T0213225

More information about this series at http://www.springer.com/series/7412

Zhen Cui · Jinshan Pan · Shanshan Zhang ·
Liang Xiao · Jian Yang (Eds.)

Intelligence Science and Big Data Engineering

Visual Data Engineering

9th International Conference, IScIDE 2019
Nanjing, China, October 17–20, 2019
Proceedings, Part I

 Springer

Editors
Zhen Cui
Nanjing University of Science
and Technology
Nanjing, China

Jinshan Pan
Nanjing University of Science
and Technology
Nanjing, China

Shanshan Zhang
Nanjing University of Science
and Technology
Nanjing, China

Liang Xiao
Nanjing University of Science
and Technology
Nanjing, China

Jian Yang
Nanjing University of Science
and Technology
Nanjing, China

ISSN 0302-9743   ISSN 1611-3349   (electronic)
Lecture Notes in Computer Science
ISBN 978-3-030-36188-4   ISBN 978-3-030-36189-1   (eBook)
https://doi.org/10.1007/978-3-030-36189-1

LNCS Sublibrary: SL6 – Image Processing, Computer Vision, Pattern Recognition, and Graphics

This Springer imprint is published by the registered company Springer Nature Switzerland AG
The registered company address is: Gewerbestrasse 11, 6330 Cham, Switzerland

Preface

The International Conference on Intelligence Science and Big Data Engineering (IScIDE 2019), took place in Nanjing, China, during October 17–20, 2019. As one of the annual events organized by the Chinese Golden Triangle ISIS (Information Science and Intelligence Science) Forum, this meeting was scheduled as the 9th in a series of annual meetings promoting the academic exchange of research on various areas of intelligence science and big data engineering in China and abroad.

We received a total of 225 submissions, each of which was reviewed by at least 3 reviewers. Finally, 84 papers were accepted for presentation at the conference, with an acceptance rate of 37.33%. Among the accepted papers, 14 were selected for oral presentations, 35 for spotlight presentations, and 35 for poster presentations. We would like to thank all the reviewers for spending their precious time on reviewing the papers and for providing valuable comments that helped significantly in the paper selection process. We also included an invited paper in the proceedings entitled "Deep IA-BI and Five Actions in Circling" by Prof. Lei Xu.

We are grateful to the conference general co-chairs, Lei Xu, Xinbo Gao and Jian Yang, for their leadership, advice, and help on crucial matters concerning the conference. We would like to thank all members of the Steering Committee, Program Committee, Organizing Committee, and Publication Committee for their hard work. We give special thanks to Prof. Xu Zhang, Prof. Steve S. Chen, Prof. Lei Xu, Prof. Ming-Hsuan Yang, Prof. Masashi Sugiyama, Prof. Jingyi Yu, Prof. Dong Xu, and Prof. Kun Zhang for delivering the keynote speeches. We would also like to thank Prof. Lei Xu for contributing a high-quality invited paper. Finally, we greatly appreciate all the authors' contributions to the high quality of this conference. We count on your continued support of the ISIS community in the future.

October 2019

Zhen Cui
Jinshan Pan
Shanshan Zhang
Liang Xiao
Jian Yang

Organization

General Chairs

Lei Xu   Shanghai Jiao Tong University, China
Xinbo Gao   Xidian University, China
Jian Yang   Nanjing University of Science and Technology, China

Program Chairs

Huafu Chen   University of Electronic Science and Technology
     of China, China
Zhouchen Lin   Peking University, China
Kun Zhang   Carnegie Mellon University, USA
Zhen Cui   Nanjing University of Science and Technology, China

Organization Chairs

Liang Xiao   Nanjing University of Science and Technology, China
Chen Gong   Nanjing University of Science and Technology, China

Special Issue Chairs

Mingming Cheng   Nankai University, China
Jinshan Pan   Nanjing University of Science and Technology, China

Publication Chairs

Shanshan Zhang   Nanjing University of Science and Technology, China
Wankou Yang   Southeast University, China

Program Committee

Mingming Gong   The University of Melbourne, Australia
Joseph Ramsey   Carnegie Mellon University, USA
Biwei Huang   Carnegie Mellon University, USA
Daniel Malinsky   Johns Hopkins University, USA
Ruben Sanchez-Romero   Rutgers University, USA
Shohei Shimizu   RIKEN, Japan
Ruichu Cai   Guangdong University of Technology, China
Shuigeng Zhou   Fudan University, China
Changdong Wang   Sun Yat-sen University, China
Tao Lei   Shaanxi University of Science & Technology, China

Lei Zhang	Chongqing University, China
Tao Wang	Nanjing University of Science and Technology, China
Changxing Ding	South China University of Technology, China
Xin Liu	Huaqiao University, China
Yang Liu	Dalian University of Technology, USA
Ying Tai	YouTu Lab, Tencent, China
Minqiang Yang	Lanzhou University, China
Guangwei Gao	Nanjing University of Posts and Telecommunications, China
Shuzhe Wu	Chinese Academy of Sciences, China
Youyong Kong	Southeast University, China
Qiguang Miao	Xidian University, China
Chang Xu	The University of Sydney, Australia
Tengfei Song	Southeast University, China
Xingpeng Jiang	Central China Normal University, China
Wei-Shi Zheng	Sun Yat-sen University, China
Yu Chen	Motovis Inc., Australia
Zebin Wu	Nanjing University of Science and Technology, China
Wei Luo	South China Agricultural University, China
Minxian Li	Nanjing University of Science and Technology, China
Ruiping Wang	Chinese Academy of Sciences, China
Jia Liu	Xidian University, China
Yang He	Max Planck Institute for Informatics, Germany
Xiaobo Chen	Jiangsu University, China
Xiangbo Shu	Nanjing University of Science and Technology, China
Yun Gu	Shanghai Jiao Tong University, China
Xin Geng	Southeast University, China
Zheng Wang	National Institute of Informatics, Japan
Lefei Zhang	Wuhan University, China
Liping Xie	Southeast University, China
Xiangyuan Lan	Hong Kong Baptist University, Hong Kong, China
Xi Peng	Agency for Science, Technology and Research (A*STAR), Singapore
Yuxin Peng	Peking University, China
Cheng Deng	Xidian University, China
Dong Gong	The University of Adelaide, Australia
Meina Kan	Chinese Academy of Sciences, China
Hualong Yu	Jiangsu University of Science and Technology, China
Kazushi Ikeda	NARA Institute of Science and Technology, Japan
Meng Yang	Sun Yat-Sen University, China
Ping Du	Shandong Normal University, China
Jufeng Yang	Nankai University, China
Andrey Krylov	Lomonosov Moscow State University, Russia
Shun Zhang	Northwestern Polytechnical University, China
Di Huang	Beihang University, China
Shuaiqi Liu	Tianjin Normal University, China

Chun-Guang Li	Beijing University of Posts and Telecommunications, China
Huimin Ma	Tsinghua University, China
Longyu Jiang	Southeast University, China
Shikui Tu	Shanghai Jiao Tong University, China
Lijun Wang	Dalian University of Technology, USA
Xiao-Yuan Jing	Wuhan University, China
Shiliang Sun	East China Normal University, China
Zhenzhen Hu	HeFei University of Technology, China
Ningzhong Liu	NUAA, China
Hiroyuki Iida	JAIST, Japan
Jinxia Zhang	Southeast University, China
Ying Fu	Beijing Institute of Technology, China
Tongliang Liu	The University of Sydney, Australia
Weihong Deng	Beijing University of Posts and Telecommunications, China
Wen Zhang	Wuhan University, China
Dong Wang	Dalian University of Technology, USA
Hang Dong	Xi'an Jiaotong University, China
Dongwei Ren	Tianjin University, China
Xiaohe Wu	Harbin Institute of Technology, China
Qianru Sun	National University of Singapore, Singapore
Yunchao Wei	University of Illinois at Urbana-Champaign, USA
Wenqi Ren	Chinese Academy of Sciences, China
Wenda Zhao	Dalian University of Technology, USA
Jiwen Lu	Tsinghua University, China
Yukai Shi	Sun Yat-sen University, China
Enmei Tu	Shanghai Jiao Tong University, China
Yufeng Li	Nanjing University, China
Qilong Wang	Tianjin University, China
Baoyao Yang	Hong Kong Baptist University, Hong Kong, China
Qiuhong Ke	Max Planck Institute for Informatics, Germany
Guanyu Yang	Southeast University, China
Jiale Cao	Tianjin University, China
Zhuo Su	Sun Yat-sen University, China
Zhao Zhang	HeFei University of Technology, China
Hong Pan	Southeast University, China
Hu Han	Chinese Academy of Sciences, China
Hanjiang Lai	Sun Yat-Sen University, China
Xin Li	Harbin Institute of Technology, Shenzhen, China
Dingwen Zhang	Northwestern Polytechnical University, China
Guo-Sen Xie	Inception Institute of Artificial Intelligence, UAE
Xibei Yang	Jiangsu University of Science and Technology, China
Wang Haixian	Southeast University, China
Wangmeng Zuo	Harbin Institute of Technology, China
Weiwei Liu	University of Technology, Sydney, Australia

Shuhang Gu	ETH Zurich, Switzerland
Hanli Wang	Tongji University, China
Zequn Jie	Tencent, China
Xiaobin Zhu	University of Science and Technology Beijing, China
Gou Jin	Huaqiao University, China
Junchi Yan	Shanghai Jiao Tong University, China
Bineng Zhong	Huaqiao University, China
Nannan Wang	Xidian University, China
Bo Han	RIKEN, Japan
Xiaopeng Hong	Xi'an Jiaotong University, China
Yuchao Dai	Northwestern Polytechnical University, China
Wenming Zheng	Southeast University, China
Lixin Duan	University of Science and Technology of China, China
Hu Zhu	Nanjing University of Posts and Telecommunications, China
Xiaojun Chang	Carnegie Mellon University, USA

Contents – Part I

Deep IA-BI and Five Actions in Circling. 1
  Lei Xu

Adaptive Online Learning for Video Object Segmentation 22
  Li Wei, Chunyan Xu, and Tong Zhang

Proposal-Aware Visual Saliency Detection with Semantic Attention 35
  Lu Wang, Tian Song, Takafumi Katayama, and Takashi Shimamoto

Constrainted Subspace Low-Rank Representation with Spatial-Spectral
Total Variation for Hyperspectral Image Restoration 46
  Jun Ye and Xian Zhang

Memory Network-Based Quality Normalization of Magnetic Resonance
Images for Brain Segmentation. 58
  Yang Su, Jie Wei, Benteng Ma, Yong Xia, and Yanning Zhang

Egomotion Estimation Under Planar Motion with an RGB-D Camera 68
  Xuelan Mu, Zhixin Hou, and Yigong Zhang

Sparse-Temporal Segment Network for Action Recognition 80
  Chaobo Li, Yupeng Ding, and Hongjun Li

SliceNet: Mask Guided Efficient Feature Augmentation
for Attention-Aware Person Re-Identification 91
  Zhipu Liu and Lei Zhang

Smoother Soft-NMS for Overlapping Object Detection in X-Ray Images 103
  Chunhui Lin, Xudong Bao, and Xuan Zhou

Structure-Preserving Guided Image Filtering. 114
  Hongyan Wang, Zhixun Su, and Songxin Liang

Deep Blind Image Inpainting 128
  Yang Liu, Jinshan Pan, and Zhixun Su

Robust Object Tracking Based on Multi-granularity Sparse Representation. . . 142
  Honglin Chu, Jiajun Wen, and Zhihui Lai

A Bypass-Based U-Net for Medical Image Segmentation. 155
  Kaixuan Chen, Gengxin Xu, Jiaying Qian, and Chuan-Xian Ren

Real-Time Visual Object Tracking Based on Reinforcement Learning
with Twin Delayed Deep Deterministic Algorithm   165
  Shengjie Zheng and Huan Wang

Efficiently Handling Scale Variation for Pedestrian Detection   178
  Qihua Cheng and Shanshan Zhang

Leukocyte Segmentation via End-to-End Learning of Deep Convolutional
Neural Networks   191
  Yan Lu, Haoyi Fan, and Zuoyong Li

Coupled Squeeze-and-Excitation Blocks Based CNN
for Image Compression   201
  Jing Du, Yang Xu, and Zhihui Wei

Soft Transferring and Progressive Learning for Human
Action Recognition   213
  Shenqiang Yuan, Xue Mei, Yi He, and Jin Zhang

Face Sketch Synthesis Based on Adaptive Similarity Regularization   226
  Songze Tang and Mingyue Qiu

Three-Dimensional Coronary Artery Centerline Extraction and Cross
Sectional Lumen Quantification from CT Angiography Images   238
  Hengfei Cui, Yong Xia, and Yanning Zhang

A Robust Facial Landmark Detector with Mixed Loss   249
  Xian Zhang, Xinjie Tong, Ziyu Li, and Wankou Yang

Object Guided Beam Steering Algorithm for Optical Phased
Array (OPA) LIDAR   262
  Zhiqing Wang, Zhiyu Xiang, and Eryun Liu

Channel Max Pooling for Image Classification   273
  Lu Cheng, Dongliang Chang, Jiyang Xie, Rongliang Ma,
  Chunsheng Wu, and Zhanyu Ma

A Multi-resolution Coarse-to-Fine Segmentation Framework with Active
Learning in 3D Brain MRI   285
  Zhenxi Zhang, Jie Li, Zhusi Zhong, Zhicheng Jiao, and Xinbo Gao

Deep 3D Facial Landmark Detection on Position Maps   299
  Kangkang Gao, Shanming Yang, Keren Fu, and Peng Cheng

Joint Object Detection and Depth Estimation in Multiplexed Image   312
  Changxin Zhou and Yazhou Liu

Weakly-Supervised Semantic Segmentation with Mean Teacher Learning. . . .   324
  Li Tan, WenFeng Luo, and Meng Yang

APAC-Net: Unsupervised Learning of Depth and Ego-Motion
from Monocular Video   336
   Rui Lin, Yao Lu, and Guangming Lu

Robust Image Recovery via Mask Matrix.   349
   Mengying Jin and Yunjie Chen

Multiple Objects Tracking Based Vehicle Speed Analysis with Gaussian
Filter from Drone Video   362
   Yue Liu, Zhichao Lian, Junjie Ding, and Tangyi Guo

A Novel Small Vehicle Detection Method Based on UAV Using Scale
Adaptive Gradient Adjustment   374
   Changju Feng and Zhichao Lian

A Level Set Method for Natural Image Segmentation by Texture
and High Order Edge-Detector   386
   Yutao Yao, Ziguan Cui, and Feng Liu

An Attention Bi-box Regression Network for Traffic Light Detection   399
   Juncai Ma, Yao Zhao, Ming Luo, Xiang Jiang, Ting Liu, and Shikui Wei

MGD: Mask Guided De-occlusion Framework for Occluded
Person Re-identification   411
   Peixi Zhang, Jianhuang Lai, Quan Zhang, and Xiaohua Xie

Multi-scale Residual Dense Block for Video Super-Resolution   424
   Hetao Cui and Quansen Sun

Visual Saliency Guided Deep Fabric Defect Classification   435
   Yonggui He, Yaoye Song, Jifeng Shen, and Wankou Yang

Locality and Sparsity Preserving Embedding Convolutional
Neural Network for Image Classification   447
   Yu Xia and Yongzhao Zhan

Person Re-identification Using Group Constraint.   459
   Ling Mei, Jianhuang Lai, Zhanxiang Feng, Zeyu Chen, and Xiaohua Xie

A Hierarchical Student's t-Distributions Based Unsupervised SAR Image
Segmentation Method   472
   Yuhui Zheng, Yahui Sun, Le Sun, Hui Zhang, and Byeungwoo Jeon

Multi-branch Semantic GAN for Infrared Image Generation
from Optical Image   484
   Lei Li, Pengfei Li, Meng Yang, and Shibo Gao

Semantic Segmentation for Prohibited Items in Baggage Inspection.   495
   Jiuyuan An, Haigang Zhang, Yue Zhu, and Jinfeng Yang

Sparse Unmixing for Hyperspectral Image with Nonlocal Low-Rank Prior . . . 506
Feiyang Wu, Yuhui Zheng, and Le Sun

Saliency Optimization Integrated Robust Background Detection
with Global Ranking 517
*Zipeng Zhang, Yixiao Liang, Jian Zheng, Kai Li, Zhuanlian Ding,
and Dengdi Sun*

Improvement of Residual Attention Network for Image Classification 529
Lu Liang, Jiangdong Cao, Xiaoyan Li, and Jane You

Nuclei Perception Network for Pathology Image Analysis 540
Haojun Xu, Yan Gao, Liucheng Hu, Jie Li, and Xinbo Gao

A k-Dense-UNet for Biomedical Image Segmentation 552
Zhiwen Qiang, Shikui Tu, and Lei Xu

Gated Fusion of Discriminant Features for Caricature Recognition 563
*Lingna Dai, Fei Gao, Rongsheng Li, Jiachen Yu, Xiaoyuan Shen,
Huilin Xiong, and Weilun Wu*

Author Index 575

Contents – Part II

Analysis of WLAN's Receiving Signal Strength Indication
for Indoor Positioning 1
    Minmin Lin, Zhisen Wei, Baoxing Chen, Wenjie Zhang,
    and Jingmin Yang

Computational Decomposition of Style for Controllable and Enhanced
Style Transfer 15
    Minchao Li, Shikui Tu, and Lei Xu

Laplacian Welsch Regularization for Robust Semi-supervised
Dictionary Learning 40
    Jingchen Ke, Chen Gong, and Lin Zhao

Non-local MMDenseNet with Cross-Band Features for Audio
Source Separation 53
    Yi Huang

A New Method of Metaphor Recognition for A-is-B Model
in Chinese Sentences 65
    Wei-min Wang, Rong-rong Gu, Shou-fu Fu, and Dong-sheng Wang

Layerwise Recurrent Autoencoder for Real-World Traffic
Flow Forecasting 78
    Junhui Zhao, Tianqi Zhu, Ruidong Zhao, and Peize Zhao

Mining Meta-association Rules for Different Types of Traffic Accidents 89
    Ziyu Zhao, Weili Zeng, Zhengfeng Xu, and Zhao Yang

Reliable Domain Adaptation with Classifiers Competition 101
    Jingru Fu and Lei Zhang

An End-to-End LSTM-MDN Network for Projectile Trajectory Prediction . . . 114
    Li-he Hou and Hua-jun Liu

DeepTF: Accurate Prediction of Transcription Factor Binding Sites
by Combining Multi-scale Convolution and Long Short-Term Memory
Neural Network 126
    Xiao-Rong Bao, Yi-Heng Zhu, and Dong-Jun Yu

Epileptic Seizure Prediction Based on Convolutional Recurrent Neural
Network with Multi-Timescale 139
    Lijuan Duan, Jinze Hou, Yuanhua Qiao, and Jun Miao

L2R-QA: An Open-Domain Question Answering Framework   151
  Tieke He, Yu Li, Zhipeng Zou, and Qing Wu

Attention Relational Network for Few-Shot Learning.   163
  Jia Shuai, JiaMing Chen, and Meng Yang

Syntactic Analysis of Power Grid Emergency Pre-plans Based
on Transfer Learning   175
  He Shi, Qun Yang, Bo Wang, Shaohan Liu, and Kai Zhou

Improved CTC-Attention Based End-to-End Speech Recognition on Air
Traffic Control   187
  Kai Zhou, Qun Yang, XiuSong Sun, ShaoHan Liu, and JinJun Lu

Revisit Lmser from a Deep Learning Perspective   197
  Wenjin Huang, Shikui Tu, and Lei Xu

A New Network Traffic Identification Base on Deep
Factorization Machine   209
  Zhenxing Xu, Junyi Zhang, Daoqiang Zhang, and Hanyu Wei

3Q: A 3-Layer Semantic Analysis Model for Question Suite Reduction   219
  Wei Dai, Siyuan Sheni, and Tieke Hei

Data Augmentation for Deep Learning of Judgment Documents   232
  Ge Yan, Yu Li, Shu Zhang, and Zhenyu Chen

An Advanced Least Squares Twin Multi-class Classification Support
Vector Machine for Few-Shot Classification   243
  Yu Li, Zhonggeng Liu, Huadong Pan, Jun Yin, and Xingming Zhang

LLN-SLAM: A Lightweight Learning Network Semantic SLAM   253
  Xichao Qu and Weiqing Li

Meta-cluster Based Consensus Clustering with Local Weighting
and Random Walking   266
  Nannan He and Dong Huang

Robust Nonnegative Matrix Factorization Based on Cosine Similarity
Induced Metric   278
  Wen-Sheng Chen, Haitao Chen, Binbin Pan, and Bo Chen

Intellectual Property in Colombian Museums: An Application
of Machine Learning   289
  Jenny Paola Lis-Gutiérrez, Álvaro Zerda Sarmiento, and Amelec Viloria

Hybrid Matrix Factorization for Multi-view Clustering.   302
  Hongbin Yu and Xin Shu

Car Sales Prediction Using Gated Recurrent Units Neural Networks
with Reinforcement Learning 312
   Bowen Zhu, Huailong Dong, and Jing Zhang

A Multilayer Sparse Representation of Dynamic Brain Functional
Network Based on Hypergraph Theory for ADHD Classification 325
   Yuduo Zhang, Zhichao Lian, and Chanying Huang

Stress Wave Tomography of Wood Internal Defects Based on Deep
Learning and Contour Constraint Under Sparse Sampling 335
   Xiaochen Du, Jiajie Li, Hailin Feng, and Heng Hu

Robustness of Network Controllability Against Cascading Failure. 347
   Lv-lin Hou, Yan-dong Xiao, and Liang Lu

Multi-modality Low-Rank Learning Fused First-Order and Second-Order
Information for Computer-Aided Diagnosis of Schizophrenia 356
   Huijie Li, Qi Zhu, Rui Zhang, and Daoqiang Zhang

A Joint Bitrate and Buffer Control Scheme for Low-Latency
Live Streaming 369
   Si Chen, Yuan Zhang, Huan Peng, and Jinyao Yan

Causal Discovery of Linear Non-Gaussian Acyclic Model
with Small Samples. 381
   Feng Xie, Ruichu Cai, Yan Zeng, and Zhifeng Hao

Accelerate Black-Box Attack with White-Box Prior Knowledge 394
   Jinghui Cai, Boyang Wang, Xiangfeng Wang, and Bo Jin

A Dynamic Model + BFR Algorithm for Streaming Data Sorting 406
   Yongwei Tan, Ling Huang, and Chang-Dong Wang

Smartphone Behavior Based Electronical Scale Validity
Assessment Framework 418
   Minqiang Yang, Jingsheng Tang, Longzhe Tang, and Bin Hu

Discrimination Model of QAR High-Severity Events Using
Machine Learning 430
   Junchen Li, Haigang Zhang, and Jinfeng Yang

A New Method of Improving BERT for Text Classification 442
   Shaomin Zheng and Meng Yang

Author Index 453

Deep IA-BI and Five Actions in Circling

Lei Xu[1,2](✉) (iD)

[1] Centre for Cognitive Machines and Computational Health (CMaCH), SEIEE,
Shanghai Jiao Tong University, Minhang, Shanghai, China
lxu@cs.sjtu.edu.cn
[2] Neural Computation Research Centre, Brain and Intelligence Sci-Tech Institute,
ZhangJiang National Lab, Shanghai, China
http://www.cs.sjtu.edu.cn/~lxu/

Abstract. Deep bidirectional Intelligence (BI) via YIng YAng (IA) system, or shortly Deep IA-BI, is featured by circling A-mapping and I-mapping (or shortly AI circling) that sequentially performs each of five actions. A basic foundation of IA-BI is bidirectional learning that makes the cascading of A-mapping and I-mapping (shortly A-I cascading) approximate an identical mapping, with a nature of layered, topology-preserved, and modularised development. One exemplar is Lmser that improves autoencoder by incremental bidirectional layered development of cognition, featured by two dual natures DPN and DCW. Two typical IA-BI scenarios are further addressed. One considers bidirectional cognition and image thinking, together with a proposal that combines theories of Hubel-Wiesel's versus Chen's. The other considers bidirectional integration of cognition, knowledge accumulation, and abstract thinking for improving implementation of searching, optimising, and reasoning. Particularly, an IA-DSM scheme is proposed for solving a doubly stochastic matrix (DSM) featured combinatorial tasks such as travelling salesman problem, and also a Subtree driven reasoning scheme is proposed for improving production rule based reasoning. In addition, some remarks are made on relations of Deep IA-BI to Hubel and Wiesel theory, Sperry theory, and A5 problem solving paradigm.

Keywords: Bidirectional · Cognition · Image thinking · Abstract thinking · Inferring · Reasoning · Topology · Optimising · Production rule

1 Deep Bidirectional Intelligence

Bidirectional intelligence (BI) recently are overviewed in Ref. [57]. As illustrated in the centre of Fig. 1(b), the bi-direction is featured by a circling similar to the

L. Xu—Supported by the Zhi-Yuan Chair Professorship Start-up Grant WF220103010 from Shanghai Jiao Tong University, and National New Generation Artificial Intelligence Project 2018AAA0100700.

Z. Cui et al. (Eds.): IScIDE 2019, LNCS 11935, pp. 1–21, 2019.
https://doi.org/10.1007/978-3-030-36189-1_1

one of ancient Chinese yIng yAng logo[1]. First, real bodies or patterns in the *Actual* world or shortly A-domain (the domain that is visible or named yAng) are mapped along the inward direction into the inner coding domain or shortly I-domain (the domain that is invisible or named yIng). This mapping transfers real body to information coding seed like a yAng or male animal (named A-mapping), performing *Abstraction* tasks such as perception, recognition, and cognition. Second, codes, concepts or symbols in the inner I-domain are mapped along the outward direction named I-mapping (a mapping from coding seed to real body in A-domain, i.e., acting like a yIng or female animal) to make *Inference* tasks that may be categorised into *image thinking* and *abstract thinking*.

The A-mapping and then I-mapping circling (or shortly A-I circling) performs each of five actions sequentially. The first action is *acquiring* data X and features that describe X. As to be further addressed in Sect. 2, the process is featured by either deep neural networks (NN) or convolutional NN (CNN) that proceeds layer by layer to perform hierarchical extraction and abstraction. The 1981 Nobel prize winners Hubel and Wiesel [24,25] developed a theory that explains how this proceeds in a manner of hierarchy as illustrated in Fig. 1(a). This H-W theory has greatly impacted the subsequent efforts in the studies of artificial intelligence and neural networks, including the recent more than decade long bloom of deep learning driven AI studies.

The second action performs *abstraction* by an inner code Y that indicates one among labels or concepts, allocates chances among candidates, gets a dimension-

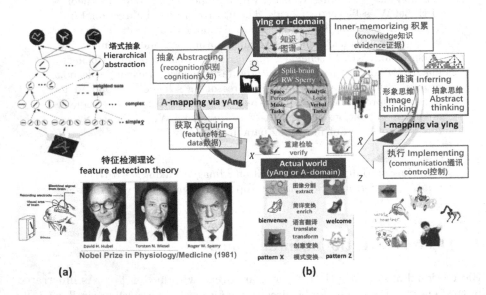

Fig. 1. Deep IA-BI (i.e., bidirectional Intelligence via YIng YAng system)

---

ally much reduced code as an inner representation of X, and even forms an icon or a subtree structure. This action not only performs perception and recognition, but also provides cognitions and evidences to the third action *Inner-memory* that accumulates knowledge and evidence.

The knowledge and evidence come from two sources. One acquires, memorizes, and organises various knowledges via education, e.g., in a formulation of knowledge graph. The other not only gets evidences to these organised and structured knowledges, but also adds in concepts and cognitions from the second action.

The fourth action *inference* may be jointly activated by the status of its previous two actions and possibly some short-cut signals as well, which performs either or both of following two manners:

Image Thinking. It is also closely related to what called concrete thinking elsewhere, and is usually referred to thought process based on dependencies among either or both of real/concrete things and their mappings/images in the I-domain of brain. The key point is specific, concrete, and detailed. Cognition via A-mapping trained by supervised deep learning is just one scenario. Another scenario happens when there is no teaching label. Whether the perceived Y by the A-mapping $X \to Y$ makes a sense is verified by checking if $\hat{X}$ generated by the I-mapping $Y \to \hat{X}$ approximates X as closely as possible. Based on the cognition by $X \to Y$, the I-mapping $Y \to \hat{X}$ performs various tasks that map X of input patterns into Z of simplified patterns, enriched patterns, and transformed patterns, as well as imaginary and creative patterns. Typical examples include language to language, text to image, text to sketch, sketch to image, image to image, 2D image to 3D image, past to future, image to caption, image to sentence, music to dance, ..., etc. All these image thinking tasks are performed by an A-I cascading and featured by an information flow that varies layer by layer in a topological preservation manner, as if displayed by an image sequence. Other details are referred to Sect. V.A in Ref. [57].

Abstract Thinking. It is also closely related to what called rational thinking, and is usually referred to thought process based on either or both of causal relations among events and logical relations among concepts in a broad, general and non-specific sense, typically described by symbolic or graphical representations. Typical examples are searching, selecting, optimising, reasoning, and planning, which are performed in a discrete space of individual or combinatoric choices. Traditionally, abstract thinking is performed by I-mapping that searches among discrete space according to knowledges and evidences accumulated and organised in the third action, which usually encounters intractable computing difficulties. Exemplified by AlphaGo [38] and AlphaGoZero [39], searching performances can be significantly improved with help of one appropriate A-mapping via deep neural networks that provides either or both of probabilistic selecting polices and heuristic values.

Following either or both of image thinking and abstract thinking, the fifth action is *implementation* of communication (verbal, writing, gesturing, posturing, etc.) and control (motoring, monitoring, steering, etc.) as desired.

As to be further addressed in the last section, the above A5 featured AI circling is actually a further development of A5 problem solving [51], which was motivated from analysing the key ingredients of randomised Hough transformation and multi-learner based problem solving [49,50]. Also, one early exemplar of IA-BI system is Bayesian Ying Yang (BYY) system [42,51]. Moreover, the principle that the A-I cascading approximates one identical mapping is just one of special cases that are addressed by Bayesian Ying Yang learning theory [42,47,48,51–54]. Recently, it is further extended to cover abstract thinking in a general thesis named BYY intelligence potential theory (BYY-IPT) [57].

In subsequent sections, further insights on IA-BI and A5 are addressed. The next section backtracks the advances on the A-I cascading from the later eighties and the early nineties to some recent studies, with insights on the layered, topology-preserved, and modularly development of bidirectional learning. Section 3 further addresses bidirectional cognition and image thinking from a perspective of Hubel-Wiesel versus Chen theories, with one combined scheme suggested. Section 4 further considers bidirectional integration of cognition, knowledge accumulation, and abstract thinking, with insights and suggestions on improving implementation of searching, optimising, and reasoning. Particularly, an IA-DSM scheme is proposed for solving those doubly stochastic matrix (DSM) featured combinatorial tasks such as travelling salesman problem. Also, a Subtree driven reasoning scheme is proposed for improving production rule based reasoning. In the last section, after a summary, remarks are made on relations of Deep IA-BI to the split-brain theory by the 1981 Nobel prize winner R.W. Sperry and A5 problem solving paradigm proposed more than two decades ago.

2 Layered, Topology-Preserved, and Modularly Developing

As mentioned above, the basic foundation of bidirectional intelligence is obtained from a bidirectional learning that makes the A-I cascading approximates one identical mapping. Early efforts along such a line can be backtracked to the later eighties and the early nineties in the last century [1,3,7,12], under the name Autoencoder (shortly AE or called auto-association). As illustrated in Fig. 2(a), AE encodes X into a vector Y in a lowered dimension by a multilayer net called encoder and decodes Y back to $\hat{X}$ by a multilayer net called decoder. The decoder shares the same structure of the encoder in a mirror architecture and acts as an inverse of the encoder.

The encoder and the decoder jointly makes $\hat{X}$ approximate X as close as possible, which may be regarded as a principle of primitive cognition (PC) that justifies the encoding $X \rightarrow Y$ is a meaningful abstraction via requiring $Y \rightarrow \hat{X}$ by a mirror architecture to perform an inverse process, such that X can be perceived or understood via Y.

Another early bidirectional learning example is the Least Mean Square Error Reconstruction (Lmser) self-organizing network that was first proposed in 1991 [40,46], which performs the PC principle by a mirror architecture similar to AE but differs in that the encoder and decoder are overlapped, resulting in several favourable characteristics, see further details in Table 3 of [57].

First, neurons per layer in the encoder and the decoder are bidirectionally connected pairwisely to the corresponding neurons on the corresponding layer, with each neuron taking a dual role both in the encoder and decoder as illustrated in Fig. 2(d), which is referred as Duality in Paired Neurons (DPN) [57]. Specifically, this duality leads to the following three extensions of autoencoder.

(a) Fast-lane Lmser: as illustrated in Fig. 2(b), neurons per layer in the encoder are directed pairwisely to their counteractions in the decoder, in a role similar to skip connections in U-net [35], ResNet [16] and DenseNet [22].

(b) Feedback Lmser: as illustrated in Fig. 2(c), neurons per layer in the decoder are directed pairwisely to their counteractions in the encoder, in a role similar to those in recurrent neural networks (RNN) for enhancing robustness.

(c) Lmser and flexible Lmser: as illustrated in Fig. 2(d), each neuron per layer j enforces the activity $v^{(j)}$ in the encoder and the activity $u_{(j)}$ in the decoder to become identical, which implies that the encoding action from the j-th layer up to the top and the decoding action from the top down to the j-th layer jointly perform an identical mapping $v^{(j)} = u_{(j)}$. In other words, in addition to seeking one global identical mapping $X = \hat{X}$ as AE does, identical mappings are also sought for distributed implementation of the PC principle, not only from the j-th layer to the top and then down to the j-th layer, but also between the j-th layer and the $j + 1$-th layer.

Second, bidirectional layered development of cognition is also considered in LMSER by another dual nature called Duality in Connection Weights (DCW), with same connecting weights between every two consecutive layers taking a

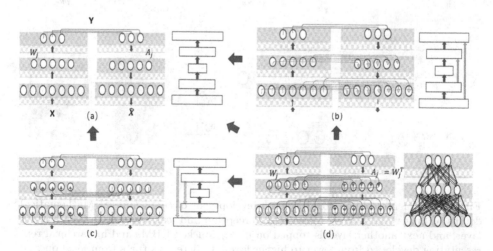

Fig. 2. Lmser differs from Autoencoder. (a) Autoencoder without the dualities DPN and DCW, (b) Fast-lane Lmser with DPN only in skip direction, (c) Feedback Lmser with DPN only in feedback direction, (d) Lmser with DPN and DCW.

dual role both in the encoder and in the decoder. From the j-th to the $j + 1$-th layer, we have $A_j = W_j^T$ in Lmser as illustrated in Fig. 2(d) while A_j is learned without the constraint $A_j = W_j^T$ as illustrated in Fig. 2(e). When W_j is an orthogonal matrix, $A_j = W_j^T$ approximately acts as its pseudo-inverse such that $A_j W_j^T = W_j W_j^T$. In other words, DCW enhances distributed implementation of the PC principle consecutively from the j-th to the $j + 1$-th layer.

Insights may also come from an incremental bidirectional layered development of cognition, as illustrated in Fig. 2(b). Perception and learning start at the bottom layer, i.e., one layer Lmser that learns templates of feature extraction, as demonstrated empirically in 1991 [40,46], with details referred to a recent reinvestigation [23] and a systematical survey [57]. Then, another layer is topped on the learned one layer Lmser, and learning is further made on the resulted two layer Lmser, ..., so on and so forth. This procedure is similar to learning stacked RBMs [18,19], as illustrated in Fig. 3(c). The importances of the DPN and DCW dualities may be interestingly backtracked from Hinton's progresses on bidirectional learning from Helmholtz machines [11,17] that also considers an architecture that locates between AE and Lmser (i.e., with DPN but without DCW) to stacked RBMs that share both the DPN and DCW dualities.

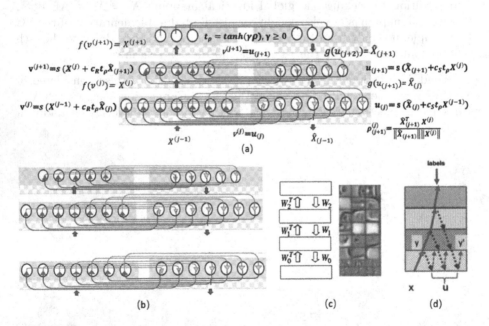

Fig. 3. Incremental bidirectional layered development of cognition. (a) Mathematical details of Lmser and flexible Lmser, (b) Perception and learning start at the bottom layer, and next another layer is topped on it, (c) Stacked RBMs and hierarchical representation developed from lower to higher layers to detect features from local ones to global ones [18,19], (d) An analog to light-ray propagation in layered media [55].

If we have samples of Y too, incremental bidirectional layered development may also be made from the top layer down by back propagation supervised learning technique layer by layer for both the direction $X \to Y$ and the direction $Y \to X$, which may be coordinated via the DCW constraints $A_j = W_j^T$ in both the directions. In general, as illustrated in Fig. 2(d), information flows of forward propagation versus backward propagation are actually coupled, in analog to light propagation via layered media [55].

Third, bidirectional layered development of cognition is also accompanied with another advantage that helps to preserve the relation of neighbourhood and topology, as well as the hierarchy of abstract concepts. It was addressed in [55] that a large number of layers help to accommodate a hierarchy and thus to preserve these relations. Also, it follows from Fig. 6 and the last paragraph of [56] that topology can be preserved by a multilayer net cascaded a typical three step structure, that is, FAN-in linear summation activates a post-nonlinear function and then FAN-out propagates to the neurons of the next layer, as long as the post-nonlinear function satisfies some nature that is held by a classic sigmoid function or a typical LUT function. Performing the PC principle, even one layer Lmser makes inner neurons become independence [40,46], which helps to preserve the relation of neighbourhood and topology. Moreover, as discussed above, multilayer Lmser performs distributed PC principle, which not only enhances preservation of neighbourhood and topology, but also facilitates to form the hierarchy of concepts via conditional independence. The importance of this preserving nature will be further addressed in the next section.

Last but not the least, the DPN and DCW dualities make modular development of cognition too, with help of checking the discrepancy either between the bottom up perception $v^{(j)}$ and the top-down reconstruction $u_{(j)}$ or between the bottom up $X^{(j)}$ and the top-down $\hat{X}_{(j+1)}$. As recently addressed by the last (10)–(14) items in Section 2 of Ref. [57], checking whether such discrepancies are bigger than some pre-specified thresholds will allocate inputs to one of multiple Lmser networks in a pipeline or a mixture.

3 Deep IA-BI Cognition and Image Thinking: From Hubel-Wiesel vs Chen to One Combined Scheme

Started from early fifties of the last century, the main stream studies on perception and cognition proceed along the direction of feature detector hypothesis on visual information process, exemplified by the 1981 Nobel Prize winners Hubel and Wiesel who developed a feature detection theory [24,25]. They found that some neurons called simple cells that fires rapidly when presented with lines at one angle, while others responded best to another angle, and also complex cells that detect edges regardless of where they were placed in the receptive field and could preferentially detect motion in certain directions. Feature detection proceeds from detecting direction at the bottom gradually up to organise into more complicated patterns in a manner of hierarchy, as illustrated in Fig. 1(a).

This H-W theory has greatly impacted the subsequent efforts on modelling of intelligence in the studies of artificial intelligence and neural networks, roughly summarised into three streams as follows:

(1) The early computing power was very limited in the later seventies and the early eighties to support the demands of AI studies. Marr and Poggio [30] proposed a simplified scheme of only three layers, which got popularised in the AI literature during the eighties of the last century. However, recent advances of deep learning for computer visions have actually abandoned this stream.

(2) Fukushima is the first who attempted to build up a computational model that is loyal to the H-W scheme, proposed Cognitron [13] and then developed into Neocognitron [14], in which equations for S-cells and C-cells are both provided with the connections for S-cells modified by learning while ones for C-cells pre-fixed [15]. Neocognitron may be regarded as a junior version of the current convolutional neural networks (CNN) [26,27]. The convolutional layers of CNN share the same point as S-cells that aims at recognising stimulus patterns based on the geometrical similarity (Gestalt), while the pooling layers of CNN like C-cells. The key difference is that Neocognitron modifies weights by self-organising while CNN modifies weights by back-propagation.

(3) There are also two roads on how feature detectors or receptive fields are developed. One considers Gabor functions or other wavelets, with parameters estimated from data. The other considers how neurons in a simple model shown in Fig. 4(a) develop receptive fields [2,6,29,36,37,40,46]. At the first stage, it has been found that receptive fields come from the evolution of weight vectors by a Hebb-type learning. A typical example is Linsker's feedforward linear network [29] with local connections from a lower layer to its next layer, which is functionally similar to a convolutional layer. Each neuron in this linear network is a special case of the one illustrated in Fig. 4(a) at simply a linear activation $s(r) = r$. It follows [36,37] that such a type of evolution may come from either of principal component analysis (PCA) and maximum information transfer.

However, learning linear neurons break coupled symmetry poorly. It was found firstly in [40] that one layer Lmser with a sigmoid nonlinearity $s(r) \neq r$ makes coupling of detectors reduced, acting like independent component analysis (ICA) that was latter further studied with improved feature detectors [2]. One layer Lmser learning rule shown in Fig. 4(c) is closely related to the learning rule used in the classical stacked RBMs as illustrated in Fig. 4(b) [18,19], sharing a common part (see the boxes in red coloured dashed lines). Also, performing learning in stacked RBMs from lower layer to higher ones developed a hierarchy of feature maps from local ones to more global ones, as illustrated in Fig. 4(c).

Rooted in Hubel-Wiesel, all the above studies are featured by local-to-global and bottom up development of cognition. In contrast, Chen believes [4,5] that perceptual process is from global to local: wholes are coded priori to analyses of their separable properties or actions, following the perceptual organisation

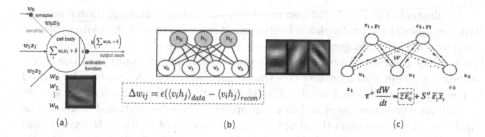

Fig. 4. Development of feature detectors or receptive fields. (a) simple neuron that develops receptive field, (b) feature detector by one layer RBM, (c) Lmser and RBM share a same structure and also a common term (i.e., two red dashed boxes) in learning. (Color figure online)

of Gestalt psychology. Proceeding far beyond the notation "whole is more than the simple sum of actions", Chen suggests that "holistic registration is priori to local analyses" and emphasises topological structure in visual perception. This 'priori' has two meanings. One implies that global organisations, determined by topology, are the basis that perception of local geometrical properties depends on. The other is that topological perception (based on physical connectivity) occurs earlier than the perception of local geometrical properties. Though there have been a number of evidences that support this global precedence, there has no computational approach to successfully implement or even illustrate, which remains an attractive challenge.

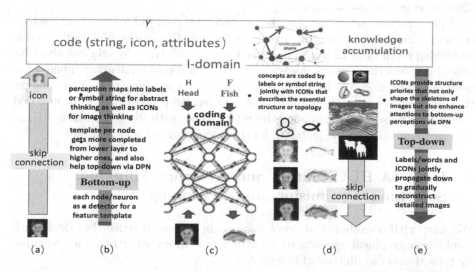

Fig. 5. IA bidirectional scheme combines local-to-global and global-to-local theses. (a) fast-lane for perceiving icon, (b) bottom up perception, (c) bidirectional implementation, (d) & (e) top-down reconstruction jointly driven by coding vector and icon, plus fast lane for top down attention.

As illustrated in Fig. 5 and also Fig. 1(b), we explore a computational scheme that combines the local-to-global bottom up development and the global-to-local top down attention. In general, each inner code consists of three ingredients. One is a label or a string of symbols that perceives the current input into the corresponding concept and a parsing tree that organises several concepts. The second is a vector of attributes that describe the input in a high dimensional feature space. These two have been widely encountered in the conventional studies. The third ingredient considers the simplest representation of spatial, structural, and topological dependence, as illustrated on the top of Fig. 5(d) and here shortly called *icon*. In the existing studies, this ingredient has been rarely considered. Instead, at least one layer of a fully connected networks is topped on CNN to output a vector as an inner code, even when we use CNN networks on images.

Following Chen's view, such icons are believed to take an important role in image thinking. As illustrated in Fig. 5(a) & (b) & (c), the bottom up perception may have a fast lane to perceive such icons, such that "holistic registration is priori to local analyses", in a coordination of certain accumulated knowledge. A rough example may be reducing a high resolution image into a much lower resolution, e.g., in a way similar to one recent study [28]. Further study may considered by adjusting icons for initialisation via bidirectional learning.

On the other hand, the top-down reconstruction of images may be jointly driven by a coding vector and an icon that acts as a structural and topological priori, again in a coordination of certain accumulated knowledge. In implementation, such icons go down directly via CNN blocks in place of a usual fully connected networks. Also, there may be a fast lane skip connections (e.g, as illustrated in Fig. 5(d) in Ref. [57]) to provide some top down attentions, as illustrated in Fig. 5(c) & (d) & (e).

This combined scheme acts as the fundamental part of IA-BI in Fig. 1, which echoes Chinese thought in term of not only the holistic Chinese philosophy that is advocated recently by efforts on machine learning [42, 47, 48, 51–54], but also Chinese characters that favours images in thinking and communication. There have been exemplified efforts by Chinese scientists in recent decades. Qian thought that image thinking plays a leading role in creative process [34]. Pan proposed a synthesis reasoning, expounded the relationship with image thinking [31], as well as Chen believed the global to local principle for cognition [4,5].

4   Deep IA-BI Cognition and Abstract Thinking: Searching, Optimising, and Reasoning

We start with considering *abstract thinking* in a general sense that deals with symbolic or graphical representations with help of discrete mathematics, featured by typical tasks as illustrated in Figs. 6 and 7.

The first task is *search vs selection*. The simple case is selecting among a finite number of individuals. The same situation can be found in making a decision or classifying/allocating among a number of choices. Beyond the simple case, search is performed by sequential decision to find a path, i.e., making selection per step as illustrated in Fig. 6(a). Typically, search is finding an optimal path in a tree or spanning a tree as illustrated on the top of Fig. 6(b), which are typically encountered in solving travelling salesman problem (TSP) and attributed graph matching (AGM), as well as alphaGo.

The second task is *satisfaction vs optimisation*. The target of searching can be not only one or more nodes but also one or more paths, such that some specifications or conditions are satisfied. Moreover, we may select an optimal one or ones among all the satisfactory ones. Traditionally, tree searching usually encounters intractable computing difficulties. Exemplified by AlphaGo [38] and AlphaGoZero [39], searching performances can be significantly improved with help of an appropriate A-mapping via deep neural networks that provides either or both of selecting probabilities and value heuristics. Furthermore, optimisation may also be made on other tasks of graph analyses, illustrated in Fig. 6(c).

Here, we sketch one IA scheme for solving a doubly stochastic matrix (DSM) featured combinatorial task, or shortly IA-DSM optimisation, e.g., the TSP task illustrated in Fig. 6(c). The solution is featured by the following problem

$$\min_{V} E(V),$$

$$V = [v_{ij}], i, j = 1, \cdots, n, \ v_{ij} = 0, 1, \ \sum_{j} v_{ij} = 1, \ \sum_{i} v_{ij} = 1, \qquad (1)$$

where $v_{ij} = 1$ indicates going from the ith city to the jth city, and there are only n ones with $v_{ij} = 1$ that defines a TSP tour. The problem is finding a shortest TSP tour such that a cost $E(V)$ is minimised, which is typically NP hard.

Since 1994, a class of neural networks algorithms (named LT algorithms) [8–10,41,43] have been developed by iteratively solving a mirror of Eq. (1) with

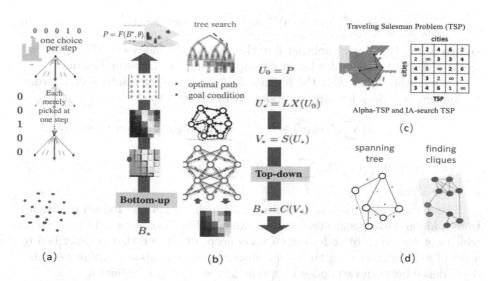

Fig. 6. Bidirectional searching. (a) simple selection and depth first search (DFS), (b) IA-DSM scheme, (c) travelling salesman problem, (d) other graph analyses.

V replaced by U, u_{ij} indicates a probability that proceeds from the ith to jth city.

For simplicity, we use

$$U^* = LX(U_0) \tag{2}$$

to denote an outcome U^* of using one of these LT algorithm started with U_0. Then we use some sharping algorithm

$$V^* = S(U^*), \ V^* = [v_{ij}^*], i, j = 1, \cdots, n, \tag{3}$$

that turns U^* into V^* that satisfies the constraints on V by Eq. (1).

Next, we further make feature enrichment on V^*

$$B^* = C(V^*), \tag{4}$$

to get a chessboard like image or configuration B as illustrated at the bottom of Fig. 6(b), with further details referred to Fig. 15 and p 886 in Ref. [57].

Following AlphaGo [38] that uses a deep neural networks to provide a policy, we use a deep neural networks or CNN to make the following mapping

$$\{P, v\} = F(B^*, \theta), \ P = [p_{ij}], i, j = 1, \cdots, n,$$
$$subject \ to \ 0 \le p_{ij} \le 1, \ \sum_j p_{ij} = 1, \ \sum_i p_{ij} = 1; \ 0 \le v \le 1, \tag{5}$$

where v is a value that indicates the goodness of P or the chance of reaching an optimal solution by $\hat{V} = S(P)$, which is proportional to $exp[-\gamma E(\hat{V})]$.

On one hand, we may let

$$U_0 = vP + (1 - v)U^* \ or \ U_0 = vP + (1 - v)V^* \tag{6}$$

and go to Eq. (2) to start another iteration, where the indices of u_{ij}^*, v_{ij}^* and p_{ij} are sorted into a same order by an appropriate number of permutations.

On the other hand, with the indices of v_{ij}^* and p_{ij} sorted into a same order, we implement a deep learning to update parameters θ by

$$\max_\theta [\sum_{i,j} v_{ij}^* \ln p_{ij} + R(\theta)], \ \text{with a regularisation term } R(\theta),$$
$$subject \ to \ 0 \le p_{ij} \le 1, \ \sum_j p_{ij} = 1, \ \sum_i p_{ij} = 1. \tag{7}$$

The third typical task is *reasoning* on which we get the following insights from a hierarchical perspective as illustrated in Fig. 7(a), especially a graph on which not only each of nodes denotes a concept of a thing that is described by a set of attributes or even their joint distribution, but also each link describes dependence between two nodes by a function or a joint distribution.

In a wide sense, reasoning can be understood as a process that *updates of one or more nodes cause a dynamic process of updating of all the rest nodes to reach a new balance.* Whether or not it is explicitly observed, there is an underlying potential $E(\xi_1, \cdots, \xi_n | \theta_R)$ that governs this dynamics. Such a dynamic process on a graph in general may not be stabilised, i.e. the reasoning may fail. A reasoning process makes a sense only when a balance is reached at one minimum point of $E(\xi_1, \cdots, \xi_n | \theta_R)$. It is an interesting problem to investigate the conditions that ensure a reasoning to make a sense, e.g., ones specified by an energy like ones either in the classic Hopfield networks [21] or Boltzmann machine [20].

We may also get a Gibbs or Boltzmann distribution $q(\xi_1, \cdots, \xi_n | \theta_R)$ from $E(\xi_1, \cdots, \xi_n | \theta_R)$. Reasoning may be understood from a perspective of uncertainty propagation, namely, given the changing of one or more marginal distribution $p(\xi_j | \theta_r)$ that describes how ξ_j varies with uncertainty specified, the task is propagating such changing all over the entire distribution $p(\xi_1, \xi_2, \cdots, \xi_n | \theta_R)$, such that all the changed marginal distributions may be obtained.

When the graph is a directed acyclic graph (DAG), i.e., each link is directed as illustrated on the second layer in Fig. 7(a), we encounter a Bayesian networks $q(\xi_1, \cdots, \xi_n | \theta_R)$, on which reasoning can be conducted by an effective propagation procedure [32,33]. Such DAGs may be further simplified by pruning links with relatively little roles and merging nodes as a big concept or thing, which provides a better understanding on main or important reasoning paths. Moreover, similar to the IA-DSM scheme shown in Fig. 6(b), we may use the I-mapping to generate a 2D or higher dimensional image X that describes the current status of reasoning, and then use a deep learning based A-mapping to update one or more nodes to adjust the reasoning.

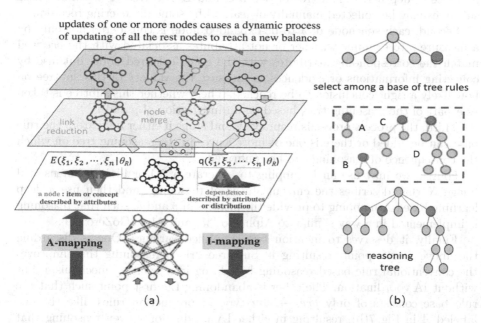

Fig. 7. IA reasoning. (a) hierarchical reasoning, (b) Sub-Tree driven reasoning. (Color figure online)

In many cases we do not have $E(\xi_1, \cdots, \xi_n | \theta_R)$ or $q(\xi_1, \cdots, \xi_n | \theta_R)$ available directly. In a classical sense, reasoning is featured by a path or subtree, starting from certain premises or preconditions to reach some expected conclusion or consequence, based on a set of reasoning rules as a rule base. One typical example is the production rule based reasoning in an expert system, which was a main theme in the traditional AI studies. At a node to be expanded in a search tree, the number of child nodes to be considered relates to both the current situation and the related rules in the knowledge base. As illustrated in Fig. 7(b), a node in a green color is considered by searching a production rule *pre* → *con* from a rule base such that attributes associated with the *pre* well match the corresponding attributes with the green coloured node. Each of production rules is featured by just one edge, e.g., the one labeled A in Fig. 7(b). Such a reasoning process is also featured by tree searching that is similar to ones in Fig. 6, and thus suffers a computationally intractable challenge still.

In addition to using a deep neural networks to provide either or both of policy probabilities and value heuristics, AlphaGo [38] tackles the challenge by MCTS search that makes a lookahead 'scouting', sharing one idea similar to the scouting-averaging technique used in CNneim-A that improves A* search [45,59]. Such an idea may also improve the production rule based reasoning.

This observation motivates the following IA Sub-tree driven reasoning scheme.

First, the knowledge base is augmented by adding those subtrees in, as illustrated in Fig. 7(b), that is, the *pre* remains same but *con* is a subtree, e.g., B is a tree of depth-1, C is a tree of depth-2, and D is a tree of depth-3. These sub-trees may be collected manually or learned by some discovering techniques.

Second, each son node of a to-be-expanded node is considered not only by a measure p to examine whether or not attributes associated with the *pre* well match the corresponding attributes with the green coloured node, but also by collecting informations or evidences to measure the value v of this subtree on how likely a right conclusion to be reached. Then, whether this subtree is taken as a part of the reasoning tree, based on an integration of p and v.

Third, the process proceeds recursively and stops if either none from the rule base can be added or there is one or more leaves in the reasoning tree on which the consequence of reasoning can be confirmed.

Fourth, we may use an I-mapping to generate a 2D or higher dimensional image X that describes the current status of reasoning, and then use a deep learning based A-mapping to provide the priories of p and v, such that reasoning is implemented in a way similar to AlphaGo [38] and AlphaGoZero [39].

Finally, it deserves to mention two simplified variants. One is discarding the above fourth point, resulting in Sub-tree driven reasoning that improves the conventional rule based reasoning by adding in a scouting mechanism, but without IA coordination. The other is abandoning the first point such that the rule base consists of only *pre* → *con* type of production rules, like the one labeled A in Fig. 7(b), resulting in either IA production driven reasoning that may be still implemented in a way similar to AlphaGo or even production-(p, v)

reasoning, which is still different from the conventional rule based reasoning that is actually a special case of letting $v = 0$.

5   Summary and Remarks on Split-Brain, IA-BI, and A5

Intelligence is featured by a bidirectional implementation in a yIng yAng system perspective, and is thus shortly named IA-BI. Its implementation is featured by A-I circling that performs each of five actions (A5) sequentially. As a basic foundation of IA-BI, bidirectional learning makes an A-I cascading approximate one identical mapping, with the natures of layered, topology-preserved, and modular development. Lmser improves autoencoder for performing bidirectional learning by incremental bidirectional layered development of cognition, featured by two dual natures. Moreover, bidirectional cognition and image thinking are addressed with a proposal that combines theories of Hubel-Wiesel versus Chen. Furthermore, bidirectional integration of cognition, knowledge accumulation, and abstract thinking are considered for improving implementation of searching, optimising, and reasoning. Particularly, an IA-DSM scheme is proposed for solving a doubly stochastic matrix (DSM) featured combinatorial tasks such as TSP, and also a Sub-tree driven reasoning scheme is proposed for improving production rule based reasoning.

It may be also interestingly to explore some link between bidirectional intelligence and Sperry's split-brain study that was awarded the Nobel Prize in Physiology or Medicine in 1981 too. On one hand, this study reveals functional specialisation of the cerebral hemispheres, and promoted the subsequent studies that explore how some cognitive functions tend to be dominated by one side or the other; that is, how they are lateralised. On the other hand, popularisations of Sperry's findings, especially in pop psychology, oversimplify the science about lateralisation, roughly by a story that presents functional differences between hemispheres as being more absolute than is actually the case. Such a story was widely held even in the scientific community for years. However, in recent years there are quite many disputes on and even an overturn of such a story.

Interestingly, from the perspective of two hemispheres together with the BI view illustrated in Fig. 1(b) (placing the head to face downward), it follows that a main percentage of popularised functions (PFs) of lateralisation still make sense but some modifications are needed. One of typical PFs to be performed in the right hemisphere is perception, which concords with the second action of BI (i.e., cognition). Also, the PFs such as music, creative writing/art, emotional, imaginative, intuitive thoughts, etc. were believed to be performed in the right hemisphere and basically tasks referred as *image thinking* that is the third action of BI in Fig. 1(b). Both the actions are indeed involved in the right hemisphere, which is consistent to the PFs story. However, modifications should be made.

Though mainly performed in the right hemisphere, perception is also aided in cooperation of the left hemisphere. Actually, perception is a foundation on which both image thinking and abstract thinking base on. As illustrated in the centre of Fig. 1(b), two typical PFs in the left hemisphere are analytic and logical ones, which belongs to the third action of BI implemented by the I-mapping in the left hemisphere. Moreover, they were performed based on this foundation in the right hemisphere, i.e., involving two hemispheres too.

The A5 formulation comes from a general problem solving paradigm [50], refined from the mechanisms embedded in Hough Transform (HT), Randomized Hough Transform (RHT) [44,58] and Multi-sets-learning [49]. It was discussed in Appendix B(2) of Ref. [51] that this A5 paradigm is consistent to and conceptually echoes the famous ancient Chinese WuXing theory that lays the foundation of TCM, as illustrated in Fig. 8(a). Here, we further discuss that the five action circling in Fig. 1(b) is another exemplar of this A5 paradigm or WuXing theory.

As illustrated in Fig. 8(b), taking line detection within one image as an example, the HT detection is featured by a circular flow of five basic mechanisms or actions. First, it starts from picking one pixel from image, which is an instance of the action named *acquisition* (shortly denoted as A-1) for sampling evidence or data from the world in observation, which corresponds the action of *acquiring* feature in Fig. 1(b).

Second, the HT maps the pixel picked into a line in its parameter space $\{a, b\}$ which is an instance of the action *allocation & assumption* (A-2) that allocates information contained in the picked pixel, featured by a distributed allocation of evidence along a line that represents a set of candidate assumptions in the parameter space, which is generalised into the action of *abstraction* in Fig. 1(b). Third, the HT quantises a window of the parameter space into a lattice on which every cell is placed with an accumulator. We add one score to those accumulators located on the candidate assumptions provided by A-2, which is an example of the action *accumulation & amalgamation* (A-3) for integrating evidences about these candidate assumptions, while its counterpart in Fig. 1(b) is the action of *inner-memorising*.

Next, we inspect the scores of all the accumulators and detect those, that pass some threshold or become local maxima, as the final candidate conclusions on detected lines, which is an instance of the fourth action *apex-seeking &*

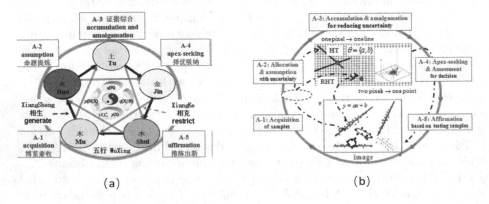

(a)   (b)

Fig. 8. The five action circling and A5 Problem Solving

assessment (A-4) that decides one or a set of final candidates with their corresponding scores either locating at peaks or becoming bigger than a threshold, which is further extended to making *inference* via image thinking and abstract thinking. Finally, the HT tests whether each of final candidates can be regarded as a detected line. In general, the job is named *affirmation* (A-5) that assesses whether each of candidate conclusions should be either discarded or identified as a final conclusion, while its counterpart in Fig. 1(b) is *implementing actuators* for outcomes as desired.

Last but not least, we continue the topic made at the end of Sect. 3, that is, image thinking versus abstract thinking. The flow of abstract thinking is illustrated by a grey coloured outer circle, representing an analytical process of induction and deduction. Observing its outward part in Fig. 9(d), the action of *Inferring* is featured by tree searching that starts with a root that represents an abstracted and broad concept deductively towards one or a few leaves with each representing a specific and concrete body or pattern, with its outcomes driving the action of *implementing* control and communication. Typically, the inferring suffers a combinatorial complexity that explodes exponentially with the depth, somewhat similar to a parsing of one western language featured with a small size of alphabets and grammar rules. Those languages used in current computers are simplified from a natural language by adding restrictions, such that searching complexity becomes manageable. Even critically, searching bases on the action of *inner-memorising* for accumulating knowledge and evidence, which was actually one fatal barrier that the classical AI study failed to pass over.

We believe that human tackles the challenge by the inward part and especially the action of *abstracting*, as illustrated in Fig. 9(a), which involves not only classify or recognise real bodies into categories or concepts together with their

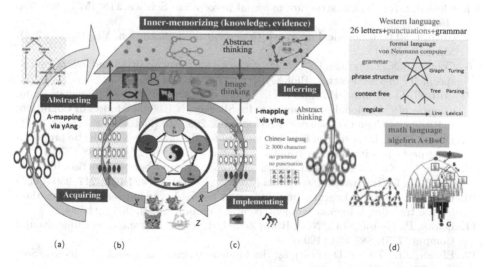

Fig. 9. Further insights on image thinking versus abstract thinking (Color figure online)

relationships but also how each concept is developed or abstracted layer by layer from the action of *acquiring*, as illustrated by a converging tree. This process involves not just an individual human intelligence, but a collective and inherited knowledge base from not only the current human society but also evolution of a long history. Current deep learning studies mostly involve mapping bodies into categories and forming concepts by clustering. It is still a long way to get the knowledge base by learning from data in comparable to human intelligence.

The flow of image thinking is illustrated by a blue coloured outer circle, describing how a specific and concrete subject or image varies from one to the other. Observing its outward part in Fig. 9(c), the action of *Inferring* is featured by mapping a small dimensional inner code into a large dimensional pattern, holistically by a deep neural networks. The scenario is somewhat similar to understand an ancient Chinese language. Being different from abstract thinking, the action of *inner-memorising* just provides a priori distribution to adjust this mapping. Also, the action of *abstracting* in the inward part is a mapping from a large dimensional pattern into a small dimensional inner code, holistically by a deep networks illustrated in Fig. 9(b). Jointly, an image thinking circle maps an input pattern X into either its reconstruction $\hat{X}$ to calibrate whether this code makes a sense but also another pattern form Z to perform various tasks.

References

1. Ballard, D.H.: Modular learning in neural networks. In: AAAI, pp. 279–284 (1987)
2. Bell, A.J., Sejnowski, T.J.: The independent components of natural scenes are edge filters. Vision Res. **37**(23), 3327–3338 (1997)
3. Bourlard, H., Kamp, Y.: Auto-association by multilayer perceptrons and singular value decomposition. Biol. Cybern. **59**(4–5), 291–294 (1988)
4. Chen, L.: Topological structure in visual perception. Science **218**(4573), 699–700 (1982)
5. Chen, L.: The topological approach to perceptual organization. Vis. Cogn. **12**(4), 553–637 (2005)
6. Cooper, L.N., Liberman, F., Oja, E.: A theory for the acquisition and loss of neuron specificity in visual cortex. Biol. Cybern. **33**(1), 9–28 (1979)
7. Cottrell, G., Munro, P., Zipser, D.: Image compression by backpropagation: an example of extensional programming. In: Sharkey, N.E. (ed.) Models of Cognition: A Review of Cognition Science, Nonvood, pp. 208–240 (1989)
8. Dang, C., Xu, L.: A barrier function method for the nonconvex quadratic programming problem with box constraints. J. Global Optim. **18**(2), 165–188 (2000)
9. Dang, C., Xu, L.: A globally convergent Lagrange and barrier function iterative algorithm for the traveling salesman problem. Neural Netw. **14**(2), 217–230 (2001)
10. Dang, C., Xu, L.: A Lagrange multiplier and hopfield-type barrier function method for the traveling salesman problem. Neural Comput. **14**(2), 303–324 (2002)
11. Dayan, P., Hinton, G.E., Neal, R.M., Zemel, R.S.: The Helmholtz machine. Neural Comput. **7**(5), 889–904 (1995)
12. Elman, J.L., Zipser, D.: Learning the hidden structure of speech. J. Acoust. Soc. Am. **83**(4), 1615–1626 (1988)
13. Fukushima, K.: Cognitron: a self-organizing multilayered neural network. Biol. Cybern. **20**(3–4), 121–136 (1975)

14. Fukushima, K.: Neocognitron: a self-organizing neural network model for a mechanism of pattern recognition unaffected by shift in position. Biol. Cybern. **36**(4), 193–202 (1980)

15. Fukushima, K., Miyake, S., Ito, T.: Neocognitron: a neural network model for a mechanism of visual pattern recognition. IEEE Trans. Syst. Man Cybern. **5**, 826–834 (1983)

16. He, K., Zhang, X., Ren, S., Sun, J.: Deep residual learning for image recognition. In: Proceedings of the IEEE Conference on Computer Vision and Pattern Recognition, pp. 770–778 (2016)

17. Hinton, G.E., Dayan, P., Frey, B.J., Neal, R.M.: The wake-sleep algorithm for unsupervised neural networks. Science **268**(5214), 1158–1161 (1995)

18. Hinton, G.E., Osindero, S., Teh, Y.W.: A fast learning algorithm for deep belief nets. Neural Comput. **18**(7), 1527–1554 (2006)

19. Hinton, G.E., Salakhutdinov, R.R.: Reducing the dimensionality of data with neural networks. Science **313**(5786), 504–507 (2006)

20. Hinton, G.E., Sejnowski, T.J., et al.: Learning and relearning in Boltzmann machines. In: Parallel Distributed Processing: Explorations in the Microstructure of Cognition, vol. 1, no. 282–317, p. 2 (1986)

21. Hopfield, J.J.: Neural networks and physical systems with emergent collective computational abilities. Proc. Natl. Acad. Sci. **79**(8), 2554–2558 (1982)

22. Huang, G., Liu, Z., Van Der Maaten, L., Weinberger, K.Q.: Densely connected convolutional networks. In: Proceedings of the IEEE Conference on Computer Vision and Pattern Recognition, pp. 4700–4708 (2017)

23. Huang, W., Tu, S., Xu, L.: Revisit Lmser and its further development based on convolutional layers. CoRR abs/1904.06307 (2019)

24. Hubel, D.H., Wiesel, T.N.: Receptive fields, binocular interaction and functional architecture in the cat's visual cortex. J. Physiol. **160**(1), 106–154 (1962)

25. Hubel, D.H., Wiesel, T.N.: Receptive fields and functional architecture of monkey striate cortex. J. Physiol. **195**(1), 215–243 (1968)

26. LeCun, Y., et al.: Handwritten digit recognition with a back-propagation network. In: Advances in Neural Information Processing Systems, pp. 396–404 (1990)

27. LeCun, Y., Kavukcuoglu, K., Farabet, C.: Convolutional networks and applications in vision. In: Proceedings of 2010 IEEE International Symposium on Circuits and Systems, pp. 253–256. IEEE (2010)

28. Li, P., Tu, S., Xu, L.: GAN flexible Lmser for super-resolution. In: ACM International Conference on Multimedia, 21–25 October 2019, Nice, France. ACM (2019)

29. Linsker, R.: Self-organization in a perceptual network. Computer **21**(3), 105–117 (1988)

30. Martin, K.A.: A brief history of the feature detector. Cereb. Cortex **4**(1), 1–7 (1994)

31. Pan, Y.: The synthesis reasonning. Pattern Recog. Artif. Intell. **9**, 201–208 (1996)

32. Pearl, J.: Fusion, propagation, and structuring in belief networks. Artif. Intell. **29**(3), 241–288 (1986)

33. Pearl, J.: Probabilistic Reasoning in Intelligent Systems: Networks of Plausible Inference. Morgan Kaufmann, San Mateo (1988)

34. Qian, X.: On thinking sciences. Chin. J. Nat. **8**, 566 (1983)

35. Ronneberger, O., Fischer, P., Brox, T.: U-Net: convolutional networks for biomedical image segmentation. In: Navab, N., Hornegger, J., Wells, W.M., Frangi, A.F. (eds.) MICCAI 2015. LNCS, vol. 9351, pp. 234–241. Springer, Cham (2015). https://doi.org/10.1007/978-3-319-24574-4_28

36. Rubner, J., Schulten, K.: Development of feature detectors by self-organization. Biol. Cybern. **62**(3), 193–199 (1990)
37. Sanger, T.D.: Optimal unsupervised learning in a single-layer linear feedforward neural network. Neural Netw. **2**(6), 459–473 (1989)
38. Silver, D., et al.: Mastering the game of go with deep neural networks and tree search. Nature **529**(7587), 484–489 (2016)
39. Silver, D., et al.: Mastering the game of go without human knowledge. Nature **550**(7676), 354 (2017)
40. Xu, L.: Least MSE reconstruction for self-organization: (i) multi-layer neural nets and (ii) further theoretical and experimental studies on one layer nets. In: Proceedings of International Joint Conference on Neural Networks-1991-Singapore, pp. 2363–2373 (1991)
41. Xu, L.: Combinatorial optimization neural nets based on a hybrid of Lagrange and transformation approaches. In: Proceedings of World Congress on Neutral Networks, pp. 399–404 (1994)
42. Xu, L.: Bayesian-Kullback coupled Ying-Yang machines: unified learnings and new results on vector quantization. In: Proceedings of the International Conference on Neural Information Process (ICONIP 1995), pp. 977–988 (1995)
43. Xu, L.: On the hybrid LT combinatorial optimization: new U-shape barrier, sigmoid activation, least leaking energy and maximum entropy. In: Proceedings of the ICONIP, vol. 95, pp. 309–312 (1995)
44. Xu, L., Oja, E., Kultanen, P.: A new curve detection method Randomized Hough Transform (RHT). Pattern Recogn. Lett. **11**, 331–338 (1990)
45. Xu, L.: Investigation on signal reconstruction, search technique, and pattern recognition. Ph.D. dissertation, Tsinghua University, December 1986
46. Xu, L.: Least mean square error reconstruction principle for self-organizing neural-nets. Neural Netw. **6**(5), 627–648 (1993)
47. Xu, L.: A unified learning scheme: Bayesian-Kullback Ying-Yang machine. In: Advances in Neural Information Processing Systems, pp. 444–450 (1996)
48. Xu, L.: BYY prod-sum factor systems and harmony learning. Invited talk. In: Proceedings of International Conference on Neural Information Processing (ICONIP 2000), vol. 1, pp. 548–558 (2000)
49. Xu, L.: Data smoothing regularization, multi-sets-learning, and problem solving strategies. Neural Netw. **16**(5–6), 817–825 (2003)
50. Xu, L.: A unified perspective and new results on RHT computing, mixture based learning, and multi-learner based problem solving. Pattern Recogn. **40**(8), 2129–2153 (2007)
51. Xu, L.: Bayesian Ying-Yang system, best harmony learning, and five action circling. Front. Electr. Electron. Eng. China **5**(3), 281–328 (2010)
52. Xu, L.: Codimensional matrix pairing perspective of BYY harmony learning: hierarchy of bilinear systems, joint decomposition of data-covariance, and applications of network biology. Front. Electr. Electron. Eng. China **6**, 86–119 (2011)
53. Xu, L.: On essential topics of BYY harmony learning: current status, challenging issues, and gene analysis applications. Front. Electr. Electron. Eng. **7**(1), 147–196 (2012)
54. Xu, L.: Further advances on Bayesian Ying Yang harmony learning. Appl. Inform. **2**(5) (2015)
55. Xu, L.: The third wave of artificial intelligence. KeXue (Sci. Chin.) **69**(3), 1–5 (2017). (in Chinese)
56. Xu, L.: Deep bidirectional intelligence: AlphaZero, deep IA search, deep IA infer, and TPC causal learning. Appl. Inform. **5**(5), 38 (2018)

57. Xu, L.: An overview and perspectives on bidirectional intelligence: Lmser duality, double ia harmony, and causal computation. IEEE/CAA J. Autom. Sin. **6**(4), 865–893 (2019)
58. Xu, L., Oja, E.: Randomized Hough transform: basic mechanisms, algorithms, and computational complexities. CVGIP Image Underst. **57**(2), 131–154 (1993)
59. Xu, L., Yan, P., Chang, T.: Algorithm cnneim-a and its mean complexity. In: Proceedings of 2nd International Conference on Computers and Applications, Beijing, 24–26 June 1987, pp. 494–499. IEEE Press (1987)

Adaptive Online Learning for Video Object Segmentation

Li Wei, Chunyan Xu$^{(\boxtimes)}$, and Tong Zhang

Key Laboratory of Intelligent Perception and Systems for High-Dimensional
Information of Ministry of Education, School of Computer Science and Engineering,
Nanjing University of Science and Technology, Nanjing, China
weilixhm@gmail.com, {cyx,tong.zhang}@njust.edu.cm

Abstract. In this work, we address the problem of video object segmentation (VOS), namely segmenting specific objects throughout a video sequence when given only an annotated first frame. Previous VOS methods based on deep neural networks often solves this problem by fine-tuning the segmentation model in the first frame of the test video sequence, which is time-consuming and can not be well adapted to the current target video. In this paper, we proposed the adaptive online learning for video object segmentation (AOL-VOS), which adaptively optimizes the network parameters and hyperparameters of segmentation model for better predicting the segmentation results. Specifically, we first pre-train the segmentation model with the static video frames and then learn the effective adaptation strategy on the training set by optimizing both network parameters and hyperparameters. In the testing process, we learn how to online adapt the learned segmentation model to the specific testing video sequence and the corresponding future video frames, where the confidence patterns is employed to constrain/guide the implementation of adaptive learning process by fusing both object appearance and motion cue information. Comprehensive evaluations on Davis 16 and SegTrack V2 datasets well demonstrate the significant superiority of our proposed AOL-VOS over other state-of-the-arts for video object segmentation task.

Keywords: Online-learning · Video object segmentation · Adaptation

1 Introduction

While moving ahead with tracking technology, the pixel-level tracking task has gradually become an important component of video analysis, which can be called as video object segmentation. Here we mainly focus on semi-supervised video object segmentation task, which is to segment the region of the specific object from the video sequence with the first frame mask as initialization. It is a challenging task in computer vision, and many related higher-level applications are widely used, including video surveillance [1], virtual reality [2], anomaly detection [3] and autonomous driving [4,5].

© Springer Nature Switzerland AG 2019
Z. Cui et al. (Eds.): IScIDE 2019, LNCS 11935, pp. 22–34, 2019.
https://doi.org/10.1007/978-3-030-36189-1_2

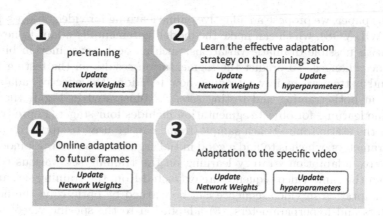

Fig. 1. The process of the adaptive online learning for video object segmentation, including pre-training stage on a single frame, learning the effective adaptation strategy on the training set, adaptation to the specific testing video sequence and online adaptation to future video frames.

Recently, there has been much previous work denoted to video object segmentation task by employing different technical strategies, such as mask propagation [21], data augmentation [23], detection cues [19,20,32], motion cues [14,15], online adaptation [28], pixel-wise metric learning [16] and multi-task learning [17,18] etc. One of the most popular VOS methods is to fine-tune the trained segmentation model with the first frame of the test video sequence [29]. Each frame of the video is tested without considering the sequence and motion information, and the segmentation result of the object is independently predicted. Although this kind of VOS methods can improve the performance of segmentation networks in the test sequence through repeated learning and understanding of the first frame information, the first frame of the sequence can not represent the whole sequence well and it is a very time-consuming process by fine-tuning the segmentation model for adapting to the test video sequence. For solving these above problems, numerous VOS methods have attempted to utilize motion information [14,15] and detection cues [19,20,32] to reduce model drifting during online learning. For example, motion cue learning and refinement process are used to improve the performance of video object segmentation (e.g., MoNet [14]). The Siamese encoder-decoder network [21] is designed to utilize mask propagation and object detection. SegFlow [17] is to predict pixel-wise object segmentation and optical flow at the same time to boost them for each other. Recently, Voigtlaender et al. [28] present an online adaptive learning method by adjusting the segmentation model according to the specificity of each frame, but with a higher computation complexity. These existing VOS methods mainly focus on the generation or simple use of those confidence cues, but can not consider how to optimize these confidence cues and establish an adaptive learning process to optimize the parameters/hyperparameters of the segmentation model.

In this paper, we propose an adaptive online learning for video object segmentation (AOL-VOS), which can make the model incrementally adapt to next new frames with possible dramatic variations, such as occlusions, motion blur and appearance changes. To predict the segmentation results on the testing videos faster and better, we mainly introduce how to adopt the effective adaptation strategy in both training and testing sets. As illustrated in Fig. 1, the adaptive online learning for object segmentation includes four stage: (1) pre-training stage on the static frames: each sample/frame can be used to updated the network parameters, which is to find a good initial segmentation network model. (2) the effective adaptation strategy learning on the training set: various training pairs from the same video sequence are extracted from the training set, and the rule of adaptation in video sequence can be learned by optimizing the network parameters and hyperparameters. (3) adaptation to the specific video: we will fast update the network parameters and hyperparameters within on or a few steps, just for making the segmentation model adapt to specific video sequence. (4) online adaptation to future frames: With the guide of confidence patterns, we can online adapt the segmentation model to each future frame of the test video sequence for better predicting the segmentation results.

To constrain/evaluate the adaptation learning strategy, specifically, we use confidence patterns which is generated by two streams of information: motion and detection cue. On the one hand, to rapidly learn the adaptive segmentation, we accumulate segmentation experience on the basis of online adaptation learning strategy and follow the idea of propagation. On the other hand, in order to reduce the error propagation in long-term video, we also jointly model the motion information and the appearance of the target by adding the guidance of confidence patterns. When the target object changes or occludes in the course of motion, motion information (optical flow) can compensate for some cues to determine the confidence patterns. On the contrary, in the case of motion blur or statical target, the appearance component will take effect. Instead of building a global model for the object appearance, we designed a part-based detector to fly on the framework to effectively detect these confidence patterns. The main contributions are summarized in three aspects:

- We propose the adaptive online learning strategy for improving the performance of video object segmentation in the testing process, which can adaptively optimize the segmentation model with only a few iteration steps.
- We introduce confidence patterns for constraining/guiding the implementation of adaptive learning process, which can be obtained by fusing both object appearance (e.g. with detection) and motion cue (e.g. optical flow) information.
- We validate the effectiveness of the proposed AOL-VOS method on Davis16 and Segtrack V2 datasets, and demonstrate its superiority when compared with existing state-of-the-art methods [14,22,30].

2   The AOL-VOS Method

We propose an adaptive online learning strategy for video object segmentation named AOL-VOS. For learning to adapt the test video segmentation, we build an adaptive learning process by training model across different videos/frames which we called adaptation-strategy. The weight of the model can be adjusted for each frame of each video sequence, and the adaptive process can be greatly accelerated by adaptive online learning. Here, adaptive online learning strategies transfer the current segmentation model to a new environment by adaptively identifying these steps. The adaptive online learning strategy provides the effective guidance for the segmentation model to develop in the right direction. Figure 2 gives an overview of our adaptive online learning strategy.

2.1   Model Foundation

The basic segmentation network in this paper is a general fully-connected network, which can be pre-trained on static images by a binary cross entropy loss. Foe reducing the computational memory cost of the intermediate characteristics and improving the training/testing speed, we adopt the powerful DeepLabv3+ [27] model as the basic network, set the output stride as 16, and employ the design of enhanced Atrous Spatial Pyramid Pooling (ASPP) [26] as well.

Let's introduce the basic knowledge of adaptation-learning in this work. Here we mainly introduce the idea of adaptation-learning and the implementation details of some key techniques includes adaptive-update and update AOL-VOS model. In the previous video object segmentation networks, the adaptive network model θ is usually updated by using a gradient descent style as $\theta - \lambda \bigtriangledown_\theta \mathcal{L}$, where the learning rate λ is fixed as a hyperparameter, and $\mathcal{L}$ is a common binary cross entropy loss function. Due to the scalar λ is humanly fixed, The learning style of network parameters might not better adapt to the incoming video frames. We expect the model update could be adaptively learnt from training/online fine-tuning sequences. For this reason, we change the learning rate λ as a tensor α to be learnt, which is with the same size as the network parameter θ. Therefore, the additional advantage is that it can better adjust the optimization direction of spatial and channel-wise representation.

At the adaptive-update step, our goal is to update the weight of the network θ by using the learned learning rate α as hyper-parameters. And the learning rate α has same size as the network parameter θ. Hence, each element in network parameter θ has its own update step size. Formally,

$$\mathcal{L} = Loss(y_{i,j}, F(\mathbf{x}_{i,j}, \theta_{k-1})), \tag{1}$$
$$\theta_k = \theta_{k-1} - \alpha \odot \bigtriangledown_\theta \mathcal{L}, \tag{2}$$

where $\mathbf{x}_{i,j}$ is the input (e.g., convolution features) for a video i and an initial frame j, $y_{i,j}$ is the expected output and $\odot$ denotes the element-wise product. We can find the update of the network weight θ_k depends on the θ_{k-1}, the hyperparameter θ and the computed gradient information $\bigtriangledown_\theta \mathcal{L}$ in the adaptively learning process. In this way, we can make the network model converge to the frames of the specified sequence in only a few iteration steps.

Then we introduce how to update the AOL-VOS model. The AOL-VOS model consists of two parts: network parameter θ and learning rate α. Here we update the AOL-VOS model by evaluating the model effect after adaptive-update. If adaptive-update works well, we don't need to adjust the learning rate α. On the contrary, if adaptive-update does not work well, we need to update the learning rate α. To optimize these parameters θ, α, we employ an alternate update strategy by computing the partial derivatives of the above objective function about them. To give a good initialization, we may assign those general trained model of video object segmentation to θ^0 and α^0, which have been adaptively learnt based on the training data. Formally,

$$\mathcal{L}_{k-1} = Loss(y_{i,j}, F(\mathbf{x}_{i,j}, \theta_{k-1})), \tag{3}$$

$$\theta_k = \theta_{k-1} - \alpha_k \odot \bigtriangledown_\theta \mathcal{L}_{k-1}, \tag{4}$$

$$\mathcal{L}_k = Loss(y_{i,j+l}, F(\mathbf{x}_{i,j+l}, \theta_k)), \tag{5}$$

$$\mathrm{grad}_{\theta_k} \leftarrow \mathrm{grad}_{\theta_k} + \nabla^2_{\theta_k} \mathcal{L}_k, \tag{6}$$

$$\mathrm{grad}_{\alpha_k} \leftarrow \mathrm{grad}_{\alpha_k} + \nabla^2_{\alpha_k} \mathcal{L}_k, \tag{7}$$

$$\theta_{k+1} \leftarrow \mathrm{Optimizer}(\theta_k, \mathrm{grad}_{\theta_k}), \tag{8}$$

$$\alpha_{k+1} \leftarrow \mathrm{Optimizer}(\alpha_k, \mathrm{grad}_{\alpha_k}) \tag{9}$$

where $\mathbf{x}_{i,j}$ and $\mathbf{x}_{i,j+l}$ are the inputs for a video i and initial frames j and $j+l$, $y_{i,j}$ and $y_{i,j+l}$ are the expected outputs and $\odot$ denotes the element-wise product. α guides the network how to adapt the online video object segmentation problem. We can find the updates of the learning rate α depends on the $\mathcal{L}_k$, which is a measure on if adaptive-update works well. We expect to find the model of dynamic changes on the learning rate by computing the partial derivatives of the learning rate α, and then we can easily optimize it as well as network weight θ. Concretely, we can perform the optimization in Eq. (9) to search the most suitable learning rate θ.

2.2   Adaptive Online Learning for VOS

Concretely, the adaptive online learning strategy mainly refers to a training and online adaptive method. Using paired video frames as training data, the model can capture the adjustment strategies of the changes between video frames, and acquire the ability to quickly adapt to the new video sequence segmentation model. The adaptive online learning strategy includes four stages: pre-training stage, the effective adaptation strategy learning on the training set, adaptation to the specific video and adaptation to future frames. Specifically,

- Pre-training stage: For a good initialization, we can learn the general segmentation model from the training data. Every sample is used to optimize the

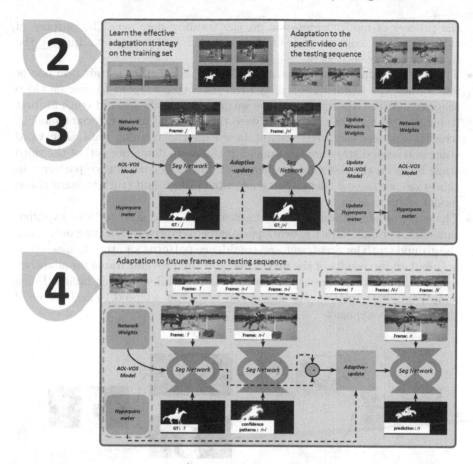

Fig. 2. The adaptive online learning strategy, including pre-training stage, the effective adaptation strategy learning on the training set, adaptation to the specific video and adaptation to future frames. We will train step by step according to the above order. Here we do not show the general pre-training process, which is similar with the general network optimization. Details of our adaptive online learning strategy will be introduced in the paper.

segmentation model on the training set, which is no different from the general model training. The main purpose is to find a good initialization network model.

- The effective adaptation strategy learning on the training set: Some training pairs are extracted from the training set. These pairs of frames from the same video sequence are trained by adaptation-learning, which updates the network weight as well as the hyperparameters. Here we first adaptively update the network model in the previous frame, and then evaluate the effect of adaptive-update model in the latter frame, so as to update the AOL-VOS model. The implementation details are shown in Fig. 2. Here we successfully model the process of model adaptation on the training data set. The purpose of this stage is to learn the rule of change in video sequence by adjusting the hyperparameter such as α.

- The stage of adaptation to the specific video: By further learning on one or few frames, the segmentation model can be better adapted to specific video sequence. At the testing stage, since we only known the ground truth for first frame, the AOL-VOS models θ, α were taken as the main invariant models, which have been adaptively learnt based on the training data. The third stage and second stage are the same, but the main difference is that the third stage focus on the test video sequence and introduce confidence patterns. In this stage, we use confidence patterns instead of ground truth to learn about model adaptation in test videos.

- The stage of adaptation to future frames: there is no need to learn hyperparameters, but only adjust network parameters adaptively. Since we only know the ground truth for first frame and confidence patterns for other frames, so we accumulate the results of the first frame and the previous frame for adaptive-update. In addition, we output the test results while adapting online. Using adaptive-update method, the model can be better adapted to each frame of the test video sequence.

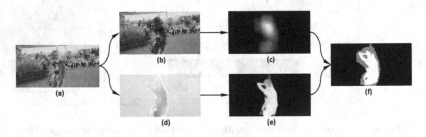

Fig. 3. Illustration of different parts for constructing AOL-seg criterion. (a) Testing frames, (b) Results of Part-based detector, (c) Bounding box mask, (d) Optical Flow, (e) Relative Motion Saliency (RMS), (f) confidence patterns for building the AOL-seg criterion.

2.3   Confidence Pattern

For achieving the adaptation to the specific video and adaptation to future frames for each test video sequence, the most important point is to construct a confidence pattern with these confidence cues, which can better provide guidelines of distinguishing objects and backgrounds in a video. Here we explore these confidence patterns of the objects to build the confidence pattern, where the confidence cues include predicted segmentation results, motion cues and object appearances can be simultaneously considered in it. The detailed confidence patterns can be shown in Fig. 3.

The overall appearance of objects in video often changes greatly, but their local information is very stable. For example, when a person turns around, the

Table 1. Comparison of segmentation performance with our AOL-VOS and existing state-of-the-art methods in terms of mean region similarity $\mathcal{J}$, mean counter accuracy $\mathcal{F}$ and global mean $\mathcal{G}$ on DAVIS16 dataset.

Methods		OnAVOS [22]	OSVOS^{S} [30]	OSVOS [29]	Lucid Tracker [23]	Mask Track [31]	RGMP [21]	MVOS [34]	MoNet [14]	AOL-VOS
DAVIS 16	$\mathcal{J}$	86.1	85.6	79.8	84.8	79.7	81.5	83.3	84.7	86.2
	$\mathcal{F}$	84.9	86.4	80.6	82.3	75.4	82.0	84.1	84.8	87.7
	$\mathcal{G}$	85.5	86.0	80.2	83.6	77.6	81.8	83.7	84.8	87.0

Table 2. Comparison of segmentation performance on the SegTrack V2 dataset

Measure	Mask Track [31]	OBJ Flow [13]	Mask RNN [32]	Lucid Tracker [23]	LSE [33]	MoNet [14]	AOL-VOS
$mIoU[\%]$	70.3	67.5	72.1	78.0	69.7	72.4	80.4

overall appearance has changed a lot, but the corresponding local information does not change much. Therefore, compared with these global-based detectors, local-based detectors which can be fine-tuned from Faster-RCNN [7] can better simulate the appearance of specific objects. When the ratio of the object pixel in the first frame is greater than 70%, we believe that the selected bounding box of the frame is a positive sample, and use it as training data to fine-turn the Faster-RCNN network. Then, we can get the detection results of all frames in the video, as can be illustrated in Fig. 3(b). Finally, considering the detection results based on parts, the boundary box mask is generated, as shown in Fig. 3(c).

In almost all video sequences, the motion of the target object is usually different from that of the background. Optical flow can be used as an important supplementary information to help us obtain the location and boundary information of the target object. In our work, we use Flownet 2.0 [6] to get motion cues, as shown in Fig. 3(d). We further calculate the average values of all optical flows outside these boundaries, which can be regarded as background motion information Ψ. Relative Motion Saliency (RMS) (seen in Fig. 3(e)) is solved by calculating the difference between background optical flow Ψ and optical flow O at each pixels. In order to make RMS more stable, we have added two norms of absolute optical flow standards, whose weight is $\kappa(\kappa = 0.5)$. The calculation formula is defined as follows:

$$RMS_{m,n} = \|O_{m,n} - \Psi\| + \kappa\|O_{m,n}\|, \tag{10}$$

where m and n denote the index of pixels in a video frame.

Finally, we integrate the RMS and bounding box mask to estimate the confidence patterns, as shown in Fig. 3(f). We believe that when the results of the three cues (RMS, bounding box bask and the predicted segmentation result) are significant, we predict them as positive patterns. Conversely, other regions/pixels can be considered negative patterns. When some is significant and the other is not, we label it as uncertain patterns. The confidence patterns, which can provide the guide for adaptation to the specific video and adaptation to future frames, refer to online adaptation rules for these confidence patterns used to determine specific objects.

3   Experiments

3.1   Datasets

On DAVIS16 [25] and SegTrack V2 [24] datasets, we evaluate the effectiveness of our AOL-VOS. The DAVIS16 dataset contains 50 high quality videos (i.e., totally 3455 frames) with all of them segmented with the pixel-level accuracy. 30 training and 20 validation videos are used in DAVIS16. It is the most credible dataset of semi-supervised learning under single object segmentation. Dense annotations given by DAVIS are very expensive, but important for the application of deep learning in video segmentation tasks. For DAVIS16 dataset, we follow the standard protocol for evaluation [25], and report region similarity in terms of intersection over union ($\mathcal{J}$), contour accuracy ($\mathcal{F}$) and the overall measure of performance $\mathcal{G}$. We test our AOL-VOS approach on the testing set and get the better performance compared with those existing video object segmentation methods.

The SegTrack V2 dataset contains 14 low-resolution video sequences which include 976 frames and 24 generic foreground objects. It is a classic dataset of similar tasks and the style and shape of the video included is different. So it's more difficult than DAVIS. For SegTrack V2 dataset, we use the same evaluation criterion (i.e., intersection over union, $mIoU[\%]$) as in [23, 33].

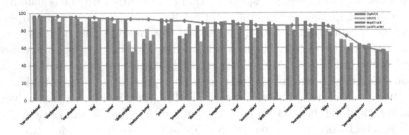

Fig. 4. Comparison of the per-sequence performance with four state-of-the-arts on the DAVIS validation dataset.

3.2   Results and Comparisons

We can find our proposed AOL-VOS and several state-of-the-art video object segmentation approaches [14, 20–23, 29–31, 34] on DAVIS16 dataset as the shown in Table 1. In the mean of region similarity $\mathcal{J}$ on DAVIS16, our AOL-VOS can outperform five existing baselines: 6.4% over OSVOS [29], 1.4% over Lucid Tracker [23], 6.5% over Mask Tracking method [31], 5.7% over RGMP [21], 2.9% over MVOS [34], 5.8% over MoNet method [14] and 0.6% over OSVOS^S [30]. And we can achieve 0.1% mean of region similarity more than the OnAVOS method [22]. For the counter accuracy $\mathcal{F}$ on DAVIS16 evaluation, our AOL-VOS achieves a very large gain, i.e., 7.1%, 1.3% and 12.3% over OSVOS^{S} [30], OSVOS [29] and Mask Tracking method [31], respectively. The improving results of our AOL-VOS from three different aspects (i.e., region similarity, counter accuracy and temporal instability of target masks) can present that the meta-seg approach we adopt can better capture these essential characteristics of the specific target in a video. Further, we compare the mean of region similarity with four state-of-the-arts for each sequence, as can be illustrated in Fig. 4. Generally, our AOL-VOS shows a higher performance than these baseline methods. We can significantly improve the performance in different situations, such as the larger changes of specific target appearance (i.e., in "drift-straight" and "motorcross-jump" videos) and the cluttered background (i.e., in "dance-twirl" video). The AOL-VOS can get comparable results even in the occlusion situations, such as "libby" and "bmx-trees" video sequences. The comparisons with the state-of-the-art methods on SegTrack V2 dataset are given in Table 2. With the intersection over union (IoU) evaluation, our AOL-VOS method can substantially outperform these baselines by 10.1% over Mask Track method [31], 12.9% over OBJ Flow algorithm [13], 8.3% over Mask RNN [32], 2.4% over Lucid Tracker [23], 10.7% over LSE [33], and 8.0% over MoNet [14], respectively.

3.3 Algorithm Analysis

Table 3 shows the results of the proposed online adaptation scheme and multiple variants on DAVIS validation set. Under the full online adaptation method, we can achieve the mean of region similarity $\mathcal{J}$ of 86.2%. Our AOL-VOS framework can lead to 4% improvement, compared with the situation without the per-training on DAVIS dataset. When disabling all learning processes on the

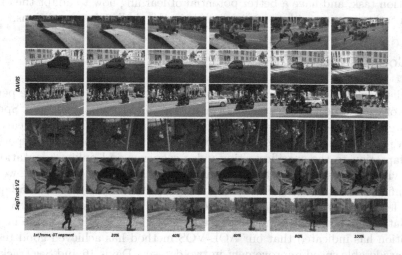

Fig. 5. The results of video sequences in DAVIS and the SegTrack V2 dataset, shows the accuracy and superiority of our method.

Table 3. Comparisons of online adaptation results on DAVIS validation set.

Methods	Measure ($\mathcal{J}$)
AOL-VOS	**86.2**
No pre-training on DAVIS	82.2
No learning on the training set	81.0
No adaptation to the specific video	76.0
No adaptation to future frames	67.7

training set, the result will decrease to 81.0%, which indicates the effectiveness of our online adaptation with the AOL-seg mechanism. We evaluate the situation without adaptation to the specific video, which degrades the performance to 76.0%. When removing adaptation to future frames, we just obtain the 67.7% performance, which significantly drops by 18.4% compared with the AOL-VOS. This demonstrates that all steps in our adaptive online learning strategy can improve its capability of learning to adapt specific patterns in VOS task.

Moreover, the online segmentation results of our AOL-VOS method on both DAVIS and low-resolution SegTrack V2 datasets are shown in the Fig. 5. Especially from the results of the DAVIS dataset, we can find that our method can achieve good results in dealing with segmenting targets from challenging situations, for example, occlusion (i.e., the video sequence of "a dog is passing through woods" in the fourth row), illumination changes (i.e., the video sequence of "a car is running on the shadow road" in the second row), the larger changes of specific target appearance and cluttered background (i.e., the video sequences of in the first and third rows). These above results indicate that our AOL-VOS method can improve the network discriminative capability for video object segmentation task, and have a better potential of learning how to adapt the online segmentation across a video sequence, even in the challenging conditions.

4   Conclusion

The adaptive online learning strategy has been proposed to improve the performance of video object segmentation by adapting to test video sequences. Specifically, we firstly pre-train the segmentation model on a single frame and learn the effective adaptation strategy on the training set. In the test process, we perform the adaptation to the specific testing video sequence and online adaptation to future video frames for predicting the final segmentation results. Besides, we also use confidence patterns in different videos and frames to help construct adaptive online learning strategies. Our Adaptive online learning strategy well balances the relationship between training speed and model accuracy. The comprehensive evaluation has indicated that our AOL-VOS method has achieved good results and considerable speed improvement in two datasets Davis 16 and SegTrack V2.

Acknowledgment. This work was supported by the National Natural Science Foundation of China (Nos. 61972204, 61906094, U1713208), Tencent AI Lab Rhino-Bird Focused Research Program (No. JR201922), the Fundamental Research Funds for the Central Universities (Nos. 30918011321 and 30919011232).

References

1. Venetianer, P.L., et al.: Video surveillance system employing video primitives. Google Patents, US Patent 9,892,606 (2018)
2. Serrano, A., Sitzmann, V., Ruiz-Borau, J., Wetzstein, G., Gutierrez, D., Masiz, B.: Movie editing and cognitive event segmentation in virtual reality video. ACM Trans. Graph. **36**, 47 (2017)
3. Li, W., Mahadevan, V., Vasconcelos, N.: Anomaly detection and localization in crowded scenes. IEEE Trans. Pattern Anal. Mach. Intell. **36**, 18–32 (2014)
4. Zhang, Z., Fidler, S., Urtasun, R.: Instance-level segmentation for autonomous driving with deep densely connected MRFs. In: CVPR, pp. 669–677 (2016)
5. Teichmann, M., Weber, M., Zoellner, M., Cipolla, R., Urtasun, R.: MultiNet: real-time joint semantic reasoning for autonomous driving. In: 2018 IEEE Intelligent Vehicles Symposium (IV), pp. 1013–1020 (2018)
6. Mayer, N., Saikia, T., Keuper, M., Dosovitskiy, A., Brox, T.: FlowNet 2.0: evolution of optical flow estimation with deep networks. In: CVPR, pp. 1647–1655 (2017)
7. Ren, S., He, K., Girshick, R., Sun, J.: Faster R-CNN: towards real-time object detection with region proposal networks. In: NIPS, pp. 91–99 (2015)
8. Ravi, S., Larochelle, H.: Optimization as a model for few-shot learning. In: ICLR, pp. 730–738 (2017)
9. Hu, Y.-T., Huang, J.-B., Schwing, A.G.: Unsupervised video object segmentation using motion saliency-guided spatio-temporal propagation. In: Ferrari, V., Hebert, M., Sminchisescu, C., Weiss, Y. (eds.) ECCV 2018. LNCS, vol. 11205, pp. 813–830. Springer, Cham (2018). https://doi.org/10.1007/978-3-030-01246-5_48
10. Li, S., Seybold, B., Vorobyov, A., Fathi, A., Huang, Q., Jay Kuo, C.-C.: Instance embedding transfer to unsupervised video object segmentation. In: CVPR, pp. 6526–6535 (2018)
11. Li, S., Seybold, B., Vorobyov, A., Lei, X., Kuo, C.-C.J.: Unsupervised video object segmentation with motion-based bilateral networks. In: Ferrari, V., Hebert, M., Sminchisescu, C., Weiss, Y. (eds.) ECCV 2018. LNCS, vol. 11207, pp. 215–231. Springer, Cham (2018). https://doi.org/10.1007/978-3-030-01219-9_13
12. Croitoru, I., Bogolin, S.-V., Leordeanu, M.: Unsupervised learning from video to detect foreground objects in single images. In: ICCV, pp. 4345–4353 (2017)
13. Tsai, Y.-H., Yang, M.-H., Black, M.J.: Video segmentation via object flow. In: CVPR, pp. 3899–3908 (2016)
14. Xiao, H., Feng, J., Lin, G., Liu, Y., Zhang, M.: MoNet: deep motion exploitation for video object segmentation. In: CVPR, pp. 1140–1148 (2018)
15. Hu, P., Wang, G., Kong, X., Kuen, J., Tan, Y.-P.: Motion-guided cascaded refinement network for video object segmentation. In: CVPR, pp. 1400–1409 (2018)
16. Chen, Y., Pont-Tuset, J., Montes, A., Van Gool, L.: Blazingly fast video object segmentation with pixel-wise metric learning. In: CVPR, pp. 1189–1198 (2018)
17. Cheng, J., Tsai, Y.-H., Wang, S., Yang, M.-H.: SegFlow: joint learning for video object segmentation and optical flow. In: ICCV, pp. 686–695 (2017)

18. Xu, M., Fan, C., Wang, Y., Ryoo, M.S., Crandall, D.J.: Joint person segmentation and identification in synchronized first- and third-person videos. In: Ferrari, V., Hebert, M., Sminchisescu, C., Weiss, Y. (eds.) ECCV 2018. LNCS, vol. 11205, pp. 656–672. Springer, Cham (2018). https://doi.org/10.1007/978-3-030-01246-5_39

19. Li, X., Loy, C.C.: Video object segmentation with joint re-identification and attention-aware mask propagation. In: Ferrari, V., Hebert, M., Sminchisescu, C., Weiss, Y. (eds.) ECCV 2018. LNCS, vol. 11207, pp. 93–110. Springer, Cham (2018). https://doi.org/10.1007/978-3-030-01219-9_6

20. Li, X., Loy, C.C.: Video object segmentation with re-identification. In: The 2017 DAVIS Challenge on Video Object Segmentation - CVPR Workshops (2017)

21. Wug Oh, S., Lee, J.-Y., Sunkavalli, K., Joo Kim, S.: Fast video object segmentation by reference-guided mask propagation. In: CVPR, pp. 7376–7385 (2018)

22. Voigtlaender, P., Leibe, B.: Online adaptation of convolutional neural networks for video object segmentation. In: BMVC, pp. 656–672 (2017)

23. Khoreva, A., Benenson, R., Ilg, E., Brox, T., Schiele, B.: Lucid data dreaming for object tracking. arXiv preprint arXiv:1703.09554 (2017)

24. Li, F., Kim, T., Humayun, A., Tsai, D., Rehg, J.M.: Video segmentation by tracking many figure-ground segments. In: ICCV, pp. 2192–2199 (2013)

25. Perazzi, F., Pont-Tuset, J., McWilliams, B., Van Gool, L., Gross, M., Sorkine-Hornung, A.: A benchmark dataset and evaluation methodology for video object segmentation. In: CVPR, pp. 724–732 (2016)

26. Chen, L.-C., Papandreou, G., Schroff, F., Adam, H.: Rethinking Atrous convolution for semantic image segmentation. arXiv preprint arXiv:1706.05587 (2017)

27. Chen, L.-C., Zhu, Y., Papandreou, G., Schroff, F., Adam, H.: Encoder-decoder with Atrous separable convolution for semantic image segmentation. arXiv preprint arXiv:1802.02611 (2018)

28. Voigtlaender, P., Leibe, B.: Online adaptation of convolutional neural networks for the 2017 DAVIS challenge on video object segmentation. In: The 2017 DAVIS Challenge on Video Object Segmentation - CVPR Workshops (2017)

29. Caelles, S., Maninis, K.-K., Pont-Tuset, J., Leal-Taixé, L., Cremers, D., Van Gool, L.: One-shot video object segmentation. In: CVPR, pp. 5320–5329 (2017)

30. Maninis, K.-K., et al.: Video object segmentation without temporal information. IEEE Trans. Pattern Anal. Mach. Intell. **41**(6), 1515–1530 (2018)

31. Perazzi, F., Khoreva, A., Benenson, R., Schiele, B., Sorkine-Hornung, A.: Learning video object segmentation from static images. In: CVPR, pp. 3491–3500 (2017)

32. Hu, Y.-T., Huang, J.-B., Schwing, A.: MaskRNN: instance level video object segmentation. In: NIPS, pp. 325–334 (2017)

33. Ci, H., Wang, C., Wang, Y.: Video object segmentation by learning location-sensitive embeddings. In: Ferrari, V., Hebert, M., Sminchisescu, C., Weiss, Y. (eds.) ECCV 2018. LNCS, vol. 11215, pp. 524–539. Springer, Cham (2018). https://doi.org/10.1007/978-3-030-01252-6_31

34. Xiao, H., Kang, B., Liu, Y., Zhang, M., Feng, J.: Online meta adaptation for fast video object segmentation. IEEE Trans. Pattern Anal. Mach. Intell.(2019)

Proposal-Aware Visual Saliency Detection with Semantic Attention

Lu Wang$^{(\boxtimes)}$, Tian Song, Takafumi Katayama, and Takashi Shimamoto

Department of Electrical and Electronic Engineering, Division of Science and
Technology Graduate School of Technology, Industrial and Social Science,
Tokushima University, Tokushima City, Tokushima 770-8506, Japan
jasmine.lulu.ll@gmail.com,
{tiansong,katayama,simamoto}@ee.tokushima-u.ac.jp

Abstract. In this paper, we propose a proposal based method for
saliency detection. Our method separates the salient proposals out by
assigning them a novel attention mechanism, semantic attention (SeA).
The attention are established based on the observation that regions
with high attention should have similarly semantic concepts with salient
objects. The SeA takes the high-level semantic features from Faster
Region-based Convolutional Neural Network (Faster R-CNN) to assist
the proposal selection in images with complex background. We select
the salient proposals according to their semantic attention probabilities.
Quantitative and qualitative experiments on four datasets demonstrate
that the proposed algorithm performs favorably against the state-of-the-
art methods.

Keywords: Saliency detection · Semantic attention · Object
proposal · Faster Region-based Convolutional Neural Networks

1 Introduction

Visual saliency detection aims to localize the most conspicuous and eye-
attracting object regions in an image. As a fundamental problem of image processing and computer vision, it has been widely used for solving many problems
including visual tracking [1], image compression [2], object detection [3] and
recognition [4]. Although numerous models have been proposed, it is still challenging to localize salient objects accurately in complex scenarios.

Object proposal technology generates a set of class-independent image
regions and assigns each region a score to indicate the probability of a region
being an object. It was first introduced to visual saliency detection by Chang
et al. [5] and boosted by [6,7]. The proposals preserve object-level knowledge
and boundary cues by segmenting all or part of objects from the input image.
However, current techniques usually generate a large number of cropped or background regions along with the salient object proposals, making it hard to train
the downstream models for saliency detection. On the other hand, the original

© Springer Nature Switzerland AG 2019
Z. Cui et al. (Eds.): IScIDE 2019, LNCS 11935, pp. 35–45, 2019.
https://doi.org/10.1007/978-3-030-36189-1_3

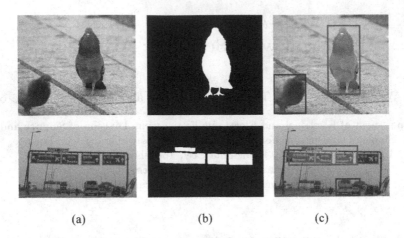

Fig. 1. (a) Input image. (b) Ground truth saliency map. (c) Object proposals generated by [8]. The red proposals have large probabilities to be an object, but they are less salient regarding to the input images. (Color figure online)

probability scores cannot be directly utilized to measure the saliency values of regions. As shown in Fig. 1, the object proposals with red rectangular bounding boxes are fragments of the object "bird" and "coach", but these proposals cannot be regarded as salient regions because they lay on the background of the image. To make full use of object proposals in visual saliency detection, an efficient proposal selection and reweighing mechanism is urgently needed to solve this problem.

Current unsupervised methods address visual saliency detection using bottom-up computation models and low-level feature representations. They usually make heuristic saliency assumptions, such as color contrast [9–12], boundary background [13–15] and center prior [16] to help saliency detection. Since most assumptions are triggered by low-level visual cues, they may obtain unsatisfying results in complex images. Recently, the deep Convolutional Neural Network (CNN) [17] has drawn lots of attention due to its effective capacity for extracting rich features from raw images. Compared with traditional hand-crafted features, the deep CNN features capture more complex semantic concepts appearing in the images. In this paper, we take the advantages of high-level deep features to predict the saliency values of object proposals and compute saliency maps.

We propose an attention mechanism to allow for salient object proposals to dynamically come to the forefront as needed. The attention module assigns each object proposal with an attention probability based on the observation that regions with high attentions usually present similar semantic concepts with the salient objects. Specifically, a Faster R-CNN [18] is trained for object detection. It predicts a large number of object bounding boxes with objectness scores. We utilize their network but train the models for the saliency detection task, which estimates the saliency scores for bounding boxes. Since the bounding boxes only provide rough locations of objects, a fine-grained segmentation is applied on the bounding boxes to separate the salient objects out. We extract the deep features of masked objects and object proposals from the fully connected layer. Instead of generating the saliency maps based on the masked objects, which lose some subtle object parts during the segmentation, we compute the attentions of object proposals according to their feature distances and overlaps with the masked object. The lost object parts can be constructed by conjoining multiple object proposals. By incorporating more human semantic knowledge on the training process, the deep features are capable of representing objects and generating more accurate attention values. We select salient object proposals according to their semantic attention probabilities. The final saliency map is calculated by the weighted average of selected object proposals.

Our work has two key new contributions. (1) We propose an attention models to solve the proposal-aware visual saliency detection. Our method dynamically separates the salient object proposals out as needed by assigning them high-level semantic attention probabilities. The saliency maps generated from selected object proposals are uniformly highlighted with clear object boundaries. (2) We compare the proposed approach with 13 state-of-the-art saliency detection methods on four benchmarks. Our method achieves the best performances under different evaluation metrics.

2   Proposed Algorithm

2.1   Generate Object Proposals

Given an input image, we first generate a set of object proposals using the geodesic object proposal (GOP) method [8]. The GOP takes an image and its boundary probability map as input and then identifies a set of foreground seeds to generate foreground and background masks. Each pair of masks specify an object proposal and generate a score to indicate its probability of being an object. One problem is that some proposals are either too small which cannot be valid segments or too large with strange aspect ratios. We refine the proposal candidate set by filter out unreasonable cases. Proposals whose overlap with the whole image under 2% or bigger than 70% or their self aspect ratio smaller than 10% are not counted. After refinement, we select top N candidates with the largest probabilities for computational efficiency and denote them as $R_i, i \in \{1 \ldots N\}$. The selected object proposals cannot be directly utilized to construct saliency maps. In existing saliency detection datasets, it is common that objects locate on the image corner and only show a small tip. Some subtle objects even scatter around the image and cannot be regarded as salient. For proposals centering on these kinds of objects, their original probabilities of being an object are not suitable to measure the saliency. We propose a semantic attention model to re-weight the object candidates.

2.2   Semantic Attention Model

We propose a semantic attention model which calculates proposals' attention probabilities by comparing their feature distances with the possible salient object. We first estimate the location and shape of salient objects by the Grab-cut segmentation method. Then we fine-tune the Faster R-CNN network on the saliency detection task to extract deep features. These features capture high-level semantic visual patterns and are more robust to complex backgrounds. Since object proposals that are close to salient objects in the feature space should be paid more attention, the semantic attention of each object proposal is predicted according to its feature distance and overlap with the possible salient object.

Training Faster R-CNN. Faster R-CNN network [18] is proposed for object detection. It takes an image as input and outputs a set of bounding boxes and their probability estimates over different object classes. Observing the efficiency and effectiveness of Faster R-CNN, we apply it to visual saliency detection. Different from object detection, which detects and discriminates objects of various categories, saliency detection aims at separating general foreground objects of all classes. In this case, a group of rectangular boxes $B_i, i = \{1, \cdots, N_B\}$ are first generated through the RPN network. During the training phrase, each box will be assigned a class label y. $y = 1$ means current rectangle is positive and its IoU with the salient objects is larger than 0.7. $y = 0$ represents negative samples with the IoUs smaller than 0.2. Rectangles with $0.2 \leqslant IoUs \leqslant 0.7$ are excluded from the training set to avoid the neural network being confused. We take the pre-trained VGG16 model [19] as backbone network and fine-tune it on the saliency detection task. The saliency detection is formulated as a binary classification problem, where the last layer of Faster R-CNN is replaced by a fully connected layer and a sigmoid function. The output is a scalar between 0 and 1 to represent the salient degree. We train neural network by minimizing the following binary cross-entropy loss,

$$L = -\frac{1}{N_B} \sum_{i=1}^{N_B} [y_i \log s_i + (1 - y_i) \log(1 - s_i)] \tag{1}$$

where y_i and s_i are the target label and predicted saliency score of B_i. N_B is the number of selected training samples.

Masked Object. Faster R-CNN can predict the attractive values of object rectangles. We fuse k most salient ones to estimate the location of objects. However, coarse bounding box locations do not contribute to fine-grained saliency detection by missing most detailed shape and contour information. On the other hand, feature representations generated by Faster R-CNN network contains plenty of background noises. As shown in the third column of Fig. 2, backgrounds occupy most area of the bounding box and objects features generated in these cases might be misguided by noisy backgrounds. To obtain objects with accurate shape and contour, one simple strategy is to segment the regions out from bounding boxes. This is realized by a GrabCut method [20], which generates binary maps

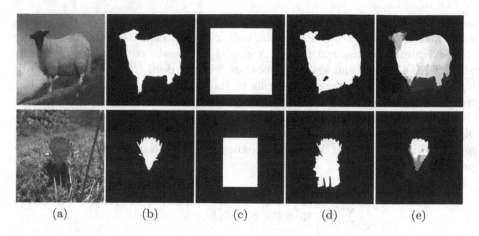

| (a) | (b) | (c) | (d) | (e) |

Fig. 2. Example maps generated in the semantic attention module. (a) Test image, (b) ground truth, (c) bounding box generated by the Faster R-CNN, (d) masked object, (e) semantic attention map.

to indicate the object and background regions. Some examples of masked objects are shown in Fig. 2(d). Compared with separating objects from the whole image directly, we first estimate the possible location of foregrounds and then perform a fine-grained segmentation. The position limitation of bounding boxes avoids GrabCut being disturbed by irrelevant image areas and improves the accuracy of salient object detection. The mask maps are further utilized to calculate semantic attentions.

Semantic Attention Probability. In this section, we evaluate the attractive probabilities of regions by measuring their semantic similarities with masked objects. Although the mask maps can indicate possible locations of salient objects, there are still some subtle parts are misclassified. Take the masked images in Fig. 2 as example, in the first case, the background grassland between sheep's two legs are mis-regarded as foreground. In addition, when objects have similar appearance with backgrounds, it is more challenging to separate them. The binary prediction deteriorates this situation by assigning absolute confidences, either 1 or 0, to misclassified regions. Instead of directly utilizing the masked map, we extract its deep features to predict the attention probabilities of object proposals $R_i, i \in \{1 \dots N\}$ generated in Sect. 2.1. The object proposals contain all or part of shape and contour information of objects. Fusing them together can avoid generated saliency maps being affected by the misclassified areas in mask maps.

Given a rectangular boxes B_i generated by Faster R-CNN, we segment out possible objects according to its GrabCut result and replace the background regions by the mean of B_i. The new generated rectangular is termed as B_i'. To extract more discriminative feature representations of regional proposals, the Faster R-CNN is fine-tuned with B_i' as training samples. B_i and B_i' have the same class labels. During the test phrase, we perform the same operation on the mask map R_M and extract its feature vector v_M through the fine-tuned Faster R-CNN. Similarly, the features of N object proposals are denoted as $v_i, i \in \{1 \dots N\}$. Different from low-level contrast feature, deep CNN features capture high-level semantic concepts appearing in the images and are more powerful in representing images. Proposals that have similar semantic understanding with the masked object are more potential to be salient. The semantic attention highlights proposals that are similar with the masked object in understanding high-level visual patterns. We formulate the semantic attention of each candidate proposal as follows:

$$a'_i = \frac{\exp(-\|v_i - v_M\|_2^2)}{\sum_{j=1}^{N} \exp(-\|v_j - v_M\|_2^2)}, \text{ for } i = 1, \cdots, N, \tag{2}$$

where $\|.\|$ denotes the norm-2 distance of two vectors. To generate the final saliency map, we sort the proposals according to their attention score in a descend order. Top-N' proposals are selected to formulate the saliency map via weighted summation.

3   Experimental Evaluation

3.1   Datasets

We evaluate the proposed method on four benchmark datasets: MSRA-5000 [21], ECSSD [12], SOD [22] and PASCAL-S [23]. The MSRA-5000 dataset [21] has 5000 images with saliency annotations, which is widely used for saliency detection. The ECSSD dataset proposed by Yan *et al.* [12] has $1,000$ images with various complex scenes from the Internet. The SOD dataset consists of 300 challenging images with multiple objects of various sizes and locations. The PASCAL-S dataset [23] contains 850 images with cluttered backgrounds and multiple objects.

3.2   Implementation Details and Evaluation Criteria

Implementation Details. The GOP method [8] generally produces $1,000$ object proposals for each image. We filter out unreasonable ones and select the top $N = 200$ proposals according to their GOP scores. In the semantic attention model, we fine-tune Faster R-CNN on the DUT-OMRON dataset [14]. The initial learning rate is 0.001 and decreases by a factor of 0.1 every $10k$ iterations. The momentum parameter and weight decay are 0.9 and 0.0005 respectively. The training process converges after $200k$ iterations with the stochastic gradient descent optimizer. We fuse the top $N' = 15$ candidates to generate the final saliency map.

Evaluation Criteria. We evaluate the performance of proposed method and other baseline approaches using the precision-recall (PR) curves, F-measure and

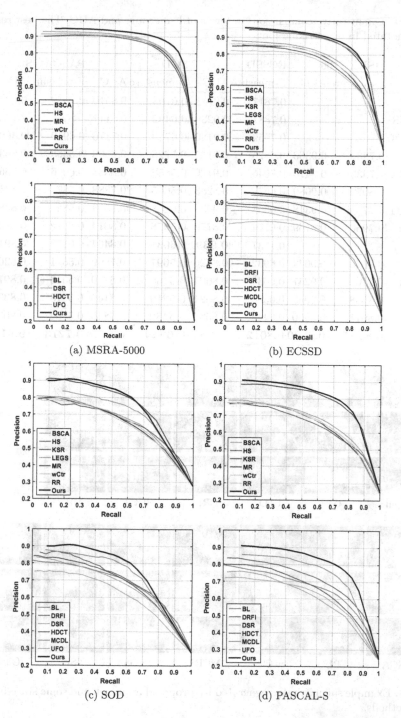

(a) MSRA-5000

(b) ECSSD

(c) SOD

(d) PASCAL-S

Fig. 3. Quantitative comparisons of proposed approach and 13 baseline methods on four saliency detection datasets. The first and second columns are PR curves of different methods and the last column is the precision and recall values under a threshold.

Table 1. Quantitative comparisons in terms of F-measure and AUC. The best results are annotated in overstriking fonts.

*	MSRA-5000		ECSSD		SOD		PASCAL-S	
	F-measure	AUC	F-measure	AUC	F-measure	AUC	F-measure	AUC
KSR	–	–	0.7817	0.9251	0.6703	0.8510	0.7039	0.8970
LEGS	–	–	0.7887	0.9239	0.6492	0.8117	–	–
MCDL	–	–	0.7959	0.9186	0.6789	0.8176	0.6912	0.8699
BL	0.7838	0.9360	0.6825	0.9147	0.5723	0.8503	0.5668	0.8633
BSCA	0.7933	0.9428	0.7046	0.9176	0.5852	0.8364	0.6006	0.8665
RR	0.8072	0.9089	0.6577	0.8283	0.5665	0.7888	0.5873	0.8251
HDCT	0.7733	0.9318	0.6897	0.9039	0.6108	0.8504	0.5824	0.8582
wCtr	0.7960	0.9169	0.6774	0.8779	0.5978	0.8014	0.5972	0.8433
DRFI	–	–	0.7337	**0.9391**	0.6031	0.8464	0.6159	0.8913
MR	0.7575	0.9044	0.6932	0.8820	0.5697	0.7899	0.5881	0.8205
DSR	0.7760	0.9247	0.6636	0.8604	0.5968	0.8210	0.5513	0.8079
HS	0.7668	0.9043	0.6363	0.8821	0.5210	0.8145	0.5278	0.8330
UFO	0.8011	0.8950	0.6442	0.8587	0.5480	0.7840	0.5502	0.8088
Ours	**0.8407**	**0.9540**	**0.8012**	0.9388	**0.6824**	**0.8514**	**0.7213**	**0.9131**

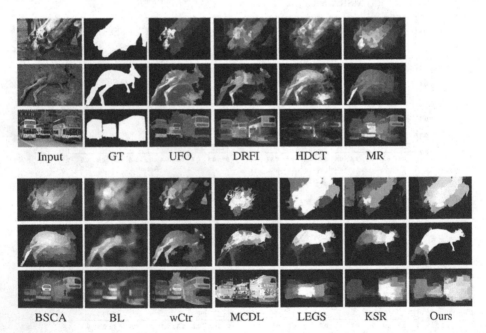

Input   GT   UFO   DRFI   HDCT   MR

BSCA   BL   wCtr   MCDL   LEGS   KSR   Ours

Fig. 4. Example saliency maps generated by proposed approach and some state-of-the-art methods.

Area Under ROC Curve (AUC). Given a saliency map, the precision and recall are computed by segmenting the map with a threshold to construct a binary map, and comparing it with the corresponding ground truth. Taking the average of precision/recall pairs of all images, we can plot the precision-recall curve at different thresholds. The F-measure is an overall performance indicator computed by the weighted harmonic of precision and recall, $F_\Upsilon = \frac{(1+\Upsilon^2)\times Precision\times Recall}{\Upsilon^2\times Precision\times Recall}$. Υ^2 is set to 0.3 to weigh more on precision as suggested in [9]. To better assess the results, we also introduce the AUC which calculates the area under the ROC curve.

3.3 Performance Comparison

We compare the proposed saliency detection model with 13 state-of-the-art methods, including MCDL [24], BL [25], BSCA [26], HS [12], UFO [7], wCtr [15], DRFI [27], DSR [28], HDCT [29], LEGS [30], MR [14], RR [31], KSR [32]. For fair comparison, the saliency maps of different methods are either provided by the authors or achieved by running available codes or softwares.

We perform comparisons of the proposed algorithm and the state-of-the-art saliency detection methods on four datasets as shown in Table 1 and Fig. 3. The F-measure and AUC shown in Table 1 demonstrate our approach achieves the best results under almost all measurements. The KSR [32], LEGS [30], MCDL [24], and DRFI [27] baselines randomly select images and ground truth from MSRA-5000 dataset for training and test, we do not report the experimental results of these four algorithms on the MSRA-5000 dataset for fair comparisons. In addition, the LEGS [30] takes about 300 images from the PASCAL-S dataset as training samples, its results on the PASCAL-S dataset are not included. In Fig. 3, each column represents the precision-recall curves on one dataset. The comparison results in terms of PR curves verify that our proposed method achieve state-of-the-art performance on four datasets. We present some example saliency maps generated by the proposed method and ten state-of-the-art algorithms in Fig. 4. Our method can locate salient regions with great accuracy and highlight the whole object regions with unambiguous boundaries.

4 Conclusion

In this paper, we propose a new framework for visual saliency detection where salient proposals are dynamically separated out according to semantic attention probabilities. High-level semantic features from Faster R-CNN are utilized to select proposals having semantic similarity with object. The semantic attention is robust to images with complex background due to the consideration of high-level object concept. We select the salient proposals by semantic attention, and combine them to generate the final saliency map. We conduct extensive experiments on four challenging datasets and demonstrate that the proposed algorithm outperforms 13 state-of-the-art methods.

References

1. Mahadevan, V., Vasconcelos, N.: Saliency-based discriminant tracking. In: Proceedings of IEEE Conference on Computer Vision and Pattern Recognition, FL, pp. 1007–1013. IEEE (2009)
2. Dhavale, N., Itti, L.: Saliency-based multifoveated MPEG compression. In: Seventh International Symposium on Signal Processing and Its Applications, Paris, vol. 1, pp. 229–232. IEEE (2003)
3. Luo, P., Tian, Y., Wang, X., Tang, X.: Switchable deep network for pedestrian detection. In: Proceedings of IEEE Conference on Computer Vision and Pattern Recognition, OH, pp. 899–906. IEEE (2014)
4. Malik, J., Shi, J.: Normalized cuts and image segmentation. IEEE Trans. Pattern Anal. Mach. Intell. **22**(8), 888–905 (2000)
5. Chang, K.-Y., Liu, T.-L., Chen, H.-T., Lai, S.-H.: Fusing generic objectness and visual saliency for salient object detection. In: Proceedings of IEEE International Conference on Computer Vision, Barcelona, pp. 914–921. IEEE (2011)
6. Li, S., Lu, H., Lin, Z., Shen, X., Price, B.: Adaptive metric learning for saliency detection. IEEE Trans. Image Process. **24**(11), 3321–3331 (2015)
7. Jiang, P., Ling, H., Yu, J., Peng, J.: Salient region detection by UFO: uniqueness, focusness and objectness. In: Proceedings of IEEE International Conference on Computer Vision, NSW, pp. 1976–1983. IEEE (2013)
8. Krähenbühl, P., Koltun, V.: Geodesic object proposals. In: Fleet, D., Pajdla, T., Schiele, B., Tuytelaars, T. (eds.) ECCV 2014. LNCS, vol. 8693, pp. 725–739. Springer, Cham (2014). https://doi.org/10.1007/978-3-319-10602-1_47
9. Achanta, R., Hemami, S., Estrada, F., Susstrunk, S.: Frequency-tuned salient region detection. In: Proceedings of IEEE Conference on Computer Vision and Pattern Recognition, FL, pp. 1597–1604. IEEE (2009)
10. Cheng, M.-M., Zhang, G.-X., Mitra, N., Huang, X., Hu, S.-M.: Global contrast based salient region detection. In: Proceedings of IEEE Conference on Computer Vision and Pattern Recognition, CO, pp. 409–416. IEEE (2011)
11. Liu, Z., Zou, W., Le Meur, O.: Saliency tree: a novel saliency detection framework. IEEE Trans. Image Process. **23**(5), 1937–1952 (2014)
12. Yan, Q., Xu, L., Shi, J., Jia, J.: Hierarchical saliency detection. In: Proceedings of IEEE Conference on Computer Vision and Pattern Recognition, OR, pp. 1155–1162. IEEE (2013)
13. Wei, Y., Wen, F., Zhu, W., Sun, J.: Geodesic saliency using background priors. In: Fitzgibbon, A., Lazebnik, S., Perona, P., Sato, Y., Schmid, C. (eds.) ECCV 2012. LNCS, vol. 7574, pp. 29–42. Springer, Heidelberg (2012). https://doi.org/10.1007/978-3-642-33712-3_3
14. Yang, C., Zhang, L., Lu, H., Ruan, X., Yang, M.-H.: Saliency detection via graph-based manifold ranking. In: Proceedings of IEEE Conference on Computer Vision and Pattern Recognition, OR, pp. 3166–3173. IEEE (2013)
15. Zhu, W., Liang, S., Wei, Y., Sun, J.: Saliency optimization from robust background detection. In: Proceedings of IEEE Conference on Computer Vision and Pattern Recognition, OH, pp. 2814–2821. IEEE (2014)
16. Jiang, Z., Davis, L.S.: Submodular salient region detection. In: Proceedings of IEEE Conference on Computer Vision and Pattern Recognition, OR, pp. 2043–2050. IEEE (2013)
17. Krizhevsky, A., Sutskever, I., Hinton, G.E.: ImageNet classification with deep convolutional neural networks. In: Advances in Neural Information Processing Systems, NV, pp. 1097–1105 (2012)

18. Ren, S., He, K., Girshick, R., Sun, J.: Faster R-CNN: towards real-time object detection with region proposal networks. In: Advances in Neural Information Processing Systems, Montreal, pp. 91–99 (2015)
19. Simonyan, K., Zisserman, A.: Very deep convolutional networks for large-scale image recognition. Computer Science arXiv:1409.1556 (2014)
20. Rother, C., Kolmogorov, V., Blake, A.: GrabCut: interactive foreground extraction using iterated graph cuts. ACM Trans. Graph. **23**(3), 309–314 (2004)
21. Liu, T., et al.: Learning to detect a salient object. IEEE Trans. Pattern Anal. Mach. Intell. **33**(2), 353–367 (2011)
22. Movahedi, V., Elder, J.H.: Design and perceptual validation of performance measures for salient object segmentation. In: Proceedings of IEEE Conference on Computer Vision and Pattern Recognition, CA, pp. 49–56. IEEE (2010)
23. Li, Y., Hou, X., Koch, C., Rehg, J.M., Yuille, A.L.: The secrets of salient object segmentation. In: Proceedings of IEEE Conference on Computer Vision and Pattern Recognition, OH, pp. 280–287. IEEE (2014)
24. Zhao, R., Ouyang, W., Li, H., Wang, X.: Saliency detection by multi-context deep learning. In: Proceedings of IEEE Conference on Computer Vision and Pattern Recognition, MA, pp. 1265–1274. IEEE (2015)
25. Tong, N., Lu, H., Ruan, X., Yang, M.-H.: Salient object detection via bootstrap learning. In: Proceedings of IEEE Conference on Computer Vision and Pattern Recognition, MA, pp. 1884–1892. IEEE (2015)
26. Qin, Y., Lu, H., Xu, Y., Wang, H.: Saliency detection via cellular automata. In: Proceedings of IEEE Conference on Computer Vision and Pattern Recognition, MA, pp. 110–119. IEEE (2015)
27. Jiang, H., Wang, J., Yuan, Z., Wu, Y., Zheng, N., Li, S.: Salient object detection: a discriminative regional feature integration approach. In: Proceedings of IEEE Conference on Computer Vision and Pattern Recognition, OR, pp. 2083–2090. IEEE (2013)
28. Li, X., Lu, H., Zhang, L., Ruan, X., Yang, M.-H.: Saliency detection via dense and sparse reconstruction. In: Proceedings of IEEE International Conference on Computer Vision, NSW, pp. 2976–2983. IEEE (2013)
29. Kim, J., Han, D., Tai, Y.-W., Kim, J.: Salient region detection via high-dimensional color transform. In: Proceedings of IEEE Conference on Computer Vision and Pattern Recognition, OH. IEEE (2014)
30. Wang, L., Lu, H., Ruan, X., Yang, M.-H.: Deep networks for saliency detection via local estimation and global search. In: Proceedings of IEEE Conference on Computer Vision and Pattern Recognition, MA. IEEE (2015)
31. Li, C., Yuan, Y., Cai, W., Xia, Y.: Robust saliency detection via regularized random walks ranking. In: Proceedings of IEEE Conference on Computer Vision and Pattern Recognition, MA, pp. 2710–2717. IEEE (2015)
32. Wang, T., Zhang, L., Lu, H., Sun, C., Qi, J.: Kernelized subspace ranking for saliency detection. In: Leibe, B., Matas, J., Sebe, N., Welling, M. (eds.) ECCV 2016. LNCS, vol. 9912, pp. 450–466. Springer, Cham (2016). https://doi.org/10.1007/978-3-319-46484-8_27

Constrainted Subspace Low-Rank Representation with Spatial-Spectral Total Variation for Hyperspectral Image Restoration

Jun Ye[✉] and Xian Zhang

School of Science, Nanjing University of Posts and Telecommunications,
Nanjing 210023, China
yj8422092@163.com

Abstract. The restoration of hyperspectral Images (HSIs) corrupted by mixed noise is an important preprocessing step. In an HSI cube, the spectral vectors can be separated into different classification based on the land-covers, which means the spectral space can be regarded as an union of several low-rank subspaces. Subspace low-rank representation (SLRR) is a powerful tool in exploring the inner low-rank structure of spectral space and has been applied for HSI restoration. However, the traditional SLRR framework only seek for the rank-minimum representation under a given dictionary, which may treat the structured sparse noise as inherent low-rank components. In addition, the SLRR framework cannot make full use of the spatial information. In this paper, a framework named constrainted subspace low-rank representation with spatial-spectral total variation (CSLRR-SSTV) is proposed for HSI restoration. In which, an artificial rank constraint is involved into the SLRR framework to control the rank of the representation result, which can improve the removal of the structured sparse noise and exploit the intrinsic structure of spectral space more effectively, and the SSTV regularization is applied to enhance the spatial and spectral smoothness. Several experiments conducted in simulated and real HSI datasets demonstrate that the proposed method can achieve a state-of-the-art performance both in visual quality and quantitative assessments.

Keywords: Restoration · Hyperspectral Image · Constrainted subspace · Low-rank · Spatial-spectral total variation

1 Introduction

Hyperspectral images (HSIs) are often degraded by mixed noise (Gaussian noise, salt-and-pepper noise, stripes and dead lines) during the acquisition process unavoidably, which will limit the efficacy of the subsequent applications, such as classification, unmixing, target detection, and so on [1–4]. Therefore, reducing the noise in HSIs is a significant preprocessing step.

J. Ye—This work is funded by the national natural science foundation of China (61771250).

Z. Cui et al. (Eds.): IScIDE 2019, LNCS 11935, pp. 46–57, 2019.
https://doi.org/10.1007/978-3-030-36189-1_4

Up to now, many HSI restoration methods have been proposed. Traditional methods treat every band of the HSI cube as a 2-D natural image, and restore them band by band.

However, this kind of restoration idea ignore the high correlation between the spectral dimension, which is quite important in HSI. Recently, the restoration methods employing both the spatial and spectral information have become the main trend. Othman and Qian [5] proposed a novel wavelet shrinkage model for HSI restoration. Rank-1 tensor decomposition [6], treating the HSI as a cube and utilizing the spatial and spectral information jointly, has also been adopted to restore HSI. Chang *et al.* [7] extend the traditional 2-D TV to 3-D, which explore the smoothness of the spectral dimension in HSI and put forward an anisotropic spatial-spectral total variation (SSTV) regularization. Low-rank (LR) matrix decomposition is a powerful tool in image processing. The LR-based HSI restoration method was proposed by Zhang *et al.* [8] at first, in which the low-rank property among the spectral space in HSIs was well explored. Another LR-based methods [9–11] put up to a good performance as well.

However, these LR-based methods have no constraint on the intrinsic structure of the spectral space. On the basis of land-covers, the spectral vectors in HSI can be separated into different classification, which means the spectral space of the HSI is composed by a union of several LR subspaces. Subspace low rank representation (SLRR) was proposed by Liu *et al.* [12] for subspace segmentation, it is pointed out that SLRR can steadily reconstruct the union structure of the data from LR subspaces by a giving dictionary. SLRR framework has also been utilized to the latest HSI restoration methods [13, 14] by treating the preliminary noise-free estimation of the degraded HSI as the redundant dictionary. Despite their good performance, as the same as these LR-based methods, the SLRR may treat the structured sparse noise, especially the dead lines in different bands located at the same place, as the inherent low-rank component. Actually, the SLRR framework only seek for the rank-minimum representation under a given redundant dictionary by minimizing the nuclear-norm, which may introduce the noise information in the dictionary. Therefore, more precise constraint on the representation results is necessary. In addition, the smoothness in both the spatial and spectral is very significant for the HSI, which is not reflected in the SLRR framework as well.

Based on above considerations, this paper proposes a novel framework, called constrainted subspace low-rank representation with spatial-spectral total variation (CSLRR-SSTV) for HSI restoration. Firstly, an artificial rank constraint is involved into the SLRR framework, termed as CSLRR, to control the rank of the representation result, which can improve the removal of the structured sparse noise and exploit the intrinsic structure of spectral space more effectively. Then, the SSTV regularization is combined into the CSLRR model to ensure the smoothness both in spatial and spectral dimension, which is effective for the Gaussian noise removal.

2 Background

2.1 SLRR Framework

The HSI data degraded by mixed noise can be denoted by $\mathcal{Y}$, and the degenerate problem can be modeled as:

$$\mathcal{Y} = \mathcal{X} + \mathcal{S} + \mathcal{N} \tag{1}$$

where $\mathcal{X}$ is the obtained clean HSI; $\mathcal{S}$ is the sparse noise (impulse noise, dead lines); and $\mathcal{N}$ denotes the Gaussian noise. All of them have the same size, i.e., $M \times N \times p$, where $M \times N$ denotes the spatial dimension and the p is the band number.

Given the matrices Y, X, S and N whose columns are vectorized by the corresponding bands of the HSI $\mathcal{Y}$, $\mathcal{X}$, $\mathcal{S}$ and $\mathcal{N}$ respectively, then the Y, X, S and N have the same size, i.e., $MN \times p$. As mentioned above, the spectral space in HSI can be indicated by a union space united by several LR subspaces. The observed HSI noisy matrix Y in SLRR framework can be modeled as:

$$Y = AZ + S + N \tag{2}$$

where A is a dictionary matrix to reconstruct the clean matrix X, Z is low rank, which denotes the high correlation of the spectral space in HSI. Then we can obtain the optimization problem modeled as below:

$$\min_{Z,S} \|Z\|_* + \lambda \|S\|_1$$
$$s.t. \ \|Y - AZ - S\|_F^2 \le \varepsilon \tag{3}$$

in which, λ is a balance parameter for the sparse noise.

2.2 SSTV Regularization

TV regularization was first proposed in Ref. [15] and has been applied to HSI processing [16, 17] widely because it can preserve the edge features and smooth the spatial information effectively. For a gray-level image u of $M \times N$, the TV norm is defined as:

$$\|u\|_{TV} = \|D_x u\|_1 + \|D_y u\|_1 \tag{4}$$

where D_x and D_y are first-order linear operators corresponding to the horizontal and vertical discrete differences respectively. Different from the 2-D natural images, HSIs have an additional information in the spectral dimension. Thus, for an observed 3-D HSI cube $\mathcal{X}$, the spatial-spectral TV (SSTV) norm can be defined as:

$$\|\mathcal{X}\|_{SSTV} = \|D_x \mathcal{X}\|_1 + \|D_y \mathcal{X}\|_1 + \tau_z \|D_z \mathcal{X}\|_1 \tag{5}$$

where D_z is the forward finite-difference operator for the spectral dimension. The gradient intensity in two spatial dimension is treated equally, and τ_z is the weight coefficient to balance the contribution of the gradient intensity in the spectral dimension. The operators D_x, D_y and D_z are defined as:

$$\begin{cases} D_x \mathcal{X} = \mathcal{X}(x+1, y, z) - \mathcal{X}(x, y, z) \\ D_y \mathcal{X} = \mathcal{X}(x, y+1, z) - \mathcal{X}(x, y, z) \\ D_z \mathcal{X} = \mathcal{X}(x, y, z+1) - \mathcal{X}(x, y, z) \end{cases} \tag{6}$$

3 Proposed Method

3.1 CSLRR Framework

For the SLRR model, a dictionary selection method given in Ref. [12] is set $A = Y$, however, in the denoising problem, Y has a serious deterioration, for which the accuracy of the represented result by Y will be reduced. SLRR-based HSI restoration methods [13, 14] have different choice of the dictionary. In our processing, we adopt a patch-based "SSGoDec" algorithm [18] to restore the noisy HSI in advance, and employ the restoration result as the dictionary, which is no doubt more close to the spectral structure of the real HSI data compared with the noisy Y.

It should be noted that the selected dictionary A is still degraded, which may bring some noise into the restoration results. Besides, as the same as these LR-based methods, the SLRR may treat the structured sparse noise, especially the dead lines in different bands locate at the same place, as the inner low-rank component. Therefore, an another constraint, $rank(Z) \leq r$, of the representation results was added to the SLRR framework, where r is a natural number with small fluctuations and mainly depends on the different HSI datasets. The proposed constrainted subspace low-rank representation (CSLRR) can be defined as follows:

$$\min_{Z,S} \|Z\|_* + \lambda \|S\|_1$$
$$s.t. \ \|Y - AZ - S\|_F^2 \leq \varepsilon, rank(Z) \leq r \tag{7}$$

in which, not only the nuclear-norm is applied as the optimal convex approximation of the rank function, an artificial rank constraint is involved to control the rank of the representation result in further, which can effectively avoid the excessive introducing of the noise in the dictionary and improve the removal of the structured sparse noise.

3.2 Constrainted Subspace Low-Rank Representation with Spatial-Spectral Total Variation

For more comprehensive utilization of the spatial-spectral features in HSI, we introduce the SSTV regularization into the CSLRR model to ensure the smoothness both in spatial and spectral dimension, named CSLRR-SSTV. Let $\mathcal{T}: \mathbb{R}^{MN \times p} \to \mathbb{R}^{M \times N \times p}$

denotes an operator to reshape the columns of the matrix AZ to the corresponding bands of the HSI. Then, the proposed CSLRR-SSTV framework can be modeled as:

$$\min_{Z,S} \|Z\|_* + \lambda\|S\|_1 + \tau\|\mathcal{T}(AZ)\|_{SSTV}$$
$$s.t. \|Y - AZ - S\|_F^2 \le \varepsilon, rank(Z) \le r \tag{8}$$

where τ is the parameter to control the contribution of the SSTV regularization, Z, S, A, λ and ε are the corresponding variables in Eq. (7).

The CSLRR-SSTV model in Eq. (8) employs both the spatial and spectral information in HSI. In which, the CSLRR is introduced to satisfy the low-rank constraint of spectral space more precisely, which can remove the sparse noise in an effective way. Then, the SSTV regularization is utilized to ensure the spatial and spectral smoothness, which can be used to reduce the Gaussian noise with advantage, and profits the decomposition of sparse noise in return.

3.3 Optimization Procedure by IALM

The inexact augmented Lagrange multiplier method (IALM) [19] is employed to solve the optimization problem of Eq. (7). Firstly, three auxiliary variables $J \in \mathbb{R}^{p \times p}$, $\mathcal{U} \in \mathbb{R}^{M \times N \times p}$ and $\mathcal{K} \in \mathbb{R}^{M \times N \times p}$ are involved, and the obtained formulation can be described as:

$$\min_{Z,S,\mathcal{K},J,\mathcal{U}} \|J\|_* + \lambda\|S\|_1 + \tau\|\mathcal{U}\|_1,$$
$$s.t. \|Y - AZ - S\|_F^2 \le \varepsilon, \mathcal{T}(AZ) = \mathcal{K}, Z = J, \mathcal{U} = D\mathcal{K}, rank(J) \le r \tag{9}$$

in which $D = [D_x, D_y, \tau_z D_z]$ denotes the SSTV operator as described in Sect. 2.2. Then minimizing the augmented Lagrangian function as below:

$$L(Z, S, \mathcal{K}, J, \mathcal{U}) = \|J\|_* + \lambda\|S\|_1 + \tau\|\mathcal{U}\|_1 + \langle\Lambda_1, Y - AZ - S\rangle + \langle\Lambda_2, \mathcal{T}(AZ) - \mathcal{K}\rangle$$
$$+ \langle\Lambda_3, Z - J\rangle + \langle\Lambda, \mathcal{U} - D\mathcal{K}\rangle$$
$$+ \frac{\mu}{2}(\|Y - AZ - S\|_F^2 + \|\mathcal{T}(AZ) - \mathcal{K}\|_F^2 + \|Z - J\|_F^2$$
$$+ \|\mathcal{U} - D\mathcal{K}\|_F^2)$$
$$s.t. rank(J) \le r \tag{10}$$

where μ is the penalty parameter, Λ_1, Λ_2, Λ_3 and $\Lambda = [\Lambda_x, \Lambda_y, \Lambda_z]$ are the Lagrangian multipliers. Besides, $\mathcal{U} = [\mathcal{U}_x, \mathcal{U}_y, \mathcal{U}_z]$, and the norm $\|\cdot\|_F^2$ denotes the sum of the squares of all elements for both 2-D and 3-D matrices in this paper.

The solution algorithm for the CSLRR-SSTV model is summarized in Algorithm 1.

Algorithm 1: CSLRR-SSTV solution algorithm

Input: Matrix Y reshaped from noisy HSI $\mathcal{Y}$, stopping criterion ϵ, regularized parameters λ, τ, τ_z and r

Output: Restored matrix AZ

Initialize: $Z = J = \Lambda_3 = 0$, $S = \Lambda_1 = 0$, $\mathcal{K} = \Lambda_2 = 0$, $\mathcal{U} = \Lambda = 0$, $\mu = 10^{-6}$,
$\quad \mu_{max} = 10^6$, $\rho = 1.5$

Repeat until convergence

Update the J, S, Z, $\mathcal{K}$ and $\mathcal{U}$

$$J = \underset{rank(J) \leq r}{\arg\min} \frac{1}{\mu} \|J\|_* + \frac{1}{2} \left\| J - \left(Z + \frac{\Lambda_3}{\mu} \right) \right\|_F^2$$

$$S = \underset{S}{\arg\min} \frac{\lambda}{\mu} \|S\|_1 + \frac{1}{2} \left\| S - \left(Y - AZ + \frac{\Lambda_1}{\mu} \right) \right\|_F^2$$

$$Z = (2A^T A + I)^{-1} \left(A^T Y - A^T S + J + A^T \mathcal{J}^{-1}(\mathcal{K}) + \frac{A^T \Lambda_1 - A^T \mathcal{J}^{-1}(\Lambda_2) - \Lambda_3}{\mu} \right)$$

$$\mathcal{K} = \mathcal{F}^{-1} \left[\frac{\mathcal{F}((\mathcal{J}(AZ) + \Lambda_2/\mu) + D^T(\mathcal{U} + \Lambda/\mu))}{1 + (\mathcal{F}(D_x))^2 + (\mathcal{F}(D_y))^2 + (\mathcal{F}(\tau_z D_z))^2} \right]$$

$$\mathcal{U} = \underset{\mathcal{U}}{\arg\min} \frac{\tau}{\mu} \|\mathcal{U}\|_1 + \frac{1}{2} \left\| \mathcal{U} - \left(D\mathcal{K} - \frac{\Lambda}{\mu} \right) \right\|_F^2$$

Update the parameter $\mu := min(\rho\mu, \mu_{max})$

Update the Lagrangian multipliers

$$\begin{cases} \Lambda_1 = \Lambda_1 + \mu(Y - AZ - S) \\ \Lambda_2 = \Lambda_2 + \mu(\mathcal{J}(AZ) - \mathcal{K}) \\ \Lambda_3 = \Lambda_3 + \mu(Z - J) \\ \Lambda = \Lambda + \mu(\mathcal{U} - D\mathcal{K}) \end{cases}$$

Check the convergence conditions

$$max\{\|Y - AZ - S\|_\infty, \|\mathcal{J}(AZ) - \mathcal{K}\|_\infty, \|Z - J\|_\infty, \|\mathcal{U} - D\mathcal{K}\|_\infty\} < \epsilon$$

---

4 Experiments

For testing the performance of the proposed CSLRR-SSTV framework in the restoration of HSI degraded by mixed noise, both simulated and real HSI data were employed in our experiments. For more thoroughly capability of the CSLRR-SSTV, three different HSI restoration methods were selected for comparison, i.e., LRMR [8], LRLSDS [20] and FSSLRL [21].

LRMR is a classical LR-based HSI restoration method, which divides the HSI cube into many over-lapped blocks and uses low rank matrix recovery (LRMR) to restore these blocks. LRRSDS and FSSLRL were proposed by Sun *et al.* successively in recent two years, under the low-rank property of the spectral space in HSI, the two methods were constructed by introducing spectral difference features and spatial similarity information separately, and have been proved to perform well in the removal of HSI mixed noise.

4.1   Simulated Data Experiments

Two HSI data sets were employed for our simulated experiments. The first is a subset (256 × 256 × 191) of Washington DC mall (WDC), and the second is a subset (256 × 256 × 103) of Pavia University (PaviaU). Before the simulated experiments, gray values of this two data sets were mapped to [0, 1].

The mean peak signal-to-noise ratio (MPSNR), the mean structural similarity index metric (MSSIM) [22] and the mean spectral angle (MSA) were applied as the quantitative evaluation. Then, three different kinds of noises were added to the simulated data sets as follows:

(1)   Zero-mean Gaussian noise was added to all the bands with $\sigma = 0.01$.
(2)   Salt-and-pepper noise with the intensity of 20% was randomly added to 10 bands.
(3)   5 structured dead lines were added to 10 bands, in which 5 bands were selected randomly from the salt-and-pepper noise bands, and 5 bands were selected randomly from the rest bands.

Figure 1 shows the restoration results of band 167 in the WDC subset. For comparison more convenient, a small block in the red box was selected to enlarge and place it under the corresponding results. The band 167 was added with all the three kinds of noise. It is clear that LRMR and LRRSDS cannot remove the Gaussian noise completely, and as presented in this Fig, LRMR handles this image excessively and LRRSDS restoration result still has some residual traces of the mixed noise. FSSLRL can remove the mixed noise effectively while keep a better value of MPSNR, which means this method is more suitable for restoring the WDC dataset. By contrast, the proposed CSLRR-SSTV can achieve more precise results in the preservation of edge features and details due to the advantages of TV. In addition, Fig. 2 displays the restoration results of band 30 of PaviaU image, which was also degraded with all the three kinds of noises. The residual traces of dead lines can be clearly distinguished in Fig. 2(c), and the restoration effects of CSLRR-SSTV are much better than LRRSDS and FSSLRL as well in this diagrammatic sketch.

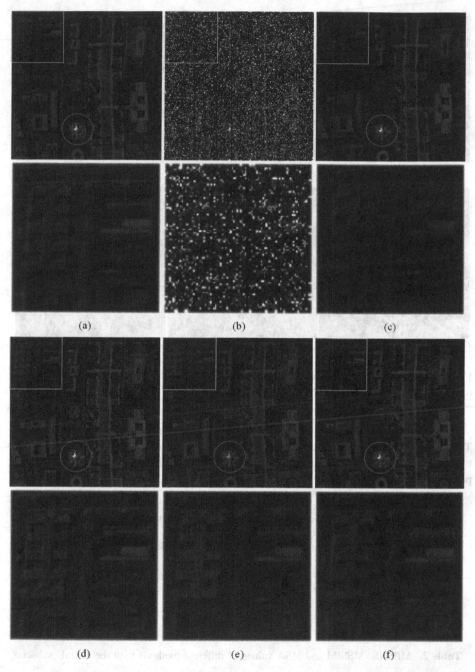

Fig. 1. Restoration results of band 167 in the WDC image: (a) Original band, (b) Noisy band, (c) LRMR, (d) LRRSDS, (e) FSSLRL, (f) CSLRR-SSTV.

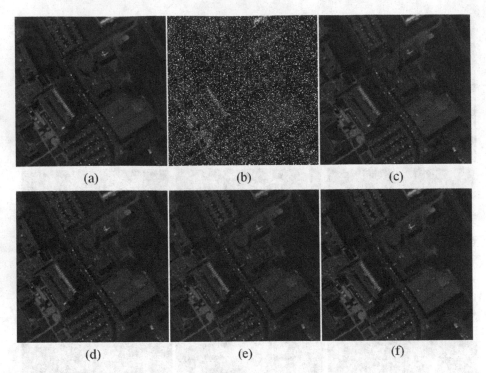

Fig. 2. Restoration results of band 30 in the PaviaU image: (a) Original band, (b) Noisy band, (c) LRMR, (d) LRRSDS, (e) FSSLRL, (f) CSLRR-SSTV.

The quantitative evaluations for the two simulated experiments are summarized in Tables 1 and 2 separately. It can be obtained that the proposed CSLRR-SSTV achieves the best values in all the quantitative indexes, especially in the WDC subset, the promotion is more evident.

Table 1. MPSNR, MSSIM and MSA values of different methods with the WDC subset.

Method	Noisy	LRMR	LRRSDS	FSSLRL	CSLRR-SSTV
MPSNR	20.1884	33.0665	32.0734	33.9622	**35.4236**
MSSIM	0.4278	0.9248	0.9082	0.9426	**0.9480**
MSA	0.4757	0.1084	0.1221	0.0912	**0.0749**

Table 2. MPSNR, MSSIM and MSA values of different methods with the PaviaU subset.

Method	Noisy	LRMR	LRRSDS	FSSLRL	CSLRR-SSTV
MPSNR	19.7145	31.9121	32.0364	33.0748	**33.2640**
MSSIM	0.3215	0.8560	0.8310	0.8903	**0.8943**
MSA	0.5742	0.1370	0.1395	0.1131	**0.1085**

For more detailed comparison of all the restoration methods, Figs. 3 and 4 present the different PSNR and SSIM values calculated from all the band of the restoration results. It is easy to recognize that CSLRR-SSTV achieves the best PSNR and SSIM values in almost all the bands, especially in those bands corrupted by mixed noises.

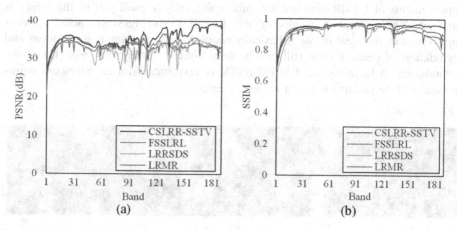

Fig. 3. PSNR and SSIM values of each band in the WDC subset of the reconstructed results on different methods. (a) PSNR values. (b) SSIM values.

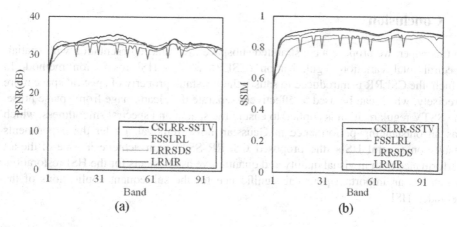

Fig. 4. PSNR and SSIM values of each band in the PaviaU subset of the reconstructed results on different methods. (a) PSNR values. (b) SSIM values.

4.2   Real Data Experiments

The HYDICE Urban (Urban) image was employed for the real data experiments. The original image is 256 × 256 × 191, and is heavily corrupted by stripes, dead lines, the atmosphere and other unknown noise in the band of 1–4, 76, 87, 101–111, 198–210, etc.

In our experiment, a subset ($150 \times 150 \times 191$) of this image with all bands were applied to demonstrate the efficiency of the LRSR-SSTV for removing the mixed noises.

Figure 5 shows the diagrammatic sketches of band 208 in the Urban subset before and after restoration. It can be clearly observed the original image of this band is polluted completely, the acquisition of texture features is almost impossible. The improvements of LRMR obtained are quite finite, only a small part of the noises is removed. The restoration results of the LRRSDS and FSSLRL achieve better improvements, the ideal image is basically restored, but the stripe noise residues and degradation of detail texture still can be seen directly. Comparing with the above methods, Fig. 5 indicates that CSLRR-SSTV is very successful in the mixed noises removal and the reconstruction of texture features.

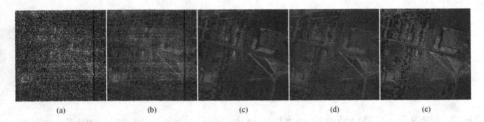

(a)   (b)   (c)   (d)   (e)

Fig. 5. Restoration results of band 208 in the Urban image. (a) Original, (b) LRMR, (c) LRLSDS, (d) FSSLRL, (e) CSLRR-SSTV.

5   Conclusion

In this paper, we propose a constrainted subspace low-rank representation with spatial-spectral total variation regularization (CSLRR-SSTV) HSI restoration method. In which, the CSLRR is introduced to satisfy the low-rank property of spectral space more precisely, which can be used to effectively separate the clean image from sparse noise, the SSTV regularization is applied to ensure the spatial and spectral smoothness, which has an outstanding performance in Gaussian noise removal. Under the experiments results on several HSIs, the proposed CSLRR-SSTV can achieve a state-of-the-art performance both in visual quality and quantitative assessments for the HSI restoration, which has an important practical significance for the subsequent applications of the degraded HSI.

References

1. Lv, F., Han, M., Qiu, T.: Remote sensing image classification based on ensemble extreme learning machine with stacked autoencoder. IEEE Access **5**, 9021–9031 (2017)
2. Bo, C., Lu, H., Wang, D.: Weighted generalized nearest neighbor for hyperspectral image classification. IEEE Access **5**, 1496–1509 (2017)
3. Yang, W., et al.: Two-stage clustering technique based on the neighboring union histogram for hyperspectral remote sensing images. IEEE Access **5**(4), 5640–5647 (2017)

4. Zhang, H., et al.: Spectral–spatial sparse subspace clustering for hyperspectral remote sensing images. IEEE Trans. Geosci. Remote Sens. **54**(6), 1–13 (2016)
5. Othman, H., Qian, S.: Noise reduction of hyperspectral imagery using hybrid spatial-spectral derivative-domain wavelet shrinkage. IEEE Trans. Geosci. Remote Sens. **44**(2), 397–408 (2006)
6. Guo, X., et al.: Hyperspectral image noise reduction based on rank-1 tensor decomposition. ISPRS J. Photogram. Remote Sens. **83**(9), 50–63 (2013)
7. Chang, Y., et al.: Anisotropic spectral-spatial total variation model for multispectral remote sensing image destriping. IEEE Trans. Image Process. **24**(6), 1852–1866 (2015)
8. Zhang, H., et al.: Hyperspectral image restoration using low-rank matrix recovery. IEEE Trans. Geosci. Remote Sens. **52**(8), 4729–4743 (2014)
9. Xue, J., et al.: Joint spatial and spectral low-rank regularization for hyperspectral image denoising. IEEE Trans. Geosci. Remote Sens. **56**(4), 1940–1958 (2018)
10. Zhao, Y., Yang, J.: Hyperspectral image denoising via sparse representation and low-rank constraint. IEEE Trans. Geosci. Remote Sens. **53**(1), 296–308 (2014)
11. Fan, F., et al.: Hyperspectral image denoising with superpixel segmentation and low-rank representation. Inf. Sci. Int. J. **397**(C), 48–68 (2017)
12. Liu, G., et al.: Robust recovery of subspace structures by low-rank representation. IEEE Trans. Pattern Anal. Mach. Intell. **35**(1), 171–184 (2013)
13. Wang, M., Yu, J., Niu, L., Sun, W.: Hyperspectral image denoising based on subspace low rank representation. In: Yuan, H., Geng, J., Bian, F. (eds.) GRMSE 2016. CCIS, vol. 699, pp. 51–59. Springer, Singapore (2017). https://doi.org/10.1007/978-981-10-3969-0_7
14. Wang, M., Yu, J., Sun, W.: LRR-based hyperspectral image restoration by exploiting the union structure of spectral space and with robust dictionary estimation. In: IEEE International Conference on Image Processing (2017)
15. Rudin, L.I., Osher, S., Fatemi, E.: Nonlinear total variation based noise removal algorithms. Physica D **60**(1–4), 259–268 (1992)
16. Addesso, P., et al.: Hyperspectral image inpainting based on collaborative total variation. In: IEEE International Conference on Image Processing (2017)
17. He, W., Zhang, H., Zhang, L.: Total variation regularized reweighted sparse nonnegative matrix factorization for hyperspectral unmixing. IEEE Trans. Geosci. Remote Sens. **55**(99), 3909–3921 (2017)
18. Zhou, T., Tao, D.: GoDec: randomized low-rank & sparse matrix decomposition in noisy case. In: Proceedings of the 28th International Conference on Machine Learning, WA, USA, pp. 33–40 (2011)
19. Lin, Z., Chen, M., Ma, Y.: The augmented Lagrange multiplier method for exact recovery of corrupted low-rank matrices. Eprint arXiv (2010)
20. Sun, L., et al.: Hyperspectral image restoration using low-rank representation on spectral difference image. IEEE Geosci. Remote Sens. Lett. **14**(7), 1151–1155 (2017)
21. Sun, L., et al.: Fast superpixel based subspace low rank learning method for hyperspectral denoising. IEEE Access **6**, 12031–12043 (2018)
22. Wang, Z., et al.: Image quality assessment: from error visibility to structural similarity. IEEE Trans. Image Process. **13**(4), 600–612 (2004)

Memory Network-Based Quality Normalization of Magnetic Resonance Images for Brain Segmentation

Yang Su[1,2], Jie Wei[2], Benteng Ma[2], Yong Xia[1,2(✉)], and Yanning Zhang[2]

[1] Research & Development Institute of Northwestern Polytechnical University in Shenzhen, Shenzhen 518057, China
yxia@nwpu.edu.cn
[2] National Engineering Laboratory for Integrated Aero-Space-Ground-Ocean Big Data Application Technology, School of Computer Science and Engineering, Northwestern Polytechnical University, Xi'an 710072, China

Abstract. Medical images of the same modality but acquired at different centers, with different machines, using different protocols, and by different operators may have highly variable quality. Due to its limited generalization ability, a deep learning model usually cannot achieve the same performance on another database as it has done on the database with which it was trained. In this paper, we use the segmentation of brain magnetic resonance (MR) images as a case study to investigate the possibility of improving the performance of medical image analysis via normalizing the quality of images. Specifically, we propose a memory network (MemNet)-based algorithm to normalize the quality of brain MR images and adopt the widely used 3D U-Net to segment the images before and after quality normalization. We evaluated the proposed algorithm on the benchmark IBSR V2.0 database. Our results suggest that the MemNet-based algorithm can not only normalize and improve the quality of brain MR images, but also enable the same 3D U-Net to produce substantially more accurate segmentation of major brain tissues.

Keywords: Medical image quality normalization · Deep learning · Magnetic resonance image · Brain tissue segmentation

1 Introduction

It has been widely acknowledged in the medical imaging community that the images of the same modality but acquired at different centers, on different scanners, using different protocols, and/or by different operators may have substantially variable quality, which brings significant difficulties not only to clinical practices, but also to computer-aided diagnosis (CAD). Since CAD systems nowadays are mostly based on deep learning, a data-driven technique, their performance is highly dependent on the quantity and quality of training images. As a result of the variable quality of medical images, a

Z. Cui et al. (Eds.): IScIDE 2019, LNCS 11935, pp. 58–67, 2019.
https://doi.org/10.1007/978-3-030-36189-1_5

deep learning model usually cannot achieve the same performance on another database as it has done on the database with which it was trained. Comparing to designing the deep learning models with a high generalization ability, normalizing the quality of medical images used in the same application may provide a more feasible and effective way to improve the generalization capabilities of CAD systems.

In this work, we use the brain magnetic resonance (MR) image segmentation [1, 2] as a case study to investigate the approach to the quality normalization of brain MR images and verify the effect of such image quality normalization on the segmentation performance. Brain MR images may have variable quality with respect to noise, soft tissue contrast, intensity non-uniformity (INU), and partial volume effect (PVE), which render a major challenge for major brain tissue segmentation using MR imaging. Many methods have been proposed in the literature to improve the quality of MR images. Yang et al. [3] employed the Wasserstein generative adversarial network (WGAN) with the perceptual loss for medical image denoising, which transfers the knowledge of visual perception to the image denoising task and is capable of not only reducing the image noise level but also trying to keep the critical information at the same time. Ran et al. [4] introduced an MR image denoising method based on the residual encoder-decoder Wasserstein generative adversarial network, which performs well in the both simulated data and real clinical data. Chang et al. [5] introduced a high-order and L_0-regularized variational model for INU correction and brain extraction, which is an efficient multi-resolution algorithm for fast computation. Feng et al. [6] proposed a local inhomogeneous intensity clustering model to estimate INU which has been extensively tested on both synthetic and real images. Zhan and Yang [7] estimated INU with histogram statistical analysis which takes full advantage of a physical property of the tissue anatomical structure and the smoothly varying property of INU. Shinohara et al. [8] presented biologically motivated normalization techniques for multi sequence brain MR images, which are simple but robust. Shafee et al. [9] classified brain voxels into major tissues by dividing a low-resolution voxel into multiple high-resolution sub-voxels to address PVE. These methods, however, consider only one image degradation factor and aim to improve the quality of each image, instead of normalizing a group of images into similar quality.

In this paper, we propose a persistent memory network (MemNet)-based algorithm to normalize the quality of MR images and adopt the 3D U-Net, a widely used deep learning model for volumetric data segmentation, to segment the gray matter (GM), white matter (WM), and cerebrospinal fluid (CSF) using the brain MR images before and after quality normalization. We evaluated the proposed algorithm on the Internet Brain Segmentation Repository (IBSR V2.0) database and achieved substantially improved segmentation accuracy.

The contributions of this work are two-fold: (1) we propose a MemNet-based algorithm to normalize the quality of brain MR image; and (2) we demonstrate that normalizing the image quality is able to improve substantially the accuracy of major brain tissue segmentation in MR images.

2 Data

The IBSR V2.0 database [12, 13] was used for this study, which consists of 18 clinical brain MRI images acquired on 18 subjects. Each image has been spatially normalized into the Talairach orientation and resliced into a dimension of $256 \times 256 \times 128$ with a voxel size of $1.0 \times 1.0 \times 1.5$ mm^3. Skull stripping has been applied to these images for the removal of non-brain tissues [13].

3 Method

Let $X \in R^{W \times H}$ represent a quality degraded MR image of size $W \times H$, and $Y \in R^{W \times H}$ represent the corresponding quality normalized MR image. The relationship between these two MR images can be formally expressed as follows

$$Y = \Phi(X) \tag{1}$$

where $\Phi : R^{W \times H} \to R^{W \times H}$ denotes the quality normalization process. Then, our target is to estimate the process Φ and use it to reconstruct the normalized high-quality MR image Y based on the observed low quality image X.

3.1 MemNet

For this study, the quality normalization process Φ is approximated by MemNet [10], which consists of the feature extraction layer, multiple stacked memory blocks, and reconstruction layer (see Fig. 1).

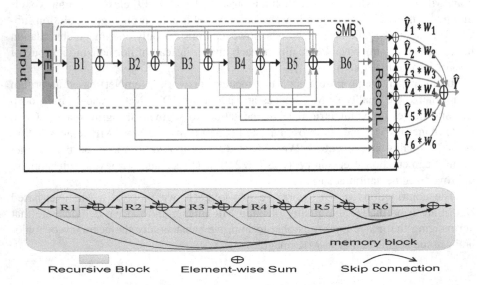

Fig. 1. Illustration of MemNet for the quality normalization of MR images. MemNet consists of feature extraction layer (FEL), multiple stacked memory blocks (SMB) and reconstruction layer (ReconL). The structure of each memory block is shown in the bottom part.

The feature extraction layer is a plain convolutional layer that extracts the features from the low-quality input image X of size 256×256, shown as follows

$$F_0 = f_{ext}(X) \tag{2}$$

where f_{ext} denotes the feature extraction function, and F_0 is the extracted feature maps which will be fed to multiple stacked memory blocks.

The multiple stacked memory blocks consists of six stacked memory blocks, denoted by B1 to B6 in Fig. 1. Each memory block contains six recursive units, denoted by R1 to R6 in Fig. 1, each further consisting of two stacked convolution layers. These six units recursively learn to generate multi-level representations under different receptive fields. Then the aggregation layer achieves persistent memory through an adaptive learning process. We use a 1×1 convolutional layer to accomplish the gating mechanism that can learn adaptive weights for different memories. As a result, the weights for the long-term memory control how much of the previous states should be reserved, and the weights for the short-term memory determine how much of the current state should be stored. Thus, the feature maps produced by the m-th memory block (denoted by M_m) can be formally expressed as follows

$$F_m = M_m(F_{m-1}) = M_m(M_{m-1}(\dots(M_1(F_0))\dots)) \tag{3}$$

The reconstruction layer is a convolutional layer, denoted by f_{rec}. Since simply using the output of the last memory block can hardly produce promising results, we feed the output of each block to the reconstruction layer, and use the weighted sum of the output to approximate the residual image. Let the reconstruction result of m-th block be formally expressed as

$$\hat{Y}_m = f_{rec}(F_m) + X \tag{4}$$

The high quality MR image estimated by MemNet is calculated as follows

$$\hat{Y} = \sum_{m=1}^{6} w_m * \hat{Y}_m \tag{5}$$

where the weights w_i can be optimized during the training. To fully use the features at different states, we add the supervision to each memory block using the same reconstruction architecture. Thus, the loss function has two parts: (1) the Euclidean loss between final output $\hat{Y}$ and the ground truth Y and (2) the Euclidean loss between each output $\hat{Y}_m$ and the ground truth, shown as follows

$$\mathcal{L}(\Theta) = \frac{\alpha}{12N} \sum_{m=1}^{6} \sum_{i=1}^{N} \left\| Y^{(i)} - \hat{Y}_m^{(i)} \right\|^2 + \frac{1-\alpha}{2N} \sum_{i=1}^{N} \left\| Y^{(i)} - \hat{Y}^{(i)} \right\|^2 \tag{6}$$

where Θ is the parameter set in MemNet, and α is the scale weight for the Euclidean loss. We empirically set the scale weight α to 0.5 for this study.

3.2 Brain MR Image Segmentation

We adopted 3D U-Net [11] to segment the brain MR images before and after quality normalization. Like the standard U-Net, 3D U-Net has an analysis and a synthesis path. In the analysis path, there are four convolution-pooling blocks. Each block contains two $3 \times 3 \times 3$ convolutional layers, each followed by a Rectified Linear Unit (ReLU) activation function, and a $2 \times 2 \times 2$ max pooling with a stride of two. In the synthesis path, there are four deconvolution-pooling blocks. Each block consists of an up convolution of $2 \times 2 \times 2$ by a stride of two in each dimension, followed by two $3 \times 3 \times 3$ convolutions, and a ReLU follows each convolution. Feature maps from higher resolutions are concatenated to up sampled feature maps by shortcut connections. The segmentation was performed on a patch-by-patch basis. Each patch has a size of $32 \times 32 \times 32$ and is extracted with a stride of 3, 9, and 3 along three directions.

4 Experiments and Results

4.1 Experimental Settings

Since the IBSR V2.0 database has only 18 brain MR images, we adopted the leave-one-out validation scheme. In each trial, we first used 17 images to train the MemNet and 3D U-Net, respectively, and then applied the trained MemNet and trained 3D U-Net to the other image.

To train the MemNet, we need a high-quality version of each training image, which we created by (1) using the segmentation ground truth to determine the volume of WM and (2) assigning the average voxel value of WM to each WM voxel. We initialized all trainable variables in MemNet using the Xavier method [16] and adopted the Adam algorithm with a batch size of 1 as the optimizer. We set the initial learning rate to 0.1 and the maximum number of epochs to 12000. The 3D U-Net was randomly initialized and optimized also by the Adam solver with a batch size of 32. We set the initial learning rate to 0.02.

The visual quality of normalized brain MR images was measured by the structural similarity index (SSIM) [15], since there are complex structures in the brain volume. Meanwhile, the performance of image quality normalization was evaluated by the improvement of the brain segmentation accuracy, which was measured by the Dice similarity coefficient (DSC) [14].

4.2 Visual Quality of Normalization Images

Figure 2 shows two example slices selected from the brain MR studies IBSR V2.0_02 and IBSR V2.0_14, respectively, their quality normalized version, the segmentation results of the slices before and after quality normalization, and the segmentation ground truth. To better visualize the difference between the image and segmentation results before and after the quality normalization, we marked a rectangular region in each image, enlarged it, and displayed it beneath the image. It shows that, comparing to the ground truth, the noise and contrast between GM and WM are obviously improved in the quality normalized slices and the segmentation results of quality normalized slices are more similar to the ground truth.

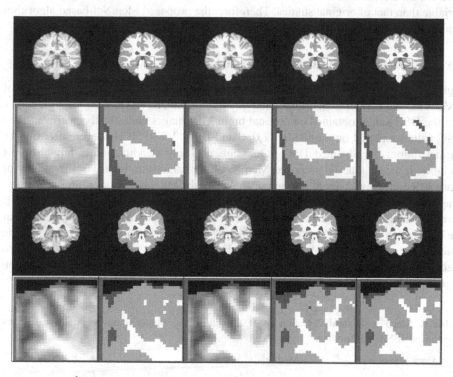

Fig. 2. The 49th coronal slice from two brain MR studies: IBSR V2.0_02 (top) and IBSR V2.0_14 (bottom). From left to right, five columns are the original MR images, segmentation of original images, quality normalized MR images, segmentation of quality normalized image, and segmentation ground truth. For better visualization, a rectangular region highlighted in red in each image was enlarged and displayed beneath each image. (Color figure online)

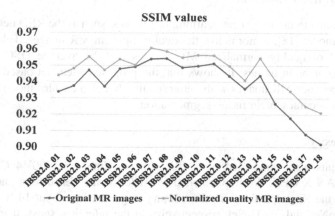

Fig. 3. SSIM values of IBSR V2.0 studies before and after quality normalization.

For a quantitative evaluation, we depicted the SSIM of each study before and after the quality normalization in Fig. 3. It reveals that, although the SSIM varies over different studies, the SSIM of quality normalized studies is steadily higher and less variable than that of original studies. Therefore, the proposed MemNet-based algorithm is able to normalize the visual quality of IBSR V2.0 studies.

4.3   Improvement of Segmentation Accuracy

Figure 4 gives the DSC values obtained by applying 3D U-Net to the segmentation of GM, WM, and CSF in each original and quality normalized IBSR V2.0 study. It shows that the DSC values obtained on original brain MR images varies badly, with very low DSC values reported on the IBSR V2.0_6 and IBSR V2.0_11. However, after quality normalization, the obtained DSC values are not only steadily higher but also less variable. It means that the proposed MemNet-based algorithm improves the quality of brain MR images and, more important, makes brain MR images have similar quality. The mean and standard deviation of obtained DSC values were displayed in Table 1. It reveals that normalizing the quality of IBSR V2.0 studies improves the average DSC of the delineation of GM, WM, and CSF by 4.13%, 3.58%, and 16.11%, respectively. The quantitative results in Fig. 4 and Table 1 demonstrate that using our MemNet-based algorithm is able to improve the segmentation accuracy even without change the segmentation method, which is consistent with the aforementioned conclusion.

Table 1. Mean ± standard deviation of DSC values (%) obtained on original and quality normalized IBSR V2.0 studies

Data	Methods	Type		
		GM	WM	CSF
Original images	3D U-Net	90.70 ± 1.61	90.74 ± 1.09	69.17 ± 5.06
Quality normalized images		**94.83 ± 0.28**	**94.32 ± 0.31**	**85.28 ± 1.02**

4.4   Comparative Evaluation

We also attempted to use other deep learning techniques, such as the RED network [17] and DnCNN model [18], to normalize the quality of brain MR images. Table 2 gives average SSIM of quality normalized MR images and the average DSC values of delineating major brain tissues. It shows that the proposed MemNet-based algorithm can not only produce MR images with better quality, but also enables the 3D U-Net to generate more accuracy brain tissue segmentation.

4.5   Complexity

Our experiments were conducted on a PC with two Intel E5-2600V4 CPU and a NVIDIA TITAN X GPU. We implemented MemNet on the Tensorflow platform and 3D U-Net on the Keras platform. In the training stage, it takes about 14 h and 4 h to train the MemNet and 3D U-Net, respectively. In the inference stage, it takes about

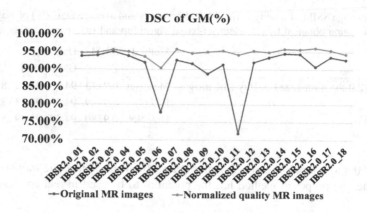

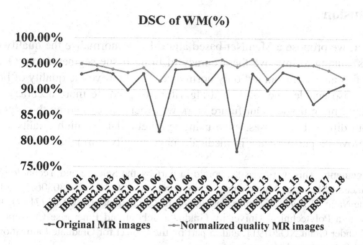

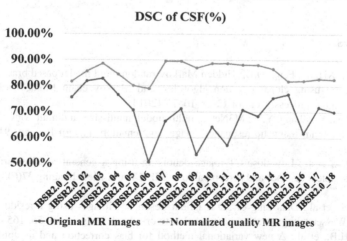

Fig. 4. DSC values obtained by applying 3D U-Net to the segmentation of GM, WM, and CSF in each original and quality normalized IBSR V2.0 study.

Table 2. Average SSIM of quality normalized MR images and average DSC (%) of major brain tissue segmentation obtained by our MemNet-based algorithm and two deep learning models.

Data	Methods	SSIM	DSC (%)		
			GM	WM	CSF
IBSR V2.0 the normalized quality MR images	MemNet	**0.9473**	**94.83**	**94.32**	**85.28**
	RED	0.9449	94.43	93.33	85.14
	DnCNN	0.9380	94.39	93.49	84.50

0.8 s and 50 s to normalize and segment a brain MR image. Since the training can be done offline, the proposed method has the potential to be used in real applications.

5   Conclusion

In this paper, we propose a MemNet-based algorithm to normalize the quality of brain MR images, aiming to improve the accuracy of brain tissue segmentation. Our results shows that this algorithm can not only normalize and improve the quality of brain MR images, but also enables the 3D U-Net algorithm to produce more accuracy segmentation of major brain tissues. Our future work will focus on extending this algorithm to images from different databases, even using synthetic data as high quality images to further improve its performance in medical image quality normalization.

Acknowledgement. This work was supported in part by the Science and Technology Innovation Committee of Shenzhen Municipality, China, under Grants JCYJ20180306171334997, and in part by the National Natural Science Foundation of China under Grants 61771397, in part by the Northwestern Polytechnical University Graduate School and Enterprise Cooperative Innovation Fund under Grant XQ201911, and in part by the Project for Graduate Innovation team of Northwestern Polytechnical University.

References

1. Zhang, T., Xia, Y., Feng, D.D.: Hidden Markov random field model based brain MR image segmentation using clonal selection algorithm and Markov chain Monte Carlo method. Biomed. Signal Process. Control **12**(1), 10–18 (2014)
2. Wei, J., Xia, Y., Zhang, Y.N.: M3Net: a multi-model, multi-size, and multi-view deep neural network for brain magnetic resonance image segmentation. Pattern Recogn. **91**, 366–378 (2019)
3. Yang, Q.S., et al.: Low-dose CT image denoising using a generative adversarial network with Wasserstein distance and perceptual loss. IEEE Trans. Med. Imaging **37**(6), 1348–1357 (2018)
4. Ran, M.S., et al.: Denoising of 3D magnetic resonance images using a residual encoder–decoder Wasserstein generative adversarial network. Med. Image Anal. **55**, 165–180 (2019)
5. Chang, H.B., et al.: A new variational method for bias correction and its applications to rodent brain extraction. IEEE Trans. Med. Imaging **36**(3), 721–733 (2017)

6. Feng, C.L., Zhao, D.Z., Huang, M.: Image segmentation and bias correction using local inhomogeneous iNtensity clustering (LINC): a region-based level set method. Neurocomputing **219**, 107–129 (2017)
7. Zhan, S., Yang, X.: MR image bias field harmonic approximation with histogram statistical analysis. Pattern Recogn. Lett. **83**, 91–98 (2016)
8. Shinohara, R.T., et al.: Statistical normalization techniques for magnetic resonance imaging. NeuroImage Clin. **6**, 9–19 (2014)
9. Shafee, R., Buckner, R.L., Fischl, B.: Gray matter myelination of 1555 human brains using partial volume corrected MRI images. NeuroImage **105**, 473–485 (2015)
10. Tai, Y., Yang, J., Liu, X.M., Xu, C.Y.: MemNet: a persistent memory network for image restoration. In: IEEE International Conference on Computer Vision, pp. 4539–4547 (2017)
11. Çiçek, Ö., Abdulkadir, A., Lienkamp, S.S., Brox, T., Rooneberger, O.: 3D U-Net: learning dense volumetric segmentation from sparse annotation. In: International Conference on Medical Image Computing and Computer-Assisted Intervention, pp. 234–241 (2015)
12. School, M.G.H.H.M.: The Internet Brain Segmentation Repository (IBSR). http://www.cma.mgh.harvard.edu/ibsr/index.html
13. Rohlfing, T.: Image similarity and tissue overlaps as surrogates for image registration accuracy: widely used but unreliable. IEEE Trans. Med. Imaging **31**(2), 153–163 (2012)
14. Bharatha, A., et al.: Evaluation of three-dimensional finite element-based deformable registration of pre- and intra-operative prostate imaging. Med. Phys. **28**(12), 2551–2560 (2001)
15. Wang, Z., Bovik, A.C., Sheikh, H.R., Simoncelli, E.P.: Image quality assessment: from error visibility to structural similarity. IEEE Trans. Image Process. **13**(4), 600–612 (2004)
16. Glorot, X., Bengio, Y.: Understanding the difficulty of training deep feedforward neural networks. In: 13th International Conference on Artificial Intelligence and Statistics (2010)
17. Mao, X., Shen, C., Yang, Y.: Image restoration using very deep convolutional encoder-decoder networks with symmetric skip connections. In: Neural Information Processing Systems (2016)
18. Zhang, K., Zuo, W., Chen, Y., Meng, D., Zhang, L.: Beyond a gaussian denoiser: Residual learning of deep CNN for image denoising. IEEE Trans. Image Process. **26**(7), 3142–3155 (2017)

Egomotion Estimation Under Planar Motion with an RGB-D Camera

Xuelan Mu$^{(\boxtimes)}$, Zhixin Hou, and Yigong Zhang

School of Computer Science and Engineering, Nanjing University of
Science and Technology, Nanjing 210094, People's Republic of China
{117106010701,zxhou}@njust.edu.cn

Abstract. In this paper, we propose a method for egomotion estimation
of an indoor mobile robot under planar motion with an RGB-D camera.
Our approach mainly deals with the corridor-like structured scenarios
and uses the prior knowledge of the environment: when at least one
vertical plane is detected using the depth data, egomotion is estimated
with one normal of the vertical plane and one point; when there are no
vertical planes, a 2-point homography-based algorithm using only point
correspondences is presented for the egomotion estimation. The proposed
method then is used in a frame-to-frame visual odometry framework. We
evaluate our algorithm on the synthetic data and show the application on
the real-world data. The experiments show that the proposed approach is
efficient and robust enough for egomotion estimation in the Manhattan-
like environments compared with the state-of-the-art methods.

Keywords: Egomotion estimation · Indoor scene · RGB-D camera ·
Planar motion · Visual odometry

1 Introduction

Egomotion estimation is an intensively discussed issue in computer vision, which
aims at understanding the six-degree-of-freedom (6-DoF) transformation (three
for the rotation and three for the translation) of the visual sensor with reference
to the input sequence of images. It has drawn a lot of attentions in numerous
applications such as augmented reality, motion control and autonomous naviga-
tion [3, 4, 14, 20]. The term visual odometry (VO) was originally presented in the
work [18] in 2004, which is the process of evaluating the egomotion of an agent
such as mobile robot with only the input of a single or multiple cameras mounted
to it. The approaches of VO can be classified into two major categories: one is
the optical flow method based on pixel information [2, 5, 15], and the other is the
vision-based method [8, 22]. Compared with the first one, the vision-based indirect method is much more robust because of its use of discernible feature points
from image. Hence, this paper focuses on the study of feature-based egomotion
estimation method which belongs to the second one.

For a calibrated camera, it needs only five points to estimate the 6-DOF pose
between two consecutive views [17] while seven or eight points [6] are needed if

© Springer Nature Switzerland AG 2019
Z. Cui et al. (Eds.): IScIDE 2019, LNCS 11935, pp. 68–79, 2019.
https://doi.org/10.1007/978-3-030-36189-1_6

the camera is not calibrated. Specifically, in the case of a 99-percent probability of success and a set of data with half-rate outliers, the linear 8-point essential matrix algorithm [7] requires about 1177 samples whereas the 5-point essential matrix algorithm [11, 13, 23] only needs 145 trials. Therefore, finding an approach with minimal points to meet the real-time requirements is necessary. To reduce the amount of needed point correspondences between frames, some reasonable hypotheses combining additional sensor data are needed. Consequently, based on the availability of low-cost RGB-D cameras, we mainly investigate the motion of the mobile robot in indoor environments where at least one plane is present in the scene. The RGB-D camera is mounted rigidly on a mobile robot and the robot is always under planar motion because of the flat indoor floor. So the pitch and roll angles remain constant during the whole process. Assuming the roll and pitch angles of the camera as known, we correct the RGB-D camera through rotating the generated point clouds so that the values of the roll and pitch angles are approximately equal to zeros. In this case, we just need to calculate a three-degree-of-freedom egomotion estimation problem, which consists of two horizontal translations and one yaw angle.

Despite of its many developments, egomotion estimation still remains somewhat challenging to be efficient and robust in structural and low-texture scenes (e.g., wall or hallway). To solve this problem, several SLAM systems using high geometric characteristics such as lines and planes have been proposed recently [1, 9, 10, 12]. While, Kim et al. [10] estimated the drift-free rotation by applying a mean-shift algorithm with the surface normal vector distribution. However this method required at least two orthogonal planes for demonstrating superior rotation estimation. Kaess [9] introduced a minimal representation with planar features using a hand-held RGB-D sensor and presented a relative plane formulation, which improved the convergence for faster pose optimization. While this method required plane extraction and matching at each frame to construct optimization function, and additional odometry sensors were utilized to perform plane matching, which increased the complexity for VO system.

In contrast to these methods, in this paper we obtain a rough plane-segmentation result using only RGB-D frame, which is fast to meet the real-time application instead of segmenting the scenes into very precise planar regions. We directly estimate the subsequent egomotion through extracting the normal in each plane from the inverse-depth-induced histograms. We will take different strategies according to the prior knowledge about the 3D scenarios. The major contributions of this work are twofold: First, if there exists at least a vertical plane, we realize the pose and location estimation with the normal of the vertical plane and one point correspondence, which is called the direction-plus-point algorithm. Second, if the plane orientation is completely not available, we propose an efficient 2-point minimal case algorithm for the homography-based method to estimate the egomotion. Compared with the classical 5-pt essential matrix estimation [17], our method just needs two matched points between views instead of five, which speeds up the process of iterative optimization for egomotion estimation. At last, we evaluate our algorithm on both synthetic and real datasets.

The rest of the paper is organized as follows. Section 2 describes two efficient algorithms for estimating the egomotion under the weak Manhattan-world assumption. Section 3 presents the performance of our solutions on synthetic and real data and compares with the classical method in a quantitative evaluation. Finally, in Sect. 4, conclusions are drawn.

2 Ego-Motion Estimation

In VO, compared with the estimation of translation which is relatively simple, the estimation of rotation deserves more attention. The majority of experimental errors are derived from the inaccurate estimation of rotation. Thus, reducing the accumulated error caused by rotation can greatly improve the performance of the algorithm. In this work, if there exists at least one vertical plane in the scene, we estimate the motion by decoupling rotation and translation so that a drift-free rotation estimation can be derived from the alignment of local Manhattan frames. Based on the accurate rotational motion, we can obtain a robust estimation of translation with 1-point RANSAC approach. In addition, we propose a new 2-pt minimal-case algorithm with the simplified motion model while no vertical planes are known.

2.1 The Plane-Plus-Point Algorithm

We design a fast plane segmentation method through using only the depth image based on a RGB-D camera, where we extract the inverse depth induced horizontal and vertical histograms to detect planes instead of segmenting the huge point clouds directly. We view the whole indoor scenery as a composition of one or several local Manhattan structures. We can recognize at least one local Manhattan coordinate frame according to the detected vertical planes at any time. And the pose estimation is simplified when knowing the vertical direction in the scenes. Then through aligning the measured vertical direction with the camera coordinate system, the y-axis of the camera is parallel to the vertical planes while the x-z-plane of the camera is parallel to the ground plane (illustrated in Fig. 1). This alignment can make relative motion reduce to a 3-DOF motion, which includes 1 DOF of the remaining rotation and 2 DOF of the translation (i.e., a 3D translation vector up to scale).

In general, we can detect only one ground plane, but multiple vertical planes which correspond to different Descartes coordinate frames may be obtained at the same time. Therefore, we need to distinguish the dominant Manhattan frame from the minor ones based on the specified Descartes coordinate frames attached to the walls. For each Descartes coordinate frame, we calculate its evaluation score according to the area of all the vertical planes in the RGB image whose normal (in the depth image) is approximately parallel or perpendicular to each other. Among them we choose the frame whose score is the largest as the dominant local Manhattan frame.

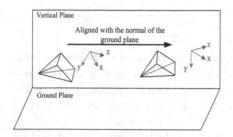

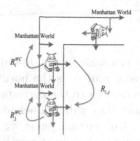

Fig. 1. Alignment of the camera with the normal of the ground plane.

Fig. 2. Rotation between two successive times.

Hence, the drift-free rotation between two successive views will be obtained as soon as the dominant Manhattan frame has been determined at each moment. As shown in Fig. 2, assuming the robot is in the same Manhattan structure (frame) denoted at two different times t_i and t_j, we can estimate only one rotation $R_{i,j}$ of C_i with respect to C_j:

$$R_{i,j} = (R_i^{wc})^T R_j^{wc} \tag{1}$$

where C_i and C_j are respectively the camera coordinate frame at the continuous time t_i and t_j and where R_i^{wc} and R_j^{wc} denote the rotations of C_i and C_j with respect to the local Manhattan coordinate.

Knowing the rotation information based on Manhattan world constraints, the translation estimation needs to be obtained through other algorithms such as a 1-point RANSAC method. We detect and match corner points from two successive RGB images and get the 3D points through the depth image. Then the translation T is estimated with one 3D point correspondence:

$$T = P' - (R_{i,j})^T P \tag{2}$$

where P and P' are respectively the current 3D points and the previous 3D points.

2.2 The 2-Point Homography-Based Algorithm

In the real-world indoor environments the vertical plane is not always available. In this case, the plane-plus-point algorithm which is proposed in the previous section does not work. Therefore we propose a new 2-point minimal case algorithm for the homography matrix based method, and we do a local refinement between the current and previous plane according to the Manhattan assumption to eliminate the drift if the vertical plane is detected again. Given $p_i = [x_i, y_i, 1]^T$ and $p_j = [x_j, y_j, 1]^T$, which are points on the ground plane in the first and second camera coordinate frames, the homography constraint is defined as:

$$\sigma p_j = H p_i \tag{3}$$

With

$$H = R - \frac{t}{d}N^T \qquad (4)$$

Where σ is a scale factor, R and t are the rotation matrix and translation vector respectively, and N is the normal vector of the 3D plane and d is the distance from the camera to the corresponding plane. Giving two 3D points in the world coordinate, they will uniquely define a virtual vertical plane and the unit normal vector of this virtual plane with respect to the i_{th} view is $n = [n_x, 0, n_z]^T$ (if they are not vertically aligned). Let $\frac{t}{d} = [t_x\ 0\ t_z]^T$. The homography induced by this virtual vertical plane can be written as:

$$H = \begin{bmatrix} \cos\theta & 0 & \sin\theta \\ 0 & 1 & 0 \\ -\sin\theta & 0 & \cos\theta \end{bmatrix} - \begin{bmatrix} t_x \\ 0 \\ t_z \end{bmatrix} \begin{bmatrix} n_x \\ 0 \\ n_z \end{bmatrix}^T, \qquad (5)$$

$$= \begin{bmatrix} \cos\theta - n_x t_x & 0 & \sin\theta - n_z t_x \\ 0 & 1 & 0 \\ -\sin\theta - n_x t_z & 0 & \cos\theta - n_z t_z \end{bmatrix}. \qquad (6)$$

There are four unknown elements of a 3×3 homography matrix in (6), therefore this matrix can be parametrized as:

$$H = \begin{bmatrix} h_1 & 0 & h_2 \\ 0 & 1 & 0 \\ h_3 & 0 & h_4 \end{bmatrix}. \qquad (7)$$

In order to eliminate the scalar factor in (3), we use cross product instead and obtain:

$$p_j \times Hp_i = 0. \qquad (8)$$

Combining (7) and (8), we have the following relation:

$$\begin{bmatrix} x_j \\ y_j \\ 1 \end{bmatrix} \times \begin{bmatrix} h_1 & 0 & h_2 \\ 0 & 1 & 0 \\ h_3 & 0 & h_4 \end{bmatrix} \begin{bmatrix} x_i \\ y_i \\ 1 \end{bmatrix} = 0. \qquad (9)$$

It gives us three linear equations, but cross product can also be expressed as a skew-symmetric matrix product, and the rank of the skew-symmetric $[p_j]_\times$ is two, only two linearly independent equations are achieved. By choosing the first two (9) can be rearranged into:

$$\begin{bmatrix} x_i y_j & y_j & 0 & 0 \\ 0 & 0 & x_i y_j & y_j \end{bmatrix} \begin{bmatrix} h_1 \\ h_2 \\ h_3 \\ h_4 \end{bmatrix} = \begin{bmatrix} x_j y_i \\ y_i \end{bmatrix}. \qquad (10)$$

One point correspondence gives two constrains and $[h_1, h_2, h_3, h_4]$ can be uniquely determined by two point correspondences if they are not vertically aligned. Note that, for different two point correspondences, the parameters $[d, n_x, n_y]$ of the plane which is determined by these two points are not the same. Once $[h_1, h_2, h_3, h_4]$ is obtained from two point correspondences, let's consider the following relations:

$$\begin{cases} n_x t_x = cos\theta - h_1, \\ n_z t_x = sin\theta - h_2, \\ n_x t_z = -sin\theta - h_3, \\ n_z t_z = cos\theta - h_4, \end{cases} \tag{11}$$

By multiplying the first equation by the forth one and multiplying the second equation by the third one, we obtain:

$$\begin{cases} n_x n_z t_x t_z = (cos\theta - h_1)(cos\theta - h_4), \\ n_x n_z t_x t_z = (sin\theta - h_2)(-sin\theta - h_3), \end{cases} \tag{12}$$

The left part of the two equations in (12) is identical. Therefore, the following relation can be obtained by associating the right parts:

$$(h_1 + h_4) \cos \theta + (h_2 - h_3) \sin \theta + h_2 h_3 - h_1 h_4 - 1 = 0, \tag{13}$$

with:

$$\sin^2 \theta + \cos^2 \theta = 1. \tag{14}$$

Using (13) and (14) we can compute $\cos \theta$ and $\sin \theta$, which have two possible solutions. The rotation of the camera motion can directly be derived from the $\cos \theta$ and the $\sin \theta$.

Then the normal vector can be obtained by dividing both sides of the first equation of (12) by the second one:

$$\frac{n_x}{n_z} = \frac{cos\theta - h_1}{sin\theta - h_2}, \tag{15}$$

with:

$$n_x^2 + n_z^2 = 1. \tag{16}$$

Finally, the translation up to scale is given by:

$$t = d \left[\frac{cos\theta - h_1}{n_x} \quad 0 \quad \frac{cos\theta - h_4}{n_z} \right]^T. \tag{17}$$

3   Experiments

To evaluate the performance of the proposed egomotion estimation method, both the synthetic data and the real-world data are used for the experiments. The synthetic data with ground truth is used to compare our 2-point algorithm with another minimal solution, the 5pt-essential method [17]. The real-world datasets are provided with two scenes, one for the laboratory building and the other for the dormitory building. Each scene that satisfies weak Manhattan constrains is captured by using a robot mounted with a Kinect v2.

3.1 Test with Synthetic Data

The synthetic data sets are generated in the following setup. The scene contains of 500 randomly sampled 3D points totally and the focal length of the camera is set to 1000 pixels with a field of view of 50°. The average distance from the first camera to the scene is set to 1 and the base line between two cameras is set to be 15% of the average scene distance. Since we focus on the indoor robot motion estimation, the second camera is rotated around y−axis with the relative rotation angle varying from $-15°$ to $15°$. The translation is set parallel to the ground and the moving direction is set into two situations, along the x-axis (sideways) and along the z-axis (forward), respectively. This is similar to Nister's test scene in [19], which has been used in [21].

As the estimated translation is up to a scalar factor, we compare the angle between the ground-truth and estimated translation vector. The errors are defined as follows:

$$\begin{cases} \xi_R = |\theta_g - \theta_e|, \\ \xi_t = arccos((t_g^T t_e)/(\|t_g\| \|t_e\|)), \end{cases} \tag{18}$$

where ξ_R is the rotation error and ξ_t is the translation error, the errors are similar to [17]. The θ_g, t_g denote the ground-truth rotation angle and translation, and θ_e, t_e are the corresponding estimated rotation angle and translation vector, respectively.

We evaluate each algorithm under the image noise (corner location) with a different standard deviation and the increased $(Roll, Pitch)$ noise. The noise can be considered as the error of the normal of the ground plane. In our experiments, we assume that there are enough feature points lying on the ground. Half of them are randomly generated on the ground plane and the rest are in the 3D space above the ground plane. We use the least square solution with all the inliers and plot the mean value of 1000 trials with different points and different transformations.

Figures 3 and 4 shows the results of the 5pt-essential matrix algorithm and our 2pt-homography method. The experiments show that our 2pt-homography matrix algorithm outperforms the state-of-art 5pt-essential method, in terms of the rotation error and the translation error. It appears that the 5pt-essential method is more sensitive to $(Pitch, Roll)$ noise while our 2pt-homography matrix algorithm is more robust. Notice, when we use this algorithm for real application, a two point RANSAC method can be used to reject outliers and the final solution is given by the least square method with all the inliers.

3.2 Performance with Real Data

In order to show the efficiency of the proposed frame-to-frame visual odometry framework, several real datasets taken in two different indoor corridor-like environments using an RGB-D camera mounted on a robot have been collected, as

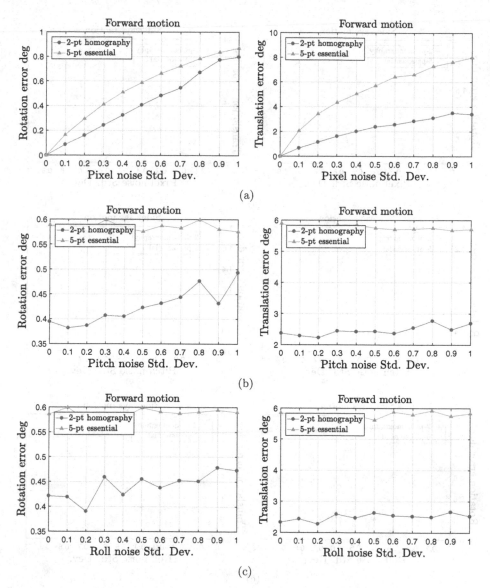

Fig. 3. Rotation and translation error for forward motion. Comparing the 5pt-essential matrix algorithm with our 2pt-homography method. Left column: Rotation error, right column: Translation error. (a) is with varying image noise. (b) is with increased Pitch noise and 0.5 pixel standard deviation image noise. (c) is with increased Roll noise and 0.5 pixel standard deviation image noise.

shown in Fig. 5. The scenes are full of low textured walls and the image resolution used is 960×540 pixels. All the experiments are run at 10 FPS on an Intel Core i5-4460 desktop computer with 3.20 GHz CPU, without GPU acceleration.

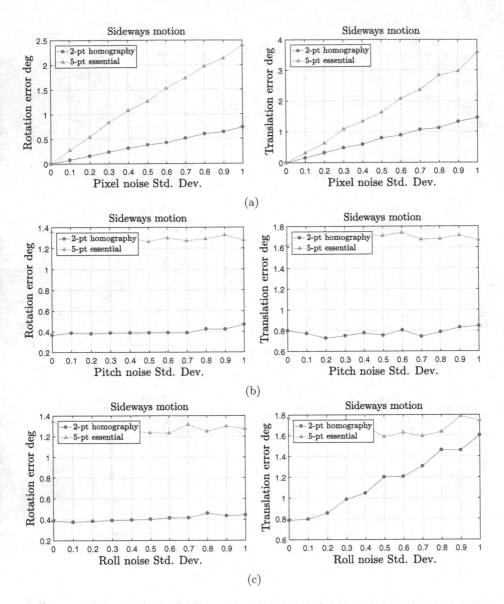

Fig. 4. Rotation and translation error for sideways motion. Comparing the 5pt-essential matrix algorithm with our 2pt-homography method. Left column: Rotation error, right column: Translation error. (a) is with varying image noise. (b) is with increased Pitch noise and 0.5 pixel standard deviation image noise. (c) is with increased Roll noise and 0.5 pixel standard deviation image noise.

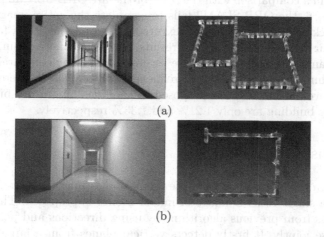

Fig. 5. (a) The laboratory building. (b) The dormitory building. First column: Example images in the school building. Second column: Reconstructed point cloud.

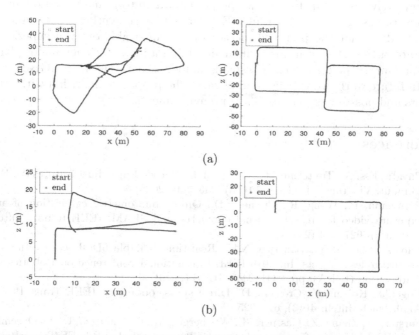

Fig. 6. Comparison between the ORB-RGBD SLAM and our proposed frame-to-frame visual odometry framework. (a) Results on the laboratory building. (b) Results on the dormitory building. First column: Trajectories of ORB-RGBD SLAM. Second column: Trajectories of our method.

We perform a comparison with the state-of-the-art ORB-SLAM2 [16]. As can be seen in Fig. 6, the ORB-SLAM2 fails to complete the entire image sequence because of lacking of features in low-textured walls and then do a false relocation. While our method achieves better results, the performance of our algorithm, using only frame-to-frame camera pose estimation, can be comparable to that of the algorithm with some non-linear refinement or loop closure detection algorithms. The overall errors of our proposed method the laboratory building and the dormitory building are only 1.21% and 1.33% respectively.

4   Conclusion

In this paper, a new method for accurate egomotion estimation of the Manhattan Frame from a single RGB-D image of indoor scenes is proposed. The proposed method differs from previous algorithms by using directions and points to estimate the pose jointly. It firstly detects vertical planes from a large number of RGB-D datasets if at least one vertical plane is available. The normal of the vertical plane is obtained directly based on the inverse-depth induced histograms and we estimate the pose through a novel 3-DOF VO. Secondly, we propose a new minimal-case algorithm to estimate the egomotion if the plane orientation is completely unknown. Finally, we propose a frame-to-frame visual odometry framework based on our algorithms. Experiments with synthetic data and real data validate that the proposed methods are comparable or even superior to the state-of-the-art algorithms while maintaining a high efficiency under planar motion. Our method is currently tested in indoor sceneries with an RGB-D camera. In future work, we will try to implement the proposed algorithm with other sensors and possibly extend to different environments.

References

1. Bay, H., Ess, A., Tuytelaars, T., Van Gool, L.: Speeded-up robust features (SURF). Comput. Vis. Image Underst. **110**(3), 346–359 (2008)
2. Bergmann, P., Wang, R., Cremers, D.: Online photometric calibration of auto exposure video for realtime visual odometry and SLAM. IEEE Robot. Autom. Lett. **3**(2), 627–634 (2018)
3. Cao, Z., Sheikh, Y., Banerjee, N.K.: Real-time scalable 6DOF pose estimation for textureless objects. In: 2016 IEEE International conference on Robotics and Automation (ICRA), pp. 2441–2448. IEEE (2016)
4. Engel, J., Koltun, V., Cremers, D.: Direct sparse odometry. IEEE Trans. Pattern Anal. Mach. Intell. **40**(3), 611–625 (2018)
5. Forster, C., Zhang, Z., Gassner, M., Werlberger, M., Scaramuzza, D.: SVO: semidirect visual odometry for monocular and multicamera systems. IEEE Trans. Robot. **33**(2), 249–265 (2017)
6. Hartley, R., Zisserman, A.: Multiple View Geometry in Computer Vision, 2nd edn. Cambridge University Press, Cambridge (2000)
7. Hartley, R.I.: In defence of the 8-point algorithm. In: Proceedings of IEEE International Conference on Computer Vision, pp. 1064–1070. IEEE (1995)

8. Hu, H., Sun, H., Ye, P., Jia, Q., Gao, X.: Multiple maps for the feature-based monocular SLAM system. J. Intell. Robot. Syst. **94**(2), 389–404 (2019)
9. Kaess, M.: Simultaneous localization and mapping with infinite planes. In: 2015 IEEE International Conference on Robotics and Automation (ICRA), pp. 4605–4611. IEEE (2015)
10. Kim, P., Coltin, B., Kim, H.J.: Visual odometry with drift-free rotation estimation using indoor scene regularities. In: 2017 British Machine Vision Conference (2017)
11. Kukelova, Z., Bujnak, M., Pajdla, T.: Polynomial eigenvalue solutions to the 5-pt and 6-pt relative pose problems. In: BMVC, vol. 2, p. 2008 (2008)
12. Le, P.H., Košecka, J.: Dense piecewise planar RGB-D SLAM for indoor environments. In: 2017 IEEE/RSJ International Conference on Intelligent Robots and Systems (IROS), pp. 4944–4949. IEEE (2017)
13. Li, H., Hartley, R.: Five-point motion estimation made easy. In: 18th International Conference on Pattern Recognition (ICPR 2006), vol. 1, pp. 630–633. IEEE (2006)
14. Li, S., Calway, A.: Absolute pose estimation using multiple forms of correspondences from RGB-D frames. In: 2016 IEEE International Conference on Robotics and Automation (ICRA), pp. 4756–4761. IEEE (2016)
15. Matsuki, H., von Stumberg, L., Usenko, V., Stückler, J., Cremers, D.: Omnidirectional DSO: direct sparse odometry with fisheye cameras. IEEE Robot. Autom. Lett. **3**(4), 3693–3700 (2018)
16. Mur-Artal, R., Tardós, J.D.: ORB-SLAM2: an open-source SLAM system for monocular, stereo, and RGB-D cameras. IEEE Trans. Robot. **33**(5), 1255–1262 (2017)
17. Nistér, D.: An efficient solution to the five-point relative pose problem. IEEE Trans. Pattern Anal. Mach. Intell. **26**(6), 0756–777 (2004)
18. Nistér, D., Naroditsky, O., Bergen, J.: Visual odometry. In: Proceedings of the 2004 IEEE Computer Society Conference on Computer Vision and Pattern Recognition, CVPR 2004, vol. 1, pp. I. IEEE (2004)
19. Nistér, D., Schaffalitzky, F.: Four points in two or three calibrated views: theory and practice. Int. J. Comput. Vis. **67**(2), 211–231 (2006)
20. Rubio, A., et al.: Efficient monocular pose estimation for complex 3D models. In: 2015 IEEE International Conference on Robotics and Automation (ICRA), pp. 1397–1402. IEEE (2015)
21. Saurer, O., Vasseur, P., Boutteau, R., Demonceaux, C., Pollefeys, M., Fraundorfer, F.: Homography based egomotion estimation with a common direction. IEEE Trans. Pattern Anal. Mach. Intell. **39**(2), 327–341 (2016)
22. Sun, H., Tang, S., Sun, S., Tong, M.: Vision odometer based on RGB-D camera. In: 2018 International Conference on Robots & Intelligent System (ICRIS), pp. 168–171. IEEE (2018)
23. Ventura, J., Arth, C., Lepetit, V.: Approximated relative pose solvers for efficient camera motion estimation. In: Agapito, L., Bronstein, M.M., Rother, C. (eds.) ECCV 2014. LNCS, vol. 8925, pp. 180–193. Springer, Cham (2015). https://doi.org/10.1007/978-3-319-16178-5_12

Sparse-Temporal Segment Network for Action Recognition

Chaobo Li[1], Yupeng Ding[1], and Hongjun Li[1,2,3,4]([envelope]) [iD]

[1] School of Information Science and Technology, Nantong University,
Nantong 226019, China
lihongjun@ntu.edu.cn
[2] State Key Laboratory for Novel Software Technology, Nanjing University,
Nanjing 210023, China
[3] Nantong Research Institute for Advanced Communication Technologies,
Nantong 226019, China
[4] TONGKE School of Microelectronics, Nantong 226019, China

Abstract. The most typical methods of human action recognition in videos rely on features extracted by deep neural network. Inspired by the temporal segment network, the sparse-temporal segment network to recognize human actions is proposed. Considering the sparse features contains the information of moving objects in videos, for example marginal information which is helpful to capture the target region and reduce the interference from similar actions, the robust principal component analysis algorithm was used to extract sparse features coping with background motion, illumination changes, noise and poor image quality. Based on different characteristics of three modal data, three parallel networks including RGB frame-network, optical flow-network and sparse feature-network were constructed and then fused through diverse ways. Comparative evaluations on the UCF101 demonstrate that three modal data contain the complementary features. Extensive experiments in subjective and objective show that temporal-sparse segment network can reach the accuracy of 94.2%, which is significantly better than several state-of-the-art algorithms.

Keywords: Action recognition · Sparse features · Temporal segment network

1 Introduction

Pattern recognition, machine learning and other methods are used in action recognition to analyze human behaviors automatically, which is a hot spot in the computer vision [1]. Human action recognition is applied on intelligent surveillance [2], video retrieval [3], human-computer interaction [4], virtual reality [5], robot [6] and other practical directions [7]. However, it is a challenging task which is influenced by many factors [8], such as different brightness, complex background, multi-view, intra-class diversity and so on.

© Springer Nature Switzerland AG 2019
Z. Cui et al. (Eds.): IScIDE 2019, LNCS 11935, pp. 80–90, 2019.
https://doi.org/10.1007/978-3-030-36189-1_7

The typical Convolutional Neural Network (CNN) achieves outstanding performance on action recognition, but it do not take into account the motion information between successive video frames [9–11]. There are several improved methods to obtain the successive information: one is based on 3D CNN and the other is based on two-stream CNN. The 3D convolution kernel is used to extract spatial and temporal features for action recognition in [12]. The Convolutional 3D (C3D) was proposed which can obtain motion information from successive RGB frames [13]. Simonyan and Zisserman [14] proposed a two-stream CNN that is divided into two parts: spatial stream convolutional network and temporal stream convolutional network. The classification results of two networks are fused to obtain the final accuracy. In order to reduce the redundancy between successive frames, Zhu et al. [15] mined decisive frames and key regions improving accuracy and efficiency. MotionNet was added before the temporal stream network which improved the performance of Optical Flow Images (OFI) [16]. Zhang et al. [17] used Motion Vector replace optical flow, which greatly accelerates the speed of two-stream CNN, but the precision is lower. In order to solve the problem that there is no interaction between spatial information and temporal information in two-stream CNN. Feichtenhofer et al. [18] fused features between spatial and temporal networks in different layers.

On the basis of two-stream CNN, the Temporal Segment Network (TSN) [19] was proposed where short-temporal motion information is extracted and fused from multiple two-stream networks at different temporal sequences. Lan et al. [20] suggested making a weighted fusion of the short-time motion information. Zhou et al. [21] proposed a temporal relational reasoning network where three full connection layers were added to learn the weight of different videos.

Among the traditional machine learning algorithms, the Improved Dense Trajectories (IDT) [22] has the better stability, but its ability to represent the action feature is limited. The algorithms based on 3D convolutional neural network are faster but less accurate than the algorithms based on two-stream network. There is a great advantage over the traditional methods in dealing with complex background and intra-class difference. The TSN algorithm solves the problem that two-stream network only pays attention to apparent characteristics and short-term motion information, but there is much redundant information if continuous RGB frames are as input of spatial stream CNN, and some useful information will miss if inputting a single frame. For the temporal stream convolution network, the motion feature contained in the optical flow is incomplete. Sparse features [23] can supplement features from single RGB images and optical flow features [24].

We proposed a Sparse-Temporal Segment Network (S-TSN) which extracts RGB, optical flow and sparse features from different temporal sequences for fusion. Several advantages of our work can be summarized below:

1. The extraction of sparse features is much quicker than optical flow, and sparse features also contain the information of moving objects in videos. It is helpful to capture the moving region and reduce the interference of similar actions.
2. Reducing redundant information between successive RGB frames and enhancing the use of video long-temporal motion information.

3. The sparse feature focuses on the foreground target in video, which complements the information contained in the RGB images and the optical flow images. It can effectively improve the performance of network.

The rest of this paper is organized as follows. First, the two-stream convolutional neural network is introduced in Sect. 2. Then we introduce the proposed sparse-temporal segment network and give a detailed explanation of its overall architecture in Sect. 3. The performance evaluation of the proposed network is conducted in Sect. 4. Finally, we conclude the paper.

2   The Two-Stream Convolutional Neural Network

The two-stream CNN is divided into spatial stream CNN and temporal stream CNN where two CNNs extract spatial information and temporal information respectively in videos. Spatial information includes the scene, objects and temporal information refers to the moving information of objects.

The input of spatial stream convolutional neural network is RGB frames. The network is similar to the typical classifier which can effectively detect the human action, such as Alexnet, GoogleNet and VGG16. It usually be pretrained on ImageNet to improve the training speed and performance of the network. The optical flow represents the correlation of movement between successive frames which can be the input of temporal stream convolutional neural network. Through the characteristic of optical flow, the human action between successive frames can be effectively recognized.

In order to match the feature dimension for fusion, the structure of temporal stream network is usually similar to the spatial stream convolution network's. The fusion method is generally divided into two types. One is the integration of results that obtained from two independent convolution networks. The other is the fusion of middle features in hidden layer.

3   Sparse-Temporal Segment Network

The Sparse-Temporal Segment Network (S-TSN) is proposed, which combines the RGB images, optical flow images and sparse features images. The spatial, temporal and sparse features are extracted at different sequences for fusion, reducing redundant information between successive frames and enhancing the efficiency of long-range temporal information. In addition, compared with RGB and optical flow, sparse features focus on the description of moving objects, which can be complementary to RGB and optical flow features and improve the performance of the network.

3.1   Extraction of Sparse Features

Many matrices have the characteristic of low rank or approximate low rank in practical applications, but the sparse error destroys the sparse characteristic of the original data. In order to recover the sparse structure, the matrix can be decomposed into the sum of two matrices, that is $D = A + E$, where A is low rank matrix and E is sparse matrix.

The Principal Component Analysis (PCA) method can be used to obtain the low rank matrix A when the E is independent and identically distributed Guassian matrix. It can be described as:

$$\min_{A,E} \|E\|_F, \ s.t. \ rank(A) \leq r, \ D = A + E \tag{1}$$

where $\|\cdot\|_F$ is Frobenius norm.

When E is sparse matrix with large noise, the PCA can not give an ideal result. However, Robust Principal Component Analysis (RPCA) can be used to obtain the optimal matrix A. The above equation can be represented as:

$$\min_{A,E} rank(A) + \lambda \|E\|_0, \ s.t. \ D = A + E \tag{2}$$

where $rank(\cdot)$ and $\|\cdot\|_0$ are not convex, which becomes the NP-hard. Because the kernel norm is the convex hull of rank function and the 1-norm is the convex hull of 0-norm, the above NP-hard problem can be transformed into the convex optimization problem:

$$\min_{A,E} \|A\|_* + \lambda \|E\|_1, \ s.t. \ D = A + E \tag{3}$$

where A is a low rank component, E is the sparse component, $\|\cdot\|_*$ denotes the kernel norm which is the sum of the singular values of the matrix, and $rank(\cdot)$ is also convex approximation; $\|\cdot\|_1$ representation l_1 norms, λ is used to balance two norms that is greater than zero. It has been proved under certain conditions that if the error matrix E is sparse enough related to A, the matrix D can be accurately divided into low rank components and sparse components, that is minimizing the weighted sum of the kernel norms and l_1 norms.

For the RPCA problem described in the formula (3), the augmented Lagrange multiplier method can be used to optimize it. The Lagrange function is:

$$L(A, E, Y, \mu) = \|A\|_* + \lambda \|E\|_1 + \langle Y, D - A - E \rangle + \frac{\mu}{2} \|D - A - E\|_F^2 \tag{4}$$

where Y is Lagrange multiplier, μ is a positive number.

Because of the correlation between successive frames, the background is approximated as a low rank component, while the foreground targets occupy only a small portion of pixels in the image, such as human body motion, which can be regarded as sparse components. By using the augmented Lagrangian multiplier method to solve the RPCA, the low rank characteristic can be obtained for the action video, as shown in Fig. 1.

Fig. 1. Comparison of three modalities data.

The first row and the last row show the RGB and Sparse Feature Images (SFI) respectively, the second and third lines show optical flow images obtained in horizontal direction and vertical direction. The RGB images represent the appearance features, including both the background and the foreground targets. The optical flow images represent the direction and velocity of the moving objects. For the horizontal direction, white means moving to the right or up and the higher the gray value, the faster the speed. The black means the moving to the left or downward, and the lower the gray value, the faster the speed. The rest of the gray area means that there is no moving object. But the sparse feature image is different from the RGB and optical flow images, it can effectively outline the moving object from foreground, removing the background to reduce the redundancy of the images and improving the training speed.

3.2 Architecture of Sparse-Temporal Segment Network

The architecture of sparse-temporal segment network is shown in Fig. 2, which operates on the entire video. During processing the training data, the whole video is divided into several snippets, the optical flow images are calculated and sparse feature images are obtained by decomposing the RGB images in each snippet. Meanwhile a RGB frame is randomly selected from each video.

The sparse-temporal segment network is composed of RGB frame-Network (R-Net), Optical flow-Network (O-Net) and Sparse features-Network (S-Net). The three nets use Inception v2 as the structural unit to build. The R-Net and S-Net take single frame RGB and sparse feature image as input respectively, while F-Net use continuous optical flow images as input. Each snippet in different sequences will be predicted its own preliminary class score. Then segmental consensus among the snippets will be derived as the predicted score of video, The three segmental consensuses are fused by weighted fusion, average fusion, and max pooling.

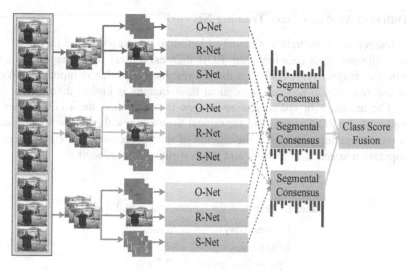

Fig. 2. Sparse-temporal segment network.

4 Experiments

In this section, we evaluate our method on publicly available dataset UCF101 [25] and compare it with the different algorithms. Our experiments were conducted on a PC laptop Intel Core i7-6800K 3.40 GHz CPU, NVIDIA GTX 1080 GPU and 16 GB RAM. Our system is mainly implemented in MATLAB, Windows 10.

4.1 Datasets

UCF101 contains 13,320 videos that categorized into 101 action classes, part of the images are shown in Fig. 3. They cover a large range of activities such as sports and human-object interaction. It is a challenging dataset as the captured videos vary in scale, illumination, background and so on. Our experiments follow the original evaluation scheme using three training or testing splits and report average accuracy over these splits.

Fig. 3. The part images of UCF101.

4.2 Different Modality Data Training Network

We use ImageNet to pre-train the model to initialize the weights, and then test model use three different input data including RGB images, optical flow images and sparse feature images respectively. Table 1 is the test results with different input data. We can see that the recognition accuracy of optical flow images is higher than that of RGB images. The accuracy of sparse feature images is lower than the RGB images, but compared with the RGB images and the optical flow images, the sparse feature images focuses on the moving target area, the amount of data processed is obviously reduced, the computation speed is improved, and the complexity is reduced.

Table 1. Test accuracy on different modal data.

Data modality	Accuracy
RGB images	84.5%
Optical flow images	87.2%
Sparse feature images	81.7%

4.3 Different Input Data

After testing on single modal data, we experiment on pairwise fusion of data from three modes, using the methods of average pooling, weighted average and max pooling on segmental consensus respectively. The results are shown in Table 2. The weights of different modal data are selected manually. The former and the latter are fused according to the weight of 1: 1.5. It is obvious that no matter which fusion method is adopted and no matter which data are fused, the result is better than the single data. In particular, the accuracy of weighted fusion of RGB images and optical flow images is 93.7%.

Table 2. Accuracy of different fusion on UCF101.

Input data	Average pooling	Weighted average	Max pooling
RGB + OFI	93.5%	**93.7%**	91.6%
RGB + SFI	89.8%	**90.0%**	89.9%
OFI + SFI	87.6%	**88.7%**	87.9%

As we can see the effect of weighted average fusion is better than average fusion and max pooling based on ImageNet, so the different weights fusion of RGB, SFI and OFI are studied. The results are shown in Table 3. On the whole, the accuracy of split 2 is higher and split 1 is lower by contrast, so we calculate the average value from three splits. It can be seen that the performance is the best when the RGB, sparse feature and the optical flow feature are fused according to the weight of 1: 0.1: 1.1. The recognition accuracy reaches 94.2%.

Table 3. The accuracy for data fusion according to different weights.

RI: SFI: OFI	Split 1 (%)	Split 2 (%)	Split 3 (%)	Avg. acc. (%)
1: 0.1: 1	93.76898	94.359785	94.125414	94.084726
1: 0.2: 1	93.709983	94.449821	94.265216	94.141673
1: 0.3: 1	93.54368	94.518715	94.297614	94.120003
1: 0.4: 1	93.383417	94.646756	94.152752	94.060975
1: 0.5: 1	93.286641	94.588097	94.096603	93.990447
1: 0.1: 1.1	**93.885609**	**94.414721**	**94.262017**	**94.187449**
1: 0.2: 1.1	93.770781	94.478574	94.296339	94.181898
1: 0.3: 1.1	93.464374	94.572357	94.278754	94.105162
1: 0.4: 1.1	93.340233	94.638898	94.138649	94.03926
1: 0.5: 1.1	93.106568	94.707863	94.144719	93.986383
1: 0.1: 1.2	93.788891	94.395527	94.366553	94.183657
1: 0.2: 1.2	93.578431	94.379099	94.340272	94.099267
1: 0.3: 1.2	93.388828	94.568935	94.361626	94.106463
1: 0.4: 1.2	93.272363	94.700797	94.366948	94.113369
1: 0.5: 1.2	93.091082	94.729839	94.273115	94.031345
1: 0.1: 1.3	93.556117	94.343969	94.428549	94.109545
1: 0.2: 1.3	93.42774	94.378431	94.386240	94.064137
1: 0.3: 1.3	93.346647	94.53615	94.439223	94.107340
1: 0.4: 1.3	93.247855	94.64062	94.448755	94.112410
1: 0.5: 1.3	93.080972	94.684158	94.436776	94.067302
1: 0.1: 1.4	93.488332	94.25593	94.50842	94.084227
1: 0.2: 1.4	93.307546	94.328871	94.521902	94.052773
1: 0.3: 1.4	93.347373	94.456709	94.395373	94.066485
1: 0.4: 1.4	93.307532	94.523247	94.380586	94.070455
1: 0.5: 1.4	93.130072	94.537164	94.363573	94.010270
1: 0.1: 1.5	93.413392	94.293434	94.441807	94.049544
1: 0.2: 1.5	93.363972	94.363405	94.445229	94.057535
1: 0.3: 1.5	93.29236	94.441562	94.333589	94.022504
1: 0.4: 1.5	93.172892	94.408261	94.423649	94.001601
1: 0.5: 1.5	93.102474	94.46605	94.386051	93.984858

Table 4. Comparison of the accuracy of different algorithms on UCF101.

Algorithms	Accuracy
Two Stream [7]	88.0%
Two Stream + LSTM [4]	88.6%
Two Stream Fusion [11]	92.5%
TSN	93.7%
Our method	94.2%

In order to further verify the effectiveness of the sparse-temporal segmentation network, our algorithm is compared with some classical algorithms on UCF101, and the results are shown in Table 4. The original two stream network gets the accuracy of 88.0%. The Long Short-Term Memory (LSTM) cells are connected to the output of the two-stream network, which can combine image information over longer time and the accuracy achieved 88.6%. Two-stream network fusing the abstract convolutional features over spatial and temporal streams further boosts performance at 92.5%. Meanwhile, temporal segmental network can acquire longer-range temporal information and fuse spatial and temporal features. Our experimental results show its accuracy is 93.7%. It can be found that our proposed sparse-temporal segmentation network that obtain accuracy of 94.2% having a certain improvement compared with other algorithms. We conjecture that the sparse features are better at capturing motion information than single apparent features which sometimes may be unstable for describing motions.

5 Conclusion

The sparse-temporal segment network is proposed for action recognition which incorporates RGB frames, optical flow and sparse features. Fusion with the different modal data takes the advantage of complementary information. It obvious that training an additional S-Net on sparse feature images is significantly better than only training R-Net and O-Net. Evaluation on challenging datasets and comparisons with the state-of-the-art algorithm demonstrate that our method achieves superior action recognition performance. And our model is more robust because of contiguous sparse features.

Acknowledgment. This work is supported by National Natural Science Foundation of China (NO. 61871241); Ministry of education cooperation in production and education (NO. 201802302115); Educational Science Research Subject of China Transportation Education Research Association (Jiaotong Education Research 1802-118); the Science and Technology Program of Nantong (JC2018025, JC2018129); Nantong University-Nantong Joint Research Center for Intelligent Information Technology (KFKT2017B04); Nanjing University State Key Lab. for Novel Software Technology (KFKT2019B15); Postgraduate Research and Practice Innovation Program of Jiangsu Province (KYCX19_2056).

References

1. Herath, S., Harandi, M., Porikli, F.: Going deeper into action recognition: a survey. Image Vis. Comput. **60**, 4–21 (2016)
2. Wu, D., Sharma, N., Blumenstein, M.: Recent advances in video-based human action recognition using deep learning: a review. In: IEEE International Joint Conference on Neural Networks, Anchorage, USA, pp. 2865–2872. IEEE (2017)
3. Ramezani, M., Yaghmaee, F.: Motion pattern based representation for improving human action retrieval. Multimedia Tools Appl. **77**(19), 26009–26032 (2018)
4. Chakraborty, B.K., Sarma, D., Bhuyan, M.K., et al.: Review of constraints on vision-based gesture recognition for human-computer interaction. IET Comput. Vis. **12**(1), 3–15 (2018)

5. Pushparaj, S., Arumugam, S.: Using 3D convolutional neural network in surveillance videos for recognizing human actions. Int. Arab. J. Inf. Technol. **15**(4), 693–700 (2019)
6. Fangbemi, A.S., Liu, B., Yu, N.H., Zhang, Y.: Efficient human action recognition interface for augmented and virtual reality applications based on binary descriptor. In: De Paolis, L.T., Bourdot, P. (eds.) AVR 2018. LNCS, vol. 10850, pp. 252–260. Springer, Cham (2018). https://doi.org/10.1007/978-3-319-95270-3_21
7. Wang, P., Liu, H., Wang, L., et al.: Deep learning-based human motion recognition for predictive context-aware human-robot collaboration. CIRP Ann. Manuf. Technol. **67**(1), 17–20 (2018)
8. Li, H.J., Suen, C.Y.: A novel Non-local means image denoising method based on grey theory. Pattern Recogn. **49**(1), 217–248 (2016)
9. Cao, C., Zhang, Y., Zhang, C., et al.: Body joint guided 3D deep convolutional descriptors for action recognition. IEEE Trans. Cybern. **48**(3), 1095–1108 (2018)
10. Ng, J.Y.H., Hausknecht, M., Vijayanarasimhan, S., et al.: Beyond short snippets: deep networks for video classification. In: IEEE Conference on Computer Vision and Pattern Recognition, Boston, USA, pp. 4694–4702. IEEE (2015)
11. Ding, Y., Li H.J., Li, Z.Y.: Human motion recognition based on packet convolution neural network. In: 2017 12th International Conference on Intelligent Systems and Knowledge Engineering, Nanjing, China, pp. 1–5. IEEE (2017)
12. Ji, S., Xu, W., Yang, M., et al.: 3D convolutional neural networks for human action recognition. IEEE Trans. Pattern Anal. Mach. Intell. **35**(1), 221–231 (2013)
13. Tran, D., Bourdev, L., Fergus, R., et al.: Learning spatiotemporal features with 3D convolutional networks. In: International Conference on Computer Vision, Santiago, Chile, pp. 4489–4497. IEEE (2014)
14. Simonyan, K., Zisserman, A.: Two-stream convolutional networks for action recognition in videos. Neural Inf. Process. Syst. **1**(4), 568–576 (2014)
15. Zhu, W., Hu, J., Sun, G., et al.: A key volume mining deep framework for action recognition. In: IEEE Conference on Computer Vision and Pattern Recognition, Las Vegas, USA, pp. 1991–1999. IEEE (2016)
16. Zhu, Y., Lan, Z., Newsam, S., et al.: Hidden two-stream convolutional networks for action recognition. arXiv preprint arXiv:1704.00389 (2017)
17. Zhang, B., Wang, L., Wang, Z., et al.: Real-time action recognition with deeply-transferred motion vector CNNs. IEEE Trans. Image Process. **27**(5), 2326–2339 (2018)
18. Feichtenhofer, C., Pinz, A., Zisserman, A.: Convolutional two-stream network fusion for video action recognition. In: IEEE Conference on Computer Vision and Pattern Recognition, Las Vegas, USA, pp. 1933–1941. IEEE (2016)
19. Wang, L., et al.: Temporal segment networks: towards good practices for deep action recognition. In: Leibe, B., Matas, J., Sebe, N., Welling, M. (eds.) ECCV 2016. LNCS, vol. 9912, pp. 20–36. Springer, Cham (2016). https://doi.org/10.1007/978-3-319-46484-8_2
20. Lan, Z., Zhu, Y., Hauptmann, A.G., et al.: Deep local video feature for action recognition. In: International Conference on Computer Vision and Pattern Recognition Workshops, Honolulu, USA, pp. 1219–1225. IEEE (2017)
21. Zhou, B., Andonian, A., Torralba, A.: Temporal relational reasoning in videos. arXiv preprint arXiv:1711.08496v1 (2018)
22. Wang, H., Schmid, C.: Action recognition with improved trajectories. In: International Conference on Computer Vision, Sydney, Australia, pp. 3551–3558. IEEE (2014)

23. Li, H.J., Suen, C.Y.: Robust face recognition based on dynamic rank representation. Pattern Recogn. **60**(12), 13–24 (2016)
24. Li, H.J., Hu, W., Li, C.B., et al.: Review on grey relation applied in image sparse representation. J. Grey Syst. **31**(1), 52–65 (2019)
25. Soomro, K., Zamir, A.R., Shah, M.: UCF101: a dataset of 101 human action classes from videos in the wild. arXiv preprint arXiv:1212.0402 (2012)

SliceNet: Mask Guided Efficient Feature Augmentation for Attention-Aware Person Re-Identification

Zhipu Liu (ID) and Lei Zhang(✉)(ID)

School of Microelectronics and Communication Engineering, Chongqing University,
Chongqing 400044, China
{zpliu,leizhang}@cqu.edu.cn

Abstract. Person re-identification (re-ID) is a challenging task since the same person captured by different cameras can appear very differently, due to the uncontrolled factors such as occlusion, illumination, viewpoint and pose variation etc. Attention-based person re-ID methods have been extensively studied to focus on discriminative regions of the last convolutional layer, which, however, ignore the low-level fine-grained information. In this paper, we propose a novel SliceNet with efficient feature augmentation modules for open-world person re-identification. Specifically, with the philosophy of divide and conquer, we divide the baseline network into three sub-networks from low, middle and high levels, which are called slice networks, followed by a Self-Alignment Attention Module respectively to learn multi-level discriminative parts. In contrast with existing works that uniformly partition the images into multiple patches, our attention module aims to learn self-alignment masks for discovering and exploiting the align-attention regions. Further, SliceNet is combined with the attention free baseline network to characterize global features. Extensive experiments on the benchmark datasets including Market-1501, CUHK03, and DukeMTMC-reID show that our proposed SliceNet achieves favorable performance compared with the state-of-the art methods.

Keywords: Person re-identification · SliceNet · Self-alignment attention

1 Introduction

The goal of person re-identification(re-ID) is to identify the query images from a large gallery across multiple cameras with non-overlapping views. It has attracted lots of research interest because of its valuable applications in video surveillance, such as multi-camera tracking, multi-camera object detection [28], and pedestrian retrieval [14,18]. In recent years, a large number of deep learning methods are proposed for person re-ID, and achieve big breakthrough. However, it is still an unsolved task due to the drastic appearance changes caused by various viewpoint, resolution, poses, background noise as well as occlusions.

Early re-ID methods [20] learn the global representation from whole-body images, but lose discriminative information lying around body parts. To capture the local discriminative information, a uniform-partition method was proposed to learn local representations from some predefined horizontal partition strips. However, person images

Z. Cui et al. (Eds.): IScIDE 2019, LNCS 11935, pp. 91–102, 2019.
https://doi.org/10.1007/978-3-030-36189-1_8

collected by automatic detectors often suffer from misalignment and the re-ID accuracy can be compromised. So horizontal stripes may be less effective when severe misalignment happens. To address this issue, pose estimation based methods [5,29] were proposed to detect key points of body such as head, foots etc. However, severe occlusions can affect the accuracy of these methods. Also, most of these methods need extra training for pose estimation, and the complexity and difficulty of feature learning are casted.

Visual attention has the ability to guide the learning toward informative and discriminative regions, and can be exploited to capture fine-grained saliency regions. That is, it can help to discover the most discriminative regions by producing attention maps containing more personal information. Consequently, a number of attention based person re-ID methods [22,30] were proposed, but most existing works only pay attention to the high-level semantic features but ignore the low-level fine-grained information such as clothing color. The conventional strategy to use the low-level information is intuitively concatenate the convolutional feature maps into the fully-connected layer. However, it ignores the training inequality of each level and may lead to unreliable features. Therefore, to achieve more reliable person re-ID, we propose to exploit the rich fine-grained features from a network, by taking into account both low-level features (e.g., shape, color etc.) and high-level semantic details (e.g., identity).

In order to augment the high-level semantic by exploiting the low-level features and make them compatible, with the philosophy of divide and conquer, we focus on the parallel learning from low-level to high-level. Additionally, we propose a self-alignment attention module in each level to enable the person feature learning. Specifically, in this paper, we propose a novel SliceNet which aims to address the open-world person re-ID from three perspectives: (1) we propose to learn each level in parallel by cutting the network into multiple sub-networks instead of the conventional layer-wise feature concatenation, such that the training is balanced level-to-level and enable reliable person re-ID; (2) we propose the mind of self-alignment attention and aim to find out the discriminative align-attention local saliency regions by learning channel and spatial attention masks from their own feature maps; (3) we leverage the mind that global whole-body features has high semantic discrimination, the high-level information is used for feature augmentation. The main contributions of this paper are three-fold:

(1) With the philosophy of divide and conquer, we propose a SliceNet comprising of three sub-networks with different depth and one global network, which aims to learn the informative features from low-level to high-level in parallel. The three sub-networks contain two, three and four stages of ResNet-50 [11], respectively. The global network is the complete ResNet-50 for whole-body feature learning.
(2) We propose a novel Self-Alignment Attention Module in each subnetwork to capture discriminative local align-attention saliency regions for addressing open-world person re-ID challenges.
(3) Exhaustive experiments on three large datasets including Market-1501, CUHK03, and DukeMTMC show that the proposed SliceNet outperform a number of state-of-the arts in person re-identification.

2 Related Work

With the development of deep learning and the availability of large datasets, deep learning based approaches have dominated the re-ID research community due to their significant superiority in discriminative feature representation. Existing deep learning based re-ID methods can be divided into two categories: (1) learning global whole-body features [12] and (2) learning local information (e.g., part, pose estimation, attention etc.) [5,22,29,30].

Specifically, for the former, Lin et al. [17] proposed a simple but effective convolutional neural network, which integrates an identity classification loss and a number of attribute classification losses. Hermans et al. [12] proposed a variant of triplet loss to focus on the hard positive and hard negative examples. Chen et al. [2] designed a quadruplet loss to supervise the training of their model. Zhong et al. [37] proposed a camera style transfer model to address the issue of image style variations caused by different cameras. Although these methods achieved improved performance, the local discriminative regions caused by background noise, pose and occlusion were ignored.

For the latter, recently, a number of parts based deep learning methods have been proposed to capture richer and finer local visual cues. These newly proposed approaches can be broadly classified into three subcategories according to the parts learning scheme, including uniformly partition methods [4,16,31], pose estimation based methods [5,29] and attention based methods [30]. The uniformly partition methods can learn local information, but can not address the issue when the same person in two images are misaligned. Although the pose estimation based methods can address the issue of misalignment, they may be compromised when body parts are occluded or missing during pedestrian detection. Attention based methods is an another subcategory that have been widely explored in various tasks, including image classification [9], object recognition [13], image captioning as well as person re-ID [22,30]. Zhao et al. [30] proposed a part-align human representation, which detects the discriminative human body regions beneficial to person matching. Sun et al. [22] proposed a Part-based Convolutional Baseline (PCB) network and a refined part pooling (RPP) method to learn discriminative part-information features for person retrieval. Si et al. [21] proposed a Dual Attention Matching network (DuATM) to learn context-aware feature sequences and perform attentive sequence comparison from a dual attention mechanism including intra-sequence and inter-sequence attention strategies.

Our attention module is inspired by the part-align method [30], which learns discriminative regions by computing the corresponding regions of a pair of probe and gallery images. Nevertheless, our work differs significantly from part-align methods in three aspects. First, the principle of mask learning is different. The mask in [30] is learned only in channel-wise while ignoring the spatial context-aware information. Our attention mind aims to learn both channel attention and spatial attention masks by a self-alignment attention module, such that more spatial context-aware information can be exploited to guide the attention maps toward informative and discriminative regions. In addition, the same person captured by different cameras can appear in various pose, as a result, the two images do not contain part-align regions, therefore we adopt a soft cross-entropy loss for slight relax on the labels. Finally, with the philosophy of divide and conquer, we propose to parallel learn the local information level by level for reliable re-ID instead of intuitively concatenate the convolutional feature maps.

3 Methodology

In this section, we present the details of the proposed SliceNet, as illustrated in Fig. 1, which includes three subnetworks embedded with a self-alignment attention module for local information learning and one global network for identity semantic feature learning.

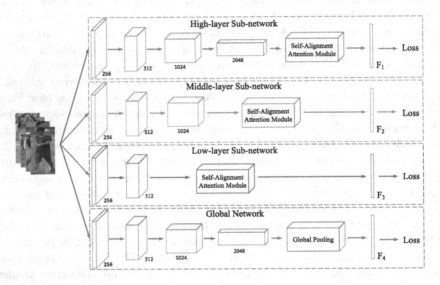

Fig. 1. The architecture of proposed SliceNet. For parallel learning each level equally, with the philosophy of divide and conquer, we divide the network into three subnetworks: high-level subnetwork, middle-level subnetwork and low-level subnetwork, and combined with the attention-free baseline Global Network. For exploiting the local information of subnetworks, a Self-Alignment Attention Module is embedded. Note that all networks are separately learned. During testing, the distance of each feature between probe image and gallery is summed as the final similarity metric.

3.1 Architecture of SliceNet

Previous attention based person re-ID approaches only focus on learning the last convolutional layer local feature, while discarding many local details contained in low-level layers. The low-level information, e.g., color and texture of attention regions, are also important clues for person re-ID. To address this problem, by cutting the baseline network level by level, a SliceNet is proposed as illustrated in Fig. 1. The baseline of our network is ResNet-50 [11], that contains four stages. Each stage (block) comprises of

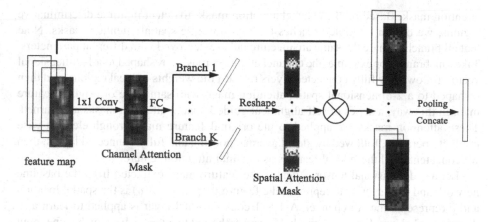

Fig. 2. The framework of the proposed self-alignment attention module. The feature map, extracted from the SliceNet, are followed by a 1×1 convolutional layer to get a channel attention mask. Then, it is reshaped to a 1-dimensional vector which is feeded into the K branches for estimate K spatial attention masks. The masks are applied to the original feature map through element-wise product operation, followed by global pooling and feature concatenation operation.

multiple convolutional layers. At the end of each stage, the feature is spatially down-sampled and fed into the next layer.

SliceNet consists of three sub-networks, which contains four, three and two stages of ResNet-50. The last global average pooling layer is replaced by the proposed Self-Alignment Attention Module (Fig. 2). The first subnetwork contains all stages of ResNet-50 and is named high-level subnetwork, and the second and third subnetwork are middle-level subnetwork and low-level subnetwork, as shown in Fig. 1. The feature map extracted from the three subnetworks are followed by the proposed self-alignment attention module. The last network is an attention-free baseline named Global Network, followed by a global average pooling layer without dimensionality reduction operation.

During training phase, each subnetwork is trained equally, independently and efficiently, which is the philosophy of divide and conquer, such that the mutual negative impacts between different layers can be eliminated and their respective local attention saliency regions can be effectively explored. During testing phases, with the feature augmentation, the distance of each feature-pair between probe image and gallery is summed together with different weights as the final similarity metric.

3.2 Self-alignment Attention Module

The Self-Alignment Attention Module, as illustrated in Fig. 3, first learns a channel attention mask which is fed into the following K branches, and K spatial attention masks can be obtained.

The input of the attention module is the feature map extracted from each subnetwork in SliceNet. First, a 1×1 convolutional layer is utilized to learn a 2-dimensional channel attention mask. Based on the channel attention mask, to detect multiple discriminative regions, we design multiple branches to learn multiple spatial attention masks. Note that all branches share the same architecture but are deployed with different parameters. Take one branch for example, the channel attention mask is reshaped to a 1-dimensional vector, followed by fully connected layers to learn the weights of local regions, and then reshaped to a 2-dimensional spatial attention mask with same size of original feature map. In this way, a series of self-alignment guided spatial attention masks are learned. These attention masks are applied to the original feature map through element-wise product operation, followed by global average pooling and fully-connected layers, then we concatenate all these local features as the final attention feature.

Let a 3-dimensional tensor $\mathbb{T}$ denote the feature maps extracted from the baseline network and use (x, y, c) to represent the feature map size. (x, y) is the spatial location and c represents the cth channel. A 1×1 convolutional layer is applied to learn a 2-dimensional channel attention mask M_c, and reshaped to a vector V_c, which is the input of the subsequent multiple branches. Each branch contains two fully-connected layers:

$$A_k = F_k(V_c) \tag{1}$$

where F_k represents the fully connected layers of the kth branch, A_k denotes spatial attention vector, followed by a reshape operator for transforming into a 2-dimensional spatial attention mask M_k with the same size of M_c:

$$M_k = R_k(A_k) \tag{2}$$

where R_k represents the reshape operator of the kth branch. Then, the spatial attention mask M_k represents the saliency weights for local regions, which is applied on the feature maps $\mathbb{T}$:

$$T_k(x, y, c) = T(x, y, c) \odot M_k(x, y) \tag{3}$$

where $\odot$ represents element-wise product. Then, by adding a global average pooling layer after T_k, i.e., $f_k = AvePooling(T_k)$, a feature vector f_k is obtained. Further, for reducing the dimension of f_k, two fully-connected layers are used to transform the feature f_k to a 128-dimensional feature f_k' in each branch. Finally, we concatenate all the local features to obtain the final attention feature of local subnetworks in Fig. 1:

$$F = [f_1' \ f_2' \ \cdots \ f_k'] \tag{4}$$

where F is the final feature of all attention regions.

3.3 Loss Function

To improve the learning ability of SliceNet, the loss functions we use to train our network is the combination of triplet loss and soft cross-entropy loss.

Triplet loss has been widely used in re-ID, which aims to learn features such that the distance between positive samples decreases and the distance between negative samples increases. Given a batch of images X, consisting of P individuals and K images per person, and the triplet loss can be represented as follows:

$$\mathcal{L}_{triplet} = \sum_{p=1}^{P} \sum_{k=1}^{K} \left[d_{pos}^{p,k} - d_{neg}^{p,k} + m \right]_+ \tag{5}$$

where $d_{pos}^{p,k} = \max_{a=1,\cdots,K} D(\phi(x_p^k), \phi(x_p^a))$ and $d_{neg}^{p,k} = \min_{q \neq p} D(\phi(x_p^k), \phi(x_q^b))$
represent the distance of the hard positives and hard negative, respectively. $D(.,.)$ represents the L2 distance between two features and $\phi(x_p^k)$ represents the feature of image k with respect to person p. m is a margin that controls the distance between positives and negatives.

A network with only triplet loss considered can easily lead to over-matching when the same person appear various poses captured in different cameras. To alleviate this problem, we add an extra soft cross-entropy loss, which is deployed after the linear layer activated by softmax probability function. The cross-entropy loss is

$$\mathcal{L}_{softmax} = -\sum_{n=1}^{N}\sum_{i=1}^{C} y_{n,i} \log p_{n,i} \tag{6}$$

where y_n denotes the one-hot encoded label vector of image x_n. Compared with traditional softmax loss, we use $y_{n,i} = 0.7$ for x_n and $0.3/(n-1)$ for other bits. C is the number of classes. N is the batch size. The probability $p_{n,i}$ is computed by softmax function, shown as

$$p_{n,i} = \frac{\exp(W_i f_{n,i})}{\sum_{i=1}^{C} \exp(W_i f_{n,i})} \tag{7}$$

where W_i is the weights of linear layer.

The total loss of our SliceNet is the combination of triplet loss and softmax loss:

$$\mathcal{L}_{total} = \lambda_1 \mathcal{L}_{triplet} + \lambda_2 \mathcal{L}_{softmax} \tag{8}$$

where λ_1 and λ_2 are the trade-off parameters. In our experiments, both are set to 1.

3.4   Implementation Details

Details of the Backbone Network. Inspired by the success of deep object detection methods [6,19], the granularity of feature can be enriched by removing the last spatial down-sampling operation in the backbone network. This was introduced to person re-ID in [22] and we follow this setting in the backbone of ResNet-50. In addition, the feature map extracted from the low-level subnetwork that only contains two stages, has a very large scale, which is difficult to learn local attention regions, so we add the spatial down-sampling operation at the end of first stage in the subnetwork. The input images are resized to 480×160 for training and testing all subnetworks in our SliceNet. We set the batch size to 128 with $P = 32$ and $K = 4$ to train our model. Our model loads the weights of backbone network ResNet-50 pre-trained on ImageNet [7]. Note that the global whole-body feature extracted from the last Global Network is a 2048-dimensional vector without further dimensionality reduction.

Details of the Self-alignment Attention Module. In this module, the output of backbone network is a 3-dimensional feature map, which is followed by a 1×1 convolutional layer and computes a channel attention mask. Then, the channel attention mask is reshaped to a vector and then feed into the K branches to further learn K spatial attention masks. In this paper, we set K to 10. In each branch, the spatial attention mask is learned by two fully-connected layers. The first fully-connected layer has 800 neural nodes followed by batch normalization, ReLU function and dropout operation with 0.5. The second layer has 300 units and is reshaped into a 2-dimensional spatial mask 30×10 with the same size as the feature map extracted from subnetworks. After element-wise product between spatial attention mask and the feature map, two fully-connected layers are followed for dimension reduction. The first layer has 1024 units followed by batch normalization operation and ReLU function, and the second layer has 128 units.

Network Training. We use Adam optimizer [8] to train our network, and update the learning rate as follows:

$$lr(t) = \begin{cases} lr_0 & , t \leq t_0 \\ l_0 0.001^{\frac{t-t_0}{t_1-t_0}} & , t_0 \leq t \leq t_1 \end{cases} \tag{9}$$

where lr_0 is the initial learning rate, set as $lr_0 = 3e - 4$. $t_0 = 300$ and $t_1 = 600$. The margin m of the triplet loss is set to 0.3.

4 Experiments

4.1 Datasets and Evaluation Metrics

To evaluate the performance of our proposed method, we conduct exhaustive experiments on three large datasets: Market-1501 [32], CUHK03 [24] and DukeMTMC [34].

Market1501 were captured from 6 different cameras and contains 12,936 training images of 751 identities and 19,732 testing images of 750 identities. The pedestrians are automatically detected by DPM-detector [10]. During test, it contains single-query and multiple-query models. The single-query model only contains 1 query image of a person and has 3368 query images. The multiple-query model use the avg- or max-pooling features of multiple images.

CUHK03 contains 13,164 images of 1467 persons captured from 6 cameras and each person is captured by 2 cameras. The bounding boxes of pedestrians contain both manually labelled and DPM detected, and we adopt the latter in this paper. The original training/testing protocol is to randomly select 100 identities for testing and the remaining ones for training. 20 random train/test splits [24] are considered, but time-consuming for deep learning.

DukeMTMC-reID is a subset of DukeMTMC captured from 8 high-resolution cameras and detected by manually labelled. It contains 16,522 training images, 17,661 gallery images and 2,228 queries from total 1404 identities.

Evaluation Protocol. In our experiments, we employ the standard cumulative matching characteristics (CMC) accuracy (Rank-1) and mean average precision (mAP) [32] on all datasets to evaluate the performance of different re-ID methods. On Market-1501 dataset, we select the single query model. To simplify the evaluation procedure on CUHK03 dataset, we adopt the new training/testing protocol proposed by [35].

Table 1. Rank-1 and mAP comparison (%) of SliceNet with other state-of-the arts on Market-1501. 'RR' represents the re-ranking operation proposed by [35].

Methods	Rank-1	mAP
part-aligned [30]	81.00	63.40
APR [17]	84.29	64.67
TriNet [12]	84.92	69.14
DaRe (R) [26]	86.40	69.30
PL-Net [27]	88.20	69.30
HA-CNN [25]	91.20	75.70
DuATM [21]	91.42	76.62
SPReID [15]	93.68	83.36
PCB+RPP [22]	93.80	81.60
SliceNet (Ours)	**95.43**	**86.86**
TriNet (RR) [12]	86.67	81.07
DaRe (R, RR) [26]	88.30	82.00
SPReID (RR) [15]	94.63	90.96
SliceNet (Ours, RR)	**96.35**	**94.44**

4.2 Comparison with State-of-the Arts

We compare our proposed method with state-of-the arts on three widely used datasets: Market-1501, CUHK03 and DukeMTMC-reID.

Comparison on Market-1501. Table 1 shows the results of our proposed method and other state-of-the-art methods on Market-1501. As shown in Table 1, our proposed method achieves rank-1 accuracy of 95.43 and mAP of 86.86, which shows competitive performance compared with all of them. By comparing with the very recent PCB+RPP [22], our method outperforms it by 1.63% in rank-1 accuracy and 5.20% in mAP. The attention based methods part-aligned [30], SPReID [17], DuATM [21] and PCB+RPP [22] achieve superior accuracy with 81.00, 91.20, 91.42 and 93.80 of Rank-1. By comparing to attention based methods, the rank-1 accuracy can be improved by 14.43, 4.23, 4.01 and 1.63, respectively, by using our SliceNet.

Comparison on CUHK03. The results on CUHK03 (Detected) is summarized in Table 2, from which we observe that our method achieves Rank-1 of 69.71 and mAP of 66.81. Our model outperforms all the compared methods by a large increment. It exceeds the state-of-the art PCB+RPP [22] by 6.0 in Rank-1 and 9.3 in mAP.

Comparison on DukeMTMC-reID. On DukeMTMC-reID datasets, our proposed method achieves the best Rank-1 accuracy of 88.7 and mAP of 76.1, as is shown in Table 2. PCB+RPP [22] achieves Rank-1 of 83.3 and mAp of 69.2, which surpasses all other compared methods of Table 2, but our methods outperforms it by 5.4 in Rank-1 and 6.9 in mAP. Therefore, the effectiveness of the proposed SliceNet is verified and the importance of self-alignment attention mask learning for multi-level fine-grained local saliency region features is obvious.

Table 2. Rank-1 and mAP Comparison (%) with state-of-the arts on DukeMTMC-reID and CUHK03 (Detected) with the same setting as [35].

Methods	CUHK03		DukeMTMC-reID	
	Rank-1	mAP	Rank-1	mAP
PAN [33]	36.3	34.0	71.6	51.5
DPFL [3]	40.7	37.0	79.2	60.6
SVDNet [23]	41.5	37.3	76.7	56.8
HA-CNN [25]	41.7	38.6	80.5	63.8
MLFN [1]	52.8	47.8	81.0	62.8
TriNet+Era [36]	55.5	50.7	73.0	56.6
PCB+RPP [22]	63.7	57.5	83.3	69.2
Ours	**69.7**	**66.8**	**88.7**	**76.1**

4.3   Discussions

Local Details Visualization in Different Layers. We know that the representations in low-level focus on learning color and texture information, but features extracted from deeper of network tend to be more abstract. As a result, the feature learned from different layers contain different semantic information, by focusing on different local details. The visualization of the learned attention saliency regions on different layers are shown in Fig. 3, in which the attention maps learned from middle-level subnetwork and high-level subnetwork are described. The first column is the original feature map extracted from baseline, and the following ten columns show ten attention maps learned from by using our SliceNet. We can see that the attention map in the two subnetworks are different, and the attention regions in middle-level subnetwork are bigger than that of high-level subnetwork. Additionally, the attention map learned from high-level subnetwork are more focused and the attention maps learned from middle-level subnetwork

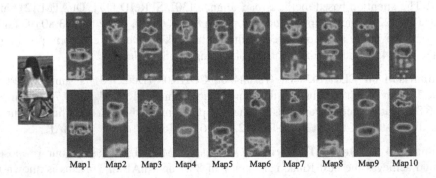

Map1   Map2   Map3   Map4   Map5   Map6   Map7   Map8   Map9   Map10

Fig. 3. Examples of visualization results learned from self-alignment attention module. The first row is the result extracted from middle-level subnetwork and the second row is from high-level subnetwork.

have obvious noises. The reason is that the semantic information learned from low-level is weaker than high-level subnetwork.

5 Conclusion

In this paper, we propose a novel trainable architecture named SliceNet, followed by self-alignment attention module to discover and exploit the multi-level discriminative local attention saliency regions. The framework is with the philosophy of divide and conquer for parallel learning of each level equally, independently, and efficiently. Compared with most existing attention based methods that only focus on learning local features of the last convolutional layer, the SliceNet not only learns high-, middle- and low- level local saliency features but also combines the global whole-body information, which can achieve more reliable person re-ID tasks. Exhaustive experiments on three widely used datasets in person re-ID show that our proposed method has superior performance over the state-of-the arts.

References

1. Chang, X., Hospedales, T.M., Tao, X.: Multi-level factorisation net for person re-identification. In: CVPR (2018)
2. Chen, W., Chen, X., Zhang, J., Huang, K.: Beyond triplet loss: a deep quadruplet network for person re-identification. In: CVPR (2017)
3. Chen, Y., Zhu, X., Gong, S.: Person re-identification by deep learning multi-scale representations. In: IEEE International Conference on Computer Vision Workshop (2017)
4. Cheng, D., Gong, Y., Zhou, S., Wang, J., Zheng, N.: Person re-identification by multi-channel parts-based CNN with improved triplet loss function. In: Computer Vision & Pattern Recognition (2016)
5. Chi, S., Li, J., Zhang, S., Xing, J., Wen, G., Qi, T.: Pose-driven deep convolutional model for person re-identification. In: ICCV (2017)
6. Dai, J., Yi, L., He, K., Jian, S.: R-FCN: object detection via region-based fully convolutional networks. In: CVPR (2016)
7. Deng, J., Dong, W., Socher, R., Li, L.J., Li, K., Li, F.F.: ImageNet: a large-scale hierarchical image database. In: IEEE Conference on Computer Vision & Pattern Recognition (2009)
8. Kingma, D.P., Ba, J.: Adam: a method for stochastic optimization. In: ICLR (2015)
9. Fei, W., et al.: Residual attention network for image classification. In: CVPR (2017)
10. Felzenszwalb, P.F., Mcallester, D.A., Ramanan, D.: A discriminatively trained, multiscale, deformable part model. In: CVPR IEEE Conference on Computer Vision & Pattern Recognition (2008)
11. He, K., Zhang, X., Ren, S., et al.: Deep residual learning for image recognition. In: CVPR (2016)
12. Hermans, A., Beyer, L., Leibe, B.: In defense of the triplet loss for person re-identification. In: CVPR (2017)
13. Ba, J., Mnih, V., Kavukcuoglu, K.: Multiple object recognition with visual attention. In: CVPR (2014)
14. Jing, X., Rui, Z., Feng, Z., Wang, H., Ouyang, W.: Attention-aware compositional network for person re-identification. In: CVPR (2018)
15. Kalayeh, M.M., Basaran, E., Gokmen, M., Kamasak, M.E., Shah, M.: Human semantic parsing for person re-identification. In: CVPR (2018)

16. Li, W., Zhu, X., Gong, S.: Person re-identification by deep joint learning of multi-loss classification. In: CVPR (2017)
17. Lin, Y., Liang, Z., Zheng, Z., Yu, W., Yi, Y.: Improving person re-identification by attribute and identity learning. In: CVPR (2017)
18. Liu, H., Feng, J., Qi, M., Jiang, J., Yan, S.: End-to-end comparative attention networks for person re-identification. IEEE Trans. Image Process. Publ. IEEE Signal Process. Soc. 26(7), 3492–3506 (2017)
19. Liu, W., et al.: SSD: single shot multibox detector. In: Leibe, B., Matas, J., Sebe, N., Welling, M. (eds.) ECCV 2016. LNCS, vol. 9905, pp. 21–37. Springer, Cham (2016). https://doi.org/10.1007/978-3-319-46448-0_2
20. Prosser, B., Zheng, W.S., Gong, S., Tao, X.: Person re-identification by support vector ranking. In: British Machine Vision Conference (2010)
21. Si, J., et al.: Dual attention matching network for context-aware feature sequence based person re-identification. In: CVPR (2018)
22. Sun, Y., Liang, Z., Yi, Y., Qi, T., Wang, S.: Beyond part models: Person retrieval with refined part pooling (and a strong convolutional baseline). In: European Conference on Computer Vision (2018)
23. Sun, Y., Zheng, L., Deng, W., Wang, S.: SVDNet for pedestrian retrieval. In: IEEE International Conference on Computer Vision (2017)
24. Wei, L., Rui, Z., Tong, X., Wang, X.G.: DeepReID: deep filter pairing neural network for person re-identification. In: Computer Vision & Pattern Recognition (2014)
25. Wei, L., Zhu, X., Gong, S.: Harmonious attention network for person re-identification. In: CVPR (2018)
26. Yan, W., Wang, L., You, Y., Xu, Z., Weinberger, K.Q.: Resource aware person re-identification across multiple resolutions. In: CVPR (2018)
27. Yao, H., Zhang, S., Zhang, Y., Li, J., Qi, T.: Deep representation learning with part loss for person re-identification. IEEE Trans. Image Process. PP(99), 1 (2017)
28. Zhang, S., Wen, L., Xiao, B., Zhen, L., Li, S.Z.: Single-shot refinement neural network for object detection. In: CVPR (2017)
29. Zhao, H., et al.: Spindle net: Person re-identification with human body region guided feature decomposition and fusion. In: IEEE Conference on Computer Vision & Pattern Recognition (2017)
30. Zhao, L., Li, X., Wang, J., et al.: Deeply-learned part-aligned representations for person re-identification. In: ICCV (2017)
31. Zheng, F., Sun, X., Jiang, X., Guo, X., Yu, Z., Huang, F.: A coarse-to-fine pyramidal model for person re-identification via multi-loss dynamic training. In: CVPR (2019)
32. Zheng, L., Shen, L., Tian, L., Wang, S., Wang, J., Tian, Q.: Scalable person re-identification: a benchmark. In: IEEE International Conference on Computer Vision (2015)
33. Zheng, Z., Liang, Z., Yi, Y.: Pedestrian alignment network for large-scale person re-identification. In: CVPR (2017)
34. Zheng, Z., Zheng, L., Yang, Y.: Unlabeled samples generated by GAN improve the person re-identification baseline in vitro. In: IEEE International Conference on Computer Vision (2017)
35. Zhong, Z., Zheng, L., Cao, D., Li, S.: Re-ranking person re-identification with k-reciprocal encoding. In: IEEE Conference on Computer Vision & Pattern Recognition (2017)
36. Zhong, Z., Zheng, L., Kang, G., Li, S., Yang, Y.: Random erasing data augmentation. In: CVPR (2017)
37. Zhong, Z., Zheng, L., Zheng, Z., Li, S., Yang, Y.: Camera style adaptation for person re-identification. In: CVPR (2017)

Smoother Soft-NMS for Overlapping Object Detection in X-Ray Images

Chunhui Lin[1(✉)], Xudong Bao[1], and Xuan Zhou[2]

[1] Lab of Image Science and Technology, School of Computer Science
and Engineering, Southeast University, Nanjing, China
220171733@seu.edu.cn
[2] School of Automation, Southeast University, Nanjing, China

Abstract. As a contactless security technology, X-ray security inspection machine is widely used in the detection of dangerous object in all kinds of densely populated public places to ensure the safety. Unlike a natural image, various objects overlapping with each other can be observed in an X-ray image for its perspectivity. It brings us a challenge that the traditional NMS (Non-maximum suppression) algorithm will suppress the less significant objects. In this paper, we propose a Smoother Soft NMS based on the difference in aspect ratios and areas of different object bounding boxes to improve the accuracy of overlapping object detection. We also propose a special data augmentation method to simulate the generation of complex samples of overlapping objects. On our dataset, we boost the mean Average Precision of ResNet-101 FPN from 89.44% to 96.67% and Cascade R-CNN from 96.43% to 97.21%. Detector trained by Smoother Soft NMS has a significant improvement in overlapping cases.

Keywords: Smoother Soft NMS · Dangerous object detection · X-ray images

1 Introduction

Object detection is fundamental for many downstream practical computer vision applications like face recognition [1, 2], video indexing [3, 4] and video surveillance [5, 6]. Recent advances on deep convolutional neural network [7–11] greatly improve the performance of object detection [12–16].

Dangerous object detection from X-ray images is also an important application as a contactless security technology and is widely used in all kinds of densely populated public places for safety. However, X-ray security inspections rely heavily on huge labor efforts. Thus, manual inspections are inefficient. Motivated by this, we propose an automatic detection algorithm instead of manual inspections to accelerate it.

Recent object detection algorithms fall into two categories: efficient one-stage detectors like YOLO [15], SSD [16] and high accuracy two-stage detectors like Faster RCNN [14] and Mask RCNN [17]. Considering that the conveyor speed of the security inspection machine is only 0.2 m/s, our work focuses on the latter one.

Limited by the dataset, we propose a special data augmentation method. We simulate the generation of complex samples of overlapping objects by adding some translucent masks surrounded or distributed around the object (Fig. 1).

© Springer Nature Switzerland AG 2019
Z. Cui et al. (Eds.): IScIDE 2019, LNCS 11935, pp. 103–113, 2019.
https://doi.org/10.1007/978-3-030-36189-1_9

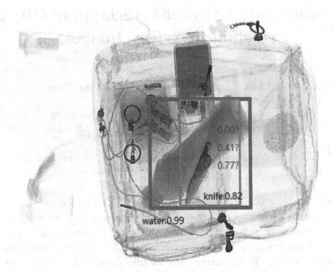

Fig. 1. This image has two confident dangerous goods water and knife (shown in red and green) which have a score of 0.99 and 0.82 respectively. The green detection box has a significant overlap with the red one. The score of the green box will be suppressed to 0 or 0.41 after NMS and Soft-NMS. Can we slow down the score decaying as much as possible and keep most of the score such as 0.77. (Color figure online)

NMS [18] (Non-maximum suppression) is essential for state-of-the-art two-stage detectors, which generates cluttered bounding boxes. During NMS, candidate bounding boxes are ranked based on classification scores. Bounding boxes of lower scores will be suppressed. However, multiple overlapping objects often appear in X-ray images. Different from the occlusion of natural images, we can observe almost all of them because of the perspectivity of the X-ray images. In this situation, object with lower scores bounding boxes will be suppressed by the NMS algorithm. Although soft NMS [19] improved this situation by giving a decayed score instead of suppress directly, the phenomenon of missing objects is still very common. Considering that the original intention of NMS is to suppress repeated bounding boxes. The characteristic of these repeated bounding boxes is that they have the similar aspect ratios and areas. In other words, it is possible that these bounding boxes with different aspect ratios and areas can be predicted as different objects.

Motivated by this, we propose a Smoother Soft NMS to make the decayed score higher if the aspect ratio and areas of the lower score box are far different from the higher score box. In this way, we greatly improved the Average Precision of the ResNet FPN [20].

The major contributions of our work are summarized as follows:

- We implement dangerous object detection for X-ray images using deep convolutional neural network algorithm instead of manual.
- We propose a special data augmentation method to simulate the generation of complex samples of overlapping objects.

- We propose a Smoother Soft NMS based on the difference in aspect ratios and areas of different object bounding boxes for overlapping object detection.

2 Related Work

There has been a significant amount of works on object detection.

Two-stage Detectors: Although one-stage detection algorithms [15, 16] are efficient, state-of-the-art object detectors are based on two-stage, proposal-driven mechanism. Two-stage detectors [14, 17, 20] will generate cluttered object proposals, which result in a large number of duplicate bounding boxes. However, during standard NMS procedure, overlapping objects' bounding boxes with lower classification scores will be discarded even they predict the true objects. Our Smoother Soft NMS tries to reserve these bounding boxes based on the difference of aspect ratios and areas.

Data Augmentation: Data Augmentation refers to random rotation, flipping, cropping, randomly setting the brightness and contrast of the image, and normalizing the data. Through these operations, we can get more sample images. In other words, the original one can be changed into multiple images, which expands the sample size, which is very helpful for improving the accuracy of the model and improving the generalization ability of the model.

OHEM [21] focuses on some complex samples that cause large loss values during model training to improve the capacity of complex sample detection. A-Fast- RCNN [22] tries to generate examples which are hard for the object detector to recognize by Generative Adversarial Networks [23]. We also propose a special data augmentation method to simulate the generation of complex samples of overlapping objects.

Non-maximum Suppression: NMS has been an essential part of computer vision for many decades. It is widely used in edge detection [18], feature point detection [24] and objection detection [14–17].

Recently, Soft NMS [19] is proposed for improving NMS results. Instead of eliminating all lower scored surrounding bounding boxes, Soft NMS decays the detection scores of all other neighbors as a continuous function of their overlap with the higher scored bounding box. Our Smoother Soft NMS further improves the Soft NMS considering the difference of aspect ratios and areas between these boxes to slow down the delay of scores.

3 Approach

In this section, we first propose a special data augmentation method to simulate the generation of complex samples of overlapping objects. Then a new NMS approach is introduced for improving the recall of the overlapping objects based on their difference of aspect ratios and areas.

3.1   Complex Sample Generation

The dual-energy X-ray baggage security inspection machine obtains high-energy and low-energy images according to the different attenuation of X-rays of different materials. According to the absorption characteristics of the two energy spectra, a colorized image representing different materials with different colors is reconstructed. However there are still some difference between the traditional natural images and the reconstructed X-rays images.

- First, the X-ray security image has rich color information, which can directly reflect the material information of the object.
- Second, the line-of-sight of the X-ray security image does not change. The same type of object does not change greatly on the scale, but the appearance will be deformed due to the difference in the placement angle.
- Third, the X-ray image is transparent, so effective texture information cannot be obtained from the X-ray image.

Based on the above characteristics, we found that we can represent different objects by generating rectangular translucent masks with different colors. For example, we can use a green rectangle to represent a book, blue for metal, and orange for liquid. In addition, we can also cover the position of the target on the original image to simulate the overlap of different objects. Besides, we can also generate some complex samples to improve the generalization ability of the model (Fig. 2).

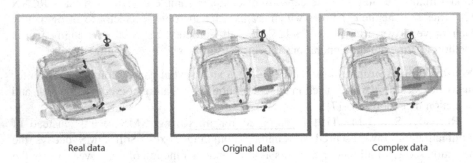

Real data   Original data   Complex data

Fig. 2. Complex samples generation. The left image is the real data that there is a knife under a book. The middle image is the original data and the right image is the complex data that add a translucent rectangular mask above the knife to simulate the left image.

In this paper, aiming at a simple scene with only one target in the image, We respectively generate five different translucent rectangular masks with the center of the target box, the upper left corner, the lower left corner, the upper right corner, and the lower right corner as the center, and the size is same as the target box according to the method above. With this complex sample generation method, we have expanded the dataset by a factor of five.

3.2 Smoother Soft NMS

As is known to us all, the purpose of NMS is to eliminate duplicate boxes in object detection tasks. According to the algorithm, if an object lies within the predefined overlap threshold, it leads to a miss. Instead of eliminating all lower scored surrounding bounding boxes, Soft NMS decays the detection scores of all other neighbors as a continuous function of their overlap with the higher scored bounding box. However there are still some problems.

We found that the duplicate boxes that NMS originally intended to suppress have the same characteristics that their aspect ratios and areas are very similar to the high score bounding box. According to above considerations, we want to retard the decay score if a box is far different from the high score bounding box in aspect ratios and area. And the greater the difference is, the less the score will be reduced.

Algorithm 1 smoother Soft-NMS

Input: $B = \{b_1, \dots b_N\}, S = \{S_1, \dots, S_N\}, N_t$

    B is the list of initial detection boxes

    S contains corresponding detection scores

    N_t is the final threshold

begin

    $D \leftarrow \{ \}$

    While $B \neq$ empty **do**

        $m \leftarrow$ argmax S

        $M \leftarrow b_m$

        $D \leftarrow D \cup M; B \leftarrow B - M$

        for b_i **in** B **do**

            if $\text{iou}(M, b_i) \geq N_t$ **then**

                $B \leftarrow B - b_i; S \leftarrow S - s_i$

            end   ⇨   NMS

            $s_i \leftarrow s_i f(\text{iou}(M, b_i))$ ⇨   Soft-NMS

            $\sigma_i^r \leftarrow \text{ror}(M, b_i)$

            $\sigma_i^a \leftarrow \text{roa}(M, b_i)$

            $s_i \leftarrow s_i f(\text{iou}(M, b_i), \sigma_i^r, \sigma_i^a)$ ⇨ Smoother Soft-NMS

        end

    end

    return D, S

end

Fig. 3. The pseudo code in red is Original NMS and blue is Soft-NMS. We improve the Soft-NMS with two line of code that we can see in green. We retard the decay score based on difference in aspect ratios and areas. (Color figure online)

As is shown in Algorithm 1, compared to Soft-NMS, we just add two lines of code to calculate the difference in aspect ratios and areas, and use them as the smoothing coefficient of the Gaussian penalty function in Soft-NMS. As a result, the decay score of the box will be retarded if there are some differences, and the box is more likely to be retained. However, boxes that are similar in aspect ratios and areas to high scoring boxes will be still well suppressed.

We calculate RoR (ratio of aspect ratio) and RoA (ratio of area) to represent the difference of the aspect ratios and areas as follows,

$$\sigma_i^r = \frac{r(b_i)}{r(b_m)} = \frac{\frac{l(b_i)}{w(b_i)}}{\frac{l(b_m)}{w(b_m)}} \tag{1}$$

Where $l(b_i), l(b_m)$ and $w(b_i), w(b_m)$ are the length and width of the box and the high score box, $r(b_i)$ and $r(b_m)$ are the aspect ratio of these boxes, σ_i^r is the ratio of aspect ratio and indicates the difference between these two boxes. σ_i^r will take his reciprocal if σ_i^r is less than 1.

$$\sigma_i^a = \frac{a(b_i)}{a(b_m)} = \frac{l(b_i) * w(b_i)}{l(b_m) * w(b_m)} \tag{2}$$

Where $l(b_i), l(b_m)$ and $w(b_i), w(b_m)$ are the length and width of the box and the high score box, $a(b_i)$ and $a(b_m)$ are the area of these boxes, σ_i^a is the ratio of area and indicates the difference in area. σ_i^a will take his reciprocal if σ_i^a is less than 1.

We further improve the Soft-NMS by taking the difference in aspect ratio and area into consideration. Therefore the Gaussian penalty function is as follows,

$$s_i = s_i e^{-\frac{iou(M,b_i)^2}{\sigma * (\sigma_i^r)^2 * (\sigma_i^a)^2}}, \forall b_i \notin D \tag{3}$$

This rule is applied in each iteration and scores of all remaining detection boxes are updated.

The Smoother Soft-NMS algorithm is formally described in Fig. 3, where RoR (M, b_i) is the ratio of aspect ratio, RoA(M, b_i) is the ratio of area and $f(iou(M, b_i), \sigma_i^r, \sigma_i^a)$ is the overlap and difference based weighting function (Fig. 4).

This algorithm improves the recall rate of the detection of overlapping objects while retaining the suppression effect on the duplicate boxes. Besides, the computational complexity for N detection boxes is $O(N^2)$, which is the same as traditional greedy-NMS and Soft-NMS.

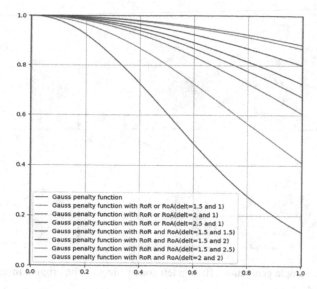

Fig. 4. Gauss penalty function with or without RoR or RoA, we can find the larger ROR and ROA are, the more scores the bounding box will retain.

4 Dataset and Experiment

4.1 Dataset

Our dataset mainly includes two kinds of dangerous goods to be tested, such as knives and liquids, as well as some objects that are easy to interfere with the detection of dangerous goods such as headphone cable, charging treasure, books and so on. There are 6 categories in the dataset, three of which are dangerous goods such as fruit knives, bottled water, and stainless steel water bottles filled with water. The other three are similar non-dangerous goods such as empty stainless steel cups, umbrellas and laptops. Our dataset has a total of 5,000 images, 80% of which are used for training and the rest is used for testing. The image in the dataset is a pseudo-color image reconstructed from a high-energy and a low-energy image.

4.2 Experiment

4.2.1. Experiment of Data Augmentation

As we can see in the figure, we augment the data using the method 3.1, and we add the training data from 4000 to 12000 and the result is shown in Table 1. The evaluation index 'AP@50' means the Average Precision when the detector has IOU threshold value 0.5. The 'AP(0.5:0.75)' means the mean Average Precision with the IOU threshold value 0.05 at intervals from 0.5 to 0.75 (Fig. 5).

Complex sample generation

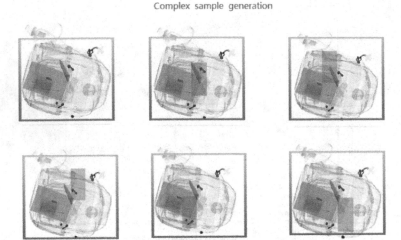

Fig. 5. Complex sample generation. The top left corner image is the original image and the rest are generated data.

Table 1. Performance comparison with FPN ResNet-101 on dataset without or with data augmentation. Best performance is marked in bold for each method.

Method	Iter	AP (0.5:0.75)	AP@50	AP@60	AP@70
Before data augmentation	100k	89.44%	96.43%	93.86%	86.12%
After data augmentation	100k	**90.12%**	**96.98%**	**95.22%**	**88.69%**

4.2.2. Experiment of Smoother Soft-NMS

First, we train a ResNet-101 FPN network with NMS. Then we apply Soft-NMS and Smoother Soft NMS with Ratio of length and width or Ratio of areas and both (Figs. 6 and 7).

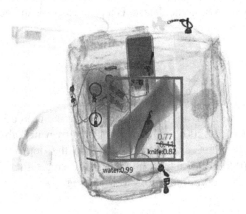

Fig. 6. The ratio of the aspect ratio of these boxes is 1.57 and the ratio of the area is 1.67 and finally the score of the knife increases from 0.41 to 0.77 using our Smoother Soft-NMS instead of Soft-NMS.

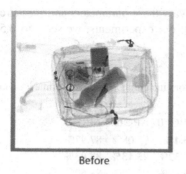
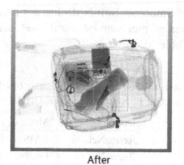

Before   After

Fig. 7. As a result, the knife can be detected and the comparison is shown in the figure.

Table 2. The contribution of each element in our detection pipeline. The baseline model is ResNet-101 FPN. Best performance is marked in bold for each method.

NMS	Soft-NMS	Smoother with RoR	Smoother with RoA	AP (0.5:0.75)	AP@50	AP@60	AP@70
✓				90.12%	96.98%	95.22%	88.69%
	✓			94.07%	97.25%	96.44%	91.89%
		✓		96.41%	97.49%	97.39%	95.85%
			✓	96.07%	97.28%	96.99%	95.27%
		✓	✓	**96.67%**	**97.84%**	**97.47%**	**96.36%**

We evaluate the contribution of each element in our detection pipeline: Soft-NMS, Smoother Soft-NMS with RoR and ROA with ResNet-101 FPN in Table 2.

As we expected, simply training with Soft-NMS instead of greedy-NMS with ResNet FPN improves the AP by 3.95%. We found that the improvement of AP@50 was not obvious while for AP@70 the improvement was more than 3%. When we training with our Smoother Soft NMS solely with RoR or RoA, the improvement is only 0.24% and 0.03% for AP@50. Further, we combine both of them, the AP is increased significantly to 2.59% in contrast to Soft-NMS and 6.54% to greedy-NMS.

Table 3. Sensitivity Analysis across multiple parameters σ for Smoother Soft-NMS using FPN. Best performance is marked in bold for each method.

σ	AP (0.5:0.75)	AP@50	AP@60	AP@70
0.1	96.21%	97.47%	97.37%	95.73%
0.3	96.34%	97.54%	97.51%	95.85%
0.5	**96.67%**	**97.84%**	97.47%	**96.36%**
0.7	96.56%	97.79%	**97.57%**	95.43%
0.9	96.43%	97.65%	97.51%	95.98%
1.1	96.35%	97.51%	97.45%	95.86%

Smoother Soft-NMS has a σ parameter and we vary these parameters and measure average precision on our dataset, see Table 3. In all our experiments, we set σ to 0.5 (Table 4).

Table 4. Comparision on popular baseline object detectors including one-stage RetinaNet and two-stage FPN and Cascade R-CNN.

Detector	Method	AP (0.5:0.75)	AP@50	AP@60	AP@70
RetinaNet	NMS	89.77%	**96.4%**	95.08%	87.97%
RetinaNet	Ours	**89.81%**	96.32%	**95.13%**	**88.03%**
FPN	NMS	90.12%	96.98%	95.22%	88.69%
FPN	Ours	**96.67%**	**97.84%**	**97.47%**	**96.36%**
Cascade R-CNN	NMS	96.43%	97.27%	96.88%	95.31%
Cascade R-CNN	Ours	**97.21%**	**98.22%**	**97.89%**	**96.87%**

We test our Smoother Soft NMS with other detectors such as one-stage detector RetinaNet and two-stage detector Cascade R-CNN [25]. The results shows us that our method is very effective on two-stage detector while has little impact on one-stage detector.

5 Conclusion

To conclude, Soft-NMS is an effective approach to improve the average precision while object overlap is a common phenomenon in the detection of dangerous goods. Further, we improve this method on account of the difference of aspect ratios and areas between two bounding boxes. We call this method with smoothing coefficient based on RoR (ratio of aspect ratio) and RoA (ratio of area) Smoother Soft-NMS. Besides, the complex sample generation method contribute to the capacity under complex conditions. Compelling results are demonstrated for both ResNet-101 FPN and Cascade R-CNN on our dataset.

References

1. Zhao, W., Chellappa, R., Phillips, P.J., Rosenfeld, A.: Face recognition: a literature survey. ACM Comput. Surv. (CSUR) **35**(4), 399–458 (2003)
2. Taigman, Y., Yang, M., Ranzato, M., Wolf, L.: DeepFace: closing the gap to human-level performance in face verification. In: Proceedings of the IEEE Conference on Computer Vision and Pattern Recognition, pp. 1701–1708 (2014)
3. Sivic, J., Zisserman, A.: Video google: a text retrieval approach to object matching in videos. In: Null, p. 1470. IEEE (2003)
4. Philbin, J., Chum, O., Isard, M., Sivic, J., Zisserman, A.: Object retrieval with large vocabularies and fast spatial matching. In: IEEE Conference on Computer Vision and Pattern Recognition, CVPR 2007, pp. 1–8. IEEE (2007)

5. Bouwmans, T., Zahzah, E.H.: Robust PCA via principal component pursuit: a review for a comparative evaluation in video surveillance. Comput. Vis. Image Underst. **122**, 22–34 (2014)
6. Ma, X., et al.: Vehicle traffic driven camera placement for better metropolis security surveillance. In: IEEE Intelligent Systems (2018)
7. Krizhevsky, A., Sutskever, I., Hinton, G.E.: ImageNet classification with deep convolutional neural networks. In: Advances in Neural Information Processing Systems, pp. 1097–1105 (2012)
8. Simonyan, K., Zisserman, A.: Very deep convolutional networks for large-scale image recognition. arXiv preprint arXiv:1409.1556 (2014)
9. Szegedy, C., Vanhoucke, V., Ioffe, S., Shlens, J., Wojna, Z.: Rethinking the inception architecture for computer vision. arXiv preprint arXiv:1512.00567 (2015)
10. Szegedy, C., Ioffe, S., Vanhoucke, V.: Inception-v4, inception-resnet and the impact of residual connections on learning. arXiv preprint arXiv:1602.07261 (2016)
11. He, K., Zhang, X., Ren, S., Sun, J.: Deep residual learning for image recognition. arXiv preprint arXiv:1512.03385 (2015)
12. Girshick, R.B., Donahue, J., Darrell, T., Malik, J.: Rich feature hierarchies for accurate object detection and semantic segmentation. In: CVPR, pp. 580–587 (2014)
13. Girshick, R.B.: Fast R-CNN. In: ICCV, pp. 1440–1448 (2015)
14. Ren, S., He, K., Girshick, R.B., Sun, J.: Faster R-CNN: towards real-time object detection with region proposal networks. In: NIPS, pp. 91–99 (2015)
15. Redmon, J., Divvala, S., Girshick, R., et al.: You Only Look Once: Unified, Real-Time Object Detection. ArXiv preprint arXiv:1506.02640
16. Liu, W., et al.: SSD: Single Shot MultiBox Detector. In: Leibe, B., Matas, J., Sebe, N., Welling, M. (eds.) ECCV 2016. LNCS, vol. 9905, pp. 21–37. Springer, Cham (2016). https://doi.org/10.1007/978-3-319-46448-0_2
17. He, K., Gkioxari, G., Dollár, P., et al.: Mask R-CNN. In: IEEE International Conference on Computer Vision, pp. 2980–2988. IEEE Computer Society (2017)
18. Rosenfeld, A., Thurston, M.: Edge and curve detection for visual scene analysis. IEEE Trans. Comput. **5**, 562–569 (1971)
19. Bodla, N., Singh, B., Chellappa, R., Davis, L.S.: Soft-NMS improving object detection with one line of code. In: 2017 IEEE International Conference on Computer Vision (ICCV), pp. 5562–5570. IEEE (2017)
20. Lin, T.Y., Dollár, P., Girshick, R., et al.: Feature Pyramid Networks for Object Detection. ArXiv preprint arXiv:1612.03144
21. Shrivastava, A., Gupta, A., Girshick, R.: Training region-based object detectors with online hard example mining. In: 2016 IEEE Conference on Computer Vision and Pattern Recognition (CVPR), pp. 761–769 (2016)
22. Wang, X., Shrivastava, A., Gupta, A.: A-Fast-RCNN: Hard Positive Generation via Adversary for Object Detection. ArXiv preprint arXiv:1704.03414
23. Goodfellow, I.J., Pouget-Abadie, J., Mirza, M., et al.: Generative adversarial nets. In: International Conference on Neural Information Processing Systems, pp. 2672–2680. MIT Press (2014)
24. Lowe, D.G.: Distinctive image features from scale-invariant keypoints. Int. J. Comput. Vision **60**(2), 91–110 (2004)
25. Cai, Z.: Nuno Vasconcelos. Cascade R-CNN: Delving into high quality object detection. ArXiv preprint arXiv:1712.00726

Structure-Preserving Guided Image Filtering

Hongyan Wang, Zhixun Su$^{(\boxtimes)}$, and Songxin Liang

School of Mathematical Sciences, Dalian University of Technology,
Dalian 116024, China
zxsu@dlut.edu.cn

Abstract. Guided filter behaves as a structure-transferring filter which takes advantage of the guidance image. Nevertheless, it is likely to suffer from structure information loss problem and artifacts would be introduced in practical tasks, e.g., detail enhancement. We in this paper propose to deal with the structure loss problem. We modify the original objective function and develop a re-weighted algorithm to proceed the filtering process iteratively. The proposed filter inherits good properties of guided filter and is more capable in avoiding structure information loss. Many vision tasks can be benefited from the proposed filter. Few applications we outline include flash/no-flash image restoration, image dehazing, detail enhancement, HDR compression, and image matting. Experimental comparisons with relative methods for these tasks demonstrate the effectiveness of the proposed filter.

Keywords: Image filter · Guided filter · Bilateral filter

1 Introduction

Image filtering has attracted many research attentions for years and been witnessed significant advances. The goal is to remove fine-scale details or textures and preserve sharp edges. In computer vision and graphics community, it is a simple and fundamental tool to extract meaningful information for understanding and analyzing images.

Most early proposed image filters are linear translation-invariant filters (e.g., Gaussian and Laplacian filters), which can be explicitly expressed by convolution operator between one input image and a specific filter kernel. They usually achieve poor performance due to their simple forms and lacking of elaborate designing. To better preserve edges, bilateral filter (BF) has been proposed in [1,30], taking both spatial and range information into consideration. By using an additional favourable image instead of the input image, bilateral filter can be extended to joint bilateral filter (JBF) [10,25]. However, one well-known limitation of (joint) bilateral filter is that it may generate gradient reversal artifacts [2,11] in detail enhancement and HDR compression.

© Springer Nature Switzerland AG 2019
Z. Cui et al. (Eds.): IScIDE 2019, LNCS 11935, pp. 114–127, 2019.
https://doi.org/10.1007/978-3-030-36189-1_10

Joint filtering techniques need an input image and a guidance image. Based on a local linear model between the guidance image and the output, the guide filter (GF) [14] is a representative structure-transferring filter and overcomes the gradient reversal limitation. Unluckily, one trouble thing is that the guidance image may be insufficient or unreliable locally, which may lead to unpleasing artifacts. Shen et al. propose the concept of mutual-structure in [28] to address the structure inconsistency problem. Relying on mutual-structures that are contained in both the input and the guidance image to generate the output, MSJF [28] is suitable to specific problems like joint structure extraction and joint segmentation. It does not has structure-transferring property.

Many methods [16, 19, 20] have been proposed to improve guided filter [14]. However, most of them work on designing various adaptive weights or regularization terms and pay little attention on the structure loss problem. Guided filter [14] applies L_2 norm distance on intensity to formulate fidelity term, leading that some meaningful structures may not be preserved well, particularly near edges. This can be illustrated in Fig. 1. As shown in Fig. 1(b), there is noticeable loss of structural information near the edge. This easily causes errors or artifacts in many applications, e.g., detail enhancement.

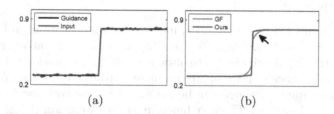

(a)   (b)

Fig. 1. Illustration of structure loss on 1D signals. The guided filter output losses structural information and blurs the edge. Our method is more capable of preserving the edge and the output edge is sharper.

We in this work propose an algorithm to improve the capability of the guided filter on avoiding structure loss. Our contribution is two-fold. First, we modify the original objective function and develop an efficient algorithm by iterative re-weighting mechanism. Second, we show that the proposed method benefits many vision applications. Experimental results compared with state-of-the-art methods demonstrate the effectiveness of our method.

2   Proposed Model and Optimization

In this section, we propose a new objective function based on the similar assumption as [14]. Then we develop a numerical algorithm and give the iterated solutions of the proposed method.

2.1   Proposed Model

Given an guidance image G, the proposed method is based on the following local model:

$$q_i = a_k G_i + b_k, \forall i \in W_k, \tag{1}$$

where q denote the expected output, and i is the pixel index in the window W_k, which is centered at pixel k. a_k and b_k are two linear transform coefficients. All pixels in W_k are assumed to share the same a_k and b_k. W_k is set to be square with $2r + 1$ pixels on the side.

For an input image p, the output q is expected to contain major structures associated with p. Details, textures, and noise are expected to be contained in $n = p - q$. Based on these assumptions, we propose the following objective function:

$$E(a_k, b_k) = \sum_{i \in W_k} \left(|a_k G_i + b_k - p_i| + \epsilon a_k^2 \right). \tag{2}$$

To deal with the structure loss, L_1 norm is employed in (2) to formulate the fidelity term.

2.2   Optimization

The data term in (2) is not quadratic, making the optimization problem not a simple linear regression problem. To solve (2), we employ iterative re-weighted least squares (IRLS) algorithm [17] to obtain the iterative solutions of a_k and b_k. IRLS solves a sequence of least square problems within an iterating framework, and every least square problem can be penalized by the reciprocal of absolute error of previous iteration. The cost function at t-th iteration ($t \geq 1, t \in N$) is defined as

$$E(a_k^t, b_k^t) = \sum_{i \in W_k} \omega_i^t (a_k^t G_i + b_k^t - p_i)^2 + \epsilon(a_k^t)^2, \tag{3}$$

where the weight at pixel i is given by $\omega_i^t = 1/\max\left\{|q_i^{t-1} - p_i|, \nu\right\}$. ν is a parameter to avoid the zero denominator. We define the q^0 as the output of the guided filter [14].

The energy function (3) is a linear regression problem [9]. By setting the derivatives of (3) with respect to a_k^t and b_k^t to zero respectively, we can obtain the iterative solutions of a_k^t and b_k^t:

$$\begin{cases} a_k^t = \dfrac{\frac{1}{|W|} \sum_{i \in W_k} \omega_i^t G_i p_i - \widetilde{G}_k^t \widetilde{p}_k^t \widetilde{\omega}_k^t}{(\widetilde{\sigma}_k^t)^2 + \epsilon}, \\ b_k^t = \widetilde{p}_k^t - a_k^t \widetilde{G}_k^t, \end{cases} \tag{4}$$

where $\widetilde{G}_k^t$ and $(\widetilde{\sigma}_k^t)^2$ denote the weighted mean and weighted variance of G in W_k, given by $\widetilde{G}_k^t = \frac{\sum_{i \in W_k} \omega_i^t G_i}{\sum_{i \in W_k} \omega_i^t}$ and $(\widetilde{\sigma}_k^t)^2 = \frac{1}{|W|} \sum_{i \in W_k} \omega_i^t (G_i - \widetilde{G}_k^t)^2$. $\widetilde{p}_k^t$ denotes the weighted mean of p and $\widetilde{\omega}_k^t$ denotes the mean of all the penalized weights ω_i^t in W_k, given by $\widetilde{\omega}_k^t = \frac{1}{|W|} \sum_{i \in W_k} \omega_i^t$ and $\widetilde{p}_k^t = \frac{\sum_{i \in W_k} \omega_i^t p_i}{\sum_{i \in W_k} \omega_i^t}$.

Similar to [14], overlapping problem appears. In each iteration, we compute $\overline{a}_i^t$ and $\overline{b}_i^t$ by averaging strategy: $\overline{a}_i^t = \frac{1}{|W|} \sum_{k:i \in W_k} a_k^t$ and $\overline{b}_i^t = \frac{1}{|W|} \sum_{k:i \in W_k} b_k^t$. The final output is calculated by $q_i^t = \overline{a}_i^t G_i + \overline{b}_i^t$.

We point out that the calculations of (4) involve $\widetilde{G}_k^t$, $\widetilde{p}_k^t$, and $\widetilde{\sigma}_k^t$, which are associated with the penalized weights $\widetilde{\omega}_i^t$. This is different from the calculations of the solutions (a_k and b_k) of guided filter [14], since the filtering process of [14] is not iterative.

(a) Input/Guidance   (b) GF (q^0)   (c) Ours

Fig. 2. Comparisons of guided filter and the proposed filter.

3   Discussions

In this section, we discuss some properties of the proposed iterative filter and provide the expressions of our iterative kernel weights. Extension to color images and limitations are also discussed.

3.1   Edge-Preserving Filter

In this section we analyze how does the proposed filter work. When $G = p$, the equations in (4) are simplified to $a_k^t = \left(\widetilde{\sigma}_k^t\right)^2 / \left((\widetilde{\sigma}_k^t)^2 + \epsilon\right)$ and $b_k^t = (1 - a_k^t) \widetilde{G}_k^t$. As ϵ is positive, for each q_i^t there are two special cases:

- If $\left(\widetilde{\sigma}_k^t\right)^2 \ll \epsilon$, then $a_k^t \approx 0$, so $q_i^t \approx b_k^t \approx \widetilde{G}_k^t$.
- If $\left(\widetilde{\sigma}_k^t\right)^2 \gg \epsilon$, then $a_k^t \approx 1$ and $b_k^t \approx 0$, so $q_i^t \approx G_i$.

For pixels located in a flat window, their intensities are approximative and we have $\widetilde{G}_k^t \approx G_i$. Then $(\widetilde{\sigma}_k^t)^2 \to 0$ and $(\widetilde{\sigma}_k^t)^2 \ll \epsilon$, we obtain $a_k^t \approx 0$ and $q_i^t \approx \widetilde{G}_k^t$. In other words, the proposed filter handles pixel in a flat window by weighted averaging to reach the goal of smoothing. On the other hand, only when $(\widetilde{\sigma}_k^t)^2 \gg \epsilon$, pixel centered at this window is preserved. Note that $(\widetilde{\sigma}_k^t)^2$ is influenced by both ω_i^t and structures in the patch W_k. This means that whether the pixel is preserved or not is determined by G, p and q^{t-1}. That is, the criterion "what is an edge" or "structures which are expected to be preserved" is no longer simply measured by the given parameter ϵ like [14]. In the t-th iteration, pixels where p and q^{t-1} are approximate are assigned large weights to reach the similar smoothing effects as guided filter, whereas pixels where p and q^{t-1} are quite different are assigned small weights for proper modifications. This explains why the proposed method is more capable of avoiding structural information loss than guided filter. Visual comparison of guided filter and the proposed filter is shown in Fig. 2.

3.2   Gradient-Preserving Filter

The proposed filter is able to avoid the gradient reversal artifacts. We take detail enhancement for example and follow the algorithm based on base-detail layers decomposition $E = B + \tau D$, where B, D, E denote the base layer, the detail layer and the enhanced image, respectively. τ is a parameter to control the magnification of details.

In practice, the base layer B is generated by filtering on the input image p, and the detail layer D can be viewed as $D = p - B$. This relationship ensures that $\partial D = \partial p - \partial B$. If B can not be consistent with the input signal p and further leads to $\partial D \cdot \partial p < 0$, the gradient reversal artifacts would appear in the enhanced signal after magnifying the detail layer D.

Theoretically, the local linear model (1) indicates that ∂B is a_k^t times of ∂p when $p \equiv G$. For a_k^t, we have $a_k^t = \tilde{\sigma}_k^2/(\tilde{\sigma}_k^2 + \epsilon)$ and it is less than 1. Then we further have $\partial D = \partial p - \partial B = (1 - a_k^t)\partial p$ and $\partial D \cdot \partial p \geq 0$.

An example of 1D signal is shown in Fig. 3. As can be seen in Fig. 3(c), our final enhanced signal avoids gradient reversal artifacts safely and does not introduce over-sharpened artifacts in the enhanced signal.

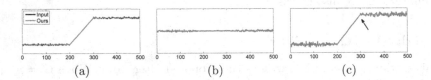

(a)   (b)   (c)

Fig. 3. The proposed filter is gradient-preserving. (a) Input signal and our filtering result. (b) Detail layer. (c) The enhanced signal. The proposed method does not generate unrealistic details near edges and further produces natural enhanced signal without over-sharpened artifacts.

3.3   Iterative Filter Kernel

The filter kernel of the proposed method varies in each iteration. The explicit expressions of kernel weights in t-th iteration can be given by

$$
W_{ij}^t = \frac{1}{|W|^2} \sum_{k:(i,j)\in W_k} M\left(\frac{(G_i - \tilde{G}_k^t)(G_j - \tilde{G}_k^t)}{(\tilde{\sigma}_k^t)^2 + \epsilon} + \frac{1}{\tilde{\omega}_k^t}\right),
\tag{5}
$$

where $M = \omega_j^t + T_j H\left(|q_j^{t-1} - p_j| - \nu\right)\left(p_j - \tilde{p}_k^t - a_k^t(G_j - \tilde{G}_k^t)\right)$, and $T_j = \text{sgn}(q_j^{t-1} - p_j)/(q_j^{t-1} - p_j)^2$. We use $\text{sgn}(\cdot)$ to denote sign function and $H(\cdot)$ to denote heaviside step function (outputting ones for positive values and zeros otherwise).

(5) can be proved by the Chain Rule and a series of careful algebraic manipulations. Here we visually show several kernels in Fig. 4.

3.4   Filtering Using Color Guidance Image

Section 2 presents the iterated filtering process for the case of a gray input with a gray guidance image. However, RGB images usually contain more information than gray images. Thus, we develop another proper algorithm for the case of color guidance image. We rewrite model (1) in vector form:

$$q_i = a_k^T G_i + b_k, \forall i \in W_k, \tag{6}$$

where $(\cdot)^T$ denotes matrix transposing operator, G_i denotes RGB intensities at pixel i, and a_k denotes coefficient vector. Note that G_i and a_k are 3×1 vectors, while q_i and b_k are still scalars. Then (2) becomes

$$E(a_k^t, b_k^t) = \sum_{i \in W_k} \omega_i^t(a_k^t G_i + b_k^t - p_i)^2 + \epsilon(a_k^t)^T a_k^t, \tag{7}$$

and the solutions of a_k^t and b_k^t can be obtained by linear regression:

$$\begin{cases} a_k^t = \left(\tilde{\Sigma}_k^t + \epsilon E\right)^{-1}\left[\dfrac{1}{|W|}\sum \omega_i^t p_i G_i - \tilde{\omega}_k^t \tilde{p}_k^t \tilde{G}_k^t\right], \\ b_k^t = \tilde{p}_k^t - (a_k^t)^T \tilde{G}_k^t \end{cases} \tag{8}$$

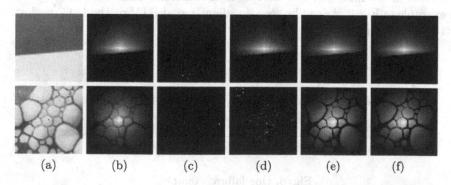

(a)   (b)   (c)   (d)   (e)   (f)

Fig. 4. Several iterative filtering kernels for two special cases. (a) Image patches. (b) Kernels of guided filter (W^0) at the pixels denoted by the red dots in (a). (c)–(f) Our iterative kernels after 1st, 4th, 7th and 10th iteration. The final kernels are capable in preserving underlying data structure.

where E is a 3×3 matrix with all one elements, and $\widetilde{G}_k^t$ is a 3×1 weighted averaged vector of G_i. $\widetilde{\Sigma}_k^t$ is a 3×3 weighted covariance matrix, expressed as $\widetilde{\Sigma}_k^t = \frac{1}{|W|} \sum_{i \in W_k} \omega_i \left(G_i - \widetilde{G}_k^t \right) \left(G_i - \widetilde{G}_k^t \right)^T$.

After dealing with the overlapping problem, the final filter output are given by $q_i^t = \frac{1}{|W|} \sum_{k:i \in W_k} \left(\left(a_k^t \right)^T G_i + b_k^t \right) = \left(\overline{a}_i^t \right)^T G_i + \overline{b}_i^t$.

We show an example in Fig. 5. By comparing the results of using gray guidance and RGB guidance visually, we can see that edges in Fig. 5(c) are preserved better than that in Fig. 5(b).

(a) Input image   (b) Gray guidance   (c) RGB guidance

Fig. 5. Comparisons on color guidance image and gray guidance image.

In addition, filtering a gray input with a color guidance image is also very useful for some vision tasks, such as dehazing and image matting. These applications can be found in Sect. 4.

3.5 Limitations

The proposed method would not work well if there is complex texture patterns contained in the image, or it is incapable of removing dense textures. We show an example in Fig. 6. As can be seen, gradient-based RTV method [33] performs better than the proposed method.

Input $r = 8, \epsilon = 0.2^2$ $r = 16, \epsilon = 0.6^2$ RTV [33]

Fig. 6. One failure example.

4   Applications and Experimental Results

The proposed method can be applied to a variety of computer vision tasks. Several tasks we outline in this section are flash/no-flash image restoration, image dehazing, detail enhancement, HDR compression, and image matting.

Parameter Settings. We state parameter settings first. Empirically, we set window radius $r < 100$ and the regularization parameter $\epsilon < 1$. ν is set to be 0.0001 in all experiments. The number of iterations is set to be 5.

Running Time. The proposed algorithm has been implemented in MATLAB on a PC with Intel Xeon E5630 CPU and 12 GB RAM. It takes about 0.4 s to process a 321×481 gray image without code optimization. For the color case, processing an image with the same size takes about 23.8 s.

4.1   Flash/No-Flash Image Restoration

Flash/No-Flash Denoising. The observed flash image can be viewed as a guidance image to facilitate denoising a noisy no-flash input. Figure 7 shows an example. The compared methods include joint BF, GF [14], and WLS [11]. As can be seen, the results shown in Fig. 7(c) (joint BF) contain noticeable gradient reversal artifacts. GF is incapable of preserving edges and can not produce sharp edges in some regions (Fig. 7(d)). The results of WLS (Fig. 7(e)) contain artifacts generated by the intensive noises. Our result is visually better than others.

Flash/No-Flash Deblurring. One common way for flash/no-flash deblurring is to generate an image which both preserves the ambiance lighting and contains clear edges and details. This process can be simply finished by the base-detail layers decomposition model mentioned in Sect. 3.2, which may save much computation compared with existing deblurring methods. The base layer B is generated by filtering on the no-flash image p guided by the flash image G in order to maintain the ambient lighting. The detail layer D is produced by $D = G - \widehat{G}$, where $\widehat{G}$ denotes a self-guidance filtered output of G. Then we can combine the base layer B and the detail layer D to generate a blur-free image.

A challenging case is that some saturated regions may appear in the blurry no-flash image, as shown in Fig. 8(a). Two representative blind image deblurring methods [22,24], fail to produce clear structure around the saturated region, as shown in Fig. 8(b)–(c). Severe artifacts can be found in these results. Even for methods [15,23], which are specifically proposed to deal with outliers, their deblurred results shown in Fig. 8(d)–(e) are still unpleasing. We also provide results of several filtering methods, including joint BF, GF [14], WLS [11], L_0 gradient minimization [32], domain transform filter (DTF) [12], RTV [33], RGF [34], and MSJF [28]. As shown in Fig. 8(g)–(o), ours is visually the best.

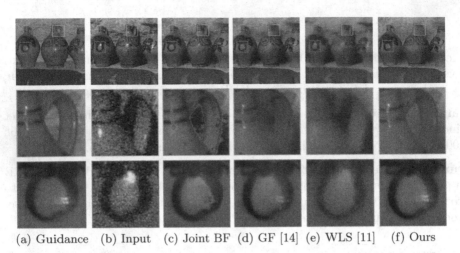

(a) Guidance   (b) Input   (c) Joint BF   (d) GF [14]   (e) WLS [11]   (f) Ours

Fig. 7. Denoising with flash/no-flash image pair. The no-flash image (input) suffers from severe noises while the structures and edges in the flash image (guidance) are quite clear. Compared with the results shown in (c)–(e), the proposed filtering method produce a noise-free result with clear edges.

4.2   Image Dehazing

We follow the widely used the hazy image formation model $I(x) = J(x)T(x) + A(x)(1 - T(x))$, where I, J, A, T denote the observed hazy image, the scene radiance, the global atmospheric light and the medium transmission, respectively. x

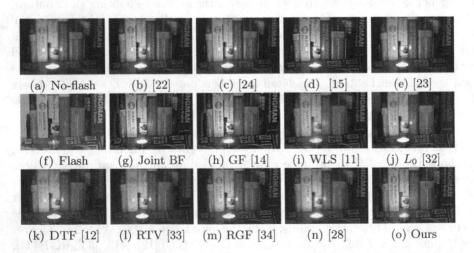

(a) No-flash   (b) [22]   (c) [24]   (d) [15]   (e) [23]

(f) Flash   (g) Joint BF   (h) GF [14]   (i) WLS [11]   (j) L_0 [32]

(k) DTF [12]   (l) RTV [33]   (m) RGF [34]   (n) [28]   (o) Ours

Fig. 8. Blur removal example with flash/no-flash image pair. Results shown in (b)–(e) are restored by deblurring methods [15, 22–24], respectively (blur kernel size: 33 × 33; running time of kernel estimation process: 465.7 s, 2020.6 s, 306.8 s, and 1933.4 s). Results shown in (g)–(o) require no kernel estimation process (running time: 1.6 s, 1.1 s, 7.4 s, 10.2 s, 78.4 s, 17.5 s, 6.4 s, 23.8 s and 7.8 s).

is the pixel index. We estimate the atmospheric light A and the raw transmission map $T^0(x)$ within the framework [13] and refine $T^0(x)$ by filtering $T^0(x)$ instead of solving the matting Laplacian matrix like [13], which is very slow. As can be seen in Fig. 9, our refined transmission map T (Fig. 9(d)) contains more meaningful structures than T^0 (Fig. 9(b)). The final dehazed result is shown in Fig. 9(e). The competitive dehazing methods include DCP [13], BCCR [21], NLD [3], DehazeNet [5], and MSCNN [26]. Our result is visually comparable to the results of conventional methods DCP, BCCR and NLD, and is a little better than that of learning-based methods DehazeNet and MSCNN.

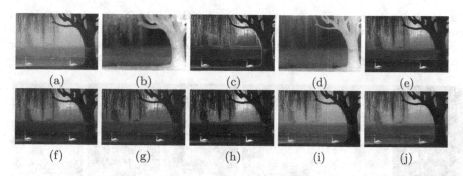

Fig. 9. An image dehazing example. (a) Input hazy image. (b) Raw transmission map T^0. (c) Dehazed by (b). (d) Our refined T. (e) Dehazed by (d). (f) DCP [13]. (g) BCCR [21]. (h) NLD [3]. (i) DehazeNet [5]. (j) MSCNN [26]. It cost about 57 s and 40 s for (f) (using matting Laplacian) and (e) (ours) to refine T^0.

4.3 Detail Enhancement and HDR Compression

Detail Enhancement. The detail enhancement algorithm has been described in Sect. 3.2. Figure 10 shows comparisons of using GF [14], LLF [29], RTV [33], mRTV [8], RoG [4] and the proposed filter on an example. The results shown in Fig. 10(b)–(f) suffer from unrealistic artifacts (See close-ups). In comparison, our method avoids to generate unrealistic artificial details (Fig. 10(g)).

HDR Compression. Different from detail enhancement, HDR compression aims to generate low dynamic range image by compressing the base layer at some rate while preserving the details. We show an example in Fig. 11 compared with some filter methods [4,11,29,32]. To display we convert the input HDR radiance to a logarithmic scale and then map the result to [0, 1] (See Fig. 11(a)). For each result, we show two close-ups of a highlighted area and a dark area. The proposed method produces clean result with natural details, whereas the other methods either suffer from aliasing or fail to compress the range properly.

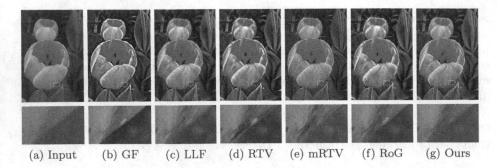

(a) Input   (b) GF   (c) LLF   (d) RTV   (e) mRTV   (f) RoG   (g) Ours

Fig. 10. A detail enhancement example compared with [4,8,14,29,33]. The guided filter generates an over-enhanced results with unrealistic artifacts due to structural information loss, as shown in (b). Our result is visually more natural with little unrealistic artifacts than (b)–(f).

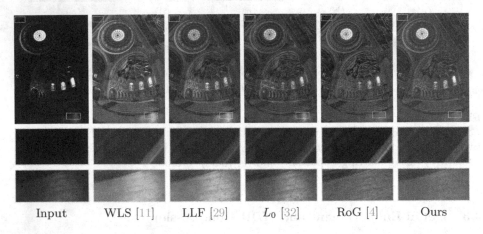

Input   WLS [11]   LLF [29] L_0 [32]   RoG [4]   Ours

Fig. 11. HDR compression. Close-ups show that the proposed method compresses the highlighted area and the dark area effectively.

4.4   Image Matting

An accurate matte can be generated from filtering a coarse binary mask with the guidance of corresponding clear image. We compare our method with image matting methods [6,7,18,27,31]. Our result shown in Fig. 12(g) is visually comparable with the results shown in Fig. 12(b)–(f). Nevertheless, the proposed method does not require another user-assisted input but a coarse binary mask. In comparison, all the competitive methods require user-assisted input (either scribbles or trimap image) for labeling.

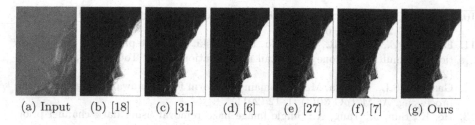

(a) Input   (b) [18]   (c) [31]   (d) [6]   (e) [27]   (f) [7]   (g) Ours

Fig. 12. An image matting example.

5 Conclusion

In this paper, we propose to modify the measurement function of fidelity term in the objective function of guided filter to address its structural information loss limitation and improve its capability on preserving structures. We then develop an efficient iterative re-weighting algorithm to solve the proposed model. We analyze the attractive properties of our method. The extension to color guidance image (with gray input) leads the proposed filter to benefit some specific tasks, e.g., image matting. We also outline other applications which can be benefited from the proposed method. We expect to apply it to more practical applications.

Acknowledgements. This work has been partially supported by National Natural Science Foundation of China (No. 61572099).

References

1. Aurich, V., Weule, J.: Non-linear Gaussian filters performing edge preserving diffusion. In: Sagerer, G., Posch, S., Kummert, F. (eds.) Mustererkennung, 17. DAGM-Symposium, pp. 538–545. Springer, Heidelberg (1995). https://doi.org/10.1007/978-3-642-79980-8_63
2. Bae, S., Paris, S., Durand, F.: Two-scale tone management for photographic look. ACM ToG **25**(3), 637–645 (2006)
3. Berman, D., Treibitz, T., Avidan, S.: Non-local image dehazing. In: CVPR, pp. 1674–1682 (2016)
4. Cai, B., Xing, X., Xu, X.: Edge/structure preserving smoothing via relativity-of-Gaussian. In: ICIP, pp. 250–254 (2017)
5. Cai, B., Xu, X., Jia, K., Qing, C., Tao, D.: Dehazenet: an end-to-end system for single image haze removal. IEEE TIP **25**(11), 5187–5198 (2016)
6. Chen, Q., Li, D., Tang, C.K.: KNN matting. IEEE TPAMI **35**(9), 2175–2188 (2013)
7. Cho, D., Tai, Y.-W., Kweon, I.: Natural image matting using deep convolutional neural networks. In: Leibe, B., Matas, J., Sebe, N., Welling, M. (eds.) ECCV 2016. LNCS, vol. 9906, pp. 626–643. Springer, Cham (2016). https://doi.org/10.1007/978-3-319-46475-6_39
8. Cho, H., Lee, H., Kang, H., Lee, S.: Bilateral texture filtering. ACM ToG **33**(4), 128:1–128:8 (2014)
9. Draper, N.R., Smith, H.: Applied Regression Analysis. Wiley series in Probability and Mathematical Statistics, 2nd edn. Wiley, New York (1981)

10. Eisemann, E., Durand, F.: Flash photography enhancement via intrinsic relighting. ACM ToG **23**(3), 673–678 (2004)
11. Farbman, Z., Fattal, R., Lischinski, D., Szeliski, R.: Edge-preserving decompositions for multi-scale tone and detail manipulation. ACM ToG **27**(3), 67:1–67:10 (2008)
12. Gastal, E.S.L., Oliveira, M.M.: Domain transform for edge-aware image and video processing. ACM ToG **30**(4), 1–12 (2011)
13. He, K., Sun, J., Tang, X.: Single image haze removal using dark channel prior. IEEE TPAMI **33**(12), 2341–2353 (2011)
14. He, K., Sun, J., Tang, X.: Guided image filtering. IEEE TPAMI **35**(6), 1397–1409 (2013)
15. Hu, Z., Cho, S., Wang, J., Yang, M.: Deblurring low-light images with light streaks. In: CVPR, pp. 3382–3389 (2014)
16. Kou, F., Chen, W., Wen, C., Li, Z.: Gradient domain guided image filtering. IEEE TIP **24**(11), 4528–4539 (2015)
17. Levin, A., Fergus, R., Durand, F., Freeman, W.T.: Image and depth from a conventional camera with a coded aperture. ACM ToG **26**(3), 70 (2007)
18. Levin, A., Lischinski, D., Weiss, Y.: A closed-form solution to natural image matting. IEEE TPAMI **30**(2), 228–242 (2008)
19. Li, Z., Zheng, J., Zhu, Z., Yao, W., Wu, S.: Weighted guided image filtering. IEEE TIP **24**(1), 120–129 (2015)
20. Liu, W., Chen, X., Shen, C., Yu, J., Wu, Q., Yang, J.: Robust guided image filtering. Computing Research Repository abs/1703.09379 (2017). http://arxiv.org/abs/1703.09379
21. Meng, G., Wang, Y., Duan, J., Xiang, S., Pan, C.: Efficient image dehazing with boundary constraint and contextual regularization. In: ICCV, pp. 617–624 (2013)
22. Pan, J., Hu, Z., Su, Z., Yang, M.: Deblurring text images via L0-regularized intensity and gradient prior. In: CVPR, pp. 2901–2908 (2014)
23. Pan, J., Lin, Z., Su, Z., Yang, M.: Robust kernel estimation with outliers handling for image deblurring. In: CVPR, pp. 2800–2808 (2016)
24. Pan, J., Sun, D., Pfister, H.: Blind image deblurring using dark channel prior. In: CVPR, pp. 1628–1636 (2016)
25. Petschnigg, G., Szeliski, R., Agrawala, M., Cohen, M.F., Hoppe, H., Toyama, K.: Digital photography with flash and no-flash image pairs. ACM ToG **23**(3), 664–672 (2004)
26. Ren, W., Liu, S., Zhang, H., Pan, J., Cao, X., Yang, M.-H.: Single image dehazing via multi-scale convolutional neural networks. In: Leibe, B., Matas, J., Sebe, N., Welling, M. (eds.) ECCV 2016. LNCS, vol. 9906, pp. 154–169. Springer, Cham (2016). https://doi.org/10.1007/978-3-319-46475-6_10
27. Shahrian, E., Rajan, D., Price, B.L., Cohen, S.: Improving image matting using comprehensive sampling sets. In: CVPR, pp. 636–643 (2013)
28. Shen, X., Zhou, C., Xu, L., Jia, J.: Mutual-structure for joint filtering. IJCV **125**(1–3), 19–33 (2017)
29. Sylvain, P., Samuel, W.H., Jan, K.: Local laplacian filters: edge-aware image processing with a laplacian pyramid. Commun. ACM **58**(3), 81–91 (2015)
30. Tomasi, C., Manduchi, R.: Bilateral filtering for gray and color images. In: ICCV, pp. 839–846 (1998)
31. Varnousfaderani, E.S., Rajan, D.: Weighted color and texture sample selection for image matting. IEEE TIP **22**(11), 4260–4270 (2013)
32. Xu, L., Lu, C., Xu, Y., Jia, J.: Image smoothing via L_0 gradient minimization. ACM ToG **30**(6), 174:1–174:12 (2011)

33. Xu, L., Yan, Q., Xia, Y., Jia, J.: Structure extraction from texture via relative total variation. ACM ToG **31**(6), 139:1–139:10 (2012)
34. Zhang, Q., Shen, X., Xu, L., Jia, J.: Rolling guidance filter. In: Fleet, D., Pajdla, T., Schiele, B., Tuytelaars, T. (eds.) ECCV 2014. LNCS, vol. 8691, pp. 815–830. Springer, Cham (2014). https://doi.org/10.1007/978-3-319-10578-9_53

Deep Blind Image Inpainting

Yang Liu[1(✉)], Jinshan Pan[2], and Zhixun Su[1]

[1] School of Mathematical Science, Dalian University of Technology, Dalian, China
lewisyangliu@gmail.com, zxsu@dlut.edu.cn
[2] School of Computer Science, Nanjing University of Science and Technology,
Nanjing, China
sdluran@gmail.com

Abstract. Image inpainting is a challenging problem as it needs to fill the information of the corrupted regions. Most of the existing inpainting algorithms assume that the positions of the corrupted regions are known. Different from existing methods that usually make some assumptions on the corrupted regions, we present an efficient blind image inpainting algorithm to directly restore a clear image from a corrupted input. Our algorithm is motivated by the residual learning algorithm which aims to learn the missing information in corrupted regions. However, directly using existing residual learning algorithms in image restoration does not well solve this problem as little information is available in the corrupted regions. To solve this problem, we introduce an encoder and decoder architecture to capture more useful information and develop a robust loss function to deal with outliers. Our algorithm can predict the missing information in the corrupted regions, thus facilitating the clear image restoration. Both qualitative and quantitative experimental demonstrate that our algorithm can deal with the corrupted regions of arbitrary shapes and performs favorably against state-of-the-art methods.

Keywords: Blind image inpainting · Residual learning · Encoder and decoder architecture · CNN

1 Introduction

Image inpainting aims to recover a complete ideal image from the corrupted image. The recovered region should either be as accurate as the original without disturbing uncorrupted data or visually seamlessly merged into the surrounding neighborhood such that the reconstruction result is as realistic as possible. The applications of image inpainting are numerous, ranging from the restoration of damaged images, videos, and photographs to the removal of selected regions. Image inpainting problem is usually modeled as

$$x(i) = \begin{cases} y(i) + n(i), & M(i) = 1, \\ u(i), & M(i) = 0. \end{cases} \tag{1}$$

where x, y and n denote the corrupted image, clear image, and noise, respectively; i denotes the pixel position; u denotes the values of all other image pixels corrupted by other factors; M is a binary indicator, in which 0 denotes missing

© Springer Nature Switzerland AG 2019
Z. Cui et al. (Eds.): IScIDE 2019, LNCS 11935, pp. 128–141, 2019.
https://doi.org/10.1007/978-3-030-36189-1_11

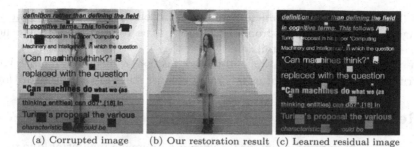

    (a) Corrupted image   (b) Our restoration result   (c) Learned residual image

Fig. 1. Image inpainting results using the proposed method. Our method directly learns the missing information in the corrupted regions (c). By plugging the learned information into the corrupted image, our algorithm can generate a realistic image.

region and 1 indicates that the data is intact but the data is usually influenced by noise n) (Fig. 1).

Recovering a clear image y from corrupted image x is a highly ill-posed problem. Most existing inpainting algorithms [5,21,23,31] usually assume that the corrupted regions are known in advance. The success of these algorithms is mainly due to the use of different image priors [7,23,31]. However, for the blind image inpainting, the corrupted regions are usually unknown in most cases, e.g., removing certain scratches from archived photographs, which accordingly increases the difficulty and fails most state-of-the-art methods.

Recent methods [13,18,20,29] use generative adversarial nets (GANs) [9] to solve image inpainting problems. However, these methods cannot be applied to blind image inpainting as they highly depend on the mask information. Thus, it is great of interest to develop an algorithm to solve the image inpainting when positions of corrupted regions are unknown.

In this paper, we propose an efficient blind image inpainting algorithm to directly restore a clear image from a corrupted image using a deep convolutional neural network (CNN). Motivated by the success of the deep residual learning algorithm [11], our deep feed-forward neural network learns the information that is lost in the corrupted regions. However, as most information is missing in the corrupted regions, directly using the neural networks in image restoration, e.g., [15], does not generate clear images. To solve this problem, we develop an encoder and decoder architecture to capture more useful information. With this architecture, our algorithm is able to predict the missing information in the corrupted regions. With the learned information, we plug it into the input image and get the final clear image.

The main contributions are summarized as follows.

- We propose an effective blind image inpainting algorithm based on a deep CNN which contains an encoder and decoder architecture to capture useful information. The proposed deep CNN is based on a residual learning algorithm which is able to restore the missing information in the corrupted regions.

- We develop a robust loss function to deal with outliers. In addition, we use image gradients computed from the inputs in the residual learning to help the estimation of image details.
- The network is trained in an end-to-end fashion, which can handle large corrupted regions and performs favorably against state-of-the-art algorithms.

2 Related Work

Recent years have witnessed significant advances in image inpainting due to the use of statistical priors on natural images and deep neural networks. We briefly review the most representative methods related to ours and put this work in proper context.

Image Inpainting. Image inpainting is firstly introduced in [5], which exploits a diffusion function to propagate low-level features from surrounding known regions to unknown regions along the boundaries of given mask. This work pioneered the idea of filling missing parts of an image using the information from surrounding regions and has been further developed by introducing the Navier-Stokes equations [3]. In [27], the normalized weighted sum of all the neighboring known pixels is used to infer the pixel to inpainting along the image gradient. Bertalmio et al. [4] first decompose an image into the sum of a structure part and a texture part and then reconstruct these two parts based on [5] separately. While these algorithms perform well on image inpainting, they are limited to deal with small objects as global information has not been considered.

Numerous image priors have been proposed to tackle image inpainting problem. In [17], hand-crafted features are introduced to help to learn image statistics for inpainting. Roth and Black [23] develop a Fields of Experts (FOE) framework based on a Markov Random Field (MRF) for learning generic image priors which encode the statistics of the entire image and could apply to general texts removal. In [31], patch-based Gaussian Mixture prior is proposed based on the EPLL framework which shows good performance in restoring the corrupted region. Barnes et al. [2] propose an efficient patch matching algorithm so the user-specified region could be filled in. These priors are effective for image inpainting. However, it difficult and expensive to utilize prior knowledge define the priors due to the variety of images.

CNN in Image Restoration. Recently, we have witnessed excellent progress in image restoration due to the use of CNN. In image restoration problem, CNN can be regarded as a mapping function, which maps the corrupted images into the clear images. The main advantage of CNN is that it can be learned by large paired training data. Thus, the restoration methods using CNN achieve better performance, such as image denoising [30], super-resolution [8,14,15,24] and deblurring [26]. One representative method is the SRCNN which is proposed by Dong et al. [8]. This method proposes a three-layer CNN to reconstruct a high-resolution image from a low-resolution input. Motivated by the success of the residual learning algorithm, Kim et al. [15] develop a much deeper CNN model with twenty layers based on residual learning which can generate the results with fine details.

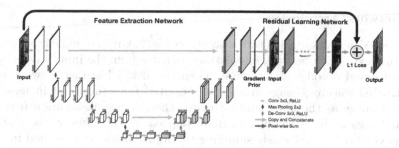

Fig. 2. Proposed network architecture. Our algorithm is based on the residual learning algorithm. The feature extraction network first encodes useful information from the input image. The following residual network aims to restore the details and structures in the corrupted regions under the guidance of the learned features and pre-extracted features. With the predicted details and structures, we can combine the input image to estimate the final clear image. Our network is jointly trained in an end-to-end manner.

Deep learning based algorithms have been made remarkable progress in image inpainting. Xie et al. [28] adopt a deep neural network pre-trained with sparse denoising auto-encoder and demonstrate its ability in blind image inpainting. Ren et al. [21] learn a three-layer convolutional neural network for non-blind image inpainting and achieve promising results given the masked image. However, this method assumes that the corrupted regions are known and it cannot be applied to the blind image inpainting problem.

Recently, GAN [9] has been used to solve image inpainting. Pathak et al. [20] employ a CNN with an adversarial loss, which performs well in predicting missing information of corrupt regions. This framework is further explored in [29], where finer details are presented. To complement face images, Li et al. [18] use a semantic face parsing loss and achieve realistic face completion results. However, this algorithm highly depends on the face domain knowledge and cannot be directly extended to natural image inpainting. In [13], Lizuka et al. aim to handle arbitrary inpainting masks. Although this method generates realistic image completion results for various scenes, it still depends on mask information to some extent.

Different from existing methods, we develop a residual learning algorithm based on CNNs to solve the blind image inpainting problem. Our neural network is based on an encoder and decoder architecture, which can preserve the details and structures of the restored images. The proposed algorithm does not require the mask information during the test stage and can be generalized on arbitrary corrupted regions well.

3   Proposed Algorithm

In this section, we describe the design methodology of the proposed method, including the network architecture, gradient prior, and loss functions.

3.1   Network Architecture

The proposed network architecture is shown in Fig. 2. Given a single input image, the feature extraction network first extract features from the input image based an encoder and decoder architecture. Then the detail learning network takes the extracted features, image gradients computed from the input images, and the input image as the input and estimates the missing information from the corrupted regions by a residual learning approach [11]. Finally, the restored image is yielded by pixel-wisely summing the input image and residual image.

The overall network has 40 layers. The feature extraction network contains 17 convolution layers and 3 de-convolution layers, while the residual learning network contains 20 convolution layers. The filter size of these layers is 3×3 pixels, each followed by a rectified linear unit(ReLU). Similar to the VGG nets [25] and VDSR [15], the filter number of the first layer is 64 which remains unchanged for the same output feature map size and is halved/doubled for the max pooling/de-convolution operation. The feature extraction network has 3 additional max pooling operations with stride 2 of size 2×2 adopted for downsampling. In addition, 3 skip link operations are used to copy and concatenate the feature maps, which can propagate the low-level features in the first several layers to the features in the last several layers, thus facilitating the missing information estimation. We apply the zero padding with padding size of 1 to ensure that the feature map in each layer has the same resolution. The whole network is jointly trained in an end-to-end manner without any post-processing.

3.2   Pre-extracted Gradient Prior

We note that image gradients are able to model the structures and details. We thus use image gradients to help the estimation of the details and structures in the corrupted regions.

Given the corrupted image, we extract its horizontal and vertical gradients. These two directional gradients are then copied and concatenated with the input image to help the detail and structure estimation. We will demonstrate the effect of the gradients in Sect. 5.2.

3.3   Loss Function

Given a training dataset $\{x^{(i)}, y^{(i)}\}_{i=1}^{N}$, a natural way to train the proposed network is to minimize the loss function

$$L(x, y) = \frac{1}{N} \sum_{i=1}^{N} \phi \left(f(x^{(i)}) - y^{(i)} \right) \tag{2}$$

where f is the mapping function which is learned by the proposed network, and $\phi(\cdot)$ denotes a robust function.

In this work, we set $\phi(\cdot)$ to be the L_1 norm instead of the L_2 norm to increase the robustness to the outliers. As the proposed network is based on the residual learning algorithm [11], the loss function used for the training is:

$$L(x, y) = \frac{1}{N} \sum_{i=1}^{N} \left\| f(x^{(i)}) + x^{(i)} - y^{(i)} \right\|_1 \tag{3}$$

We will show that the loss function based on L_1 norm is more robust to outliers thereby leading to higher accuracy results in Sect. 5.3.

4   Experimental Results

In this section, we evaluate our algorithm on several benchmark datasets and compare it with state-of-the-art methods. Similar to existing methods [21,28], we only train our model on grayscale images, and all the PSNR and SSIM values in this paper are computed by gray-scale images. We use color images for visualization, where each channel of the color image is processed independently. The trained models and source code will be made available to the public.

4.1   Datasets

We create a binary mask image dataset for training and testing, which mainly contains text patterns and block patterns. We use 20 text images to generate the training dataset and another 10 mask images for generating test dataset. These text patterns are of different font sizes (10 to 50 pt) and randomly distributed formats. In addition, randomly distributed blocks with the size of 5 to 40 pixels are added. Note that the texts and blocks indicate the corrupted regions which are unknown in advance in our blind inpainting task. Although the datasets mainly contains text patterns and block patterns, we will show that the proposed algorithm trained on this dataset generalized well.

Training Dataset. We adopt the dataset by [19] which is widely used in image super-resolution and related problems. When synthesizing data, we take the data augmentation, e.g., scale and rotation, to diversify the training samples for training. The input size of each patch for training is 64×64 pixels and 100,000 image patches are used as the training data. Our algorithm trained with such data is able to converge well and generates good results as presented in the experiments.

Test Dataset. Four datasets are employed for evaluating the proposed algorithm, i.e., Set5, Set14, Urban100 [12] and BSDS500 [1], which have been widely used as test datasets in image restoration problems. The corrupted test images are generated by taking the same pixel-wise multiplication with the created mask images, containing randomly distributed texts and blocks, which are identical for each test method.

Table 1. Quantitative evaluations on the benchmark datasets (Set5, Set14, Urban100, and BSDS500) in terms of PSNR and SSIM. Our method generates the results with the highest PSNR/SSIM than other methods including non-blind inpainting methods.

Dataset	FoE [23] PSNR/SSIM	EPLL [31] PSNR/SSIM	ShCNN [21] PSNR/SSIM	U-Net [22] PSNR/SSIM	VDSR [15] PSNR/SSIM	CE [20] PSNR/SSIM	Lizuka [13] PSNR/SSIM	Ours PSNR/SSIM
Set5	24.02/0.9322	28.27/0.9383	20.13/0.8817	31.47/0.9509	31.08/0.9432	21.02/0.7764	28.44/0.9198	**32.52/0.9569**
Set14	25.52/0.9346	28.57/0.9393	20.98/0.8768	30.07/0.9405	29.55/0.9346	19.78/0.7826	28.03/0.9222	**30.70/0.9462**
Urban100	28.16/0.9741	30.83/**0.9771**	23.93/0.9233	32.04/0.9683	31.32/0.9655	22.27/0.9186	29.38/0.9670	**32.95**/0.9748
BSDS500	25.53/0.9300	28.79/0.9329	20.28/0.8769	30.27/0.9404	30.18/0.9346	19.70/0.7529	28.08/0.9179	**30.79/0.9440**

4.2   Network Training

Our implementation is based on PyTorch. We train all networks on an NVIDIA 1080 Ti GPU using the created 100,000 image patches of the size 64×64 pixels. The weights of our proposed network are initialized by the method proposed by He et al. [10], which is confirmed to favor the performance of the rectified linear unit (ReLU). We use the Adam [16] algorithm with $\beta = (0.9, 0.999)$ and set the weight decay to be 0 to optimize the network. The number of the optimal training epoch is 60 (47100 iterations in total with batch size 128). The network is first trained with 50 epochs at a learning rate of 10^{-3}, then another 10 epochs at a lower rate of 10^{-4}. We empirically find that more iterations do not significantly improve the results.

4.3   Comparisons with State-of-the-Art Methods

We present experimental evaluations of the proposed algorithm against several state-of-the-art methods, including the conventional inpainting methods (FoE [23], EPLL [31]) and deep learning based algorithms (ShCNN [21], CE [20], Lizuka 2017 [13]), and use the publicly available codes for fair comparisons. As our algorithm contains an encoder and decoder architecture which is similar to the U-Net [22] and residual learning algorithm [15] (VDSR), we also retrain these networks for fair comparison.

Table 1 summarizes the quantitative evaluations on several benchmark datasets. Our method performs favorably against state-of-the-art methods by a large margin in term of PSNR and SSIM.

Visual Comparisons. Figure. 3 shows several visual comparisons with state-of-the-art methods. The corrupted regions in these examples are complex and some important features (e.g, boundaries of the leaves and the eye of the parrot) of the input images are missing. State-of-the-art methods fail to generate clear images. In contrast, our method generates much clearer images.

Although our network is trained with text patterns and blocks patterns, our algorithm can also generalize to other patterns, e.g., scratches. Figure 4 shows an example from [5], where the corrupted regions are arbitrary scratches. Our method generates realistic images.

Fig. 3. Visual comparisons with state-of-the-art methods. The results by prior methods contain considerable artifacts, while the proposed algorithm generates much clearer images with higher PSNR/SSIM values.

Evaluations on the Examples with Significant Noise. Our algorithm is robust to image noise. We examine the effect on the corrupted images with significant noise, i.e., salt and pepper noise. The noise level ranges from 0% to 40%. Figure 5 shows the results under different noise density, where our algorithm is

(a)   (b)   (c)   (d)   (e)   (f)   (g)   (h)   (i)

Fig. 4. Visual comparisons with state-of-the-art methods on an old photograph. (a) Input. (b) FoE [23]. (c) EPLL [31]. (d) ShCNN [21]. (e) U-Net [22]. (f) VDSR [15]. (g) CE [20]. (h) Lizuka 2017 [13]. (i) Ours. Our algorithm does not require the positions of corrupted regions and generates realistic results.

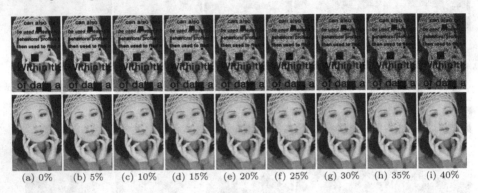

(a) 0%   (b) 5%   (c) 10%   (d) 15%   (e) 20%   (f) 25%   (g) 30%   (h) 35%   (i) 40%

Fig. 5. Evaluations of our algorithm on the images with salt and pepper noise of different density. Our algorithm performs well even when the noise density is high.

Table 2. Average running time (seconds) of the evaluated methods on Set5.

Methods	FoE [23]	EPLL [31]	ShCNN [21]	CE [20]	Lizuka [13]	Ours
Avg. Running Time	947.90	2451.82	6.42	0.33	2.91	0.52

Table 3. Quantitative evaluations for the different part of our network on the benchmark datasets (Set5, Set14, Urban100, and BSDS500) in terms of PSNR and SSIM. Our method achieves higher PSNR/SSIM values than other baselines.

Dataset	ours PSNR/SSIM	Baseline (i) PSNR/SSIM	Baseline (ii) PSNR/SSIM	w/o gradient PSNR/SSIM	w/ L2 loss PSNR/SSIM
Set5	**32.52/0.9569**	31.91/0.9532	31.57/0.9525	32.08/0.9532	31.96/0.9509
Set14	**30.70/0.9462**	30.09/0.9434	30.13/0.9437	30.53/0.9441	30.17/0.9410
Urban100	**32.95/0.9748**	32.41/0.9729	32.38/0.9732	32.85/0.9728	32.66/0.9704
BSDS500	**30.79/0.9440**	30.53/0.9418	30.51/0.9415	30.69/0.9424	30.62/0.9407

able to generates realistic images. We also evaluate the state-of-the-art methods with the same experiment. Since these methods need the guidance of mask information which is unavailable, they fail to generate satisfactory results.

Running Time. We benchmark the running time of all methods on dataset Set5 (RGB color space) with a machine of an Intel Core i7-6850K CPU and an NVIDIA GTX 1080 Ti GPU. Table 2 shows that our method compares favorably against competing inpainting methods.

5   Analysis and Discussion

In this section, we further analyze and discuss the proposed network, including an ablation study to analyze the effect of each part of the proposed network, convergence properties, additional baselines to exam the architectural choices and the limitations.

5.1   Effect of Feature Extraction Network

We examine the effect of feature extraction network by replacing it with different networks. Baseline (i): ours without feature extraction network; Baseline (ii): the feature extraction network is replaced by a 20-layer network which is of the same setting to the residual learning network, except that the first and last layer both contain 64 filters of the size $3 \times 3 \times 64$.

Table 3 shows the quantitative comparison with the aforementioned baselines. The PSNR/SSIM of our method is significantly higher than that of baseline (i) or (ii). A visual comparison is presented in Fig. 6. The results of baseline (i) or (ii) contain some artifacts. Although baselines (i) and (ii) can still remove texts well, they fail to recover sharp edge, suggesting the effectiveness of the feature extraction network.

We note that the skip connection operations [22] used in the feature extraction network is able to fuse both low-level and high-level features to help the final restoration. The comparisons with baselines (i) and (ii) further show the effectiveness of the skip connection operations.

5.2   Effect of Pre-extracted Gradient

The pre-extracted gradients are used for helping the details and structures estimation. We train a network without pre-extracted gradient to evaluate its effect. Table 3 summarizes the quantitative evaluations of these two networks in terms of PSNR and SSIM. The results show that the pre-extracted gradient is able to facilitate image restoration. Figure 7 shows some visual comparisons. The method using the pre-extracted gradients is able to recover the main structures and generate sharper edges (Fig. 7(c)).

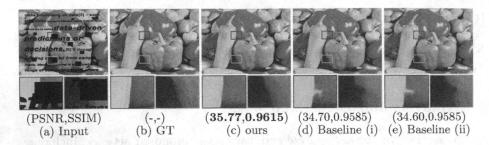

(PSNR,SSIM)   (-,-)   (**35.77,0.9615**) (34.70,0.9585)   (34.60,0.9585)
   (a) Input   (b) GT   (c) ours   (d) Baseline (i) (e) Baseline (ii)

Fig. 6. Visual comparisons of the basline methods. Feature extraction network can help the estimation of structures and details.

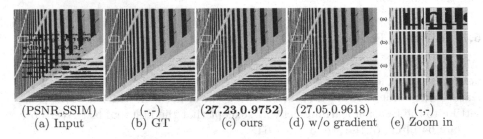

(PSNR,SSIM)   (-,-)   (**27.23,0.9752**) (27.05,0.9618)   (-,-)
   (a) Input   (b) GT   (c) ours   (d) w/o gradient   (e) Zoom in

Fig. 7. Visual comparisons of using pre-extracted gradient prior or not. The network trained with image gradient is able to recover fine detailed structures while the network trained without gradient fails to fill the useful information in the corrupted regions.

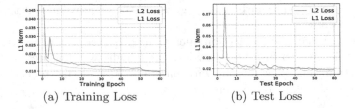

(a) Training Loss   (b) Test Loss

Fig. 8. The effect of different loss functions. The method using L1 loss function has lower errors and better convergence property in both training and test stages.

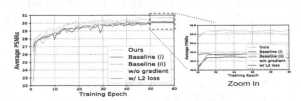

Fig. 9. Quantitative evaluations of the convergence property on Set14. The proposed networks converge well after 50 epochs.

5.3 Effect of the Loss Function

We note that L_2 norm based loss function is widely used in CNN to restore image [8]. However, as most information is missing in the corrupted regions, lots of outliers may be introduced. As the loss function based on L_2 norm is less effective for outliers, the generated image obtained by the loss function based on L_2 norm would not generate good results.

Figure 8 shows the values of the training loss functions based on L_1 norm and L_2 norm. The method with the loss function based on L_1 norm converges better than that with the loss function based on L_2 norm, suggesting that the loss function based on L_1 norm is more robust to outliers. Quantitative evaluations w.r.t. loss functions on four benchmark test datasets are summarized in Table 3. The method with the loss function based on L_1 norm generates better results.

5.4 Convergence Property

We quantitatively evaluate the convergence properties of our method against the aforementioned baseline networks on the benchmark dataset Set14. As presented in Fig. 9, the proposed algorithms start converging after 50 epochs. Moreover, our method achieves higher PSNR values than other baselines.

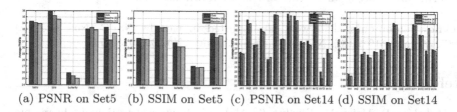

(a) PSNR on Set5   (b) SSIM on Set5   (c) PSNR on Set14 (d) SSIM on Set14

Fig. 10. Quantitative evaluations on the benchmark datasets Set5 and Set14. Our method achieves higher PSNR/SSIM values that those of baseline (iii) or baseline(v).

5.5 Additional Baselines

To further validate the effect of the proposed network architecture, we also consider several additional baselines. Baseline (iii): note that the input image itself is injected into the first layer of the residual learning network, so we remove it in this baseline; Baseline (iv): we inject the input image into each layer similar to the cascaded refinement network from Chen and Koltun [6]; Baseline (v): instead of using residual learning, we directly mask out the known part and only predict the unknown regions.

We evaluate these baselines on benchmark datasets Set5 and Set14. As shown in Fig. 10, our original network achieves higher PSNR/SSIM results than that of baseline (iii) or baseline (v). There is no quantitative results with baseline (iv), this is because this network is unstable during training. We adjust it with different learning rate, i.e., 10^{-3} to 10^{-7}, but gradient explosion still occurs, indicating that this configuration might not be suitable for the inpainting task.

The experiments with baseline (iii) indicate that removing the input image is unfavorable to the restoration. As the input image itself contains the most basic features, more context information could been used to facilitate the recovery of structure and texture via injecting the input image into the network.

In baseline (v), masking out the known part and only predicting the unknown regions needs to know the positions of the corrupted regions. Also, this method will reduce to the non-blind image inpainting task, which is different from the task that we solve. As presented in Fig. 10, this method is less effective compared to the proposed method.

6   Conclusion

In this paper, we have proposed an efficient blind image inpainting algorithm that directly restores a clear image from a corrupted input. We introduce an encoder and decoder architecture to capture useful features and fill semantic information based on the residual learning algorithm. We develop L_1 norm based loss function in the proposed work and show that it is more robust to outliers. Extensive qualitative and quantitative experiments show that the proposed algorithm performs favorably against state-of-the-art methods.

Acknowledgements. This work is supported in part by NSFC (Nos. 61572099, 61872421, and 61922043), NSF of Jiangsu Province (No. BK20180471).

References

1. Arbeláez, P., Maire, M., Fowlkes, C., Malik, J.: Contour detection and hierarchical image segmentation. IEEE TPAMI **33**(5), 898–916 (2011)
2. Barnes, C., Shechtman, E., Finkelstein, A., Goldman, D.B.: Patchmatch: a randomized correspondence algorithm for structural image editing. ACM TOG **28**(3), 24 (2009)
3. Bertalmio, M., Bertozzi, A.L., Sapiro, G.: Navier-stokes, fluid dynamics, and image and video inpainting. In: CVPR, pp. 355–362 (2001)
4. Bertalmio, M., Vese, L., Sapiro, G., Osher, S.: Simultaneous structure and texture image inpainting. IEEE TIP **12**, 882–889 (2003)
5. Bertalmio, M., Sapiro, G., Caselles, V., Ballester, C.: Image inpainting. In: SIGGRAPH, pp. 417–424 (2000)
6. Chen, Q., Koltun, V.: Photographic image synthesis with cascaded refinement networks. In: ICCV, pp. 1511–1520 (2017)
7. Dong, B., Ji, H., Li, J., Shen, Z., Xu, Y.: Wavelet frame based blind image inpainting. ACM TOG **32**(2), 268–279 (2012)
8. Dong, C., Loy, C.C., He, K., Tang, X.: Image super-resolution using deep convolutional networks. IEEE TPAMI **38**(2), 295–307 (2016)
9. Goodfellow, I.J., et al.: Generative Adversarial Networks. arXiv (2014)
10. He, K., Zhang, X., Ren, S., Sun, J.: Delving deep into rectifiers: surpassing human-level performance on imagenet classification. In: ICCV, pp. 1026–1034 (2015)

11. He, K., Zhang, X., Ren, S., Sun, J.: Deep residual learning for image recognition. In: CVPR, pp. 770–778 (2016)
12. Huang, J.B., Singh, A., Ahuja, N.: Single image super-resolution from transformed self-exemplars. In: ICCV, pp. 5197–5206 (2015)
13. Iizuka, S., Simo-Serra, E., Ishikawa, H.: Globally and locally consistent image completion. ACM Trans. Graph. **36**(4), 1–14 (2017)
14. Johnson, J., Alahi, A., Fei-Fei, L.: Perceptual losses for real-time style transfer and super-resolution. arXiv (2016)
15. Kim, J., Lee, J.K., Lee, K.M.: Accurate image super-resolution using very deep convolutional networks. In: CVPR, pp. 1646–1654 (2016)
16. Kingma, D.P., Ba, J.: Adam: a method for stochastic optimization. arXiv (2014)
17. Levin, A., Zomet, A., Weiss, Y.: Learning how to inpaint from global image statistics. In: ICCV, vol 1, pp. 305–312 (2003)
18. Li, Y., Liu, S., Yang, J., Yang, M.H.: Generative face completion. In: CVPR, pp. 3911–3919 (2017)
19. Martin, D., Fowlkes, C., Tal, D., Malik, J.: A database of human segmented natural images and its application to evaluating segmentation algorithms and measuring ecological statistics. In: ICCV, pp. 416–423 (2001)
20. Pathak, D., Krähenbühl, P., Donahue, J., Darrell, T., Efros, A.A.: Context encoders - feature learning by inpainting. In: CVPR, pp. 2536–2544 (2016)
21. Ren, J.S.J., Xu, L., Yan, Q., Sun, W.: Shepard convolutional neural networks. In: NIPS, pp. 901–909 (2015)
22. Ronneberger, O., Fischer, P., Brox, T.: U-Net: convolutional networks for biomedical image segmentation. arXiv (2015)
23. Roth, S., Black, M.J.: Fields of experts. IJCV **82**(2), 205–229 (2009)
24. Shi, W., et al.: Real-time single image and video super-resolution using an efficient sub-pixel convolutional neural network. In: CVPR, pp. 1874–1883 (2016)
25. Simonyan, K., Zisserman, A.: Very deep convolutional networks for large-scale image recognition. arXiv (2014)
26. Sun, J., Cao, W., Xu, Z., Ponce, J.: Learning a convolutional neural network for non-uniform motion blur removal. arXiv (2015)
27. Telea, A.: An image inpainting technique based on the fast marching method. J. Graph. Tools **9**(1), 23–34 (2004)
28. Xie, J., Xu, L., Chen, E.: Image denoising and inpainting with deep neural networks. In: NIPS, pp. 341–349 (2012)
29. Yang, C., Lu, X., Lin, Z., Shechtman, E., Wang, O., Li, H.: High-resolution image inpainting using multi-scale neural patch synthesis. arXiv (2016)
30. Zhang, K., Zuo, W., Gu, S., Zhang, L.: Learning deep CNN denoiser prior for image restoration. arXiv (2017)
31. Zoran, D., Weiss, Y.: From learning models of natural image patches to whole image restoration. In: ICCV, pp. 479–486 (2011)

Robust Object Tracking Based on Multi-granularity Sparse Representation

Honglin Chu[1], Jiajun Wen[1,2,3(✉)], and Zhihui Lai[1,2]

[1] College of Computer Science and Software Engineering, Shenzhen University, Shenzhen 518060, People's Republic of China
chuhonglin2018@email.szu.edu.cn,
enjoy_world@163.com, lai_zhi_hui@163.com
[2] Guangdong Key Laboratory of Intelligent Information Processing, Shenzhen University, Shenzhen 518060, People's Republic of China
[3] The National Engineering Laboratory for Big Data System Computing Technology, Shenzhen University, Shenzhen 518060, People's Republic of China

Abstract. Even though recent advances in object tracking have shown notable results in tracking efficiency, many of these algorithms are not powerful enough for its adaptation to appearance changes caused by intrinsic and extrinsic factors. In this paper, a robust object tracking method based on multi-granularity sparse representation has been proposed to exploit not only the effectiveness of holistic and local features but also make use of the representation ability of multiple patches under different granularity. For the first part, contour templates have been introduced to combine with PCA basis vectors and square templates to enhance the observation model's ability to resist the appearance changes of the target. For the second part, a novel block-division scheme is designed for multi-granularity sparsity analysis, which takes into account the joint representation ability of the target patches with different sizes. At last, in order to reduce tracking model's drift phenomenon due to model update, an adaptive update mechanism is designed by combining occlusion ratio and incremental HOG feature. Both qualitative and quantitative evaluations have been conducted on OTB-2013 datasets to demonstrate that the proposed tracking algorithm outperforms several state-of-the-art methods.

Keywords: Object tracking · Multi-granularity · PCA · Sparse representation · HOG feature

1 Introduction

As one of the research hotspots in computer vision, object tracking has been widely applied to the fields of video surveillance, virtual reality, human-computer interaction, etc. From a historical point of view, the study of the tracking methods undergoes three major phases, including the classical mean-shift, particle filter in phase one, the sparse representation-based or subspace-based methods in phase two, and correlation filter-based or deep learning-based methods in phase three. Accordingly, the competition on several major tracking datasets has been very intense in the recent years. However, the

© Springer Nature Switzerland AG 2019
Z. Cui et al. (Eds.): IScIDE 2019, LNCS 11935, pp. 142–154, 2019.
https://doi.org/10.1007/978-3-030-36189-1_12

challenges such as occlusion, deformation, illumination variation, motion blur, background clutter, etc. remain exist for designing a robust and efficient tracker which works well in various situations. In this work, a robust tracking method based on sparse representation is proposed. Therefore, our discussion mainly focuses on the key issues related to the sparse representation-based methods.

The sparse representation-based tracking methods have made great influence in the community of object tracking in the past decade [1–7, 10–12, 15–17, 24]. Mei et al. [1] applied sparse representation to the tracking problem through finding the best candidate target among all the samples with the most similar sparse coefficients and minimal reconstruction error. However, this method is time consuming due to its demand in solving L1 norm minimization problem. To improve the tracking performance, Bao et al. [3] used accelerated proximal gradient approach to optimize the L1 regularized least square problem with good tracking accuracy and high efficiency. Liu et al. [10] used a local sparse appearance model to handle appearance changes in a mean-shift framework. Jia et al. [11] proposed an adaptive structured local sparse appearance model based on alignment-pooling method to combine local and spatial structure information for enhancing the discriminant ability of the tracker. Zhang et al. [17] introduced the idea of multi-task learning into the sparse tracking framework, in which the tracking of each sample is treated as a separated learning task. Qi et al. [16] developed a structure-aware local sparse coding method which imposes the restrictions on global and local spatial structure to encode a candidate. Although these methods are robust to target occlusion, they largely depend on the effectiveness of template updating. Once the dynamic background occurs, such a tracking model will easily lost the target.

For another technology roadmap, the subspace-based tracking method has received much attention [8, 18–20] in recent years. Ross et al. [8] proposed an incremental subspace visual tracking method by learning and updating the PCA subspace representation model online, which can effectively handle appearance changes due to rotation, illumination transformation, and posture changes. However, it is not robust enough in complex scenes, especially for occlusion. Wang et al. [6] developed sparse prototypes method by incorporating the advantages of IVT [8] and L1 tracker [1] aiming at reducing the redundancy among the tracking data. However, the L1 norm regularization ignores the weak sparsity of the obtained coefficients, leading an acceptable tracking accuracy but not a perfect one. Inspired by the successful use of L2 norm in target recognition [21, 22], Xiao et al. [7] used PCA basis vectors and square templates to model the tracking target and the error term, respectively. Although the tracking speed is greatly improved, the L2 norm is less restrictive to the error term, resulting in the drift phenomenon of the tracking model.

Motivated by the above research works, we introduce contour templates to the sparse representation model, which can bring more fine features compared with the method used in [7]. To fully exploit the representation ability of the target patches with different sizes, a multi-granularity-based method is proposed to improve the tracking robustness against the influence of occlusion. In order to speed up the model's reaction to template update, an incremental HOG dictionary is developed to detect prompt appearance changes, and a novel update strategy by taking the advantages of HOG feature and square template is designed to greatly alleviate drift phenomenon. To sum up, the main contributions of our paper are as follows:

(1) Contour templates are introduced to sparse representation model in case of a coarse to fine analysis of the tracking model.
(2) A multi-granularity-based method is designed to fully exploit the representation ability of the target patches with different sizes.
(3) A reasonable update strategy by combining incremental HOG feature and error term is proposed to reduce the possibility of model drift.

The rest of this paper is organized as follows: Sect. 2 review the closely related work. Section 3 demonstrates our tracking framework and further explains the rationality of our method. Section 4 compares and analyzes test results to validate our methods compared with other state-of-the-art methods on challenging video datasets. Section 5 summarizes our work and draws the conclusions.

2   Related Works

The recent advance in subspace-based online learning tracking methods can be found in [1, 6, 7]. In this section, we mainly focus on the review of studies that are closely related with the work of our paper.

Based on the L1 tracker [1] and IVT [8], Xiao et al. combined the advantages of incremental subspace learning and sparse representation by casting object tracking as L2 norm minimization problem with less computational complexity and competitive tracking performance. Unlike L1 tracker, L2 tracker uses PCA basis vectors and square templates as dictionary to model the target appearance, which is robust to illumination, scale, rotation and deformation changes. Considering the spatial continuity of occluded pixels, they used square templates [22] to handle the partial occlusion problem. Accordingly, the candidate target can be modeled as

$$y = Bc_b + Ec_e = [B, E]\begin{bmatrix} c_b \\ c_e \end{bmatrix} = Dc \tag{1}$$

where $B \in R^{d \times k}$ denotes a target template matrix consisted of PCA basis vectors, $E \in R^{d \times n}$ represents a occlusion template, matrix consisted of square template vectors, d and k represent dimension of the PCA basis vector and its quantity, n represents the total number of square templates, $D = [B, E]$, and $c = \begin{bmatrix} c_b \\ c_e \end{bmatrix}$.

The sparse coefficients of candidate targets are obtained by solving

$$c* = \arg\min_{c} \|y - Dc\|_2^2 + \lambda \|c\|_2^2 \tag{2}$$

In [7], $P = (D^T D + \lambda I)^{-1} D^T$ as projection matrix. At t-th frame, $\{y^i\}_{i=1}^N$ represent N samples or candidate targets. Then P can be pre-calculated for all the candidate targets, which saves more computations when handling one candidate target.

Finally, the likelihood function is finally formulated as

$$p(y^i|x^i) \propto \exp(-||y^i - Bc_b^i - Ec_e^i||_2^2 - \delta||Ec_e^i||_1) \tag{3}$$

Although L2 tracker is more efficiency and robust than L1 tracker, it maybe unstable under drastic appearance changes caused by occlusion, deformation, rotation, and background clutter. In addition, the error terms usually can hardly reflect the true occlusion situation as square templates can represent any pixel of the image, leading an inaccurate occlusion estimation result. Therefore, the update scheme based on occlusion ratio will cause model drift or even a failure of object tracking.

3 The Proposed Tracking Framework

The proposed tracking algorithm consists of three main components: motion model, observation model, observation model update. The details of each part are presented as follows.

3.1 Motion Model

In the view of Markov theory, object tracking can be regarded as a Bayesian inference task [1, 6–8]. Given a series of target observation vectors $Y_t = \{y_1, y_2, \dots y_t\}$, the estimation problem of the target state x_t can be converted into a Bayesian formula to solve

$$p(x_t|Y_t) \propto p(y_t|x_t) \int p(x_t|x_{t-1})p(x_{t-1}|Y_{t-1})dx_{t-1} \tag{4}$$

where $p(x_t|x_{t-1})$ and $p(y_t|x_t)$, respectively, represent the motion model and observation model (i.e., likelihood function), and then the optimal estimation of x_t can be obtained.

3.2 Observation Model

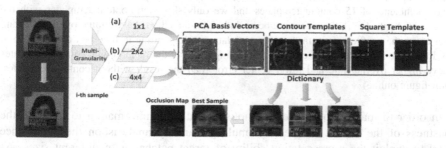

Fig. 1. Illustration for the proposed tracker model.

In this paper, contour templates are introduced for building the model of L2RLS [7]. A novel contour extraction method is designed to reduce noise while extracting smooth contours. Let C denotes the sobel gradient map of the target. Then, we normalize the gradient values to the range from 0 to 1 by normalizing equation

$$C_{norm} = \frac{C}{\max(C(i,j))} \tag{5}$$

where $C(i,j)$ represents the value of the i-th row and j-th column. The magnitude of each pixel's gradient value determines its importance to the contour description. We use each gradient values as weights to reinforce pixels near the edge and weaken pixels away from the edges. Finally, our contour template can be obtained by matrix dot production operation.

As shown in Fig. 2, we can find that most coefficients of the contour templates are much larger than PCA basis and square templates. It means the contours templates work to enhance the model's robustness to occlusion.

$$Contour = C_{norm} \cdot T \tag{6}$$

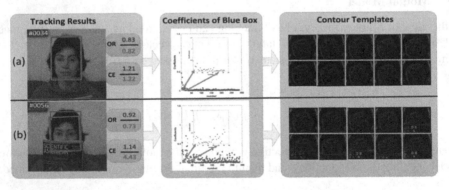

Fig. 2. OR and CE denote overlap rate and center error respectively. The red and blue boxes, respectively, represent the tracking results of L2 tracker without contour templates and with contour template. Note that in x coordinate, the coefficient values range from 17 to 31 correspond to the coefficients of 15 contour templates and we only show ten contour templates with the largest coefficients. In (a), the face is not be occluded, both methods have similar overlap rate and center error, the coefficients of the contour templates corresponding to blue box are all close to zero. In (b), the face is partial occluded, the tracking result of the method with contour templates achieve higher overlap rate and lower center error than the result without contour templates. (Color figure online)

In order to make good use of the holistic and local information to improve the robustness of the model, we perform multi-granularity analysis on the appearance model to exploit the representation ability of target patches with different sizes. As shown in Fig. 1, each candidate target is evaluated by fusing the representation results at three different granularities with different weights. Take Fig. 1(b) as an example, for a candidate target, it is divided equally into four non-overlapping sub-patches. We assume that each patch from the target local region can be linearly represented by the corresponding local patches from dictionary templates (i.e., PCA basis vectors, contour templates, square templates). Thus, the j-th sub-patch of the i-th candidate target under the l-th level of granularity can be represented as

$$y_j^{il} = B_j^l c_{bj}^{il} + G_j^l c_{gj}^{il} + E_j^l c_{ej}^{il} = [B_j^l, G_j^l, E_j^l] \begin{bmatrix} c_{bj}^{il} \\ c_{gj}^{il} \\ c_{ej}^{il} \end{bmatrix} = D_j^l c_j^{il} \qquad (7)$$

where $D_j^l = [B_j^l, G_j^l, E_j^l]$ denotes local dictionary and $c_j^{il} = \begin{bmatrix} c_{bj}^{il} \\ c_{gj}^{il} \\ c_{ej}^{il} \end{bmatrix}$ represents sparse

coefficients of local patches, G denotes a matrix with column contour template vectors, j and l corresponds to the serial number of each patch inside the candidate target (i.e., for Fig. 1(b), $j = 1, 2, 3, 4$) and the level of granularity (i.e., for Fig. 1(b), $j = 2$) respectively.

In the case of each level granularity, we obtain confidence (or named weight) of each patch by using likelihood function (3) to evaluate each sub-patches. The confidence can be calculated by

$$w_j^{il} = \exp(-||y_j^{il} - D_j^l c_j^{il}||_2^2 - \delta ||E_j^l c_{ej}^{il}||_1) \qquad (8)$$

To obtain overall weight

$$w^{il} = \frac{1}{N} \sum_{j=1}^{N} w_j^{il} \qquad (9)$$

At 1 level granularity, we get only one weight which corresponds to the current entire candidate target area. At 2th and 3th level granularity, the total number of weight N are four and sixteen, respectively. We design an adaptive weighted fusion scheme by

$$\alpha_t^l = \frac{w_{t-1}^l}{\sum_{l=1}^{L} w_{t-1}^l} \qquad (10)$$

We assume the optimal candidate target in the $t-1$-th frame is x_{t-1}^{opt}, and w_{t-1}^1, w_{t-1}^2, w_{t-1}^3 represent estimates of x_{t-1}^{opt} at different granularities respectively. If w_{t-1}^2 is larger than w_{t-1}^1 and w_{t-1}^3, we believe that the appearance model performs best at 2-th level granularity. So we hope that the contribution of 2-th level granularity to the evaluation of optimal state in the t-th frame will increase. In this way, we improve the robustness of the model.

3.3   Model Update

The error term can be represented by trivial or square templates. Based on occlusion ratio γ computed by error term, the researchers in [6] and [7] adopt one of the three model update schemes, including fully, partial and no update, to handle occlusion effectively during tracking. But the update mechanism is sensitive to drastic appearance caused by deformation and background clutters. The occlusion ratio often exceeds a high threshold $trh = 0.6$ even when no occlusion occurs [7], which causes the update of appearance model ceases for such a long time that leads to model drift or tracking failure.

Recently, HOG feature has been widely used in image recognition, especially in pedestrian detection, as it maintains good invariance to both geometric and optical deformation of the image. Motivated by them, we designed HOG-based dictionary discrimination mechanism to solve the update problem [6, 7]. As shown in Fig. 3, for the t-th frame, the optimal candidate target or sample x_t^{opt} is evaluated by our observation model. Then, we use the method in [13] to extract the HOG feature from x_t^{opt}. The x_t^{opt} is divided into small connected regions (e.g. cells), and the histogram of oriented gradients of each cell is collected, then these histograms are combined to obtain the feature descriptor.

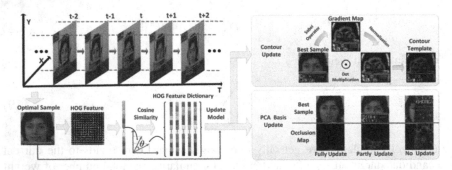

Fig. 3. The proposed model update strategy.

Let $H_t = [h_t^1, h_t^2, \cdots, h_t^n] \in R^{d \times n}$ denote a dictionary with a set of HOG feature vectors corresponding to the optimal state from the previous frames, n represent total number of the HOG vectors, $h_t^i \in R^d$. We calculate the cosine similarity between h_t an h_t^i by

$$S_i = \frac{h_t \cdot h_t^i}{\|h_t\| \|h_t^i\|}, s.t. 1 \leq i \leq n \tag{11}$$

The closer the value of S_i gets to 1, the smaller the appearance of the target changes. Then, we can obtain

$$S_{\max} = \arg \max_{1 \leq i \leq n} [S_i] \tag{12}$$

We use $S_{\max}$ to represent the extent of appearance deformation and design a HOG dictionary dynamic update strategy to make sure that the dictionary can reflect the changes of the target in time. If $S_{\max} \geq \xi$ (e.g., $\xi = 0.9$ in our method), it indicates that the deformation of current target is negligible, our HOG dictionary H_t does not need to be updated. Otherwise, the h_t^i with the smallest S_i will be replaced by h_t. We proposed a more reasonable model update mechanism by combining cosine similarity S and occlusion ratio γ [7]. If $\gamma < thl$ (e.g., in our method), the optimal candidate target is fully used to update PCA basis. If $thl \leq \gamma \leq thh$ (e.g., $thh = 0.6$ in our method), the optimal candidate target is partly used by replacing the occluded part with their corresponding part of the observation to update PCA basis. If $\gamma > thh$ and $S > \xi$, the optimal candidate target is still partly used for update. If $\gamma > thh$ and $S < \xi$, there is no update.

4 Experiment

The proposed algorithm is implemented in MATLAB 2017a on a PC with Intel® Xeon (R) CPU E5-2650 v4@2.20 GHz × 24 with 128 GB memory and runs at an average speed of 30 frames per second. All experimental parameters were set as follows,for each image,we set the total number of samples $N = 600$, and each target or candidate target (e.g., sample) was normalized to a resolution of 32 × 32. Our observation model consists of 16 PCA basis, 15 contour templates and 256 square templates. Penalty parameter λ was set to $5e - 6$. The model update period was set to 5 frames. Our HOG feature dictionary has five HOG vectors. In order to validate the effectiveness of our algorithm, we compare our algorithm with nine state-of-the-art algorithms in OTB-2013. These algorithms are L1APG [3], SP [6], L2RLS [7], IVT [8], Struck [9], ASLA [11], SCM [12], KCF [14], MTT [17].

4.1 Validation of Our Parameter

In order to determine the effective value of our cosine similarity threshold ξ, we first test our algorithm on twelve image sequences by setting different values of ξ. We set ξ from 0.8 to 0.95 in step of 0.025. The results are evaluated by calculating the overlap rate between the tracking results and the ground-truth boxes. Setting the tracking result (e.g., bounding box) as R_T and the ground-truth bounding box as R_{GT}, the overlap rate can be calculated by

$$OR = \frac{area(R_T \cap R_{GT})}{area(R_T \cup R_{GT})} \tag{13}$$

The detailed results are reported in Table 1. When $\xi = 0.9$, our algorithm can achieve the best performance, so we set $\xi = 0.9$.

Table 1. Overlap rate for each image sequence at different ξ

$\xi =$	0.800	0.825	0.850	0.875	0.900	0.925	0.950
boy	0.602	0.741	0.760	0.786	0.813	0.814	0.780
car4	0.713	0.795	0.749	0.882	0.921	0.850	0.887
deer	0.637	0.579	0.563	0.611	0.647	0.632	0.622
jumping	0.575	0.537	0.689	0.745	0.744	0.720	0.714
football	0.728	0.763	0.817	0.805	0.825	0.760	0.781
carDark	0.716	0.715	0.784	0.833	0.854	0.865	0.783
david3	0.731	0.674	0.767	0.801	0.781	0.763	0.747
walking2	0.729	0.692	0.736	0.774	0.810	0.771	0.736
faceocc1	0.602	0.787	0.801	0.852	0.927	0.910	0.878
faceocc2	0.704	0.603	0.678	0.725	0.760	0.754	0.727
singer1	0.712	0.662	0.732	0.806	0.817	0.720	0.796
girl	0.759	0.748	0.814	0.778	0.835	0.801	0.783
Average	0.684	0.683	0.741	0.783	**0.811**	0.778	0.770

4.2 Quantitative Evaluation

We use the precision and success rate of OPE (e.g., one-pass evaluation) defined in [23] to quantitatively evaluate the tracking results of all the algorithms. In Fig. 4, All trackers are ranked according to precision scores at 20 pixels and average overlap scores, respectively.

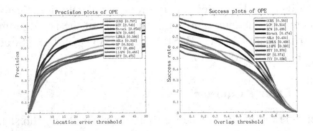

Fig. 4. The precision and success plots of OPE

Figures 5 and 6 show that the proposed method works well on the videos with occlusion, background clutter, scale variation, illumination variation and motion blur attributes. The results show that the overall performance of the our method can achieve the competitive performance compared with other state-of-the-art algorithms.

4.3 Qualitative Evaluation

Occlusion and Background Clutter: In the david3 sequence, at frame 190, OURS, SP, L2RLS and KCF perform well when the target is severely occluded by tree. It is

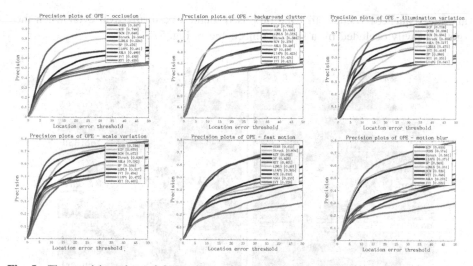

Fig. 5. The precision plots of OPE with occlusion, background clutter, illumination variation, scale variation, fast motion and motion blur attributes based on precision metric.

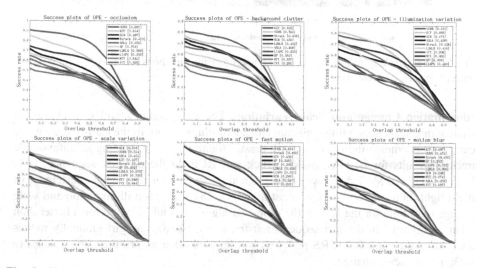

Fig. 6. The success plots of OPE with occlusion, background clutter, illumination variation, scale variation, fast motion and motion blur attributes based on overlap rate metric.

worth noting that OURS and KCF still track the target well in case of target deformation and the influence of background cluster at the 224-th frame. For sequence faceocc1 whose target is partially occluded by a book, ASLA, IVT and L1APG gradually drift away from the target, meanwhile other algorithms are robust to the interference of the book. The faceocc2 sequence is challenging as it contains rotation, and partial occlusion. In this case, the proposed algorithm achieves higher precision and overlap rate compared with other algorithms. The tracking results in football and walking2 sequences validate that the proposed method performs well even when the target is heavily occluded by a similar object, see in Fig. 7.

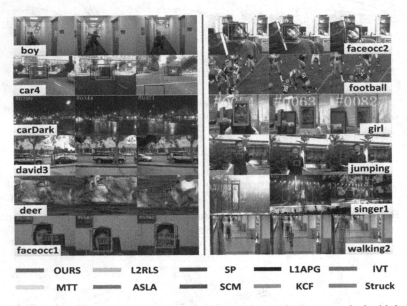

Fig. 7. Some tracking results of our method and other nine methods are marked with boxes of different colors, respectively.

Illumination Change and Scale Variation: The car4, carDark and singer1 are the challenging sequences with significant change of illumination. In the car4 sequence, the car goes underneath the overpass with drastic lighting change, leading to unfavorable tracking results for L1APG, SCM, Struck and KCF. In the carDark sequence, IVT, Struck, KCF and MTT drift away from the real target position due to the interference of night light. In the singer1 sequence, at frame 82, 181, 278, both illumination and scale change over time for the singer, the proposed algorithm and SCM perform better than other algorithms. In the girl sequence, at frame 54, 68, 82, the girl gradually moves away from the camera, OURS performs well and all other algorithms cannot adapt to such a rapid scale change.

Fast Motion and Motion Blur: In the boy, deer and jumping sequences, the target with fast motion will cause a cluster of blur area. In this case, the target is not easy to track for some trackers. In the boy sequence, MTT, SCM, L1APG, ASLA and IVT all lost target at frame 269, 281, 573, but OURS, L2RLS, SP, KCF and Struck perform well. In the jumping and deer sequences, OURS, L2RLS, SP and Struck still perform well and achieve higher overlap rate and precision compared with other algorithms.

5 Conclusion

In this paper, we present a robust tracking algorithm based on multi-granularity sparse representation. In order to enhance the robustness of the tracking model to appearance changes caused by intrinsic and extrinsic factors, we comprehensively consider the ability of different size sub-patches to represent the target by integrating the decision at different granularities. The contour template further enhances the model's robust to deformation. In addition, a more reasonable model update mechanism is designed by introducing an incremental HOG dictionary to reduce the possibility of drift due to model update based on occlusion ratio. Both qualitative and quantitative evaluations of tracking results on challenging image sequences demonstrate that the proposed algorithm performs well against several state-of-the-art algorithms. The potential disadvantage of this algorithm is that the number of samples is too large which needs much computation. In addition, the model update frequency is set to a fixed value. Our future work will focus on how to improve the sample computing efficiency, and exploring the dynamic adjustment mechanism of model update frequency.

Acknowledgements. This work was supported in part by the Natural Science Foundation of China under Grant 61703283, 61703169, 61806127, in part by the China Postdoctoral Science Foundation under Project 2016M590812, Project 2017T100645 and Project 2017M612736, in part by the Guangdong Natural Science Foundation under Project 2017A030310067, Project 2018A030310450 and Project 2018A030310451, in part by the National Engineering Laboratory for Big Data System Computing Technology, in part by the Guangdong Laboratory of Artificial-Intelligence and Cyber-Economics (SZ), in part by the Scientific Research Foundation of Shenzhen University under Project 2019049 and Project 860-000002110328, in part by the Research Foundation for Postdoctor Worked in Shenzhen under Project 707-00012210.

References

1. Mei, X., Ling, H.: Robust visual tracking using $\ell1$ minimization. In: 12th International Conference on Computer Vision, pp. 1436–1443 (2009)
2. Mei, X., Ling, H., Wu, Y., et al.: Minimum error bounded efficient $\ell1$ tracker with occlusion detection. In: CVPR, pp. 1257–1264 (2011)
3. Bao, C., Wu, Y., Ling, H., et al.: Real time robust L1 tracker using accelerated proximal gradient approach. In: IEEE Conference on Computer Vision and Pattern Recognition, pp. 1830–1837 (2012)
4. Liu, B., Yang, L., Huang, J., Meer, P., Gong, L., Kulikowski, C.: Robust and fast collaborative tracking with two stage sparse optimization. In: Daniilidis, K., Maragos, P., Paragios, N. (eds.) ECCV 2010. LNCS, vol. 6314, pp. 624–637. Springer, Heidelberg (2010). https://doi.org/10.1007/978-3-642-15561-1_45
5. Mei, X., Ling, H.: Robust visual tracking and vehicle classification via sparse representation. IEEE Trans. Pattern Anal. Mach. Intell. **33**(11), 2259–2272 (2011)
6. Wang, D., Lu, H., Yang, M.H.: Online object tracking with sparse prototypes. IEEE Trans. Image Process. **22**(1), 314–325 (2012)
7. Xiao, Z., Lu, H., Wang, D.: L2-RLS-based object tracking. IEEE Trans. Circuits Syst. Video Technol. **24**(8), 1301–1309 (2013)

8. Ross, D.A., Lim, J., Lin, R.S., et al.: Incremental learning for robust visual tracking. Int. J. Comput. Vision **77**(1–3), 125–141 (2008)

9. Hare, S., Golodetz, S., Saffari, A., et al.: Struck: structured output tracking with kernels. IEEE Trans. Pattern Anal. Mach. Intell. **8**(10), 206–2109 (2015)

10. Liu, B., Huang, J., Yang, L., et al.: Robust visual tracking with local sparse appearance model and k-selection. In: CVPR, vol. 1, no. 2, p. 3 (2011)

11. Jia, X., Lu, H., Yang, M.H.: Visual tracking via adaptive structural local sparse appearance model. In: IEEE Conference on Computer Vision and Pattern Recognition, pp. 1822–1829 (2012)

12. Zhang, T., Liu, S., Ahuja, N., et al.: Robust visual tracking via consistent low-rank sparse learning. Int. J. Comput. Vision **111**(2), 171–190 (2015)

13. Felzenszwalb, P.F., Girshick, R.B., McAllester, D., et al.: Object detection with discriminatively trained part-based models. IEEE Trans. Pattern Anal. Mach. Intell. **32**(9), 1627–1645 (2009)

14. Henriques, J.F., Caseiro, R., Martins, P., et al.: High-speed tracking with kernelized correlation filters. IEEE Trans. Pattern Anal. Mach. Intell. **37**(3), 583–596 (2014)

15. Nai, K., Li, Z., Li, G., et al.: Robust object tracking via local sparse appearance model. IEEE Trans. Image Process. **27**(10), 4958–4970 (2018)

16. Qi, Y., Qin, L., Zhang, J., et al.: Structure-aware local sparse coding for visual tracking. IEEE Trans. Image Process. **27**(8), 3857–3869 (2018)

17. Zhang, T., Ghanem, B., Liu, S., et al.: Robust visual tracking via multi-task sparse learning. In: IEEE Conference on Computer Vision and Pattern Recognition, pp. 2042–2049 (2012)

18. Zhou, Y., Li, S., Zhang, D., et al.: Seismic noise attenuation using an online subspace tracking algorithm. Geophys. J. Int. **212**(2), 1072–1097 (2017)

19. Narayanamurthy, P., Vaswani, N.: Provable dynamic robust PCA or robust subspace tracking. IEEE Trans. Inf. Theory **65**(3), 1547–1577 (2018)

20. Vaswani, N., Bouwmans, T., Javed, S., et al.: Robust subspace learning: robust PCA, robust subspace tracking, and robust subspace recovery. IEEE Signal Process. Mag. **35**(4), 32–55 (2018)

21. Zhang, L., Yang, M., Feng, X.: Sparse representation or collaborative representation: which helps face recognition? In: International Conference on Computer Vision, pp. 471–478 (2011)

22. Shi, Q., Eriksson, A., Van Den Hengel, A., et al.: Is face recognition really a compressive sensing problem? In: CVPR, pp. 553–560 (2011)

23. Wu, Y., Lim, J., Yang, M.H.: Online object tracking: a benchmark. In: Proceedings of the IEEE Conference on Computer Vision and Pattern Recognition, pp. 2411–2418 (2013)

24. Zhang, T., Xu, C., Yang, M.H.: Robust structural sparse tracking. IEEE Trans. Pattern Anal. Mach. Intell. **41**(2), 473–486 (2018)

A Bypass-Based U-Net for Medical Image Segmentation

Kaixuan Chen, Gengxin Xu, Jiaying Qian, and Chuan-Xian Ren[⊠]

School of Mathematics, Sun Yat-sen University, Guangzhou 510275,
People's Republic of China
rchuanx@mail.sysu.edu.cn

Abstract. U-Net has been one of the important deep learning models applied for biomedical image segmentation for a few years. In this paper, inspired by the way how fully convolutional network (FCN) makes dense predictions, we modify U-Net by adding a new bypass for the expansive path. Before combining the contracting path with the upsampled output, we connect with the feature maps from a deeper encoding convolutional layer for the decoding up-convolutional units, and sum up the information learned from both sides. Also, we have implemented this modification to recurrent residual convolutional neural network based on U-Net as well. The experimental results show that the proposed bypass-based U-Net can gain further context information, especially the details from the previous convolutional layer, and outperforms the original U-Net on the DRIVE dataset for retinal vessel segmentation and the ISBI 2018 challenge for skin lesion segmentation.

Keywords: U-Net · Medical image segmentation · Retinal vessel segmentation · Skin lesion segmentation

1 Introduction

Automatic segmentation is a fundamental and important task in medical image analysis [1–5]. For instance, retinal vessel segmentation plays a critical role in the early diagnosis of diabetic retinopathy (DR) which is the leading cause of visual impairment in the working-age population in developed countries [6]. Besides, melanoma is also one of the deadly forms of skin cancer [7]. The early detection is based on dermoscopic images and the accuracy means great benefits for the survival rate.

Deep learning (DL) has provided state-of-art performance for medical image processing involved in different modalities including computer tomography (CT), ultrasound, X-ray and magnetic resonance (MR) in the last few years. Since 2012, several deep convolutional neural networks (CNN) have been proposed such as AlexNet [8], GoogleNet [9], VGG-Net [10], Residual Net [11] and so on; however, these models are mostly applied for image classification since their outputs are a single label or probability. To produce a pixelwise prediction map, Jonathan Long, et al. built FCN [12] replacing the fully connected layers with convolutional layers by using 1×1 convolutional filters first and then combining semantic information from a deep, coarse layer with appearance information from a shallow, fine layer. Although the prediction of FCN

© Springer Nature Switzerland AG 2019
Z. Cui et al. (Eds.): IScIDE 2019, LNCS 11935, pp. 155–164, 2019.
https://doi.org/10.1007/978-3-030-36189-1_13

is not accurate enough as the upsampling may blur the feature maps and overlook the details, it has brought us a brand new situation for image segmentation. The defined fully convolutional net for segmentation that combines layers of the feature hierarchy and refines the spatial precision of outputs has shown the magic power of "fully convolutional" networks, based on which Ronneberger et al. proposed U-Net [13].

So far, U-Net has become another typical approach for medical image segmentation in deep learning field. A simple diagram of U-Net is shown in Fig. 1. It has a symmetric u-shaped architecture with the left part of the network capturing context information while the right part merging the upsampled featured maps with high resolution features from the contracting path. This architecture is designed to learn to assemble a more precise output and it shows great advantages for segmentation tasks: first, it can work with very few training images and yields more precise results. Second, it allows the network to learn both the context and localization of the object at the same time and can be trained end-to-end.

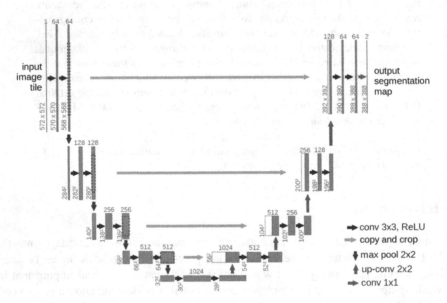

Fig. 1. A diagram of U-Net architecture with a contracting path and a symmetric expansive path taking image as input and outputting a segmentation map sharing the same size with the input [13].

The success of U-Net has led to different variants of U-Net proposed such as Attention U-Net [14] and R2U-Net [15]. Based on U-Net, Attention U-Net focuses on target structures and suppresses irrelevant regions by combining the U-Net's decoding units with a novel attention gate to highlight the useful and salient features. The function of each cooperator is clear-out. While the U-Net clarifies the objects and their locations, the attention gate helps to learn objectively with a faster speed. The recurrent residual convolutional neural network based on U-Net (R2U-Net) substitutes recurrent convolutional units with residuals (RRCU) for ordinary convolutional units in U-Net. The architecture of the RRCU is shown in Fig. 2. It refers to residual units that can overcome the crisis of the decreasing efficiency as the depth of the networks is growing and ensure the neural network to go deeper and the recurrent convolutional operations preserving the context from the same block and helping connect with the information gained by the previous steps in the same convolutional block.

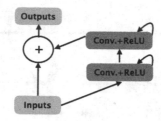

Fig. 2. A diagram of recurrent residual units [15].

We have witnessed that these deep neural networks have performed well for medical image classification, segmentation, detection and so on. Additionally, we have learned that the information gained by the shallow, fine layers is able to yield great benefits for segmentation outputs. Therefore, in this paper, we build our model upon U-Net, which we call a bypass-based U-Net, not only concatenating the feature maps from the corresponding convolutional layer, but also adding the up-convolutional units with further feature maps captured by the deeper convolutional encoding units from the next layer.

The next section explains our bypass-based U-Net with details. And the following section introduces the datasets we use for experiments, and represents the framework of our experiments. In Sect. 4 we compare the results of our proposed models with the aforementioned U-Nets. Finally, a conclusion is drawn in Sect. 5.

2   Bypass-Based U-Net

The architecture of our bypass-based U-Net (BpU-Net) is shown in Fig. 3. We have named each output of the convolutional operations, maxpooling operators and up-convolutional layers with a beginning of c, p and u respectively.

To explain the modification we have put forward, we concatenate the feature maps of c3 and u7 for example. Before concatenation, we carry out an extra convolutional operation on the feature maps of c4 and then up-sample these maps with stride 2 to keep the maps be the same size as the up-convolutional units in u7, from which we get u4. We sum up u7 and u4 and then concatenate with c3. Similarly, we apply the same adaptation to each the following block. Since we add once more convolution to c4, c3 and c2 and combine the information with the corresponding up-convolutional units, the sum we get may record further context information and make contributions to the final segmentation. Hence, we are able to assemble a more precise segmentation output. Moreover, we have implemented this modification the same way on R2U-Net as well.

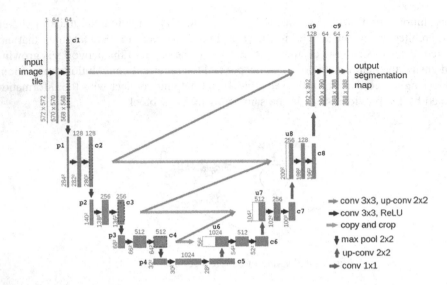

Fig. 3. The architecture of the BpU-Net combining feature maps from two previous layers with the up-convolutional units.

3 Experimental Evaluation

To evaluate the efficacy of our proposed BpU-Net, we test our models on two datasets consisting of different modalities of medical image, the DRIVE dataset [16] for retinal vessel segmentation and the ISBI 2018 skin lesion dataset [17] for lesion area segmentation. The proposed model illustrated by Fig. 3 will be applied to these datasets in three steps: (1) image preprocessing, (2) feeding and then training BpU-Net, (3) testing and segmentation result reconstruction.

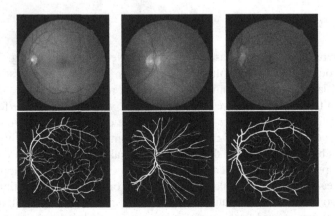

Fig. 4. Some examples from the DRIVE dataset: the first row shows the original images and the second rows shows the ground truth given by the 1st expert.

3.1 Datasets

The DRIVE dataset with 40 color fundus retinal images is shown in Fig. 4. Each image shares a same size of 565 × 584 pixels and is manually annotated by two experts. It has been split equally into training and testing sets. The other dataset is taken from the ISBI 2018 challenge on skin lesion analysis towards Melanoma detection [17–19], which is shown in Fig. 5. It is equipped with 2594 dermoscopic images of skin with one individual mask obtained by several professional and international institutions. We split it into training, validation and testing sets by 6:1:3.

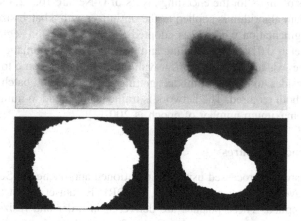

Fig. 5. Some examples from the ISBI 2018 skin lesion dataset: the first row shows the original images and the second rows gives the ground truth.

3.2 Image Pre-processing

Since the DRIVE dataset has only 20 images for training, which is very few images for deep learning, we randomly pick a pixel in the image as a center of patch and extract 9500 patches with a size of 48 × 48 pixels from each retinal image allowing overlapping, from which we obtain 190000 tiles from 20 training image in total. We take 20% of the total patches to build a dataset for validation and the rest 80% for training.

As shown in Fig. 4, the color fundus images are inhomogeneous, some extracted patches may suffer low contrast and noise. Therefore, before titling images, we use the contrast limited adaptive histogram equalization algorithm (CLAHE) and gamma adjusting algorithm to improve the global contrast, making the bright fields of view brighter and otherwise darker. These two algorithms are able to suppress noise as well. Also, we convert the RGB images into gray images to avoid the influence of hue, light and saturation and reduce the amount of calculation for the convolutional networks.

As for the ISBI 2018 skin lesion analysis dataset, on the one hand, we implement the approaches described in the preceding paragraph, the parameter in the gamma adjusting algorithm for the DRIVE dataset is set as 0.5 which can make dark regions lighter because of its low contrast; however, that for the skin lesion segmentation is set as 1.5 to make the shallows darker so that it can enhance the contrast of the whole image for the dermoscopic images. On the other hand, rather than extract patches from the original image, we resize the images into 128 × 128 pixels due to the limitation of memory and computational complexity.

3.3   Experimental Setup

To demonstrate the performance of BpU-Net and BpR2U-Net, we implement the proposed models based on Keras and TensorFlow frameworks. The experiments are all accomplished on a computer equipped with Inter Core i7-6700 CPU and NVIDIA GeForce GTX 1060 6 GB GPU.

The numbers of filters for the encoding layers of U-Net are 16, 32, 64, 128 and 256 respectively. Adam algorithm is applied as the optimizer with a batch size of 64 for the retinal vessel segmentation while 32 is adopted for the skin lesion segmentation. The learning rate is set as 0.01 at the beginning and it would adjust automatically along with the change of the loss and accuracy of the validation till it has come to the minimum value 0.0001. Besides, to avoid our models getting into overfitting, batch normalization and dropout are both required. The network is trained end-to-end to minimize the cross entropy and the maximum number of epoch is 200.

3.4   Performance Measures

Testing images are pre-processed using the mentioned approaches in Sect. 3.1 at first. Similarly, we have to extract patches from the DRIVE dataset but at this time overlapping is not applied. These patches are delivered in order and resized to 48 × 48 pixels. Then the tiles from the same image are fed into the trained model as a batch. Hence, it is convenient for us to predict the probability map for each tile and combine them back into a whole image. To test the ISBI 2018 skin lesion dataset we resize the images to 128 × 128 pixels. A threshold 0.5 is used for both datasets to generate the final truth map.

The assessment of the proposed networks involves five commonly used metrics for semantic segmentation, which are average specificity, sensitivity, the overall accuracy, precision representing the portion of the right classification to all the positive outcomes, and area under the receiver operating characteristics (ROC) curve (AUC). To calculate their values, we use variables, True Positive (TP), True Negative (TN), False Positive (FP) and False Negative (FN), from the Confusion Matrix. Their calculation equations are given in the following Eqs. (1)–(4).

$$Specificity = TN/(TN + FP) \tag{1}$$

$$Sensitivity = TP/(TP + FN) \tag{2}$$

$$Accuracy = (TP + TN)/(TP + TN + FP + FN) \tag{3}$$

$$Precision = TP/(TP + FP) \tag{4}$$

4 Experimental Results and Discussion

Table 1 shows the performance of several models including U-Net, Attention U-Net, R2U-Net, BpU-Net and BpR2U-Net on the DRIVE dataset. It reports that our modification of the U-Net combining more information from deeper layer improves the performance of the original U-Net in case of specificity and precision to a certain extent, especially for the R2U-Net. The sensitivity value of BpR2U-Net is the highest among the models on the DRIVE and it is 1.41% higher than original R2U-Net. To prove the modification is able to help capture more detailed information, we inspect the output truth maps for the segmentation of the vessels. And the results in Fig. 6 shows that BpR2U-Net performs the best in keeping the completeness and consistency of the vessels.

Table 1. Experimental results of the proposed models for DRIVE dataset.

Methods	Specificity	Sensitivity	Accuracy	Precision	AUC
U-Net	0.9827	0.7617	0.9632	0.8093	0.9621
Attention U-Net	0.9824	0.7513	0.9620	0.8045	0.9613
R2U-Net	0.9828	0.7537	0.9626	0.8085	0.9547
BpU-Net	**0.9866**	0.6823	0.9600	**0.8330**	0.9605
BpR2U-Net	0.9827	**0.7678**	**0.9637**	0.8094	**0.9627**

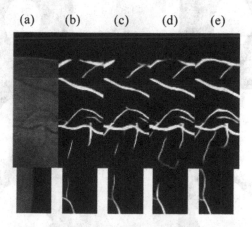

(a)   (b)   (c)   (d)   (e)

Fig. 6. Some test patches from the DRIVE dataset. Column (a) is the original image, column (b)–(e) are the results of the U-Net, Attention U-Net, R2U-Net and BpR2U-Net.

The qualitative analysis on the ISBI 2018 skin lesion dataset are reported in Table 2. The proposed models show better performance when compared to the equivalent U-Net and R2U-Net in the way of sensitivity which represents the portion of correct positive labels delivered by the model to the total positive samples. The value increases 1.19% for BpU-Net versus U-Net and is 3.48% higher than the original R2U-

Net for BpR2U-Net. Several experimental outputs of BpR2U-Net compared with the given ground truth are displayed in Fig. 7, from which we infer that the margin produced by BpR2U-Net is reliable and fair.

Table 2. Experimental results of the proposed models for ISBI2018 Skin lesion dataset.

Methods	Specificity	Sensitivity	Accuracy	Precision	AUC
U-Net	0.9709	0.8583	0.9464	0.8782	0.9862
Attention U-Net	0.9681	0.8673	0.9466	0.8790	0.9800
R2U-Net	**0.9836**	0.8252	0.9451	**0.9054**	0.9870
BpU-Net	0.9707	**0.8702**	0.9479	0.8768	0.9848
BpR2U-Net	0.9757	0.8600	**0.9493**	0.8937	**0.9896**

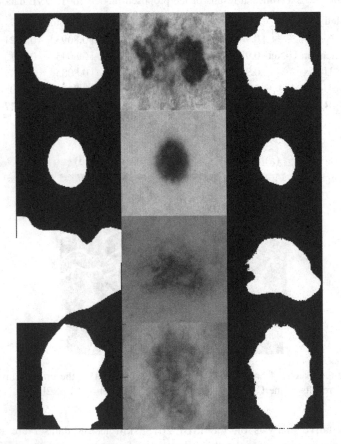

Fig. 7. Experimental outputs for the ISBI 2018 skin lesion analysis dataset. The left column shows the given ground truth, the middle column gives the original dermoscopic images and the right column shows the segmentation results of BpR2U-Net.

Additionally, the modification of the network helps to train faster relatively and take fewer epochs. For both the datasets, it takes approximately 40 epochs for the original U-Net while BpU-Net needs only 30 epochs. And for R2U-Net it needs 48 epochs at least which cost more than 12 h to train on GeForce GTX 1060 6 GB GPU while BpR2U-Net takes 7 epochs less.

5   Conclusion and Future Works

In this paper, we have presented a bypass-based U-Net where the encoder sub-networks are hierarchically connected with the feature maps produced by deeper encoding layers to capture more detailed information learned by the neural network. Also we have implemented this modification to the R2U-Net. The proposed models are called as BpU-Net and BpR2U-Net respectively. The performance of our models are demonstrated by the tests on the retinal vessel segmentation dataset DRIVE and the skin lesion dataset provided by ISBI 2018 challenge. The results explicitly show that this approach could gain better performance on both datasets especially when combining it with the R2U-Net and take less time for training as well.

In the future, we would adjust our models to improve the sensitivity and precision and explore a more effective architecture to meet the needs of better performance. We may also consider to improve the generalization and broaden the range of application.

Acknowledgments.   This work is supported in part by the National Natural Science Foundation of China under Grant 61976229, 61906046, 61572536, 11631015, U1611265 and in part by the Science and Technology Program of Guangzhou under Grant 201804010248.

References

1. Arganda-Carreras, I., Turaga, S., Berger, D., et al.: Crowdsourcing the creation of image segmentation algorithms for connectomics. Front. Neuroanat. **9**, 142 (2015)
2. Chen, H., Qi, X., Cheng, J., Heng, P.: Deep contextual networks for neuronal structure segmentation. In: Proceedings of the Thirtieth AAAI Conference on Artificial Intelligence, Phoenix, Arizona, USA, pp. 1167–1173. AAAI Press (2016)
3. Havaei, M., Davy, A., Warde-Farely, D., et al.: Brain tumors segmentation with deep neural networks. Med. Image Anal. **35**, 18–31 (2017)
4. Yu, L., Yang, X., Chen, H., Qin, J., Heng, P.: Volumetric convnets with mixed residual connections for automated prostate segmentation from 3D MR images. In: Proceedings of the Thirty-First AAAI Conference on Artificial Intelligence, California, USA, pp. 66–72. AAAI Press (2017)
5. Zheng, Y., Jiang, Z., Zhang, H., Xie, F., et al.: Histopathological whole slide image analysis using context-based CBIR. IEEE Trans. Med. Imaging **37**, 1641–1652 (2017)
6. Saker, S.: Diabetic retinopathy: in vitro and clinical studies and mechanisms and pharmacological treatments. University of Nottingham (2016)
7. Jernal, A., Siegel, R., Ward, E., Hao, Y., Xu, J., Thun, M.: Cancer statistics. CA Cancer J. Clin. **59**(4), 225–249 (2009)

8. Krizhevsky, A., Sutskever, I., Hinton, G.: ImageNet classification with deep convolutional neural networks. In: NIP'S Proceedings of the 25th International Conference on Neural Information Processing Systems, Lake Tahoe, Nevada, LNCS, vol. 1, pp. 1097–1105 (2012)
9. Szegedy, C., Liu, W., Jia, Y., et al.: Going deeper with convolutions. In: 2015 IEEE Conference on Computer Vision and Pattern Recognition (CVPR), Boston, MA, USA, pp. 1–9. IEEE (2015)
10. Simonyan, K., Zisserman, A.: Very deep convolutional networks for large-scale image recognition. In: Computer Science (2014)
11. He, K., Zhang, X., Ren, S., Sun, J.: Deep residual learning for image recognition. In: 2016 IEEE Conference on Computer Vision and Pattern Recognition (CVPR), Las Vegas, NV, pp. 770–778 (2016)
12. Long, J., Shelhamer, E., Darrel, T.: Fully convolutional networks for semantic segmentation. In: 2015 IEEE Conference on Computer Vision and Pattern Recognition (CVPR), Boston, MA, USA, pp. 3431–3440. IEEE (2015)
13. Ronneberger, O., Fischer, P., Brox, T.: U-Net: convolutional networks for biomedical image segmentation. In: Navab, N., Hornegger, J., Wells, W.M., Frangi, A.F. (eds.) MICCAI 2015. LNCS, vol. 9351, pp. 234–241. Springer, Cham (2015). https://doi.org/10.1007/978-3-319-24574-4_28
14. Oktay, O., et al.: Attention U-Net: learning where to look for the pancreas. In: 1st Conference on Medical Imaging with Deep Learning (MIDL 2018), Amsterdam, The Netherlands (2018)
15. Alom, M., Hasan, M., Yakopcic, C., Taha, T., Asari, V.: Recurrent residual convolutional neural network based on u-net (R2U-net) for medical image segmentation (2018). https://arxiv.org/abs/1802.06955
16. Web page of the digital retinal images for vessel extraction. http://www.isi.uu.nl/Research/Databases/DRIVE/. Accessed 11 Mar 2019
17. Web page of the ISBI 2018: Skin lesion analysis towards melanoma detection. https://challenge.kitware.com/#challenge/5aab46f156357d5e82b00fe5. Accessed 20 Apr 2019
18. Codella, N., Rotemberg, V., Tschandl, P., Emre Celebi, M., et al.: Skin lesion analysis toward melanoma detection 2018: a challenge hosted by the international skin imaging collaboration (ISIC) (2018). https://arxiv.org/abs/1902.03368
19. Tschandl, P., Rosendahl, C., Kittler, H.: The HAM10000 dataset, a large collection of multi-source dermatoscopic images of common pigmented skin lesions. Sci. Data 5, 180161 (2018)

Real-Time Visual Object Tracking Based on Reinforcement Learning with Twin Delayed Deep Deterministic Algorithm

Shengjie Zheng and Huan Wang[✉]

Nanjing University of Science and Technology, Nanjing, People's Republic of China
wanghuanphd@njust.edu.cn

Abstract. Object tracking as a low-level vision task has always been a hot topic in computer vision. It is well known that Challenges such as background clutters, fast object motion and occlusion et al. affect a lot the robustness or accuracy of existing object tracking methods. This paper proposes a reinforcement learning model based on Twin Delayed Deep Deterministic algorithm (TD3) for single object tracking. The model is based on the deep reinforcement learning model, Actor-Critic (AC), in which the Actor network predicts a continuous action that moves the target bounding box in the previous frame to the object position in the current frame and adapts to the object size. The Critic network evaluates the confidence of the new bounding box online to determine whether the Critic model needs to be updated or re-initialized. In further, in our model we use TD3 algorithm to further optimize the AC model by using two Critic networks to jointly predict the bounding box confidence, and to obtain the smaller predicted value as the label to update the network parameters, thereby rendering the Critic network to avoid excessive estimation bias, accelerate the convergence of the loss function, and obtain more accurate prediction values. Also, a small amount of random noise with upper and lower bounds are added to the action in the Actor model, and the search area is reasonably expanded in offline learning to improve the robustness of the tracking method under strong background interference and fast object motion. The Critic model can also guide the Actor model to select the best action and continuously update the state of the tracking object. Comprehensive experimental results on the OTB-2013 and OTB-2015 benchmarks demonstrate that our tracker performs best in precision, robustness, and efficiency when compared with state-of-the-art methods.

Keywords: Visual object tracking · Reinforcement learning ·
Actor-Critic model · Twin Delayed Deep Deterministic algorithm

1 Introduction

Contemporarily, as one of fundamental problems in computer vision, visual tracking has made remarkable progress, which has been used in many practical applications such as robot, transportation, Unmanned vehicle and electronic

© Springer Nature Switzerland AG 2019
Z. Cui et al. (Eds.): IScIDE 2019, LNCS 11935, pp. 165–177, 2019.
https://doi.org/10.1007/978-3-030-36189-1_14

surveillance equipment. Meanwhile, as a difficult topic in computer vision, visual object tracking faces many challenges such as occlusion, motion blur, fast motion, background clutters and scale variation. With the development of deep learning, the object tracking problem is gradually recognized as an object detection problem being composed of feature learning, candidate object extraction and identification in a continuous video sequence. deep neural networks can greatly improve the tracking precision by learning rich appearance feature of an tracked object compared to traditional methods. However, when facing with a complicated situation, deep neural networks still suffer frequent tracking drift and pool localization precision due to the big gap between training and test scenarios. To overcome this problem, deep reinforcement learning based tracking methods have employed to accurately locate the object area by forcing deep neural networks to learn a tracking behavior, i.e. learning to select continuous or discrete actions to move current tracking box to best fit the tracked object.

Our work aims to improve the precision and robustness of existing deep reinforcement learning based trackers. Its contributions are twofold:

(1) Our proposed method locates the tracking object in each frame of the sequences by a continuous action, which obtains more accurate action to locate the tracked object.
(2) Twin Delayed Deep Deterministic algorithm is a first attempt to be used in visual tracking which effectively improves the robustness of the tracker in various complex situations. Also, we use supervised training and greedy strategy to train Twin Delayed Deep Deterministic to improve the performance of the deep deterministic policy gradient algorithm.

2   Related Work

The traditional tracking methods mainly rely on designing various appearance model to model the object area given in the first frame of an video sequence, and the areas that best fit the constructed object appearance model in the subsequent frames are considered the position of the tracked object. The appearance model is consistently updated online to prevent tracking drift. These tracking methods are referred to as generative methods. The correlation filters, which searches the best filter template applied to the next frame, and the tracked object is the region with the largest response. In contrast, discriminative tracking methods are also broadly investigated. They view the true target regions as positive samples and collect negative samples in the background region. Afterwards, an online regressive classifier is online trained and maintained by those samples collected and applied to infer the confident or probability of being the true object over each location. The location with the largest confidence is regarded as the tracked object. The classifier and learning strategy are the most critical for the discriminative tracking methods. Besides the conventional classifiers such as Support vector machine, Adaboost as well as BP network, the deep neural network, with stronger representation ability, can better distinguish the background and the tracked object.

Deep Neural Network. CNN-based methods have been proposed to learn semantic future for the visual tracking task. HCF, HDT and other methods [6–8] make full use of the convolution characteristics of each layer in the convolutional neural network, and further improve the tracking performance by combining multi-level convolution features on the basis of correlation filtering. However, the goals of the tracking task and classification task are different, so a pre-trained network on ImageNet [5] may not be optimal for visual tracking. Nam [9] designs a multi-domain convolutional neural network specifically trained on tracking video sequences, which achieves excellent performance, but faces the issue of low processing speed.

Reinforcement Learning. Reinforcement learning is a principled paradigm to learn how to make decisions and select actions online. Due to the combination of reinforcement learning and deep learning, deep reinforcement learning has a good memory and inference ability, and deep reinforcement learning has increasingly been applied to visual tracking [11–13]. Yun et al. [11] propose a policy network, which gets a series of optimal actions according to different states, finally moves to object location by these actions. In [12], an Actor-Critic framework is introduced during offline training, where the Actor model predicts a continuous action to locate the tracked object, and the Critic model predicts the Q-value to help update the Actor and Critic networks, so that in the tracking process, the Actor-Critic framework becomes more robust to face various complex scenes. And the method has achieved good results with real-time performance in the popular benchmarks.

3 Proposed Method

3.1 Problem Description

Reinforcement learning usually consists of five basic elements: environment, state, action, reward, and agent. If one addresses the visual tracking problem using a reinforcement learning framework, the environment is considered a single frame of an video sequence, and the state represents the image patch within a bounding box predicted by a tracker. Following an action, the bounding box in the previous frame moves to the location of the tracked object in the current frame. The reward is defined as intersection over union ratio between the predicted bounding box and the true bounding box of the tracked object. The agent is responsible for getting the corresponding reward according to the environment and action.

State. In our work, the state is the observation image patch within the bounding box $b = [x, y, h, w]$, where (x, y) denotes the center coordinates and h and w are the height and width, respectively. we obtain a state s by defining a pre-processing function that crop the image patch within the bounding box b in an image F and resize it to fit the input size of our deep network.

Action and State Transition. To accelerate the tracking speed of our model, we choose a continuous action, which moves the bounding box to the location of the tracked object in the current frame directly. Here, we define the action $a = [\Delta x, \Delta y, \Delta s]$ to denote the relative motion of the tracked object. In detail, Δx and Δy denote the relative horizontal and vertical translations and Δs stands for the relative scale change. Considering the temporal continuity in the tracking problem, we made some range restrictions on the action a: $-1 \leq \Delta x \leq 1$, $-1 \leq \Delta y \leq 1$ and $-0.05 \leq \Delta s \leq 0.05$. After performing an action, the original bounding box $b = [x, y, h, w]$ moves to the new bounding box $b' = [x', y', h', w']$,

$$\begin{cases} x' = x + \Delta x \times w \\ y' = y + \Delta y \times h \\ w' = w + \Delta s \times w \\ h' = h + \Delta s \times h \end{cases} \tag{1}$$

Then, the next state s' is obtained by the state transition process $s' = f(s, a)$, In order to fit the size of our network, we define a pre-processing function $s = \phi(b, F)$ to crop the image patch within the bounding box b in a given frame F and resize it to fit the input size of the deep work. In this work, we use the Actor model to infer a continuous action $a = u(s)$, and we prevent the Critic network from overfitting by adding a small amount of random noise ϵ to the Critic network, $u(s') + \epsilon$.

3.2 Overview of Our Tracking Framework

The goal of visual tracking is to pursue the location of the tracked object and obtain the target bounding box in each subsequent frame, given the object position in the first frame. In this work, we apply an effective tracking method based on Twin Delayed Deep Deterministic algorithm of deep reinforcement learning in offline training. The Actor model predicts a continuous action, which moves the bounding box in the previous frame directly to the location of the tracked object. In offline training, the Critic model predicts the probability of target based on the image within the current bounding box to obtain a more accurate Q-value, and the Q value guides the Actor and Critic models to learn jointly, and obtains more accurate prediction values from the updated two networks. During the tracking process, the Critic model also receives the action given by the Actor model and the patch processed by the pre-processing function to determine whether to reinitialize tracking. We present our architectures in Fig. 1.

3.3 Offline Training

Twin Delayed Deep Deterministic Algorithm Framework. Q-learning is a classic algorithm for reinforcement learning. In the state s, an optimal action is obtained by predicting the future benefit Q-value of each action. When updating the Q-value of the state and action pair, the Q-value is generated by the Bellman equation, and the formula for updating $Q(s, a)$ is as follows:

$$Q(s, a) \leftarrow Q(s, a) + \alpha[r + \gamma max_{a'} Q(s', a') - Q(s, a)] \tag{2}$$

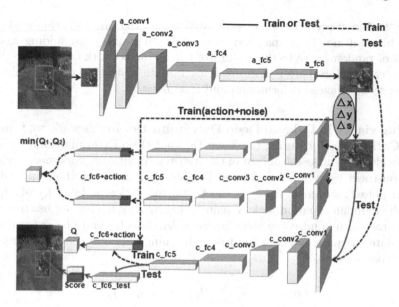

Fig. 1. The pipeline of the proposed tracking algorithm. The black dotted line represents the training process and the green solid line indicates the test process. (Color figure online)

where α is the learning rate, and γ is reward attenuation coefficient, r represents the reward obtained by the action in the current state, $Q(s, a)$ represents the Q-value of action a in state s, s' is the next state of state s, a' is the action performed in state s. In the object tracking process of the deep reinforcement learning algorithm based on the Actor-Critic model, we may overestimate the true Q-value due to the use of negative samples and repeated samples to train the Critic network. Sometimes this overestimated value will be very large, resulting in Actor and Critic have poor convergence and inaccurate prediction values. In order to reduce the excessive estimation bias, the Clipped Double Q-Learning method proposes two Q-learning networks to predict the future benefit $Q_{\theta 1}$ and the future benefit $Q_{\theta 1}$ respectively, and select a smaller predictive value to suppress the Q-value. By taking the minimum of the two Q-learning network prediction values, the best label l_Q is obtained, as shown in the following formula:

$$l_Q = r_i + \gamma \min_{j=1,2} Q_{\theta j}(s_i', u(s_i') + \epsilon), \quad \epsilon \sim clip(N(0,\sigma), -c, c) \tag{3}$$

where s_i' is the new patch obtained after moving in state s_i, $u(s_i')$ denotes the optimal action predicted under state s_i', $\min_{j=1,2} Q_{\theta j}(s_i', u(s_i') + \epsilon)$ represents the minimum Q-value of the two Q-learning network predictions, ϵ indicates the random noise, $N(0,\sigma)$ represents a normal distribution with 0 the mean and σ the variance, and c is the maximum of noise amplitude, $clip$ represents truncation of the generated random value. By using the Clipped Double Q-learning method, the Critic network does not produce excessively high estimated bias. Although this method may predict a lower estimated value, the invalid action evaluated by the bias will not be propagated to the update strategy. By adding a small amount of random noise to the actions of the Actor network output, the Actor network can explore better motions. This update method can effectively face complex scenes such as deformation and fast motion.

Training via Twin Delayed Deep Deterministic. In this work, we train the Actor-Critic network by using the Twin Delayed Deep Deterministic algorithm. Its main idea is to make the output of the Actor and Critic network more accurate and robustness by using the Clipped Double Q-Learning method and adding random noise. As shown in Eq. 4, the Critic model learns label l_Q of $Q(s,a)$ using the Bellman equation with N pairs of (s_i, a_i, r_i, s_i'). Then we use the mean square error to minimize the error between l_Q and $Q_{\theta 1}$ and $Q_{\theta 2}$, respectively, and update the parameters of the two Q-learning networks, namely the Critic network parameters.

$$Loss_{Q_{\theta 1}} = \frac{1}{N} \sum_i (l_Q - Q_{\theta 1}(s_i, a_i))^2 \tag{4}$$

$$Loss_{Q_{\theta 2}} = \frac{1}{N} \sum_i (l_Q - Q_{\theta 2}(s_i, a_i))^2 \tag{5}$$

where $l_Q = r_i + \gamma \min_{j=1,2} Q_{\theta j}(s_i', u(s_i') + \epsilon)$ is the $Q(s_i, a_i)$ label and i is the i-th frame in a video sequence. The way the Actor model parameters are updated is determined by the gradient direction of the Critic network update parameters, because the larger the reward, the better the action. So use the chain rule to update the Actor network parameters:

$$\nabla_{\theta^u} J \approx \frac{1}{N} \sum_i \nabla_a Q(s, a_i | \theta^Q)|_{s=s_i, a=u(s_i)} \nabla_{\theta^u} u(s|\theta^u)|s = s_i \tag{6}$$

Where $\nabla_a Q(s, a_i | \theta^Q)|_{s=s_i, a=u(s_i)}$ represents the gradient of the Critic network parameter, $\nabla_{\theta^u} u(s|\theta^u)|s = s_i$ represents the gradient of the Actor network parameter. During offline train, We use ImageNet [5] to train both two networks by randomly selecting a piece of training sequences $[F_k, F_{k+1}, ..., F_{k+n}]$ with corresponding ground truth$[G_k, G_{k+1}, ..., G_{k+n}]$ from the training data. The Actor network obtains action a_i, obtains the image patch s_i according to the pre-processing function $\phi(b_i, F_i)$, and then obtains a new bounding box b_i' by executing the action a_i in the initial bounding box b_i. Therefore, the IoU value of b_i' and G_i is calculated as the reward r_i of the current frame, and it gets a new image patch $s_\phi'(b_i', F_i)$ through the pre-processing function, and then the training pairs (s_i, a_i, r_i, s_i') that make up the selected sequence train the Critic and Actor network parameters (Fig. 2).

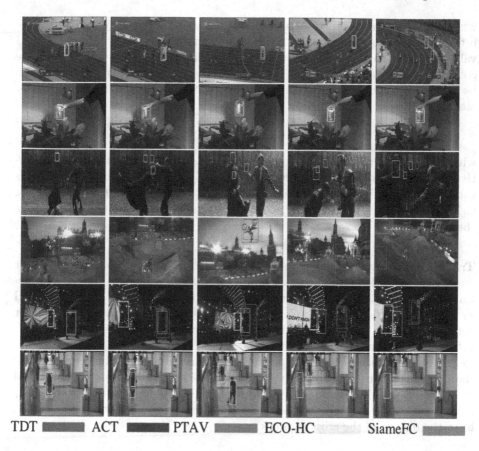

TDT ▬▬▬ ACT ▬▬▬ PTAV ▬▬▬ ECO-HC ▬▬▬ SiameFC ▬▬▬

Fig. 2. Qualitative results of our proposed method and other trackers on some challenging sequences (Blot, Coke, Matrix, MotorRolling, Singer2, Walking2)

3.4 Online Tracking

Network Initialization. We initial the parameters of the Actor and the Critic network with the ground truth in the first frame. Firstly, add a Gaussian noise on the ground-truth bounding box of the first frame to randomly generate a bounding box, and then randomly select n bounding boxes $b_1, ..., b_n$ from these bounding boxes, the action a_n is the distance from the sample bounding box to the ground-truth bounding box. Then, we obtain the state s_n by extracting the image observation for bounding box b_n using pre-processing function $s_n = \phi(b_n, F)$. Therefore, the Actor network adjusts the parameters by minimizing the $L2$ loss function

$$\min_{\theta^u} \frac{1}{N} \sum_{n=1}^{N} [u(s_n|\theta^u) - a_n]^2 \tag{7}$$

where θ^u is the parameters of Actor network. It assigns a label to the image in the n-th bounding box generated by the Gaussian method as above. When the IoU value is greater than 0.7, the label is set to 1, otherwise -1. By this method, we collect a set of training samples $\{s_n, l_n\}$. Furthermore, the Critic network can be trained by minimizing the loss function by the stochastic gradient descent method,

$$\underset{\theta^Q}{argmin} - \sum_{s \in S_+} P_+(s|Q; \theta^Q) - \sum_{s \in S_-} P_-(s|Q; \theta^Q), \tag{8}$$

where S_+ and S_- represent the positive and negative training sample set, respectively, Q represents the Critic network function $Q(s, u(s) + \epsilon)$, and θ^Q represents the Critic network parameter. For a trained Critic network, given a state s, Critic determines the probability P that the image within the bounding box is the target.

Tracking via Actor-Critic Model. For online tracking, we combine the Actor and Critic networks. Initially, we calculate the state s_t using pre-processing function $\phi(b'_{t-1}, F_t)$ (b'_{t-1} denotes the optimal bounding box in the $t - 1$ frame and F_t is the image frame). Then, We will put state s_t into the Actor network to get action $a_t = u(s_t)$. Thus, we calculate the new bounding box b'_t and image observation s'_t in current frame with the action a_t and bounding box b'_{t-1}. Therefore, we utilize the Critic network to verify the observation s'_t. If the score is large than 0, the action a_t is correctly selected and the b'_t is considered a optimal location in the t-th frame. If the Critic fails, we add Gaussian noise to the b'_{t-1} in previous frame to find the optimal location b'_t which has the highest score from the output of the critic network.

Network Update. An simple but effective update strategy is applied to improve the balance between the robustness and efficiency of our model. Specifically, when the score through the Critic output is less than zero, we think the Critic does not fit well the change of object appearance in the current environment, and then the positive and negative samples collected in the previous N frames (N is empirically set as 10 in our experiments) to update the network based on Eq. 8.

3.5   Implementation Details

Samples Generation. We train the networks in both offline and online tracking stages by obtaining 64 samples $X_t^i = (x_t^i, y_t^i, z_t^i)$ (x and y are horizontal and vertical translations; z denotes the scale) with a Gaussian distribution centered at the object centroid in frame $t - 1$. We set the covariance of all components in X_t^i as a diagonal matrix $diag(0.09d^2, 0.09d^2, 0.5^2)$, where d stands for the mean of the width and height of the tracked object.

Offline Training. We train the Actor network and Critic network on 768 video sequences from the ImagetNet Video. Consecutive twenty to forty frames in a sequence are randomly selected as training samples each time to enhance network robustness. At the beginning of each training iteration, we randomly select 64 samples as positive samples in the first frame image. These samples are obtained by adding Gaussian noise on the ground truth, and their IoU is larger than 0.7, used to initialize the Actor network parameters, where the learning rate of the Actor and Critic network is set to 10^{-4}. Each frame of the image is tracked in turn, and after the Actor network predicts the action, we use the greedy strategy to decide the action that is the value predicted by the Actor or the distance between the bounding box of the previous frame and the ground truth of the current frame to train the Actor network, the initial value of the greedy strategy is set to $\phi = 0.5$ and reduced by 5% after every ten thousand iterations. As the number of training increases, the probability that greedy strategies choose the value of the Actor predicted is increasing. We train a total of 250,000 iterations of these two networks and update parameters of the target network every ten thousand iterations. In Actor and Critic network, we set the learning rate to 10^{-6} and 10^{-5}, respectively, and the replay buffer size is set to 10^4.

Online Tracking. In online tracking, we randomly select 500 positive samples and 5000 negative samples with ground truth in first frame to train the Actor and Critic network. And the Actor network only uses positive samples to train parameters. The learning rate is set to 10^{-4} in the Actor and Critic network. The Actor and Critic network set the batch size to 64 and 128, respectively. When the Critic network obtains the score of the current sate and the score is less than 0, it means our tracker fails, we add Gaussian noise to the previous bounding box to obtain the optimal s', and its score is highest among them. At the same time, we also take the samples of 10 recent successful tracking frames for retraining. We will put 50 positive samples and 150 negative samples from each successful tracking frame in the container to use when tracking failure.

4 Experiment

4.1 Parameter Setting

We implement our tracker in python3 with the TensorFlow framework, and it runs at 23 fps on a PC which has a 3.6 GHz CPU with 32G memory and a GeForce GTX 1080 GPU. We evaluate our tracker on the popular Object Tracking Benchmark (OTB).

4.2 Datasets

As the well-known visual tracking benchmark, OTB-2015 contains 100 fully annotated sequences to evaluate tracking algorithms. And the OTB-2013 has 50 sequences from the OTB-2015 which contains some of the challenging video

---

Algorithm 1. offline training the Actor network

Input: Training frame in sequence [F] and their corresponding ground truths[G]
Output: Trained weights for Actor network

    Initial critic networks Q_{θ_1}, Q_{θ_2}, and the Actor network u_ϕ with random parameters θ_1, θ_2, ϕ

    Initialize target networks $\theta_1' \leftarrow \theta_1, \theta_2' \leftarrow \theta_2, \phi \leftarrow \phi'$

    Initialize replay buffer β

  repeat

    Randomly select a number of consecutive frames $[F_k, F_{k+1}, ..., F_{k+n}]$ with their ground truth $[G_k, G_{k+1}, ..., G_{k+n}]$

    Receive initial observation state s_k according to F_k and G_k

    Train the Actor network for an iteration utilizing s_1

  for each t=2, n+1 **do**

    Obtain state s_t according to state s_{t-1} and F_{k-1+t}

    Select action with exploration noise $a \sim u_\phi(s_t) + \epsilon, \epsilon \sim N(0, \sigma)$, and observe reward r and new state s' store transition tuple (s, a, r, s') in β

    Sample a random mini-batch of N transitions (s, a, r, s') from β

    $\tilde{a} \leftarrow u_{\phi'} + \epsilon, \epsilon \sim clip(N(0, \tilde{\sigma}), -c, c)$

    $y \leftarrow r + \gamma \min_{i=1,2} Q_{\theta_i'}(s', \tilde{a})$

    Update critics $\theta_i \leftarrow arg\,min_{\theta_i} N^{-1} \sum (y - Q_{\theta_i}(s, a))^2$

    if t mod d **then**

      Update ϕ by the deterministic policy gradient:

      $\nabla_\phi J(\phi) = N^{-1} \sum_i \nabla_a Q_{\theta_i}(s, a)|_{a=u_\phi(s)} \nabla_\phi u_\phi(s)$

      Update target networks:

      $\theta_i' \leftarrow \tau\theta + (1 - \tau)\theta_i'$

      $\phi_i' \leftarrow \tau\phi + (1 - \tau)\phi_i'$

    end if

  end for

  until Reward become stable

---

frame properties in modern object tracking domain, including 11 attributes: Illumination Variation, scale variation, occlusion, deformation, motion blur, fast motion, in-plane rotation, out-of-plane rotation, out-of-view, background clutters, low resolution.

4.3 Metrics

Precision Plots. the Euclidean distance between the center point of the target position estimated by the tracking algorithm and the center point of the ground-truth, which is less than the percentage of the video frame of the given threshold. Different thresholds result in different percentages, and the general threshold is set to 20 pixels. The disadvantage of this evaluation method is that it does not reflect changes in the size and scale of the object.

Success Plots. The bounding box is obtained by the tracking algorithm (denoted as a), and the box is given by ground-truth (denoted as b), the intersection ratio of the two (IoU) is $overlap = |a \cap b|/|a \cup b|$, $|.|$ indicates the number of pixels in the area. When the overlap of a certain frame is greater than the set threshold, the frame is regarded as successful, and the percentage of the total successful frame in all frames is the success rate. The value of overlap ranges from 0 to 1, and the general threshold is set to 0.5.

FPS. The time taken by the tracker to track the target in the video sequence, and the real-time performance of the tracking algorithm is measured by calculating the number of video frames processed per second.

4.4 Methods for Comparison

The precision and success plots over 50 sequences of OTB-2013 are showed in Fig. 3 where our proposed algorithm is denoted as TDT, we compare our tracker with 14 other trackers. The results show that our proposed method achieves the best performance compared to other trackers, and our tracker also reaches real-time performance. Due to the superior performance of CNN features in recognition, our tracker effectively identifies the tracked object which updates the model when the tracker fails. In comparison with SimaFC, our tracker proposes an accurate continuous action to locate tracked objects, which achieves better performance and enable our models to achieve real-time results. Due to using the Actor and Critic network, the ACT has good performance, and our proposed method reduces the loss error of the Actor and Critic network, which gets more accurate action and identification of the tracked object. The ECO-HC and PTAV also achieve good performance due the improved correlation filter technique or the explicit combination of a tracker and a verifier, But due to too many filters, the FPS drops. As shown in Fig. 4, compared with different trackers, our proposed work is also superior to its accuracy and success plots in OTB-2015.

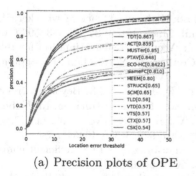

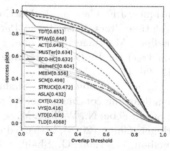

(a) Precision plots of OPE   (b) Success plots of OPE

Fig. 3. The precision and success plots of different trackers on OTB-2013 dataset

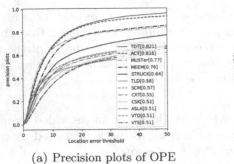
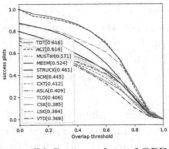

(a) Precision plots of OPE   (b) Success plots of OPE

Fig. 4. The precision and success plots of different trackers on OTB-2015 dataset

5 Conclusion

In this paper, we apply the twin delayed deep deterministic algorithm to the Actor-Critic reinforcement learning framework for visual tracking, which moves a bounding box and evaluates the patch within the bounding box simultaneously. We make the Actor network move the bounding box to optimal location in current frame by predicting continuous action. Also, the Clipped Double Q-learning method is introduced to the Critic network, leading to correctly distinguishing whether the patch in a bounding box is the tracking object in complex situations such as deformation or motion blur. Experiments on benchmark show that our method achieves better robustness and accuracy compared to the state of the art reinforcement learning based tracking methods.

Acknowledgement. This work is supported by National Science Foundation of China (Grant No. 61703209 and 61773215).

References

1. Henriques, J.F., Caseiro, R., Martins, P., et al.: High-speed tracking with kernelized correlation filters. IEEE Trans. Pattern Anal. Mach. Intell. **37**(3), 583–596 (2014)
2. Danelljan, M., Häger, G., Khan, F., et al. : Accurate scale estimation for robust visual tracking. In: British Machine Vision Conference, Nottingham, 1–5 September 2017 (2014)
3. Danelljan, M., Robinson, A., Shahbaz Khan, F., Felsberg, M.: Beyond correlation filters: learning continuous convolution operators for visual tracking. In: Leibe, B., Matas, J., Sebe, N., Welling, M. (eds.) ECCV 2016. LNCS, vol. 9909, pp. 472–488. Springer, Cham (2016). https://doi.org/10.1007/978-3-319-46454-1_29
4. Danelljan, M., Bhat, G., Shahbaz Khan, F., Felsberg, M.: ECO: efficient convolution operators for tracking. In: CVPR, pp. 6638–6646 (2017)
5. Russakovsky, O., et al.: ImageNet large scale visual recognition challenge. Int. J. Comput. Vis. **115**(3), 211–252 (2015)
6. Ma, C., Huang, J.B., Yang, X., et al.: Hierarchical convolutional features for visual tracking. In: Proceedings of the IEEE International Conference on Computer Vision, pp. 3074–3082 (2015)

7. Bertinetto, L., Valmadre, J., Henriques, J.F., Vedaldi, A., Torr, P.H.S.: Fully-convolutional Siamese networks for object tracking. In: Hua, G., Jégou, H. (eds.) ECCV 2016. LNCS, vol. 9914, pp. 850–865. Springer, Cham (2016). https://doi.org/10.1007/978-3-319-48881-3_56

8. Ma, C., Yang, X., Zhang, C., Yang, M.-H.: Long-term correlation tracking. In: CVPR, pp. 5388–5396 (2018)

9. Nam, H., Han, B.: Learning multi-domain convolutional neural networks for visual tracking. In: CVPR, pp. 4293–4302 (2016)

10. Fan H., Ling H.: Parallel tracking and verifying: A framework for real-time and high accuracy visual tracking. In: Proceedings of the IEEE International Conference on Computer Vision, pp. 5486–5494(2017)

11. Yun, S., Choi, J., Yoo, Y., Yun, K., Young Choi, J.: Action-decision networks for visual tracking with deep reinforcement learning. In: CVPR, pp. 2711–2720 (2017)

12. Chen, B., Wang, D., Li, P., Wang, S., Lu, H.: Real-time 'Actor-Critic' tracking. In: ECCV, pp. 318–334 (2018)

13. Bibi, A., Mueller, M., Ghanem, B.: Target response adaptation for correlation filter tracking. In: Leibe, B., Matas, J., Sebe, N., Welling, M. (eds.) ECCV 2016. LNCS, vol. 9910, pp. 419–433. Springer, Cham (2016). https://doi.org/10.1007/978-3-319-46466-4_25

Efficiently Handling Scale Variation for Pedestrian Detection

Qihua Cheng[1] and Shanshan Zhang[1,2(✉)]

[1] PCA Lab, Key Lab of Intelligent Perception and Systems for High-Dimensional Information of Ministry of Education, and Jiangsu Key Lab of Image and Video Understanding for Social Security, School of Computer Science and Engineering, Nanjing University of Science and Technology, Nanjing, China
{cs_cheng_qh,shanshan.zhang}@njust.edu.cn
[2] Science and Technology on Parallel and Distributed Processing Laboratory (PDL), Changsha, China

Abstract. Pedestrian detection is a popular yet challenging research topic in the computer vision community. Although it has achieved great progress in recent years, it still remains an open question how to handle scale variation, which commonly exists in real world applications. To address this problem, this paper presents a novel pedestrian detector to better classify and regress proposals of different scales given by a region proposal network (RPN). Specifically, we have made the following major modifications to the Adapted FasterRCNN baseline. First, we divide all proposals into small and large pools according to their scales, and deal with each pool in a separate classification network. Also, we employ two auxiliary supervisions to balance the effect of two parts of proposals on the back propagation. It is worth noting that the proposed new detector does not bring extra computational overhead and only introduces very few additional parameters. We have conducted experiments on the CityPersons, Caltech and ETH datasets and achieved significant improvements to the baseline method, especially on the small scale subset. In particular, on the CityPersons and ETH datasets, our method surpasses previous state-of-the-art methods with lower computational costs at test time.

Keywords: Pedestrian detection · Scale variation · Convolutional neural networks

1 Introduction

Pedestrian detection is a popular research topic as it has wide applications, such as self-driving vehicles, video surveillance and robotics [3,4]. In recent years, pedestrian detection has achieved a great success by using Convolutional Neural Networks (CNNs) [8,10,11,14]. However, we notice that the performance decreases significantly as scale goes down. It indicates that current detectors do not well handle scale variation, which is a major challenge for pedestrian detection.

Z. Cui et al. (Eds.): IScIDE 2019, LNCS 11935, pp. 178–190, 2019.
https://doi.org/10.1007/978-3-030-36189-1_15

Recently, some works have devoted to solve the scale variation problem. SAF R-CNN [30] uses two classification networks for representation learning for large-scale and small-scale proposals, respectively, and the final results are produced by fusing the predictions from both classification networks. Song *et al.* [16] propose to fuse multi-scale feature representations and regress the confidence of topological elements. In addition, other works [18,24,25] use the upsampling strategy to increase the resolution of small objects. The common disadvantage of the above methods is high computational costs during inference, induced by more parameters or a larger image scale.

In this paper, we propose a new and efficient detector based on Adapted FasterRCNN [19] framework to alleviate the scale variation problem. We split the whole proposal pool from RPN into two groups: small proposals and large proposals, which are treated differently in both feature extraction and classification. For feature extraction, we allow small proposals obtain larger feature maps from a lower layer than larger proposals; for classification, two groups of proposals are processed by two separate RCNNs. Please note each proposal only goes through one RCNN at test time, thus no additional computation is needed. Moreover, we add two auxiliary supervisions to balance the impact of two groups of proposals on the optimization of different convolutional layers. Finally, we use multi-scale training for data augmentation, to compensate the number of training samples for each RCNN stream.

In summary, the contributions of this paper are as follows:

- We propose a novel pedestrian detection method, which handles the scale variation problem by processing proposals of different scales using multi-stream RCNNs and auxiliary supervision.
- We have conducted experiments on CityPersons, Caltech and ETH datasets. Our proposed method not only significantly improves over the Adapted FasterRCNN baseline, but also obtains state-of-the-art performance, especially for small scales. Also, our method is more efficient since it uses a smaller single scale for testing.
- We re-implement the top-performing multi-scale method for generic object detection SNIP [20], and observe its generalization ability for the pedestrian detection task. We find SNIP achieves remarkable improvement especially for small scales using a multi-scale test setting; while our method obtains comparable performance at lower computational costs with single scale testing.

2   Related Work

Since our method is based on the Adapted FasterRCNN architecture [19], and focuses on solving scale variation problem for pedestrian detection, we review recent works on pedestrian detection with convolutional neural networks (CNNs), and deep multi-scale pedestrian detection respectively.

2.1 Pedestrian Detection with CNNs

Recently, pedestrian detection is dominated by the CNN-based methods [1,12,13,15,19,24,25,28]. Ouyang *et al.* [1] introduce a part deformation layer into deep models to infer part visibility. Daniel *et al.* [28] utilize the heatmaps of semantic scene parsing to promote detector training. Brazil *et al.* [12] use an independent deep CNN to replace the classifier of RCNN and add an segmentation loss to supervise the detection. Zhou *et al.* [15] propose to train part detectors by jointly learn and reduce the computational cost. Specifically, since Faster-RCNN [8] had achieved great success in generic object detection, most recent methods are built on top of it. Zhang *et al.* [19] make some modifications to the vanilla FasterRCNN so as to reach the state-of-the-art performances. RepLoss [24] and OR-CNN [25] design two regression loss functions to tackle occlusion. Zhang *et al.* [13] utilize channel-wise attention mechanism to represent various occlusion patterns.

2.2 Deep Multi-scale Pedestrian Detection

Although the overall performance has been greatly improved, scale variation still remains a challenging problem in pedestrian detection. In order to better detect small-scale object, several works [16,20,21,26,30] have been devoted to solve the scale variation problem. SAF R-CNN [30] uses two classification networks for representation learning for large-scale and small-scale proposals, respectively, and the final results are produced by fusing the predictions from both classification networks. Other methods, like MS-CNN [21] and SDP [26], independently predict objects from features with different spatial resolution without combining features or scores. Some methods propose to fuse multi-scale feature representations and regress the confidence of topological elements [16]. This approach is time-consuming and requires more computational resources. More recent methods make use of an image pyramid to predict objects at multiple resolutions images, *e.g.* SNIP [20]. This kind of approach generates the final results by testing at multiple resolutions images, which makes the whole inference process more cumbersome and time-consuming.

3 Our Method

3.1 Overview of Proposed Framework

Our method is based on Adapted FasterRCNN [19] detection framework due to its superior performance on pedestrian detection. We show the architecture of our method in Fig. 1. First, we modify the base net (*e.g.* VGG16) by removing the third max-pooling and keeping the forth max-pooling. As SE Block [27] is able to select more informative features while suppressing less useful ones, two additional SE Blocks are used to re-weigh conv4 and conv5 to conv4r and conv5r respectively. Next, RPN is used to generate a pool of proposals from conv5r.

These proposals are divided into small proposals and large proposals based on the height. At test time, the features of small proposals and large proposals are pooled from conv4r and conv5r and then are processed by RCNN1 and RCNN2 respectively. In contrast, at the training time, we add two auxiliary supervision to balance the effect of small proposals and large proposals on base net. The features of small proposals and large proposals are also be pooled from conv5r and conv4r and are then processed by RCNN3 and RCNN4 respectively.

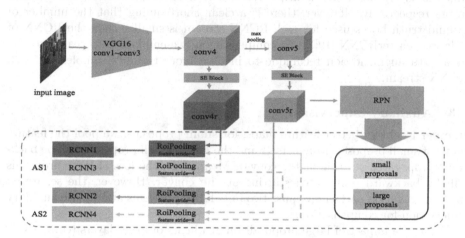

Fig. 1. The architecture of our method. AS1 and AS2 are two auxiliary supervision indicated by gray, and only exist in the training phase. The SE Block is a self-attention sub-network [27].

3.2 Multi-stream RCNNs

Scale variation is a fundamental challenge in pedestrian detection. Based on the Adapted FasterRCNN [19], we propose a multi-stream RCNNs to handle the above problem, which is applied to the RCNN stage. Intuitively, pedestrians at different scales may have different features. One batch consisted of different scales of proposals are difficult for one single RCNN to classify accurately. Generally, those pedestrians fall in the height range of [50,75] pixels are considered to be more difficult as the performance drops significantly in this range. Hence, we consider [50,75] as small scale, and [75, inf] as large scale. On the other hand, one proposal with an intersection over union (IoU) value above 0.5 to any ground truth box, will be considered as a positive sample at training time. In order to include all possible proposals for each RCNN steam, we enlarge the above range and define [0,128] as small scale and [53, inf] as large scale for proposals.

For pedestrian detection, the recall rate is relatively high while the performance is poor. Therefore, in the training phase, all ground truth boxes are still used to assign labels to anchors. However, when selecting the proposal batch from RPN, we just use the ground truth with height falls within a valid height range to assign the labels to proposals between a specified height range and ignore the ones out of that range.

As stated in [14, 19], fine-grained features are beneficial for small scale pedestrian detection. To obtain finer-grained features while ensuring the high-level semantic, we remove the third max-pooling layer from VGG16 to reduce the stride of the conv4 layer to 4 pixels. Therefore, for small proposals, we pool the features from conv4r. Due to the semantic level gap between the features of different layers, we design multiple RCNNs to deal with features from different layers respectively. However, there is a clear shortcoming that the number of ground truth boxes used for each RCNN stream is smaller than the RCNN of Adapted FasterRCNN [19]. As a compensation, we employ multi-scale training as a data augmentation technique to involve more training samples for each RCNN stream.

3.3 Auxiliary Supervision

In order to obtain better feature maps, for small proposals, we pool the feature maps of small proposals from conv4r and the loss of small proposals will be backward propagated starting at the conv4r. Meanwhile, the loss of large proposals will be backward propagated starting at the conv5r. However, the separated back propagation paths may cause two imbalance problems and we use auxiliary supervision for compensation.

First, the impact of large proposals on gradients of lower layers (conv1–3) is relatively smaller than that of small proposals, as the path is longer and gradients may diminish. To address this problem, we also pool the feature maps for large proposals from conv4r and then add a new supervision for large proposals at conv4r. It helps to balance their gradients in the layers lower than conv4. We call this auxiliary supervision as AS1.

Second, the small proposals are isolated to the optimization of conv5. This would cause a negative effect for RPN to recognize small pedestrians. To avoid this problem, we pool the feature maps of small proposals from conv5r and then add a new supervision for small proposals to help the optimization of conv5. We call this auxiliary supervision as AS2.

Our method is trained in an end-to-end manner, where each RCNN has the loss functions for classification and bounding box regression. Following Faster-RCNN [8], the cross-entropy loss is used for classification, and the loss function for bounding box regression is smooth $L1$ loss. Finally, our method can be optimized by the combined loss function:

$$
\begin{aligned}
L = & L_{RPN\_cls} + L_{RPN\_reg} \\
& + L_{RCNN1\_cls} + L_{RCNN2\_cls} + L_{RCNN3\_cls} + L_{RCNN4\_cls} \\
& + L_{RCNN1\_reg} + L_{RCNN2\_reg} + L_{RCNN3\_reg} + L_{RCNN4\_reg},
\end{aligned}
\tag{1}
$$

where L_{RPN_cls} and L_{RPN_reg} are the loss functions for classification and bounding boxes regression in RPN. The remaining terms are the loss functions for classification and bounding boxes regression in RCNN1, RCNN2, RCNN3, and RCNN4 respectively.

4   Experiments

For pedestrian detection, CityPersons [19], ETH [2] and Caltech [5] datasets are generally used to demonstrate the effectiveness of the proposed method and compare with other state-of-the-art methods. Therefore, we conduct several experiments on these three popular datasets, and more experimental analyses on our method are further given on the CityPersons validation set.

4.1   Datasets

The CityPersons dataset was built upon the Cityscapes dataset [9] to provide a new pedestrian dataset. It recordes the multiple cities and countries across Europe with different seasons and various weather conditions. The dataset includes training, validation and testing subset, which consists of 2,975, 500 and 1,525 images respectively. The Caltech dataset contains approximately 10 h of 30 Hz video. It consists of 11 videos, where the first 6 videos are used to train and the remaining 5 videos are used to test. The dataset includes training and testing set, which consists of 42,782 and 4,024 images respectively. The ETH dataset [2] was captured in the city center and consists of three sequences for testing.

4.2   Evaluation Metrics

The standard log miss rate averaged over the false positive per image range of $[10^{-2}, 10^0]$ (MR^{-2}) is used to measure the detection performance (lower is better) of all experiments in this paper. We evaluate the results on different test subsets:

1. Reasonable (**R**): height $\in$ [50, inf], visibility $\in$ [0.65, inf];
2. Reasonable small (**RS**): height $\in$ [50, 75], visibility $\in$ [0.65, inf];
3. Heavy occlusion (**HO**): height $\in$ [50, inf], visibility $\in$ [0.2, 0.65];
4. All (**All**): height $\in$ [20, inf], visibility $\in$ [0.2, inf];

4.3   Implementation Details

We use the ImageNet pre-trained VGG16 model for initialization. We optimize the network using the Stochastic Gradient Descent (SGD) algorithm with 0.9 momentum and 0.0005 weight decay. Our network is trained on one Titan X GPU with the batch size of one image. We use the scales of x0.8, x1 (original image) and x1.25 to train our method on CityPersons. For CityPersons, we train the network with the initial learning rate 10^{-3} for 6 epochs, and decay it to 10^{-4} for another 3 epochs. We use the scales of x1.2, x1.5 and x1.875 to train our method on Caltech. For Caltech-USA, we finetune from the CityPersons model, and set the initial learning rate as 10^{-4}, which is decayed to 10^{-5} after the first epoch. At test time, only one single scale is used to speed up.

Table 1. Pedestrian detection results on the CityPersons validation set. [†]indicates using multi-scale training. The runtime is tested on a Titan X GPU.

Method	Backbone	Test scale	R	RS	HO	All	Time(s)
Zhang et al. [19]	VGG16	x1	15.1	26.2	54.7	44.4	-
	VGG16	x1.3	12.8	-	-	-	-
Repulsion [24]	ResNet-50	x1	13.2	22.3	56.9	44.6	-
	ResNet-50	x1.3	11.6	16.4	54.8	39.8	-
OR-CNN [25]	VGG16	x1	12.8	16.2	55.4	42.3	-
	VGG16	x1.3	11.0	13.0	51.9	39.4	-
SNIP[†] [20]	VGG16	x0.5, x1, x2	10.5	13.2	47.6	38.9	1.02
Baseline	VGG16	x1	14.3	20.0	55.7	44.4	0.31
Ours[†]	VGG16	x1	11.2	13.7	50.8	41.0	0.36
	VGG16	x1.1	11.2	11.8	51.1	40.1	0.42

For SNIP [20], we use the Adapted FasterRCNN [19] as the base detector. SNIP is a method of multi-scale training and testing, where only the object instances that have a similar resolution to the pre-training samples are used to for training. On CityPersons, we use the scales of x0.5, x1 and x2 for training and testing SNIP. The valid height ranges of ground truth are [150, inf], [90, 180] and [50, 120] respectively on three scales. We split the range [75, 512] into 10 quantile bins (equal amount of samples per bin), and use the resulting 11 endpoints as RPN scales to generate proposals.

4.4   Comparisons with State-of-the-Art Methods

CityPersons. We compare our method with other state-of-the-art methods on the CityPersons validation set in Table 1. Compared to the baseline, our method using the original scale for testing already reduces the MR^{-2} by 3.1 points and 6.3 points on **R** and **RS** subsets respectively. Using a small upsampling factor of x1.1, our method further achieves 11.8 MR^{-2} on **RS** subset, which is 1.4 points and 8.2 points better than the second best detector [25] and baseline respectively. These results demonstrate that our method is effective for handling the scale variation problem. Compared to other state-of-the-art methods [24, 25], our method achieves comparable results at a smaller testing scale. Compared to SNIP, our method underperforms by 0.5 points on **R** subset, but our method runs more than two times faster. In addition, it is worth noting that our method outperforms SNIP on **RS** subset by 1.4 points at x1.1 scale. The above results demonstrate the superiority of the our method on pedestrian detection.

In additional, we also evaluate the proposed method on the CityPersons test set and report the results in Table 2. Our method achieves the best results at x1 or x1.1 scales on **R**, **RS** and **HO** subsets. In particular, our method achieves 11.11 MR^{-2} on **RS** subset, which is 3.01 points better than the second best detector. In addition, our method uses a smaller scale during inference compared to other state-of-the-art methods, and thus is also more efficient.

Table 2. Comparisons to state-of-the-art methods on the CityPersons test set.

Method	Backbone	Test scale	R	RS	HO	All
Adapted FasterRCNN [19]	VGG16	x1.3	12.97	37.24	50.47	43.86
Repulsion loss [24]	ResNet-50	x1.5	11.48	15.67	52.59	39.17
OR-CNN [25]	VGG16	x1.3	11.32	14.19	51.43	40.19
Ours	VGG16	x1.0	10.97	12.78	49.71	42.29
	VGG16	x1.1	10.68	11.11	48.90	40.57

Table 3. Comparison with baseline on the Caltech test set.

Method	Backbone	Test scale	R	RS
Adapted FasterRCNN [19]	VGG16	x1.875	9.18	11.20
Baseline	VGG16	x1.5	10.55	13.50
Ours	VGG16	x1.5	8.99	11.09
	VGG16	x1.6	8.56	10.67

Caltech. We compare our method with the baseline in Table 3. Our method achieves 8.99 MR^{-2} at x1.5 scale. Compared to the baseline, our method improves about 2.5 points on **RS** subset. Under x1.6 scale, our method further achieves 8.56 MR^{-2} and 10.67 MR^{-2} on **R** and **RS** subsets, improving over the baseline by 1.99 points and 2.83 points respectively. The comparison results between Adapted FasterRCNN and our method validate the effectiveness of our method on the Caltech dataset.

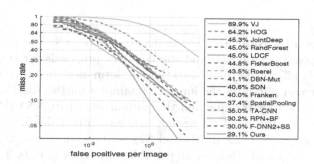

Fig. 2. Comparisons with the state-of-the-art methods on the ETH dataset.

ETH. To verify the generalization ability of our method, we directly apply our CityPersons model on the ETH dataset. We compare our method with other state-of-the-art methods in Fig. 2. We can see that we obtain the state-of-the-art performance. It not only demonstrates the effectiveness of our method, but also validates its generalization ability to other scenarios.

4.5 Ablation Study

Step-by-Step Modifications. To analyze the impact of each modification on Adapted FasterRCNN, we report the results in Table 4. We gradually add each modification on top of Adapted FasterRCNN [19]: First, we modify the backbone network to makes it more suitable for small scales. We remove the third max-pooling layer from VGG16 and keep the forth max-pooling layer. It improves 2.5 points on **RS** subset. Second, we introduce multi-stream RCNNs to better classify and regress proposals of different scale. It improves the miss rate from 20.0 to 18.0 on **RS** subset. Third, we employ two auxiliary supervisions to balance the effect of two groups of proposals on the back propagation. It improves the performance from 14.1 MR^{-2} to 12.9 MR^{-2} on **R** subset. Results on **R**, **RS** and **All** subsets are consistently improved. Finally, we use multi-scale training to compensate the number of training samples of each RCNN. It brings another 1.8 points and 3 points improvement on **R** and **RS** subsets respectively. These results demonstrate the effectiveness of each modification.

Table 4. Impact of different modifications to the baseline. Results are reported on the CityPersons validation set.

Detector aspect	**R**	**RS**	**HO**	**All**
baseline	14.7	22.5	56.1	45.2
+modify backbone	14.3	20.0	55.7	44.4
+multi-stream RCNNs	14.2	18.0	53.4	44.3
+auxiliary supervision	12.9	16.7	52.2	42.7
+multi-scale training	11.2	13.7	50.8	41.0

Scale Ranges of Proposals. For multi-stream RCNNs, we study how the range of the positive proposals and the negative proposals for training multi-stream RCNNs affects the final detection performance. In order to train the network faster and minimize the impact of samples on the RCNN, multi-scale training is not used for training and the parameters of RCNN1 (RCNN2) and RCNN3 (RCNN4) are shared. The results are shown in Table 5.

As mentioned in Sect. 3.2, in M_1 and M_2 settings, their ranges of small and large positive proposals both cover all possible positive proposals. Therefore, the results of M_1 and M_2 are very similar. Compared to M_2, results of M_3 on each subset are consistently improved, which demonstrates our strategy for selecting proposals is beneficial to train an optimal detector. While comparing M_3 and M_4, we find that ignoring partial positive proposals causes a performance degradation. However, compared to M_2, the performance of M_3 does not drop, but improves 0.7 points on **RS** subset. The above results demonstrate that the proposal batch composed of different scales of proposals is hard for RCNN to optimize.

Table 5. Comparisons of using different valid height ranges of proposals. Results are reported on the CityPersons validation set. The columns of Pos and Neg indicate valid height ranges of positive proposals and negative proposals.

Setting	Proposals	Pos	Neg	R	RS	HO	All
M_1	Small	[0, inf]	[0, inf]	12.1	16.7	52.6	42.1
	Large	[0, inf]	[0, inf]				
M_2	Small	[0, 128]	[0, inf]	12.4	16.6	52.7	42.7
	Large	[53, inf]	[0, inf]				
M_3	Small	[0, 128]	[0, 128]	11.7	15.8	51.8	41.5
	Large	[53, inf]	[53, inf]				
M_4	Small	[0, 75]	[0, 75]	12.3	15.9	51.3	43.0
	Large	[75, inf]	[75, inf]				

Table 6. Effect of using auxiliary supervision in our method. Results are reported on the CityPersons validation set.

AS1	AS2	R	RS	HO	All
		14.1	18.1	55.2	44.9
✓		12.7	17.7	52.7	42.4
	✓	13.7	17.2	54.8	44.4
✓	✓	12.9	16.7	52.2	42.7

Fig. 3. Visualization of detection results from our method (top row) and baseline (bottom row) at FPPI = 0.1. The images are selected from CityPersons validation set. Compared to the baseline, our method achieves a higher recall for small scale pedestrians. We show ground truth annotations in green and detection results in red. (Color figure online)

Auxiliary Supervisions. We conduct some experiments to study the impact of each auxiliary supervision, as shown in Fig. 3. Here we do not use multi-scale training considering speed. The results in Table 6 show the auxiliary supervision significantly improve the detection performance.

Compared to the baseline without auxiliary supervision, AS1 reduces the MR^{-2} by 1.4 points on **R** subset and AS2 reduces the MR^{-2} by 0.9 points on **RS** subset, which validate the effectiveness of AS1 and AS2 respectively. When we simultaneously add AS1 and AS2, it reduces the MR^{-2} by 1.2 points and 1.4 points on **R** and **RS** subsets respectively. It demonstrates that AS1 and AS2 can balance the gradient flow of small proposals and large proposals in the network.

5   Conclusion

In the paper, we propose a new method to deal with various sizes of instances in pedestrian detection. This work is motivated by the finding that small scale proposals are more difficult to classify compared to larger ones. It is difficult for one single RCNN to deal with proposals of different scales. Therefore, we design a multi-stream RCNN architecture to handle different scales, and employ auxiliary supervisions to enhance gradient back propagation.

The proposed method shows significant improvements over the baseline Adapted FasterRCNN detector. Specifically, our method achieves state-of-the-art performance on the CityPersons, Caltech and ETH datasets, especially for small scales. These results demonstrate that our method better handles scale variation for pedestrian detection.

Encouraged by the promising results, we plan to extend the proposed method to generic object detection in the future.

Acknowledgements. This work is supported by National Natural Science Foundation of China (Grant No. 61702262), Funds for International Cooperation and Exchange of the National Natural Science Foundation of China (Grant No. 61861136011), Natural Science Foundation of Jiangsu Province, China (Grant No. BK20181299), CCF-Tencent Open Fund (RAGR20180113), "the Fundamental Research Funds for the Central Universities" (No. 30918011322) and Young Elite Scientists Sponsorship Program by CAST (2018QNRC001).

References

1. Ouyang, W., Wang, X.: Joint deep learning for pedestrian detection. In: ICCV, pp. 2056–2063 (2013)
2. Ess, A., Leibe, B., Van Gool, L.: Depth and appearance for mobile scene analysis. In: ICCV, pp. 1–8 (2007)
3. Li, W., Zhao, R., Xiao, T., Wang, X.: DeepReID: deep filter pairing neural network for person re-identification. In: CVPR, pp. 152–159 (2014)
4. Wang, X., Wang, M., Li, W.: Scene-specific pedestrian detection for static video surveillance. PAMI **36**(2), 361–374 (2014)

5. Dollar, P., Wojek, C., Schiele, B., Perona, P.: Pedestrian detection: an evaluation of the state of the art. PAMI **34**(4), 743–761 (2011)
6. Zhang, S., Benenson, R., Omran, M., Hosang, J., Schiele, B.: Towards reaching human performance in pedestrian detection. PAMI **40**(4), 973–986 (2017)
7. Chen, D., Zhang, S., Ouyang, W., Yang, J., Tai, Y.: Person search via a mask-guided two-stream CNN model. In: ECCV, pp. 734–750 (2018)
8. Ren, S., He, K., Girshick, R., Sun, J.: Faster R-CNN: towards real-time object detection with region proposal networks. In: NIPS, pp. 91–99 (2015)
9. Cordts, M., et al.: The cityscapes dataset for semantic urban scene understanding. In: CVPR, pp. 3213–3223 (2016)
10. Girshick, R.: Fast R-CNN. In: ICCV, pp. 1440–1448 (2015)
11. Liu, W., et al.: SSD: single shot multibox detector. In: Leibe, B., Matas, J., Sebe, N., Welling, M. (eds.) ECCV 2016. LNCS, vol. 9905, pp. 21–37. Springer, Cham (2016). https://doi.org/10.1007/978-3-319-46448-0_2
12. Brazil, G., Yin, X., Liu, X.: Illuminating pedestrians via simultaneous detection & segmentation. In: ICCV, pp. 4950–4959 (2017)
13. Zhang, S., Yang, J., Schiele, B.: Occluded pedestrian detection through guided attention in CNNs. In: CVPR, pp. 6995–7003 (2018)
14. Redmon, J., Farhadi, A.: YOLO9000: better, faster, stronger. In: CVPR, pp. 7263–7271 (2017)
15. Zhou, C., Yuan, J.: Multi-label learning of part detectors for heavily occluded pedestrian detection. In: ICCV, pp. 3486–3495 (2017)
16. Song, T., Sun, L., Xie, D., Sun, H., Pu, S.: Small-scale pedestrian detection based on somatic topology localization and temporal feature aggregation. In: arXiv preprint. arXiv:1807.01438 (2018)
17. Hosang, J., Omran, M., Benenson, R., Schiele, B.: Taking a deeper look at pedestrians. In: CVPR, pp. 4073–4082 (2015)
18. Zhang, S., Benenson, R., Omran, M., Hosang, J., Schiele, B.: How far are we from solving pedestrian detection? In: CVPR, pp. 1259–1267 (2016)
19. Zhang, S., Benenson, R., Schiele, B.: Citypersons: a diverse dataset for pedestrian detection. In: CVPR, pp. 3213–3221 (2017)
20. Singh, B., Davis, L.S.: An analysis of scale invariance in object detection snip. In: CVPR, pp. 3578–3587 (2018)
21. Cai, Z., Fan, Q., Feris, R.S., Vasconcelos, N.: A unified multi-scale deep convolutional neural network for fast object detection. In: Leibe, B., Matas, J., Sebe, N., Welling, M. (eds.) ECCV 2016. LNCS, vol. 9908, pp. 354–370. Springer, Cham (2016). https://doi.org/10.1007/978-3-319-46493-0_22
22. Kong, T., Sun, F., Yao, A., Liu, H., Lu, M., Chen, Y.: Ron: reverse connection with objectness prior networks for object detection. In: CVPR, pp. 5936–5944 (2017)
23. Lin, T.-Y., Dollár, P., Girshick, R., He, K., Hariharan, B., Belongie, S.: Feature pyramid networks for object detection. In: CVPR, pp. 2117–2125 (2017)
24. Wang, X., Xiao, T., Jiang, Y., Shao, S., Sun, J., Shen, C.: Repulsion loss: detecting pedestrians in a crowd. In: CVPR, pp. 7774–7783 (2018)
25. Zhang, S., Wen, L., Bian, X., Lei, Z., Li, S.Z.: Occlusion-aware R-CNN: detecting pedestrians in a crowd. In: ECCV, pp. 637–653 (2018)
26. Yang, F., Choi, W., Lin, Y.: Exploit all the layers: fast and accurate CNN object detector with scale dependent pooling and cascaded rejection classifiers. In: CVPR, pp. 2129–2137 (2016)
27. Hu, J., Shen, L., Sun, G.: Squeeze-and-excitation networks. In: CVPR, pp. 7132–7141 (2018)

28. Daniel Costea, A., Nedevschi, S.: Semantic channels for fast pedestrian detection. In: CVPR, pp. 2360–2368 (2016)
29. Girshick, R., Donahue, J., Darrell, T., Malik, J.: Rich feature hierarchies for accurate object detection and semantic segmentation. In: CVPR, pp. 580–587 (2014)
30. Li, J., Liang, X., Shen, S., Xu, T., Feng, J., Yan, S.: Scale-aware fast R-CNN for pedestrian detection. IEEE Trans. Multimedia **20**(4), 985–996 (2018)

Leukocyte Segmentation via End-to-End Learning of Deep Convolutional Neural Networks

Yan Lu[1], Haoyi Fan[2], and Zuoyong Li[3(✉)]

[1] College of Mathematics and Computer Science, Fuzhou University,
Fuzhou 350108, China
[2] School of Computer Science and Technology, Harbin University of Science
and Technology, Harbin 150080, China
[3] Fujian Provincial Key Laboratory of Information Processing and Intelligent
Control, Minjiang University, Fuzhou 350108, China
fzulzytdq@126.com

Abstract. Identification and analysis of leukocytes (white blood cells, WBC) in blood smear images play a vital role in the diagnosis of many diseases, including infections, leukemia, and acquired immune deficiency syndrome (AIDS). However, it remains difficult to accurately segment and identify leukocytes under variable imaging conditions, such as variable light conditions and staining degrees, the presence of dyeing impurities, and large variations in cell appearances, e.g., size, color, and shape of cells. In this paper, we propose an end-to-end leukocyte segmentation algorithm that uses pixel-level prior information for supervised training of a deep convolutional neural network. Specifically, a context-aware feature encoder is first introduced to extract multi-scale leukocyte features. Then, a feature refinement module based on the residual network is designed to extract more discriminative features. Finally, a finer segmentation mask of leukocytes is reconstructed by a feature decoded based on the feature maps. Quantitative and qualitative comparisons of real-world datasets show that the proposed method achieves state-of-the-art leukocyte segmentation performance in terms of both accuracy and robustness.

Keywords: Leukocytes · Cell segmentation · Feature refinement · Deep convolutional neural networks

1 Introduction

Leukocytes are a significant indicator for the detection and treatment of several diseases, including infections, leukemia, and acquired immune deficiency syndrome (AIDS) [1, 2]. To diagnose these diseases, we usually analyze the characteristics of leukocytes by observing blood smears. Blood smears are complex scenes composed of leukocytes, red blood cells, platelets, and backgrounds. Thus, the cells should be segmented from the image before analysis. It is time-consuming and unrealistic to rely on manual segmentation. Moreover, there are different kinds of leukocytes with

© Springer Nature Switzerland AG 2019
Z. Cui et al. (Eds.): IScIDE 2019, LNCS 11935, pp. 191–200, 2019.
https://doi.org/10.1007/978-3-030-36189-1_16

different appearances, and in blood smears, some leukocytes also adhere to red blood cells. This makes it difficult to segment the leukocytes.

In recent years, the development of computer-aided methodology has made the analysis of medical images simpler and more intuitive. Computer-aided segmentation methods are primarily divided into two types: supervised learning and unsupervised learning. Unsupervised learning methods primarily involve thresholds [3] and clustering, such as k-means clustering [4], and fuzzy c-means clustering (FCM) [5]. These methods do not require manually labeled training data and thus, are more adaptable. However, unsupervised learning methods do not perform well when there is little difference between the region of interest of the cell and the region of non-interest of the cell. Supervised learning methods usually divide the problem into multiple categories, for example, the k-nearest neighbors algorithm [6] and support vector machines (SVM) [7]. Supervised learning comprises two main steps: feature extraction and classification. Manually tagged data is required for training, and the accuracy of this data determines the accuracy of the results.

Convolutional neural networks (CNN) have developed rapidly over the past few years, especially in the field of computer vision. Many methods based on CNN improvements are been widely implemented. Among them, the fully convolutional network (FCN) [8] achieves pixel-level segmentation of images. U-net [9] is a network with an encoder-decoder structure that is based on an FCN. The above CNN-based methods cannot achieve satisfactory results for leucocyte segmentation due to the complex background of blood smears. The FCN inputs an image and output is also an image. We use a fully convolutional network which implements pixel-to-pixel mapping. The FCN combines the pooling layer with the feature map to reduce the loss of information. We refer to the information fusion technology of FCN. Our network is improved based on U-net that we use the same encoder-decoder structure as U-net.

The main contribution of this paper is as follows:

(1) An end-to-end white-cell segmentation algorithm based on a deep CNN is proposed to achieve pixel-to-pixel mapping.
(2) For feature extraction, we used dilated convolution [10] to reduce over-fitting of the data.
(3) Resnet [11] refined feature extraction was added to the feature map to achieve a quick convergence and to prevent the gradient from disappearing.

The remainder of this paper is structured as follows. In Sect. 2, we present our proposed method in detail. In Sect. 3, three different datasets are analyzed and the results of the analysis with our algorithm are compared with the results obtained by other algorithms. A summary is provided in Sect. 4.

2   Proposed Method

We aim to automatically obtain the entire WBC from the complete background of the blood smear. U-net can be regarded as a network of coding and decoding structures; the coding layer performs feature extraction and the decoding layer restores object detail space dimensions. At the coding level, the feature map space dimension is degraded;

the decoding layer up-samples the feature map using deconvolution. We improved an algorithm which is based on U-net and divided the segmentation task into three parts: context-aware feature encoder, feature refinement module, and feature decoder. Figure 1 shows the structure of our network.

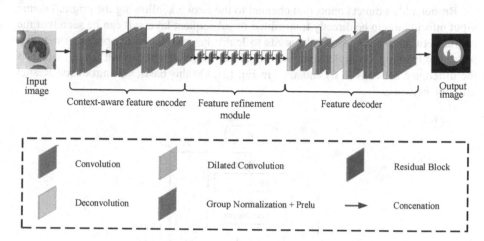

Fig. 1. Architecture of the proposed network

2.1 Context-Aware Feature Encoder

Convolution is used for feature extraction in many different applications. We used convolution and dilated convolution to extract the image features. Dilated convolution involves injecting holes into the convolution map, thereby increasing the receptive field so that each convolution output contains a wider range of information. First, the input image is convolved three times; the convolution kernels are 3×3, 5×5, and 7×7, respectively. Then, group normalization is used to make small batch neural network training more stable; the activation function (PReLU) is used [12]. Finally, dilated convolution is used on the feature map to improve the image resolution of the intermediate feature map.

2.2 Feature Refinement Module

Convolutional neural networks typically extract more complete features by deepening the number of network layers. As the number of network layers increases, the difficulty of training increases, and the problem of gradient degradation occurs, that is, the deeper the training layer, the lower the training accuracy; Resnet solves this problem well. In a traditional convolutional network, the relational formula of input and output is:

$$x_{i+1} = F(x_i). \tag{1}$$

x_i is the input of the ith layer, x_{i+1} is the output of the ith layer, and $F(\cdot)$ is the input transformation operation, including convolution operations and group normalization operations [13]. Figure 2 shows the structure of the residual block in our network. The input and output formula is:

$$x_{i+1} = Prelu(F(x_i) + I(x_i)). \tag{2}$$

Resnet adds a direct connection channel to the network, allowing the original signal input information to be directly transmitted to subsequent layers. It can be seen that the neural network of this layer only needs to learn the residual of the previous network instead of learning a complete output, the difference in Resnet learning is $I(x_i)$ which is the difference between $F(x_i)$ and x_{i+1} in Eq. (2). On this basis, we introduced Resnet into our network.

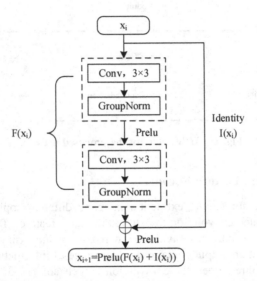

Fig. 2. Architecture of Residual block in the proposed method

2.3   Feature Decoder

In the feature extraction stage, the operation of feature convolution results in small scales of feature maps. Therefore, in the last part of our network, we used up-sampling and deconvolution to restore the feature map to the original image size. In our network, the input is the image, and the output is also the image, we perform pixel-level classification, implement pixel-to-pixel mapping and convert the segmentation problem into classification. Directly up-sampling or deconvoluting the feature map results in the loss of a lot of information. Deep learning-based FCN and U-net solve this problem by combining the previous feature maps, they will use up-sampling the feature map to the same size before fusion. We also use the fusion of the previous feature map in the part of the context-aware feature encoder to reduce the loss of information.

2.4 Loss Function

The loss function used in the proposed method is a weighted loss function based on Tversky index [14]. The definition is as follows:

$$Loss = \sum_c (1 - TI_c),\tag{3}$$

where TI_c is the Tversky similarity score and defined as follows:

$$TI_c = \frac{\sum_i^N p_{ic}g_{ic} + \varepsilon}{\sum_i^N p_{ic}g_{ic} + \alpha \sum_i^N p_{ic}g_{\bar{c}} + \beta \sum_i^N p_{\bar{c}}g_{ic} + \varepsilon},\tag{4}$$

where p_{ic} is the probability that pixel i belong to leukocyte class c, $p_{i\bar{c}}$ is the probability that pixel i belong to non-leukocyte class $\bar{c}$, and g_{ic} is the ground truth training label which is 1 for leukocyte pixel i and 0 for non-leukocyte pixel i. α and β are two parameters that can be tuned to shift the emphasis to improve recall in the case of large class imbalance; for all experiments in this paper, they are set as 0.3 and 0.7, respectively.

3 Results

In this section, the dataset and evaluation metrics are first introduced. Then, the data augmentation techniques that generate more data samples are discussed, and finally, the results are compared to other recent methods.

3.1 Dataset and Evaluation Methods

We used three datasets to evaluate the proposed method, Dataset1, Dataset2, and Dataset3. Dataset1 is from China Jiangxi Tekang Technology Co., Ltd. It consists of 300 individual leukocyte images of size 120×120 (176 neutrophils, 22 eosinophils, 1 basophil, 48 monocytes, and 53 lymphocytes). Dataset2, published in CellaVision, consists of 100 individual leukocyte images of size 300×300 (30 neutrophils, 12 eosinophils, 3 basophils, 18 monocytes, and 37 lymphocytes). Dataset3 was collected with the help of the Third People's Hospital of Fujian Province, and comprises a total of 60 individual leukocytes.

For the evaluation metrics, we used four measures commonly used in segmentation methods: misclassification error (ME) [15], false positive rate (FPR), false negative rate (FNR) [16], and kappa index (KI) [17]. The formula definitions are as follows:

$$ME = 1 - \frac{|G_b \bigcap P_b| + |G_f \bigcap P_f|}{|G_f| + |G_b|},\tag{5}$$

$$FPR = \frac{|G_b \bigcap P_f|}{|G_b|},\tag{6}$$

$$FNR = \frac{|G_f \cap P_b|}{|G_f|}, \tag{7}$$

$$KI = \frac{2|G_f \cap P_f|}{|G_f| + |P_f|}, \tag{8}$$

where P_f and P_b are the foreground and background of the predicted results of the leukocyte segmentation algorithm; G_f and G_b are the foreground and background of the ground truth. Lower ME, FPR, and FNR values, and higher KI values, indicate better segmentation.

3.2   Implement Details

In Dataset1 and Dataset2, we selected 90% and 10% of each dataset, respectively, as the training and test sets, while for Dataset3, 80% and 20% of the dataset were selected as the training and tests sets, respectively, as there was less data. Because of the small amount of training sample data, in order to make the model more stable and robust, we performed data augmentation during training.

4   Results and Analysis

The proposed method is compared with the SVM algorithm [18], combined graph segmentation algorithm (CGS) [19] and based on deep learning U-net [9] on the three datasets: Dataset1, Dataset2, Dataset3.

4.1   Qualitative Results

As shown in Figs. 3, 4, and 5, our algorithm performed significantly better than the other algorithms. In Fig. 3, it can be seen that SVM and U-net segmentation resulted in background impurities in the cytoplasm while CGS resulted in mis-segmentation of the entire cell as the nucleus in Dataset1. As seen in Fig. 4(b), SVM on Dataset2 mistook red blood cells for leukocytes, and like CGS and U-net, background impurities were misclassified into cytoplasm. Figure 5 shows that SVM divided red blood cells into the cytoplasm and CGS incorrectly divided the cytoplasm.

4.2   Quantitative Results

As shown in Table 1, our algorithm had a lower ME and FPR for Dataset1, and a higher KI, compared to the other algorithms. For Dataset2, the ME, FPR, and FNR of our algorithm were lower than the other algorithms, and the KI was higher than other algorithms. For Dataset3, ME and FPR for our algorithm were lower than the other methods, and KI was higher than the other methods.

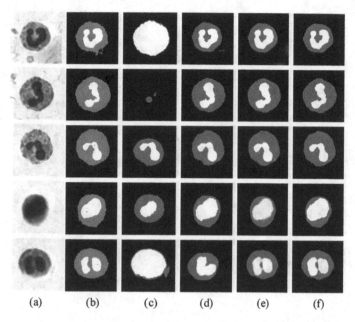

Fig. 3. Example segmentation results of different methods on Dataset1. (a) The original images, (b) Results of SVM, (c) Results of CGS, (d) Results of U-Net, (e) Results of the proposed method, (f) Ground truth.

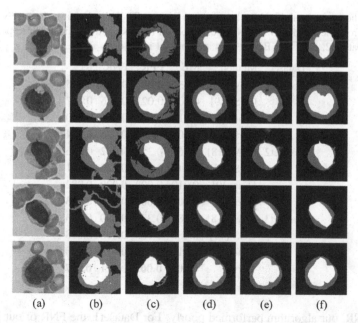

Fig. 4. Example segmentation results of different methods on Dataset2. (a) The original images, (b) Results of SVM, (c) Results of CGS, (d) Results of U-Net, (e) Results of the proposed method, (f) Ground truth.

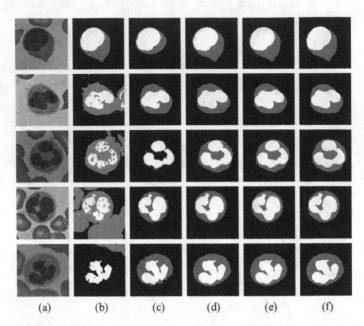

(a)   (b)   (c)   (d)   (e)   (f)

Fig. 5. Example segmentation results of different methods on Dataset3. (a) The original images, (b) Results of SVM, (c) Results of CGS, (d) Results of U-Net, (e) Results of the proposed method, (f) Ground truth.

Table 1. The average value of Misclassification Error (ME), False Negative Rate (FNR), Kappa index (KI) from the SVM, CGS, U-Net, our proposed method, and ground Truth respectively on Dataset1, Dataset2, Dataset3. Best results are marked in bold.

Dataset	Method	ME	FPR	FNR	KI
Dataset1	SVM [18]	0.034	0.05	**0.001**	0.945
	CGS [19]	0.103	0.023	0.306	0.727
	U-net [9]	0.011	0.012	0.011	0.980
	Ours	**0.009**	**0.003**	0.026	**0.984**
Dataset2	SVM [18]	0.266	0.353	0.032	0.676
	CGS [19]	0.147	0.121	0.220	0.751
	U-net [9]	0.010	0.002	**0.029**	0.982
	Ours	**0.009**	**0.001**	**0.029**	**0.983**
Dataset3	SVM [18]	0.285	0.371	0.090	0.663
	CGS [19]	0.080	0.065	0.098	0.868
	U-net [9]	0.012	0.011	**0.014**	0.980
	Ours	**0.009**	**0.002**	0.021	**0.987**

For FNR, our algorithm performed poorly. For Dataset1, the FNR of our proposed method was higher than the U-net and SVM algorithms, and for Dataset 3, the FNR of our proposed method was higher than U-net. FNR is defined as the ratio of the number of positive samples predicted to be negative to the actual number of positive samples. The lower FPR values of other algorithms primarily occur because these other methods exhibit some over-segmentation; they mistake the surrounding environment for white blood cells. Our algorithm tends to have a higher FNR value because there is less over-segmentation.

Table 2. The average value of Misclassification Error (ME), False Negative Rate (FNR), Kappa index (KI) from our proposed method without feature refinement module, without dilated conv-layer, without group normalization and prelu, and our proposed method respectively on Dataset1. Best results are marked in bold.

Method	ME	FPR	FNR	KI
Without feature refinement module	0.011	0.007	**0.023**	0.981
Without dilated conv-layer	0.011	**0.003**	0.033	0.980
Without group normalization and prelu	0.307	0.084	0.897	0.155
Ours	**0.009**	**0.003**	0.026	**0.984**

As shown in Table 2, our algorithm had a lower ME and FPR for Dataset1, and a higher KI, compared to the other algorithms. For the FNR, our method without dilated conv-layer performs better than our proposed, because we have some cases in which the backgrounds are misclassified into cytoplasm.

5 Conclusion

In this paper, we studied the problem of leukocyte segmentation, which is vital for an automatic diagnosis system. Leukocyte segmentation is challenging because of variable imaging conditions, such as variable light conditions and staining degrees, the presence of substantial dyeing impurities, and large variations in cell appearance, e.g., size, color, and shape of cells. To solve the above-mentioned issues, we proposed an end-to-end leukocyte segmentation method that implements pixel-level mapping. Firstly, a context-aware feature encoder was introduced for the extraction of multi-scale leukocyte features. Then, a feature refinement module based on the residual network was designed to extract more discriminative features for leukocyte segmentation. Finally, a finer segmentation mask of leukocytes was reconstructed based on the feature maps from previous modules using a feature decoder. Qualitative analysis and quantitative analysis demonstrated that the proposed method works well. In future work, we will also collect more image data to achieve automated positioning and analysis of cells.

Acknowledgements. This work is partially supported by National Natural Science Foundation of China (61972187, 61772254), Key Project of College Youth Natural Science Foundation of Fujian Province (JZ160467), Fujian Provincial Leading Project (2017H0030).

References

1. Zheng, X., Wang, Y., Wang, G., et al.: A novel algorithm based on visual saliency attention for localization and segmentation in rapidly-stained leukocyte images. Micron **56**, 17–28 (2014)
2. Zheng, X., Wang, Y., Wang, G., et al.: Fast and robust segmentation of white blood cell images by self-supervised learning. Micron **107**, 55–71 (2018)
3. Huang, D.C., Hung, K.D., Chan, Y.K.: A computer assisted method for leukocyte nucleus segmentation and recognition in blood smear images. J. Syst. Softw. **85**(9), 2104–2118 (2012)
4. Zhang, C., Xiao, X., Li, X., et al.: White blood cell segmentation by color-space-based k-means clustering. Sensors **14**(9), 16128–16147 (2014)
5. Theera-Umpon, N.: White blood cell segmentation and classification in microscopic bone marrow images. In: Wang, L., Jin, Y. (eds.) FSKD 2005. LNCS (LNAI), vol. 3614, pp. 787–796. Springer, Heidelberg (2005). https://doi.org/10.1007/11540007_98
6. Kong, H., Gurcan, M., Belkacem-Boussaid, K.: Partitioning histopathological images: an integrated framework for supervised color-texture segmentation and cell splitting. IEEE Trans. Med. Imaging **30**(9), 1661–1687 (2011)
7. Song, Y., Cai, W., Huang, H., et al.: Region-based progressive localization of cell nuclei in microscopic images with data adaptive modeling. BMC Bioinform. **14**(1), 173 (2013)
8. Long, J., Shelhamer, E., Darrell, T.: Fully convolutional networks for semantic segmentation. In: Proceedings of the IEEE Conference on Computer Vision and Pattern Recognition, pp. 3431–3440 (2015)
9. Ronneberger, O., Fischer, P., Brox, T.: U-Net: convolutional networks for biomedical image segmentation. In: Navab, N., Hornegger, J., Wells, William M., Frangi, Alejandro F. (eds.) MICCAI 2015. LNCS, vol. 9351, pp. 234–241. Springer, Cham (2015). https://doi.org/10.1007/978-3-319-24574-4_28
10. Yu, F., Koltun, V.: Multi-scale context aggregation by dilated convolutions. arXiv preprint arXiv:1511.07122 (2015)
11. He, K., Zhang, X., Ren, S., et al.: Deep residual learning for image recognition. In: Proceedings of the IEEE Conference on Computer Vision and Pattern Recognition, pp. 770–778 (2016)
12. He, K., Zhang, X., Ren, S., et al.: Delving deep into rectifiers: surpassing human-level performance on ImageNet classification. In: Proceedings of the IEEE International Conference on Computer Vision, pp. 1026–1034 (2015)
13. Fan, H., Zhang, F., Xi, L., et al.: LeukocyteMask: an automated localization and segmentation method for leukocyte in blood smear images using deep neural networks. J. Biophotonics, e201800488 (2019)
14. Tversky, A.: Features of similarity. Psychol. Rev. **84**(4), 327 (1977)
15. Yasnoff, W.A., Mui, J.K., Bacus, J.W.: Error measures for scene segmentation. Pattern Recognit. **9**(4), 217–231 (1977)
16. Fawcett, T.: An introduction to ROC analysis. Pattern Recognit. Lett. **27**(8), 861–874 (2006)
17. Fleiss, J.L., Cohen, J., Everitt, B.S.: Large sample standard errors of kappa and weighted kappa. Psychol. Bull. **72**(5), 323 (1969)
18. Zheng, X., Wang, Y., Wang, G.: White blood cell segmentation using expectation-maximization and automatic support vector machine learning in Chinese. Data Acquis. Process. **28**, 614–619 (2013)
19. Dong, G.G.C.: Flexible combination segmentation algorithm for leukocyte images. Chin. J. Sci. Instrum. **9**, 1977–1981 (2008)

Coupled Squeeze-and-Excitation Blocks Based CNN for Image Compression

Jing Du, Yang Xu$^{(\boxtimes)}$, and Zhihui Wei

School of Computer Science and Engineering,
Nanjing University of Science and Technology, Nanjing, China
xuyangth90@njust.edu.cn

Abstract. Recent researches have shown that deep convolutional neural networks (CNN) have achieved promising results in the field of image compression. In this paper, we propose an end-to-end image compression framework based on effective attention modules. In the proposed method, two channel attention mechanisms are employed jointly. The first is the Squeeze-and-Excitation block (SEblock) in the encoder. The other is the novel inversed SEblock (ISEblock) placed in decoder. These blocks, named coupled SEblocks, are placed behind the convolutional layer in both encoder and decoder. By using SEblocks, the encoder learns the interdependencies between different channels and the feature maps can be better distributed after entropy coding. In decoder, the inversed SEblock is employed which adaptively learns the weights and divides weights between the channels to supplement information compressed from the encoder. The whole network is trained as a joint rate-distortion optimization by using a subset of the ImageNet dataset. We evaluate our method on public Kodak test set. At low bit rates, our approach outperforms the existing Ballè's, JPEG, JPEG2000 and WebP on multi-scale structural similarity (MS-SSIM) and gets good visual qualities for all images at test set.

Keywords: Convolutional neural networks · Image compression · Coupled SEblocks · Channel attention

1 Introduction

Efficient image and video compression algorithms have always been one of the hot research directions in academia and industry. The purpose of image compression is to reduce the redundant information and irrelevant information of the image, and store or transmit at a low bit rate. The most famous image compression engineered codecs are JPEG [1], JPEG2000 [2] and WebP [3]. These methods, however, perform poor at low bit rates, and cause visual artifacts, e.g., blurring, ringing, and blocking.

The recently proposed deep learning methods have greatly improved the visual effect of decompressed image. Autoencoder which learns a more powerful latent representation than JPEG and JPEG2000 via stacking convolutional neural network (CNNs), is widely used in the latest works. Similar to the traditional methods, CNNs framework contains four components, i.e. encoder, quantizer, entropy encoder and

Z. Cui et al. (Eds.): IScIDE 2019, LNCS 11935, pp. 201–212, 2019.
https://doi.org/10.1007/978-3-030-36189-1_17

decoder. In general, lossy image compression is regarded as joint rate-distortion optimization to learn the encoder, quantizer, and decoder in an end-to-end manner.

There are also many problems with deep learning methods. Even if the encoder and decoder can be optimized by back propagation, the learning of quantizer is still a challenging problem because of its non- differentiability. Furthermore, the entropy rate estimation defined on the network is also a discrete function. In order to solve these problems, several works have been proposed within converting these to continuous approximations.

Besides, the processing power of the image spatial domain and channel domain are particularly prominent. Unlike traditional compression algorithms, we begin to focus on channel-relationship understanding of the deep learning network. We trained a weight vector that high-focus channels have higher weights and reduce other channels weights in the feature map. Therefore, the representational power of the network is improved and the feature map function is adjusted selectively.

In this paper, we propose a coupled Squeeze-and-Excitation block image compression framework which uses a variant of the SEblocks. This framework is based on autoencoder and uses channel attention mechanism to learn the relationship in feature maps. The decoder and encoder should be considered differently when dealing with image compression problems. Thus, we use SEblocks in encoder and apply inversed SEblocks, which replaced product by division in scale part, in decoder. This way ensures the integrity of the information at the decoder. Applying coupled SEblocks to the image compression network learns the better salient feature maps and gets a more concentrated distribution of feature maps. Because of the concentrated the distribution, our network gets the lower the compression ratio. Our model is trained by a joint rate-distortion optimization, and gets good performance at low bit rates. Compared with Ballè's method, JPEG, JPEG2000 and WebP, our network has better performance in Kodak data set in terms of MS-SSIM.

2   Related Work

The traditional image coding standards, e.g., JPEG [1] and JPEG2000 [2], transform the spatial domain information into the frequency domain. Then, the information of the image is further separated into low frequency components and high frequency components. The main contents are preserved in low frequency, and details are saved in high frequency. Further, quantization and lossless entropy coding are used to minimize the compression rate.

Several methods focus on improving the quality of JPEG, eliminating the blocking artifacts and other post-processing operations. [4] proposed a DCT algorithm in which the shape of the block will change adaptively with the image structure. [5, 6] used the similarity of non-local overlapping blocks as a prior to reduce compression artifacts. [7] introduced a constrained nonconvex low rank prior, and solved the deblocking of compressed image as an optimization problem.

Recently, deep learning has achieved great success in computer vision. SRCNN is designed for image super-resolution, and it is the first time applied deep learning to pixel-level image problems [8]. Dong et al. [9] applied deep learning to the restoration of compressed images and designed the network ARCNN. Based on the SRCNN, ARCNN adds a convolution layer that denoises the feature map and enhances the feature's signal-to-noise ratio. Toderici *et al.* [10] presented a variable rate image compression algorithm based on long short-term memory networks (LSTM). The algorithm reduces the image size and adjusts the number of features to achieve image compression with a 32 × 32 image into the network. The algorithm leads to better visual effects than JPEG when the image is blurred and the compression ratio is high. Then Toderici *et al.* [11] introduced a recurrent neural network (RNN) for full-resolution image compression. This framework includes RNN-based encoders and decoders and a neural network-based entropy coder. It uses perceptual errors and proposes a new deep structure based on GRU and ResNet.

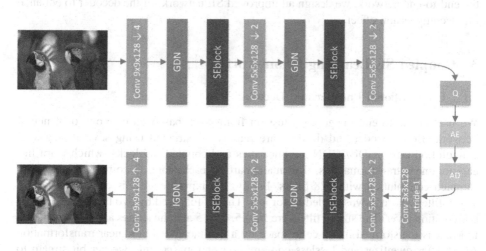

Fig. 1. Network architecture of the Coupled SE model. Upper part of the image shows the structure of encoder, and the lower part corresponds to the decoder that uses a symmetrical structure. Q represents quantization, and AE, AD represent arithmetic encoder and arithmetic decoder, respectively. The parameters of the convolution layer are interpreted as: number of kernel × kernel × filters. ↑ denotes upsampling and ↓ denotes downsampling where the following number indicates the stride.

Ballè *et al.* [12] applied convolutional neural networks (CNN) to achieve image compression. This network consists of three parts: analysis transformation structure, quantization structure and synthetic transformation structure. These structures are mainly composed of convolutional layer, downsampling layer, GDN normalization layer. At low bit rates, Ballè's method obtain higher visual quality than JPEG and JPEG2000. In Theis' work, to effectively estimate the distribution and bit rate of the coding coefficients, a GSM (Gaussian scale mixtures) is introduced to control the compression ratio based on the recoverable image [13]. Due to the dissimilarity of the local content of the image, Li *et al.* [14] proposed a content-based adaptive quantization method by introducing the importance map. Mentzer *et al.* [15] represented the entropy estimation by using a 3D-CNN which mainly learns a conditional probability.

Attention plays an important role in human perception. In these years, attention based DNNs develop quickly. Hu *et al.* [16] have proposed a compact "Squeeze-and-Excitation" (SE) block to demonstrate the relationship between feature maps. It learns channel-wise feature responses, which can be regarded as an attention mechanism applied upon channel axis. However, the spatial axis is ignored, which is also an important factor in inferring accurate attention map. Woo *et al.* [17] design a Bottleneck Attention Module (BAM). This model infers a parallel attention architecture along channel and spatial which shows consistent improvement in classification and detection performances. Inspired by this, we proposed coupled SEblocks framework applying convolutional neural networks of image compression. Considering the particularity of the end-to-end network, we design an improved SE network for the decoder to enhance the decompression effect.

3 Coupled SE Learning Framework

3.1 Convolutional Encoder and Decoder

We use an end-to-end image compression framework based on convolutional neural network. Both encoder and decoder are mainly constructed using several stages of convolution layer, GDN/IGDN nonlinearities and coupled SEblocks, which learn linear–nonlinear transformations. The main part of each stage is composed of a subsampled convolution with 128 filters. Firstly, the input image x is convolved with a 9×9 filter whose downsampled factor is 4. Then it followed by two layers of the same convolution layer. The size of filters are $5 \times 5, 5 \times 5$ and the strides are 2. The original image x is transformed into a code space by a parameterized nonlinear transformation $E(x)$. After quantizer and lossless compression entropy coding, we get bit stream to save. The process of reconstructing the compressed image generally mirrors the encoding transform in reverse order. The decoder $D(c)$ are made up of four square filters with sizes $3 \times 3, 5 \times 5, 5 \times 5, 9 \times 9$. In addition, results are upsampled by factors of $\{1, 2, 2, 4\}$, respectively.

Instead of using common nonlinear activation function, we use a generalized divisive normalization (GDN) joint nonlinearity that is inspired by local gain control behaviors observed in biological visual systems, and has proven effective for density modeling and image compression [12]. We proposed a coupled SEblocks to enhance the relationship between feature maps in both encoder and decoder. A complete network framework is shown in Fig. 1.

3.2 Coupled SEblocks

The Squeeze-and-Excitation (SE) block is a computational unit [16]. Through the research and comparison of image compression networks, we can find that most existing image compression frameworks contain many feature maps of bottleneck after encoder. Not all feature maps are full of information and each channel can be considered as a feature detector. We insist each channel of feature maps contains a specific response. In order to solve the problem of channel dependence, SEblocks are used in encoder. Considering our Image compression network is training by an end-to-end mode, and we add inversed SEblocks (ISEblock) in the decoding side for improving the quality of the decompressed image. SEblocks and inversed SEblocks (ISEblock) are collectively called coupled SEblocks.

To model the channel internal dependencies of encoder, we add a SEblock after each convolutional layer. Let $X \in \mathbb{R}^{H \times W \times C}$ denotes the input tensor, the refined feature map $\hat{X}$ is computed as:

$$\hat{X} = F_{scale}(X, M_s(X)) = X \otimes M_s(X) \tag{1}$$

where $\otimes$ refers to channel-wise multiplication between the feature maps.

We introduce channel attention mechanism, and learn a specific feature response of each channel. We take global average pooling on the feature map X and produce a channel vector $X_c \in \mathbb{R}^{1 \times 1 \times C/r}$ by fully connected layer. This vector aggregates the global information of channels softly. To estimate attention across channels from the channel vector X_c, we use a nonlinear activation function ReLU [18] and another fully connected layer. To meet our criteria, we employ a simple gating mechanism with a sigmoid activation. In short, the channel attention is computed as:

$$M_s(X) = \sigma(fc(\delta(fc(gap(X))))) \tag{2}$$

where δ and σ correspond to ReLU and sigmoid function respectively, $X_c \in \mathbb{R}^{1 \times 1 \times C/r} = fc(gap(X))$, and the reduction ratio r is set to 16.

As for decoder, we use inversed SEblocks (ISE) which are mirrored structure of SEblocks. We are working on the same arrangement of squeeze and excitation parts. The input tensor passes through the global average pooling layer, the fully connected layer, ReLU function, another fully connected layer and sigmoid function.

$$M_I(X) = \sigma(fc(\delta(fc(gap(X))))) \tag{3}$$

However, the biggest difference between the SEblocks and the inversed SEblocks is the way of scaling. In encoder, we adopt SEblocks to learn a weight vector that indicates the relationship between feature maps. The values of this vector are in the range of 0–1 due to the sigmoid function. After the SEblocks in encoder, it learns that different feature maps have different degrees of importance. Useful features are enhanced and useless features are suppressed. Nevertheless, we take the inverse scaling method in decoder. The goal of the decoder is to recover a good quality decompressed image. Considering the end-to-end learning mode, we use the channel-wise division instead of multiplication. This can reduce the suppression of the feature maps by the previous SEblocks, and enlarge the feature maps information. ISEblock is symmetric to that of the SEblocks. The ISEblock' scale is defined as:

$$\hat{X}_I = F_{I-scale}(X, M_I(X)) = X \cdot / M_I(X) \tag{4}$$

where $\cdot/$ refers to channel-wise division between the feature maps.

It is the first time that the coupled SEblocks are applied in image compression framework. Coupled SEblocks are designed to minimize the effects of noise channels. In fact, the feature maps, as the output of the convolutional layer, is a combination of the feature maps of the input. If there is too much noise in the input channels, the output is definitely not ideal. The framework we propose makes the strong feature maps stronger and the weak feature maps weaker, effectively improving decompressed image quality. The complete SEblocks and ISEblocks are shown in Fig. 2.

3.3   Quantization

Quantization is the main factor causing distortion in image coding. The basic principle of the quantization layer is to round the float number of the feature map to an integer number, that is $\hat{y} = round(y_E)$. Most of the quantization methods directly quantify each feature map in the forward propagation. However, the derivative of the quantization function itself is mostly zero, which is non-differentiable at other locations. If the gradient is directly calculated using the round function and applied to the network, the gradient cannot be transmitted to the next layer through the quantization layer. Therefore, the quantization layer needs to be approximated by a continuous function. In the case of backpropagation, we replace the quantizer with an additive *i.i.d.* uniform noise, where Δx is a random noise, which has the same width as the quantization bins.

$$y_Q = y_E + \Delta y \tag{5}$$

where y_E represents the coefficient value of feature map after encoder.

By training a discrete probability mass with weights equal to the probability mass function, we can get the marginal density of y_Q [12].

$$P_{y_Q}(y_Q = n) = \int_{n-0.5}^{n+0.5} P_{y_E}(t)dt \tag{6}$$

It is can be easily found that the differential entropy of $y_Q\psi$ approximates the entropy of $\hat{y}$. We use $\hat{y} = round(y_E)$ as the quantization operation when we test the model and the rates can be estimated precisely as well.

3.4   Model Formulation

In general, image compression problems based on DNNs can be formulated as a rate-distortion optimization problem. Our goal is to minimize the sum of the distortion loss and rate loss. Let λ be the tradeoff parameter for balancing distortion and bit rates. Therefore, we define the loss as:

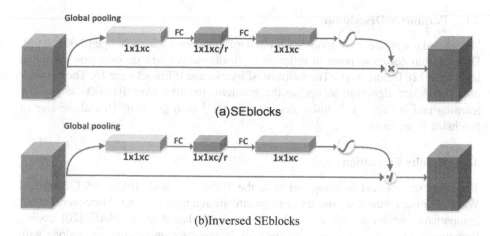

Fig. 2. Illustration of Coupled SEblocks. (a) the default architecture of SEblocks. (b) the Inversed SEblocks that we proposed. $W \times H \times C$ indicates the size of input feature map. r is the ratio of SEblocks and ISEblocks.

$$L = \sum_{x \in X} \{L_D(y_n, x_n) + \lambda L_R\} \tag{7}$$

where x_n is the input image, y_n is the decompress image, L_D denotes the distortion loss and L_R denotes the rate loss.

Distortion loss commonly uses the squared L_2 norm to evaluate the error between the original image and the decompressed image. The formula is as follows:

$$L_D(y_n, x_n) = \|y_n - x_n\|_2^2 \tag{8}$$

The actual number of bits stored depends on the entropy rate of the feature map after the quantizer. The rate loss L_R is defined as [12]:

$$L_R = -E[\log_2 P_y] \tag{9}$$

Drawing on Ballè's method [12], we apply a piecewise linear functions for approximate the discrete P_y to ensure it continuous and differentiable.

4 Experiments and Results

In the training phase, we randomly select a subset of the ImageNet. The subset contains about 6500 images in total and the extracted images are cropped into patches of size $256 \times 256 \times 3$. These patches need to be normalized to a range of 0–1. The test set is the popular Kodak [19] data set consisting of 24 high quality images. We use MS-SSIM as a distortion perceptual metric for training and bits per pixel (bpp) is used to evaluate the compression rate.

4.1   Parameter Description

We trained a separate set of models with tradeoff parameter λ in the range [0.005, 0.05]. Different bitrates correspond to different λ. Both encoder and decoder convolutional layers use 128 feature maps. The ratios of SEblocks and ISEblocks are 16. The network uses the Adam algorithm to update the gradient, iterating over 1000000 steps. The learning rate is from 10^{-4} slowly decaying to 10^{-7} by a factor of 10 and the size of batchsize is set to 16.

4.2   Results Evaluation

The proposed method is compared with the Ballè's method, JPEG, JPEG2000 and WebP. For our method, we use uniform quantization during training. Moreover, for fair comparison, we implement a simple entropy code based on CABAC [20] coding framework. In the testing phase, our network is separated and can be used alone with encoders and decoders. Figure 3 shows the average MS-SSIM over all 24 Kodak images of these five methods. We represent raw MS-SSIM (ms) in dB scale by using $-10\log_{10}(1 - ms)$ and the results are shown in Fig. 4.

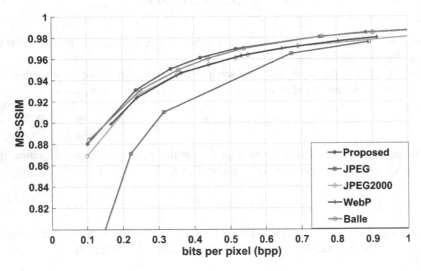

Fig. 3. Performance of our approach on Kodak Dataset, where we outperform Ballè, JPEG, JPEG2000 and WebP in MS-SSIM.

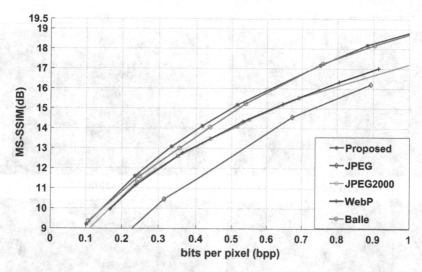

Fig. 4. Evaluated using MS-SSIM (dB) on Kodak, our approach has better performance than other four methods.

From Fig. 3, we can see our curve is above the other methods. Specifically, when the bit rates is between 0.2bpp–0.5bpp, our method is significantly higher than the other four methods. Fig. 5 shows the decompressed images of the compared methods at relatively low bit rates. Since WebP and JPEG2000 have similar performances on the Kodak dataset, and WebP is slightly better than JPEG2000, Fig. 5 only provides the visual effects of Ballè's, WebP and JPEG. From Fig. 5, we can find the appearance of the compressed image differs in quality and is improved compared to JPEG and WebP at low bit rates.

In addition, the proposed method maintains the smoothness of the outline and improves the clarity of images. In contrast, images produced by JPEG and WebP have visual interference blocking, ringing artifacts and aliasing that respond to underlying basis functions. Besides, the color of the image is distorted at low bit rates. Table 1 gives the experimental results of some images of the Kodak test set at low bit rate. In particular, compared to the Ballè's, our method has a lower bit rate under the same λ. This is because the coupled SEblocks improves the expression ability of feature maps. Coupled SEblocks make the weight of the significant feature maps larger, and the weight of the insignificant feature maps are smaller, which is beneficial to learn a more concentrated distribution in the entropy coding and effectively reduce compression ratio. From Table 1, it can be seen that our method maintains the structural information and good quality of the image while having a lower bit rate than Ballè's.

Fig. 5. Visual Quality Comparison of example Images in Kodak Dataset (Kodim07 and Kodim13)

Table 1. Comparison of partial image experiment results of Kodak dataset

Image	Evaluation	Proposed	Ballè	WebP	JPEG
Kodim01	BPP	**0.3120**	0.3230	0.3587	0.3473
	MS-SSIM	**0.9279**	0.9273	0.9260	0.9069
Kodim04	BPP	**0.2671**	0.2773	0.27112	0.3049
	MS-SSIM	**0.9379**	0.9350	0.9361	0.9160
Kodim07	BPP	**0.1023**	0.1063	0.1264	0.1223
	MS-SSIM	**0.9323**	0.9311	0.9186	0.8370
Kodim08	BPP	**0.3954**	0.4205	0.4136	0.4076
	MS-SSIM	**0.9513**	0.9512	0.9385	0.9209
Kodim09	BPP	**0.0858**	0.0892	0.0917	0.1070
	MS-SSIM	**0.9325**	0.9275	0.9188	0.8328
Kodim13	BPP	**0.3920**	0.4387	0.4169	0.4064
	MS-SSIM	**0.9021**	0.8991	0.8912	0.8843
Kodim18	BPP	**0.2759**	0.2904	0.3197	0.2901
	MS-SSIM	**0.9146**	0.9115	0.9090	0.8752
Kodim23	BPP	**0.2326**	0.2438	0.2576	0.2551
	MS-SSIM	**0.9670**	0.9655	0.9660	0.9274

5   Conclusion

In this paper, we proposed a complete end-to-end image compression framework based on coupled SEblocks. Our compression method applies attention-based feature refinement. We use SEblocks in the encoder to determine the weights of feature maps. In the decoder, we improve the scale part of the SEblocks by using channel-wise division which magnifies the response of the feature map. Our final mode learns what to emphasize or suppress and refines intermediate features effectively. To verify its efficacy, we use the classic dataset of image compression - Kodak dataset as a test set, comparing with other four methods including Ballè's, JPEG, JPEG2000 and WebP. From the experimental results, our approach is higher than these four methods at low bit rates. Moreover, compared with the traditional compression method JPEG and WebP, our method retains more image texture details with no artifacts and block effects, and the visual effect is clearly improved.

References

1. Wallace, G.K.: The jpeg still picture compression standard. IEEE Trans. Consum. Electron. **38**(1), xviii–xxxiv (1992)
2. Skodras, A., Christopoulos, C., Ebrahimi, T.: The jpeg 2000 still image compression standard. IEEE Signal Process. Mag. **18**(5), 36–58 (2001)
3. Google.: WebP: Compression techniques (2017). http://developers.google.com/speed/webp/docs/compression. Accessed 30 Jan 2017

4. Foi, A., Katkovnik, V., Egiazarian, K.: Pointwise shape-adaptive DCT for high-quality denoising and deblocking of grayscale and color images. IEEE Trans. Image Process. **16**(5), 1395–1441 (2007)
5. Zhang, X., Xiong, R., Ma, S., Gao, W.: Reducing blocking artifacts in compressed images via transform-domain non-local coefficients estimation. In: IEEE International Conference on Multimedia and Expo (ICME), pp. 836–841 (2012)
6. Zhang, X., Xiong, R., Fan, X., Ma, S., Gao, W.: Compression artifact reduction by overlapped-block transform coefficient estimation with block similarity. IEEE Trans. Image Process. **22**(12), 4613–4626 (2013)
7. Zhang, X., Xiong, R., Zhao, G., Zhang, Y., Ma, S., Gao, W.: CONCOLOR: Constrained non-convex low-rank model for image deblocking. IEEE Trans. Image Process. **25**(3), 1246–1259 (2016)
8. Dong, C., Loy, C.C., He, K., Tang, X.: Learning a deep convolutional network for image super-resolution. In: European Conference on Computer Vision (ECCV), pp. 184–199 (2014)
9. Dong, C., Loy, C.C., He, K., Tang, X.: Compression artifacts reduction by a deep convolutional network. In: IEEE International Conference on Computer Vision (ICCV), pp. 576–584 (2015)
10. Toderici, G., et al.: Variable rate image compression with recurrent neural networks. In: International Conference on Learning Representations (ICLR). arXiv: 1511.06085 (2015)
11. Toderici, G., et al.: Full resolution image compression with recurrent neural networks. In: IEEE Conference on Computer Vision and Pattern Recognition (CVPR), pp. 5435–5443 (2017)
12. Ballé, J., Laparra, V., Simoncelli, E.P.: End-to-end optimized image compression. In: International Conference on Learning Representations (ICLR). arXiv: 1608.05148 (2016)
13. Theis, L., Shi, W., Cunningham, A., Huszár, F.: Lossy image compression with compressive autoencoders. In: International Conference on Learning Representations (ICLR). arXiv: 1703.00395 (2017)
14. Li, M., Zuo, W., Gu, S., Zhao, D., Zhang, D.: Learning convolutional networks for content-weighted image compression. In: IEEE Conference on Computer Vision and Pattern Recognition (CVPR), pp. 3214–3223 (2018)
15. Mentzer, F., Agustsson, E., Tschannen, M., Timofte, R., Van Gool, L.: Conditional probability models for deep image compression. In: IEEE Conference on Computer Vision and Pattern Recognition (CVPR), pp. 4394–4402 (2018)
16. Hu, J., Shen, L., Albanie, S., Sun, G., Wu, E.: Squeeze-and-excitation networks. In: IEEE Conference on Computer Vision and Pattern Recognition (CVPR), pp. 7132–7141 (2018)
17. Park, J., Woo, S., Lee, J.Y., Kweon, I.S.: BAM: Bottleneck Attention Module. In: The British Machine Vision Conference (BMVC). arXiv:1807.06514 (2018)
18. Hinton, G. E.: Rectified linear units improve restricted boltzmann machines. In: International Conference on International Conference on Machine Learning (ICML), pp. 807–814 (2010)
19. Eastman Kodak.: Kodak Lossless True Color Image Suite (2012). http://r0k.us/graphics/kodak. Accessed Oct 2012
20. Marpe, D., Schwarz, H., Wiegand, T.: Context-based adaptive binary arithmetic coding in the h.264/avc video compression standard. IEEE Trans. Circuits Syst. Video Technol. **13**(7), 620–636 (2003)

Soft Transferring and Progressive Learning for Human Action Recognition

Shenqiang Yuan, Xue Mei$^{(\boxtimes)}$, Yi He, and Jin Zhang

School of Electrical Engineering and Control Science, NanJing Tech University,
Nanjing 211816, China
seraph_mx@163.com

Abstract. In action recognition, many different network structures of spatiotemporal features extractions has been proposed, and performed well on several mainstream datasets. Inevitably, one question occurs to us: is there any transferable characterizes between different models? In this paper, we discuss such problem by introducing a cross-architecture transferring learning scheme, dubbed soft transferring learning, aiming to overcome the limitation of divergence of different network structures. A multi-stage semi-supervision training procedure is conducted to keep the consistency internally between two different models from bottom to top. To this end, we introduce two kinds of cross-structure metric strategy to compute the mismatch value in different features level, together with entropy classification loss, integrated to be a three-stage supervision method. We additionally design a new network to learn from the supervisors, which have been trained on large-scale datasets. We fine-tune supervision model and train our new model on UCF101 and HMDB51 datasets, experiment results demonstrate the feasibility of soft transferring method, extend transfer learning to a broader sense, and show the flexibility of deploying of existing models. Our method is designed easily generalized to different networks in other computer vision task.

1 Introduction

The abundance of available data of multimedia, like images, voices and videos, made these CNN-based methods correlated to data analysis a large resurgence and proliferation. Especially in still image recognition, detection and segmentation tasks, many novel and creative ideas, such as residual learning, dense connection, diverse and small kernels, have broken through the thinking inertia in traditional network designing, bringing neural networks to a deeper and stronger era.

Video classification, holding input as a sequence of images, benefits a lot from those skills though. Specific to human action recognition field, deep networks (I3D [8], P3D [10], R(2+1)D [7], ECO [11], TSN [9], etc.) have exceeded hand-craft features both in accuracy and processing speed, and as more large-scale video datasets occurring, this superiority will be further enhanced. However, there is a problem, do they have some common features since they are useful for recognition totally even though more or less differences structurally? So, instead of discussing those models' pros and cons, we explore the off-shoot question: what the models learned from videos [4] and

Z. Cui et al. (Eds.): IScIDE 2019, LNCS 11935, pp. 213–225, 2019.
https://doi.org/10.1007/978-3-030-36189-1_18

whether there is any invariant and general features representation hidden in off the shore learnt knowledge from existing networks. We name it Soft Transferring Learning problem (STL).

In order to prove this view and answer the question, a new network is designed based on 3D DenseNet [5] blocks, called MDN (Modified DenseNet), to generate the same feature space with reference models like I3D [8] or R(2+1)D [7], which we choose as supervisors to provide knowledge of how to represent a video clip with distinguishability but robustness at the mean time. A multi-stage optimization process is utilized for progressive training to prompt MDN learn from supervisors, inspired by GoogleNet [6]. Specifically, three stages of supervision with different loss functions are designed: we measure cosine similarity of low level features at the early stage; introduce multiple kernel maximum mean discrepancy [21] metric at top feature space; and finally add a regular classifier with classification entropy loss. With this approach, knowledge can be shared and reused between heterogeneous models. If successful, a significant usage is that STL can be expected to bridge lighter models to heavy models, allowing lighter model inherit knowledge from complicated counterparts without limitations caused by different structures. Before this general exploration, we instead to limit our experiments in scope of equivalent models complexity, more in-depth research is worth looking forward to in the feature. It also should be noted that we do not try to challenge the basic view that structures play key roles in CNN models, in spite of the natural structurally independent trait of STL, our experiments also emphasize the importance of architecture itself.

To summarize, the main contributions of our paper are:

- An interesting cross-structure transferring learning problem is raised, and a three-stage supervision soft transferring method is proposed;
- Two auxiliary loss metrics are introduced to compel student model learning from supervisor model to keep consistency from bottom to top;
- A new network model MDN is designed and tested through STL method, extensive evaluations on available public activity recognition datasets demonstrate the feasibility.

2 Related Works

Recently, human action recognition base on videos becomes more realistic, many difficult data modeling works in several years ago were overcome by data driving now. Those improvements come mainly from two aspects: bigger labelled datasets and efficient convolutional network (CNN) architectures.

Video Datasets. For datasets, HMDB51 [2] and UCF101 [3] were released in 2011 and 2012, containing 6766 clips of 51 classes and 13320 clips of 101 classes respectively. Those videos are trimmed temporally, with only one action instance per clip. It is appropriate to do research about action classification and usually introduced to assess algorithms as benchmarks. Starting from 2012, since deep CNN got a great success on images classification, larger video datasets like MIT, Sports-1M, kinetics, AVA (more details are reported on[1]) and etc. are created and released successively as well. As a result, many deeper and more complicate networks can be trained to converge and generalized to other unspecific works. However, large dataset is difficult to use inevitably, thousands of video segments consume huge memory footprint, must be equipped with large groups of GPUs for parallel training. We are curbed to do effective experiments on these datasets, so here we do not intend to discuss them thoroughly.

Models Structures. In terms of methods of video processing, most of recent works are enfolded in deep CNN, We briefly review these methods as follows. At beginning, researchers attempt to explore the potency of 2D-CNN in processing video data, Li-FeiFei [12] build a 2D convolutional network with seven layers to access the single-frame feature and extended the connectivity in time by some later or early fusion strategies. Shallow model layers and insufficient temporal information summary only lead to the modest improvements compared to single frame model. After that, a lot of modified varieties were proposed from these two perspectives. Donahue [1] used deeper VGGNet [13] and introduced a recurrent structure, long short term memory (LSTM), to aggregate temporal information, achieved greater promotion on video classification and description. Two-stream [25] model was put forward to use not only RGB images but warped optical flow modality, opening a big branch of detached spatial-temporal features extracting. TSN [9], ECO [11] apply more efficient sample steps and feature-fusion mechanism, speeding up the processing of long-rang video, accomplishing a real-time running mode. The famous 3D CNN architecture C3D [14] born in 2015 from facebook. Without any preprocessing, no need for optical flow, C3D [14] distilled the compact video representation from raw frame-sequences, demonstrating a strong modeling capabilities on videos. Many later researches used C3D as a backbone for features extracting. Some techniques originally used in 2D CNN have been naturally applied to 3D CNN, like residual connection, dense connection and inception blocks, correspondingly, 3D ResNet [15, 22, 23], 3D DenseNet, and I3D were put forward. Those models perform well, whereas 2D CNN is surprisingly hard on the heels. Xie [16] come up with a top-heavy structure S3D combined the 2D and 3D ConvNets; Tran [7], author of C3D, reconsider the relationship between spatial and temporal of 3D kernel, factored 3D convolutional filters to be (2+1)D based on 3D ResNet and further underscored the necessity of temporal information in action recognition; These models have greatly improved over previous versions, and are leading a clear future for action recognition.

Transferring Learning. Currently transferring learning usually means the reuse of certain domain knowledge in a new dataset or new task [26], in many cases of CNN models, parameter-based transferring requires completely unanimous structure for containing those parameters, causing strict restrictions especially when we want to rebuild a lighter network on device of lower compute capability and still benefit from those learnt knowledge. Some representative works [21, 27–29] on transfer learning

---

[1] http://www.actionrecognition.net/.

used almost the same model structure for different domains as well, just insert extra adaptation layers on top feature space, which are categorized as structure-dependent method. Diba [17] proposed a cross-architecture approach intended to transfer knowledge from 2D CNN model to 3D CNN networks, the author evaluated correspondence of outputs of 2D and 3D models with inputs of positive/negative pairs. Even both 2D and 3D model in [17] shared the same DenseNet structure, for supervision, the 2D stream could be replaced by arbitrary models in form actually, this is very promising but not paid enough attention. As for transferring metric strategies, metric approximation and distribution approximation are usually used for homogeneous transfer [30], which demands a same feature space of source and target domain. By minimizing the feature distribution distance of different domains, distribution bias is gradually eliminated and eventually domain adaption accomplished [31–33]. Essentially, those metrics is general for data fitting works only if the data embedded in same feature spaces, however simplest European distance is usually difficult to cope with the drastic changes in data distribution. So the most important issue for transferring learning is to find an effective metric function of high adaptiveness for describing disparity between distributions.

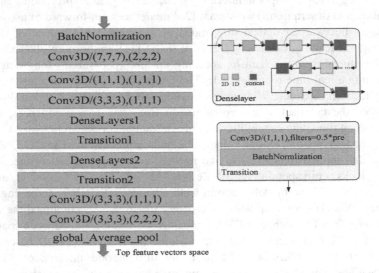

Fig. 1. Outlook of MDN structure: (k_1, k_2, k_3), (s_1, s_2, s_3) denote kernel sizes and sliding strides. More details about our MDN please refer to Table 1.

3 Proposed Method

In this section, we first introduce the structure of our MDN model, explain the calculation details about disparity measurements across features of different levels from models, Then we formalize our soft transferring method as an optimization problem of several loss functions. And finally we descript the transferring procedure from a computational perspective.

3.1 Modified DenseNet Structure

Residual connections is proved to be a useful discipline, correspondingly 3D-ResNet [24] has considerable potency for action recognition, and many works paves this way. Here in this work, we go further to adopt dense connections to utilize data along the whole depth of network. Different from the original DenseNet, inspired by [10], we detach 3D convolutional kernels into 2D and 1D in basic dense-blocks and reduce dense layers to make model lighter but efficient. For adapting variable-length, fully convolution and global pooling technologies are used to generate feature vectors with fixed size. Figure 1 depicts the summary of our MDN model. Every convolution layer is followed by a batch-normalization [18] layer and relu activation function, the top feature vectors is encoded by fully connections for generating classification vectors, this kind of details are not shown in the Fig. 1. More explicit information is displayed in Table 1.

Table 1. Structures of MDN: Input size is set to (25,112,112,3). DenseLayer1 and DenseLayer2 contain 4, 6 dense blocks respectively, all strides are (1,1,1) by default if not explicitly specified. Bias is adopted only for fully connected.

	Name	Output size	Kernel/(stride)	Flops
Layer1	Conv3d_1a_7x7	$13 \times 56 \times 56 \times 64$	$64 \times 3 \times 7 \times 7 \times 7/(2,2,2)$	19.23G
Layer2	Maxlooling0	$13 \times 28 \times 28 \times 64$	$64 \times 64 \times 1 \times 3 \times 3/(1,2,2)$	0
Layer3	Conv3d_1b_1x1	$13 \times 28 \times 28 \times 64$	$64 \times 64 \times 1 \times 1 \times 1$	0.04G
Layer4	Conv3d_1c_1x1	$13 \times 28 \times 28 \times 256$	$192 \times 64 \times 3 \times 3 \times 3$	3.15G
Layer5	Maxpooling1	$13 \times 14 \times 14 \times 256$	$256 \times 256 \times 1 \times 3 \times 3/(1,2,2)$	0
Layer3	DenseLayer1: Dense block1 $\sim$ 4	$13 \times 14 \times 14 \times 384$	$128 \times 256 \times 1 \times 3 \times 3$ $32 \times 128 \times 3 \times 1 \times 1$	3.44G
Layer4	Transition1_1x1:	$7 \times 14 \times 14 \times 480$	$480 \times 384 \times 3 \times 3 \times 3$ $480 \times 480 \times 2 \times 1 \times 1$	11.81G
Layer5	DenseLayer2: Dense block1 $\sim$ 6	$7 \times 14 \times 14 \times 672$	$128 \times 480 \times 1 \times 3 \times 3$ $32 \times 128 \times 3 \times 1 \times 1$	5.04G
Layer6	Transition2_1x1:	$7 \times 7 \times 7 \times 512$	$512 \times 672 \times 1 \times 3 \times 3$ $192 \times 192 \times 1 \times 2 \times 2$	5.19G
Layer8	Conv3d_4d_3x3	$7 \times 7 \times 7 \times 1024$	$1024 \times 832 \times 1 \times 1 \times 1$	0.54G
Layer9	Global average pool	$1 \times 1 \times 1 \times 1024$	$1024 \times 1024 \times 7 \times 7 \times 7$	0
Layer10	Fully connected	101	101×1024	0.099G

3.2 Multi-stage Supervision

Random initialization makes vacant model deviate too far to be redressed back especially when the model has different structure from its supervisor. Just inspecting on final features at top layer maybe not sufficient, Multi-stage supervision is needed to restrict model to a rational jitter scope. We intuitively set the first observation point over the output of transition1 layer in MDN, correspondingly the output of Mixed_3c layer of I3D is picked out for similarity measure. Second stage supervision is performed on the output of global average pooling layer displayed in Fig. 1, which is a feature space with spatial and temporal resolution both shrunk to 1. Every video clip corresponds to a highly integrated vector of space-time information, category information is reflected implicitly by the joint distribution of vector components. A mean discrepancy method is introduced to measure distance of this joint distribution to the output space of last mixed block of I3D. Finally we add a routine cross-entropy to make a final class prediction for action recognition.

Cosine Similarity Loss Function. In order to quantify mismatches of early layers of MDN and I3D, we need to compute the distance of the two tensors in high dimension ($13 \times 14 \times 14 \times 480$) supposed input video clip has shape of ($25 \times 112 \times 112 \times 3$), which will incur the expensive cost of explicitly working in these spaces if a pure Euclidean distance is applied. In addition, Euclidean distance constraint may be too severe as for middle supervision, likely making the non-convergence of loss. For those reasons, we first down sample the feature maps by MaxPooling in depth, height and weight without extra parameters and compute inner product along channels, depicted as Fig. 2. Every spatiotemporal grid is represented as a vector of 1×480, easy to analogize this representation to interest point description in space-time like STIP algorithm, but only replace Harr and Forstner corner point detection with CNN. We scale this cosine similarity through negative logarithmic function to formulate a loss function defined by formula (1).

$$\textit{Similarity}_{loss} = -\frac{1}{N}\sum_{j=0}^{N-1}\sum_{i=0}^{C-1} log\langle g_i^{t_j}, g_i^{s_j}\rangle$$
$$= -\frac{1}{N}log\prod_{j=0}^{N}\prod_{i=0}^{C-1}\langle g_i^{t_j}, g_i^{s_j}\rangle \tag{1}$$

N is batch-size, C is the total number of spatiotemporal grid cells ($L' \times H' \times W'$), and $g_i^t \in$ grid^t, $g_i^s \in$ grid^s, like depicted in Fig. 2.

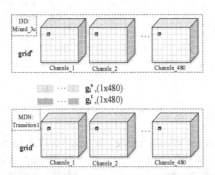

Fig. 2. Middle supervision from I3D: the output of mixed_3c in I3D model has tensor shape of (clip_length/2^1, 14, 14, 480), every space-time unit was presented as a vector of 480 dimensions.

$$MMD = ||E(\varphi(\boldsymbol{x})) - E(\varphi(\boldsymbol{y}))||_{\mathcal{H}}^2$$

$$\approx \frac{1}{N_x(N_x-1)} \sum_i^{Nx} \sum_{j,j\neq i}^{Nx} k(\boldsymbol{x}_i, \boldsymbol{x}_j)$$

$$- \frac{2}{N_x N_y} \sum_i^{Nx} \sum_j^{Ny} k(\boldsymbol{x}_i, \boldsymbol{y}_j) \tag{2}$$

$$+ \frac{1}{N_y(N_y-1)} \sum_i^{Ny} \sum_{j,j\neq i}^{Ny} k(\boldsymbol{y}_i, \boldsymbol{y}_j)$$

Maximum Mean Discrepancy (MMD). MMD loss is very popular in domain adaption field recently, it maps source and target domain features to a reproducing kernel Hilbert spaces through particular kernels and compute mean discrepancy between those feature distributions [19, 20], unbiased estimation of Maximum Mean Discrepancy defined like formula (2), which reveals the distance of two distributions in original feature space. $\kappa(\boldsymbol{x}, \boldsymbol{y}) = <\varphi(\boldsymbol{x}), \varphi(\boldsymbol{y})>$ means kernel operation, ascending dimensions and measuring distance, $\boldsymbol{x}$ and $\boldsymbol{y}$ denote feature vector. N_x and N_y indicate number of samples of source domain and target domain respectively. Specify to our applications, we need a metric for measuring the difference between top feature spaces out from MDN and I3D, an improved multiple kernel maximum mean discrepancy (named MK-MMD [21]) approach is introduced, whose author applied multi Gaussian kernels

$$MK = \begin{bmatrix} \sum_{k=1}^{K} e^{\frac{-1}{\sigma_k}||x_1-y_1||_2} & \sum_{k=1}^{K} e^{\frac{-1}{\sigma_k}||x_2-y_1||_2} & \cdots & \sum_{k=1}^{K} e^{\frac{-1}{\sigma_k}||x_{nx}-y_1||_2} \\ \sum_{k=1}^{K} e^{\frac{-1}{\sigma_k}||x_1-y_1||_2} & \sum_{k=1}^{K} e^{\frac{-1}{\sigma_k}||x_2-y_2||_2} & \cdots & \sum_{k=1}^{K} e^{\frac{-1}{\sigma_k}||x_{nx}-y_2||_2} \\ \cdots & \cdots & \cdots & \cdots \\ \sum_{k=1}^{K} e^{\frac{-1}{\sigma_k}||x_1-y_{ny}||_2} & \sum_{k=1}^{K} e^{\frac{-1}{\sigma_k}||x_2-y_{ny}||_2} & \cdots & \sum_{k=1}^{K} e^{\frac{-1}{\sigma_k}||x_{nx}-y_{ny}||_2} \end{bmatrix}_{Ny \times Nx} \tag{3}$$

$$MKMmd_{loss} = \frac{1}{Ny \times Nx} \sum_{i=1}^{Nx} \sum_{j=1}^{Ny} MK(i,j)$$

$$\arg\min_{\Theta} F(\Theta) = J(\Theta) + \gamma_1 Similarity_{loss} + \gamma_2 MKMmd_{loss}$$
$$where\ J(\Theta) = Entropy_{loss}(\Theta, clip, label) \tag{4}$$

for building Hilbert spaces on multiple basis functions. We compute MMD loss by formula (3), $N_x = N_y = N$ equal to batch_size, x_i is a feature vector out of MDN, y_j is an aligned harmonious vector from I3D, k = 1, 2, ... K, indicates index of kernels. Finally, we format the STL into an optimization problem by formula (4), where J is entropy loss function, $\Theta = \{w, b\}$.

3.3 Implementation

We use data-parallel structure for training models on multi-GPUs with SGD optimizer. For saving computation, supervision features is pre-extracted through fine-tuned I3D and R2.5D models on UCF101 and HMDB51 datasets instead of online style. Resampling strategy is included to make the most of video dataset. Specifically, we fix the clip length at 25 frames, every video is sampled with non-overlapping more than once with interval 32, consequently 70364 clips for UCF101 are collected, we divide them into training, validation, test with proportion of 0.5,0.25,0.25. During iterations, input video clip is set as $25 \times 112 \times 112 \times 3$ which is randomly cropped from resized images of 168×224 pixels.

4 Experiment Results

In this part, we show out the training procedure and report the result of several groups of contrast experiments for analysis.

4.1 Structure Correlations

As mentioned above, we consider several kinds of network structures for mutual learning, including two sota (state-of-the-art) structures and our self-building MDN. Not only serve as supervisors, R2.5D and I3D are used to perform extra soft transferring learning experiment between themselves. Here we re-implement R2.5D-18 consistent with the original version proposed in [7]. As for I3D model, a few modifications are applied for channels matching.

Some evidence suggests that a careful designed structure is necessary even though STL is structure-irrelevance formally. We have used some advanced techniques for constructing our new model, but there is still an obvious gap between MDN and these sota models. As shown in Fig. 3(a) (b) (c), the final classification loss converges quickly for I3D model, inclined to saturated after 20k steps of iteration. Comparatively shown as Fig. 3(d) (e) (f), this is not the easy case for training MDN, it arrives at a modest situation till 60k iterations, with higher value of entropy loss and lower accuracy. One possible explanation could be that rich and diverse kernels endow I3D with strong modeling ability, but MDN is subject to limitations with shallow layers (15 layers, which cannot play its due role) by comparison.

Another concern about model structure with STL is that similar structure gets much benefits than models of great structure difference. For example, learning from R2.5D is easier than learning from I3D for MDN, it takes 80k iterations arriving at 0.8 top1 validate accuracy with I3D supervision but 50k iterations with supervision from R2.5D and validate accuracy eventually converges to 0.85, nearly 5% points higher. Think it over, we find that decomposition kernels and skipping residual connections used in MDN are similar with R2.5D factually, no much effort is need to explore supervisor's structure knowledge. However, this is a hard-core but also the hardest part of STL. Most extremely, models with completely same structure would meet the best results, which degenerates STL to parameter-based transferring learning problem.

4.2 Loss Function Efficacy

To make clear the efficacy of each stage, we set three groups of super parameters in formula (4): $(\gamma_1, \gamma_2) = (0.03, 0.0), (0.0, 0.05), (0.03, 0.05)$ respectively, see Table 2. $(\gamma_1, \gamma_2) = (0.0, 0.0)$ means training from scratch using neither middle nor top

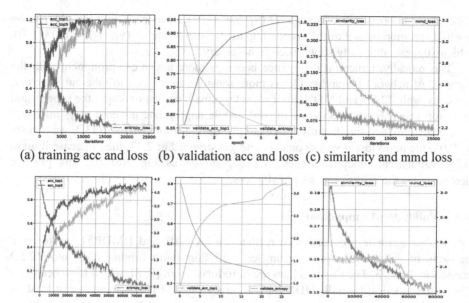

(a) training acc and loss (b) validation acc and loss (c) similarity and mmd loss

(d) training acc and loss (e) validation acc and loss (f) similarity and mmd loss

Fig. 3. Different models training and validation on ucf101: (a) (b) (c) are results of I3D+R2.5D (STL); (d) (e) (f) are results of MDN+I3D(STL).

Table 2. Accuracy gains of several different training schemes. * means neither middle nor top supervision information used. No pre-trained parameters are restored for all models. Trained and validated on UCF101 splits1.

	γ_1	γ_2	top1_val
MDN*	0.0	0.0	62.1
I3D*	0.0	0.0	82.5
MDN+I3D	0.0	0.05	59.3
MDN+I3D	0.03	0.0	63.0
MDN+I3D	0.03	0.05(0.75)	80.3
MDN+R2.5D	0.03	0.05	85.4
I3D+R2.5D	0.03	0.75	94.7

Table 3. Evaluation results of different models on UCF101 and HMDB51. Fine tuning means using pre-trained model on kinetics or Sports-1M.

	Pre-trained dataset	UCF101	HMDB51	Params (MB)	Flops (GB)
		top1/top5	top1/top5		
I3D(fine tuning)	Kinetics	95.1/98.4	74.3/84.6	11.58	28.9
R2.5D(fine tuning)	Sports-1M	96.8/97.0	74.5/86.2	31.63	76.8
I3D+R2.5D	None	94.7/98.1	73.1/85.0	11.58	28.9
MDN+I3D	None	80.3/91.0	54.9/66.1	16.40	48.7
MDN+R2.5D	None	85.4/92.7	55.4/67.3	16.40	48.7

supervision information. When we set $(\gamma_1, \gamma_2) = (0.0, 0.05)$ with only top supervision applied, promotion is limited comparing to the condition of $(\gamma_1, \gamma_2) = (0.03, 0.0)$, which suggests that early supervision is necessary.

Theoretically, structural divergence becomes more and more prominent with the increase of network hierarchy. I3D is more sensitive to γ_2, due to inherent structural disparity with R2.5D and MDN. So if we set γ_2 small, correspondingly the supervision from top layer becomes too weak to work, while appropriate increase of γ_2 improves model performance significantly. For example, during fitting MDN to I3D, we first set γ_2 0.05, the mmd loss decrease quickly and get stranded at 2.47, then increase it up to 0.75, validate accuracy continues to raise, shown as Fig. 3(e) (f), this practice demonstrates the effectiveness of MMD loss.

4.3   Validation Comparison

We further evaluate proposed STL method on UCF101 and HMDB51, make comparisons of validate accuracy, parameters and flops with original I3D and R2.5D models in Table 3. The first two lines are introduced from [7, 8], which have been pre-trained on kinetics. Overall, MDN top1 accuracy on UCF101 is increased at least by 20% under extra two supervision stages compared with MDN* reported in Table 2. Transferring from R2.5D to I3D make I3D almost reach the same level with fine-tuning method. Similarly, results on HMDB51 are always inferior to that on UCF101, this is determined by the characteristic of datasets.

We have verified that a new network indeed benefits from the pre-trained model even though they are different in structures. The proposed method achieved cross-structure learning formally, and student models are expected to inherit latent attributions from supervisors, which is shown in Fig. 4(a) (b) by a supplement experiment: we use a small fraction (1k) videos of Kinetics to retrain MDN after soft transferring on UCF101. By 20k iterations, we extract the top features of 512 dimensions and then reduce the dimensions to 2 by t-SNE algorithm for visualizations, the features are clearly scattered in two-dimensional plane, while in sharp contrast with the case of training from scratch without any supervisions, features are still gathering together like a mess. This phenomenon suggests that MDN has reached the general rules of the supervisor for modeling on Kinetics.

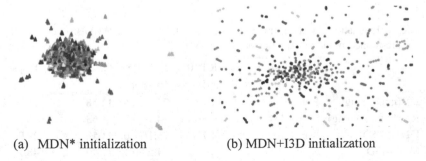

(a) MDN* initialization   (b) MDN+I3D initialization

Fig. 4. Features visualizations of 500 videos from Kinetics.

5 Conclusions

This paper explore the feasibility of mutually learning across architectures by introducing a STL method, and demonstrate the transferability of I3D and R2(+1)D models to the target model experimentally. We discuss the influence of structural differences on knowledge transferring qualitatively, find that stronger inherent action modeling ability allows a relaxed restrictions between model structures, simultaneously models with same or similar structures will have an easy convergence process. This result means that slapdash model may not benefits from those well designed and trained models, which is also compatible with common sense. Basically, STL brings positive effects to the models learning, which makes transferring learning across architecture workable and is very promising for mining of the essence of features for neural networks.

Acknowledgements. This work is highly supported by the open source models of I3D and R(2 +1)D, thanks for those pre-trained models, thanks tensorflow community for technical support. Additionally, the authors would like to thank the Beijing Advanced Innovation Center for Intelligent Robots and Systems for the support of project No. 2018IRS20.

References

1. Donahue, J., et al.: Long-term recurrent convolutional networks for visual recognition and description. IEEE Trans. Pattern Anal. Mach. Intell. **39**(4), 677–691 (2017)
2. HMDB. http://serre-lab.clps.brown.edu/resource/hmdb-a-large-human-motion-database/
3. UCF101. https://www.crcv.ucf.edu/data/UCF101.php
4. Feichtenhofer, C.: What have we learned from deep representations for action recognition? In: IEEE Conference on Computer Vision and Pattern Recognition (2018)
5. Huang, G., Liu, Z., van der Maaten, L., Weinberger, K.Q.: Densely connected convolutional networks. In: 2017 IEEE Conference on Computer Vision and Pattern Recognition (CVPR) (2017)
6. Szegedy, C., et al.: Going deeper with convolutions. In: 2015 IEEE Conference on Computer Vision and Pattern Recognition (CVPR), Boston, MA (2015)
7. Tran, D., Wang, H., Torresani, L., Ray, J., LeCun, Y., Paluri, M.: A closer look at spatiotemporal convolutions for action recognition. In: 2018 IEEE/CVF Conference on Computer Vision and Pattern Recognition, Salt Lake City, UT (2018)
8. Carreira, J., Zisserman, A.: Quo vadis, action recognition? A new model and the kinetics dataset. In: 2017 IEEE Conference on Computer Vision and Pattern Recognition (CVPR), Honolulu, HI (2017)
9. Wang, L., et al.: Temporal segment networks: towards good practices for deep action recognition. In: Leibe, B., Matas, J., Sebe, N., Welling, M. (eds.) ECCV 2016, Part VIII. LNCS, vol. 9912, pp. 20–36. Springer, Cham (2016). https://doi.org/10.1007/978-3-319-46484-8_2
10. Qiu, Z., et al.: Learning spatio-temporal representation with pseudo-3D residual networks. In: IEEE International Conference on Computer Vision (2017)
11. Zolfaghari, M., Singh, K., Brox, T.: ECO: efficient convolutional network for online video understanding. In: Ferrari, V., Hebert, M., Sminchisescu, C., Weiss, Y. (eds.) ECCV 2018, Part II. LNCS, vol. 11206, pp. 713–730. Springer, Cham (2018). https://doi.org/10.1007/978-3-030-01216-8_43

12. Karpathy, A., Toderici, G., Shetty, S., Leung, T., Sukthankar, R., Fei-Fei, L.: Large-scale video classification with convolutional neural networks. In: 2014 IEEE Conference on Computer Vision and Pattern Recognition, Columbus, OH (2014)
13. Simonyan, K., Zisserman, A.: Very deep convolutional networks for large-scale image recognition. arXiv preprint: arXiv:1409.1556 (2014)
14. Tran, D., Bourdev, L., Fergus, R., Torresani, L., Paluri, M.: Learning spatiotemporal features with 3D convolutional networks. In: 2015 IEEE International Conference on Computer Vision (ICCV), Santiago (2015)
15. He, K., Zhang, X., Ren, S., Sun, J.: Deep residual learning for image recognition. In: Proceedings of the IEEE Conference on Computer Vision and Pattern Recognition (CVPR) (2016)
16. Xie, S., Sun, C., Huang, J., Tu, Z., Murphy, K.: Rethinking spatiotemporal feature learning: speed-accuracy trade-offs in video classification. In: Ferrari, V., Hebert, M., Sminchisescu, C., Weiss, Y. (eds.) ECCV 2018, Part XV. LNCS, vol. 11219, pp. 318–335. Springer, Cham (2018). https://doi.org/10.1007/978-3-030-01267-0_19
17. Diba, A., Fayyaz, M., Sharma, V., et al.: Temporal 3D ConvNets: new architecture and transfer learning for video classification. arXiv preprint: arXiv:1711.08200 (2017)
18. Ioffe, S., Szegedy, C.: Batch normalization: accelerating deep network training by reducing internal covariate shift. arXiv preprint: arXiv:1502.03167 (2015)
19. Borgwardt, K.M., Arthur, G., Rasch, M., et al.: Integrating structured biological data by Kernel Maximum Mean Discrepancy. Bioinformatics 22(14), e49–e57 (2006)
20. Smola, A., Gretton, A., Song, L., Schölkopf, B.: A Hilbert space embedding for distributions. In: Hutter, M., Servedio, Rocco A., Takimoto, E. (eds.) ALT 2007. LNCS (LNAI), vol. 4754, pp. 13–31. Springer, Heidelberg (2007). https://doi.org/10.1007/978-3-540-75225-7_5
21. Long, M., Cao, Y., Wang, J., Jordan, M.I.: Learning transferable features with deep adaptation networks. In: ICML (2015)
22. Xie, S., Girshick, R., Dollár, P., Tu, Z., He, K.: Aggregated residual transformations for deep neural networks. In: Conference on Computer Vision and Pattern Recognition (CVPR) (2017)
23. Zagoruyko, S., Komodakis, N.: Wide residual networks. In: Proceedings of the British Machine Vision Conference (2016)
24. Hara, K., Kataoka, H., Satoh, Y.: Can spatiotemporal 3D CNNs retrace the history of 2D CNNs and ImageNet? In: Conference on Computer Vision and Pattern Recognition (CVPR) (2018)
25. Simonyan, K., Zisserman, A.: Two-stream convolutional networks for action recognition in videos. In: Advances in Neural Information Processing Systems (NIPS) (2014)
26. Pan, S.J., Yang, Q.: A survey on transfer learning. IEEE Trans. Knowl. Data Eng. 22(10), 1345–1359 (2010)
27. Long, M., Cao, Y., Cao, Z., Wang, J., Jordan, M.: Transferable representation learning with deep adaptation networks. IEEE Trans. Pattern Anal. Mach. Intell. 41(12), 3071–3085 (2018)
28. Donahue, J., Jia, Y., Vinyals, O., Hoffman, J., Zhang, N., et al.: DeCAF: a deep convolutional activation feature for generic visual recognition. In: ICML (2014)
29. Long, M., Wang, J., Jordan, M.I.: Unsupervised domain adaptation with residual transfer networks. In: NIPS (2016)
30. Yong, L., Wen, Y., Duan, L.-Y., Tao, D.: Transfer metric learning: algorithms, applications and outlooks. arXiv preprint: arXiv:1810.03944 (2018)
31. Geng, B., Tao, D., Xu, C.: DAML: domain adaptation metric learning. IEEE Trans. Image Process. 20(10), 2980–2989 (2011)

32. Xu, Y., Pan, S.J., Xiong, H., et al.: A unified framework for metric transfer learning. IEEE Trans. Knowl. Data Eng. **29**(6), 1158–1171 (2017)
33. Hu, J., Lu, J., Tan, Y.-P.: Deep transfer metric learning. In: IEEE Conference on Computer Vision and Pattern Recognition (CVPR) (2015)

Face Sketch Synthesis Based on Adaptive Similarity Regularization

Songze Tang[1](✉) and Mingyue Qiu[2]

[1] Department of Criminal Science and Technology,
Nanjing Forest Police College, Nanjing 210023, China
ts198708@163.com
[2] Department of Information and Technology, Nanjing Forest Police College,
Nanjing 210023, China

Abstract. Face sketch synthesis plays an important role in public security and digital entertainment. In this paper, we present a novel face sketch synthesis method via local similarity and nonlocal similarity regularization terms. The local similarity can overcome the technological bottlenecks of the patch representation scheme in traditional patch-based face sketch synthesis methods. It improves the quality of synthesized sketches by penalizing the dissimilar training patches (thus have very small weights or are discarded). In addition, taking the redundancy of image patches into account, a global nonlocal similarity regularization is employed to restrain the generation of the noise and maintain primitive facial features during the synthesized process. More robust synthesized results can be obtained. Extensive experiments on the public databases are carried out to validate the generality, effectiveness, and robustness of the proposed algorithm.

Keywords: Face sketch synthesis · Local similarity · Nonlocal similarity

1 Introduction

Generally, it is difficult to directly get the frontal face photo of the criminal suspect in the actual investigation. The suspect in the video surveillance may intentionally hide or elude the face. Then, the obtained limited information cannot be used to identify the suspect even using conventional state-of-the-art face recognition methods [1, 2]. However, we could get a sketch drawn by an artist according to the clues from the video surveillance. Then this sketch could be a substitute for identifying the suspect. Face sketch synthesis was born at the right moment, which refers to transform a face photo into a sketch. After several decades of development, it plays an important role in both digital entertainment and video surveillance-based law enforcement. Currently, face sketch synthesis methods can be roughly divided into two categories: image-based methods and patch-based methods.

© Springer Nature Switzerland AG 2019
Z. Cui et al. (Eds.): IScIDE 2019, LNCS 11935, pp. 226–237, 2019.
https://doi.org/10.1007/978-3-030-36189-1_19

1.1 Prior Works

The image-based methods treat the input photo image as a whole, and obtain the sketch image with some models. Wang et al. presented an automatic image-based approach for converting greyscale images to pencil sketches, in which strokes followed the image features [3]. Li et al. proposed a simple two-stage framework for face photo-sketch synthesis [4]. These methods cannot well mimic the sketch style and their reconstructed sketches are more like photos. Due to the breakthrough of deep learning [5], it has been widely applied in image style transfer [6, 7] and other image processing problems [8–10]. Face sketch synthesis can be seemed as one of the image transformation problems. Zhang et al. [11] proposed a fully convolutional network (FCN) to learn the end-to-end mapping from photos to sketches. Recently, generative adversarial network (GAN) [12] attracts growing attentions due to its ability to generate face images from natural images. To be capable of inferring photo-realistic natural images, Ledig et al. proposed a perceptual loss function which consists of an adversarial loss and a content loss [8]. Based on the perceptual similarity metrics, a class of loss functions were introduced to GAN to generate images [13]. Later, Wang et al. firstly introduced the GAN for face sketch synthesis and proposed the back projection strategy to further improve the quality of synthesized sketches [14]. Because the GAN-based methods learn a loss that adapts to the data, they can be applied to a multitude of tasks and achieved impressive results.

Three different categories of patch-based methods are often proposed, i.e., subspace learning based methods, Bayesian inference based methods, and sparse representation based methods.

The subspace learning framework mainly includes linear subspace-based methods and nonlinear subspace-based methods. The seminar work of linear face sketch synthesis was Eigen-Transformation method [15]. Considering the complexity of human faces, the linear relationship may not always be held, Liu et al. [16] proposed to characterize the nonlinear process of face sketch synthesis according to the locally linear embedding (LLE) idea [17]. Inspired by the image denoising method, Song et al. [18] proposed a spatial sketch synthesis approach, which explored K surrounding spatial neighbors for face sketch synthesis. Instead of online searching neighbors, Wang et al. randomly sampled some patches offline and used these patches to reconstruct the target sketch patch. This random sampling strategy greatly sped up the synthesis process. It was named Fast-RSLCR [19].

Sparse representation has been successfully applied to many other related inverse problems in image processing [20–22]. Chang et al. introduced sparse coding [22] to face sketch synthesis. Based on the assumption that the face photo patch takes the same sparse representation coefficients with the corresponding sketch patch, they learned two coupled dictionaries (sketch patch dictionary and photo patch dictionary) firstly. Then, a test photo patch was decomposed on the photo patch dictionary, weighted by the sparse representation coefficients. From the linear combination of the atoms of the sketch patch dictionary, the target sketch patch was produced at last.

Bayesian inference methods are proposed to explore the constraints between neighbor image patches. Gao et al. [23] employed embedded hidden Markov models to describe the dependency relationship between neighbor pixels. This kind method generated some facial deformations and cannot synthesize sketch patches not existing in the training dataset. To address this issue, Zhou et al. [24] introduced the linear combination into the MRF model (namely Markov weight field, MWF). Peng et al. [25] adaptively combined multiple representations to represent an image patch to improve the robustness.

1.2 Motivation and Contributions

As we know, face images have strong structural similarity in local regions (the mouth is not similar with the nose) [26, 27]. If photo patches and sketch patches are apart from their corresponding small regions, their similarity will come down. Thus, a local similarity constraint can be employed to search the best matching neighbors from the training samples and discard the patches that are far from the input patch. In addition, inspired by nonlocal means denoising method [28], a nonlocal similarity regularization is also introduced to improve the quality of sketch synthesis further. In brief, we propose a simple yet effective algorithm for neighbor selection using local and nonlocal similarity regularization, which provides great favor to improve the matching accuracy of nearest neighbor and synthesized performance. The flowchart of the proposed method is presented in Fig. 1.

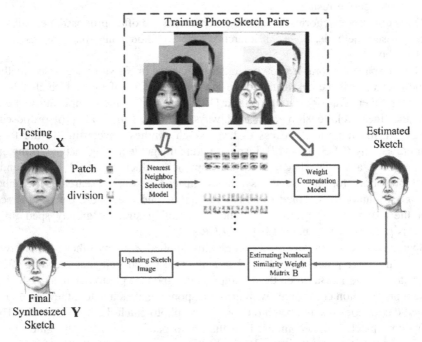

Fig. 1. The graphical outline of the proposed face sketch synthesis method.

2 Face Sketch Synthesis Based on Adaptive Similarity Regularization

In the face sketch synthesis, considering a training set with M face photo-sketch pairs, we divide each training image into N small overlapping patches. Let $\mathbf{X}$ and $\mathbf{Y}$ be an input test photo and the final synthesized sketch image, which are divided into N small overlapping patches $\{\mathbf{x}_1, \mathbf{x}_2, \ldots, \mathbf{x}_N\}$ and $\{\mathbf{y}_1, \mathbf{y}_2, \ldots, \mathbf{y}_N\}$ by the same way, respectively. It should be noted that each photo and the corresponding sketch image are the same size. Thus, each photo patch $\{\mathbf{x}_i\}_{i=1}^N$ and the corresponding sketch patch $\{\mathbf{y}_i\}_{i=1}^N$ has the same size and corresponds to the same position in $\mathbf{X}$ and $\mathbf{Y}$, respectively. For each test photo patch $\mathbf{x}_i$, we can reconstruct it using K nearest photo patches $\mathbf{P}_{i,K} = \{\mathbf{p}_{i,k}\}_{k=1}^K$ from the training dataset with corresponding weight vector $\mathbf{w}_{i,K} = \{w_{i,k}\}_{k=1}^K$. Thus, the corresponding sketch image patch $\mathbf{y}_i$ can be synthesized by the corresponding K nearest sketch patches $\{\mathbf{s}_{i,k}\}_{k=1}^K$ with the above obtained weight vector $\mathbf{w}_{i,K}$. After all target sketch patches are generated, they are assembled into a whole sketch image $\mathbf{Y}$ by averaging overlapping pixel intensities.

2.1 Adaptive Regularization by Local Similarity

The LLE method calculated the linear combination coefficients vectors without appropriate constraints. Biased solutions can be produced when the dimension of the patch is smaller than the size of the training set. Due to the large quantization errors, similar test patches might be very different, as illustrated in Fig. 2(a). Besides, the LLE process ignores the relationships between different training patches. As shown in Fig. 2 (b) in Fast-RSLCR, each test patch was more accurately represented by capturing the correlations between different training patches in Fast-RSLCR method. However, not all patches play positive roles in the final synthesized result. We introduce the local similarity regularization term to the neighbor selection model, thus leading to (i) a stable solution to the least squares problem and (ii) discriminant synthesized results (see Fig. 2(c)).

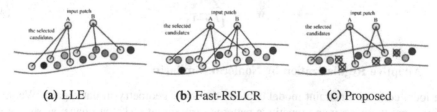

| (a) LLE | (b) Fast-RSLCR | (c) Proposed |

Fig. 2. Comparison between LLE, Fast-RSLCR and the proposed.

In our local similarity constraint model, we only consider the most relevant patches in the training set as effective samples. For each patch $\mathbf{x}_i$ in the test photo, the optimal weights are obtained by minimizing the local similarity regularized reconstruction error:

$$argmin_{\mathbf{w}_{i,K}} \|\mathbf{x}_i - \mathbf{P}_{i,K}\mathbf{w}_{i,K}\|_2^2 + \lambda_1 \|\mathbf{D}_{i,K}\mathbf{w}_{i,K}\|_2^2, \qquad s.t.\ \mathbf{1}^T\mathbf{w}_{i,K} = 1 \tag{1}$$

Where $\mathbf{D}_{i,K}=\begin{bmatrix} d_{i,1} & & & 0 \\ & d_{i,2} & & \\ & & \ddots & \\ 0 & & & d_{i,K} \end{bmatrix}$, $d_{i,j}=\exp\left(\dfrac{\|\mathbf{x}_i-\mathbf{p}_j\|_2^2}{\sigma}\right)$ represents the entries of

the exponential locality adaptor and σ is a positive value. K sampled training photo patches constitute the marix $\mathbf{P}_{i,K}$. $\mathbf{w}_{i,K}$ is the weight representation, λ_1 balances the reconstruction error and the locality constraint. To preserve the data structure, the exponential function is considered to improve the representation. Since $d_{i,j}$ grows exponentially with $\|\mathbf{x}_i - \mathbf{p}_j\|_2^2 \big/ \sigma$, the exponential locality adaptor is very large when $\mathbf{x}_i$ and $\mathbf{p}_j$ are far apart. This property is useful when we want to stress the importance of data locality. (Since $d_{i,j}$ is the weight of $w_{i,j}$ in (6), a large value of $d_{i,j}$ causes $w_{i,j}$ to be small.)

To determine the solution $\mathbf{w}_{i,K}$ in (6), we consider the Lagrange function $L(\mathbf{w}_{i,K}, \beta)$, which is defined as

$$L(\mathbf{w}_{i,K}, \lambda, \beta)=\|\mathbf{x}_i - \mathbf{P}_{i,K}\mathbf{w}_{i,K}\|_2^2 + \lambda_1 \|\mathbf{D}_{i,K}\mathbf{w}_{i,K}\|_2^2 + \beta(\mathbf{w}_{i,K}\mathbf{1}^T-1) \tag{2}$$

where β is a lagrange multiplier. (7) can be reformulated as

$$
\begin{aligned}
L(\mathbf{w}_i, \lambda, \beta) &= \|\mathbf{x}_i - \mathbf{P}_{i,K}\mathbf{w}_{i,K}\|_2^2 + \lambda_1 \|\mathbf{D}_{i,K}\mathbf{w}_{i,K}\|_2^2 + \beta(\mathbf{1}^T\mathbf{w}_{i,K} - 1) \\
&= \|(\mathbf{x}_i\mathbf{1}^T - \mathbf{P}_{i,K})\mathbf{w}_{i,K}\|_2^2 + \lambda_1 \|\mathbf{D}_{i,K}\mathbf{w}_{i,K}\|_2^2 + \beta(\mathbf{1}^T\mathbf{w}_{i,K} - 1) \\
&= \mathbf{w}_{i,K}^T \mathbf{Z}\mathbf{w}_{i,K} + \beta(\mathbf{1}^T\mathbf{w}_{i,K} - 1)
\end{aligned}
\tag{3}
$$

where $\mathbf{1}$ is a column vector of ones. $\mathbf{Z} = (\mathbf{x}_i\mathbf{1}^T - \mathbf{P}_{i,K})^T(\mathbf{x}_i\mathbf{1}^T - \mathbf{P}_{i,K}) + \lambda_1 \mathbf{D}_{i,K}^T\mathbf{D}_{i,K}$. We can get the final solution

$$\mathbf{w}_{i,K} = \frac{\mathbf{Z}^{-1}\mathbf{1}}{\mathbf{1}^T\mathbf{Z}^{-1}\mathbf{1}} \tag{4}$$

2.2   Adaptive Regularization by Nonlocal Similarity

The local context constraint model exploits the local geometry in data space. We also noticed that there are many repetitive patterns throughout a sketch image, as shown in Fig. 3. Therefore, we extend local similarity constraint to find the nonlocal self-similarity. Generally, for each extracted patch $\mathbf{y}_i$ from the estimated sketch image $\mathbf{Y}$, we search its L similar patches $\{\mathbf{y}_i^l\}_{l=1}^L$ in $\mathbf{Y}$. Then, $\mathbf{y}_i$ can be represented as a linear combination of the L similar patches

Fig. 3. Nonlocal similarity in the sketch images.

$$\mathbf{y}_i = \sum_{l=1}^{L} \mathbf{y}_i^l b_i^l \tag{5}$$

The nonlocal similarity weight b_i^l is inversely proportional to the distance between patches $\mathbf{y}_i$ and $\mathbf{y}_i^l$ and it is set as

$$b_i^l = \exp\left(-\|\mathbf{y}_i - \mathbf{y}_i^l\|_2^2 \big/ h\right) \tag{6}$$

where h is a pre-determined control factor of the weight. Let $\mathbf{b}_i$ be the column vector containing all the weights b_i^l and $\boldsymbol{\beta}_i$ be the column vector containing all $\mathbf{y}_i^l$. (14) can be rewritten as:

$$\mathbf{y}_i = \mathbf{b}_i^T \boldsymbol{\beta}_i \tag{7}$$

By incorporating the nonlocal similarity regularization term into patch aggregation, we have:

$$\mathbf{Y}^* = \arg\min_{\mathbf{Y}} \left\{ \sum_i \|\mathbf{R}_i\mathbf{Y} - \mathbf{y}_i\|_2^2 + \lambda_2 \sum_i \|\mathbf{y}_i - \mathbf{b}_i^T\boldsymbol{\beta}_i\|_2^2 \right\} \tag{8}$$

where $\mathbf{R}_i$ is an operator which extracts a patch from an image and represents the patch as a column vector by lexicographic ordering.

$$\mathbf{Y}^* = \arg\min_{\mathbf{Y}} \left\{ \sum_i \|\mathbf{R}_i\mathbf{Y} - \mathbf{y}_i\|_2^2 + \lambda_2 \sum_i \|(\mathbf{I} - \mathbf{B})\mathbf{Y}\|_2^2 \right\} \tag{9}$$

where $\mathbf{I}$ is the identity matrix and

$$\mathbf{B}(i,j) = \begin{cases} b_i^l, & \text{if } \mathbf{y}_i^l \text{ is an elment of } \boldsymbol{\beta}_i, \ b_i^l \in \mathbf{b}_i \\ 0, & \text{otherwise} \end{cases} \tag{10}$$

(17) is a convex optimization problem, and the recovered image is updated by minimizing the objective function using the equation

$$\mathbf{Y}^* = \left\{ \sum_i \mathbf{R}_i^T\mathbf{R}_i + \lambda_2(\mathbf{I} - \mathbf{B})^{-1}(\mathbf{I} - \mathbf{B}) \right\}^{-1} \left(\sum_i \mathbf{R}_i^T\mathbf{y}_i \right) \tag{11}$$

3 Experimental Results and Analysis

In this section, a lot of experiments are performed to evaluate the generality, effectiveness, and robustness of the proposed method for face sketch synthesis. We also compare our proposed method with some recent state-of-the-art methods. Since face recognition is the ultimate goal, feature matching is very important. Feature similarity index metric (FSIM) [29] is adopted as the evaluation criterion to objectively assess the quality of synthesized sketches under different experimental settings and methods. We validate our method on the Chinese University of Hong Kong (CUHK) face sketch database (CUFS) and CUHK face sketch FERET database (CUFSF). CUFS database contains the CUHK student (CUHKs) database, the AR database, and the XM2VTS database. All face images are cropped into the size of 250×200.

3.1 Parameter Analysis

There are some parameters (i.e. the number of selected neighbors and the trade-off parameters λ_1 and λ_2) in the proposed method.

(1) The influence of different regularization parameters: Our algorithm has two free regularization parameters λ_1 and λ_2, which balance the different contribution of regularization terms. We perform a set of parametric experiments to validate the effectiveness of the proposed regularization priors. We tune the local similarity parameter λ_1 (from 0 to 0.3 with step 0.02) and the nonlocal similarity parameter λ_2

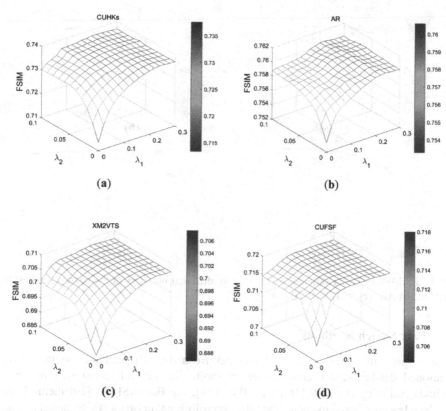

Fig. 4. Effects of regularization parameters λ_1 and λ_2 on different databases. (a) CUHKs; (b) AR; (c) XM2VTS; (d) CUFSF.

(from 0 to 0.1 with step 0.01) carefully. Figure 4 shows the surfaces of FSIM variations. It can be clearly observed that the synthesized performance is stable in terms of FSIM with regards to $\lambda_1 \in [0.18, 0.26]$ and $\lambda_2 \in [0.05, 0.07]$.

(2) The influence of the nearest neighbor number: Similar to other LLE-related methods, the synthesized performance by our method is highly correlated with the number of the nearest neighbors. We conduct some experiments on the mentioned four databases by changing the numbers. The curved lines of the FSIM values varying with the numbers of the nearest neighbors are plotted in Fig. 5. When the value of K is equal to the number of training photos, our proposed method can't get the best performance. Larger K doesn't mean better performance. This verifies the significance of selecting several similar patches for robust representation. As shown in Fig. 5, the values of FSIM increase stably with the increase of the nearest neighbors number. Nonetheless, when K reaches one suitable value, the performance of our proposed method will be steady. We also observe that the performance decreases stably with the increase the value of K (larger than 80% of the number of training photos). In view of this, to achieve the optimal or nearly optimal performance, we recommend that K is 80% of the number of training photos.

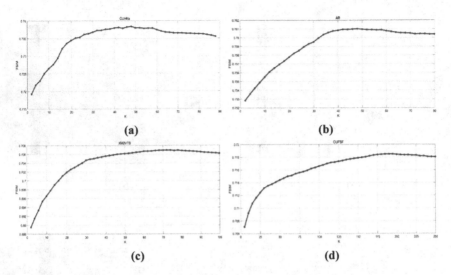

Fig. 5. FSIM score of the different databases with different numbers of the nearest patches. (**a**) CUHKs; (**b**) AR; (**c**) XM2VTS; (**d**) CUFSF.

3.2   Face Sketch Synthesis

Figure 6 presents some synthesized sketches from different methods on the above-mentioned databases. We compare our method with several recent state-of-the-art methods including the LLE [16], the MWF [24], the Fast-RSLCR [19] method, and two deep learning based methods, the fully convolutional network (FCN) based method [11] and GAN based method [14]. Generally speaking, the proposed method could generate much more details in comparison to the other five popular methods. This is because the selected candidate patches in our proposed methods are effective. In addition, the nonlocal similarity regularization also plays an important role in maintaining primitive facial features.

We also provide the quantitative comparisons of different methods in terms of average FSIM in Table 1. It can be seen that the proposed method obtains the good performance for all the test databases.

Table 1. Average FSIM score of different methods on different databases

Methods	LLE	MWF	FCN	GAN	Fast-RSLCR	Proposed
CUFS	0.7180	0.7294	0.6936	0.6899	0.7142	*0.7352*
CUFSF	0.7043	0.7029	0.6624	0.6814	0.6775	*0.7181*

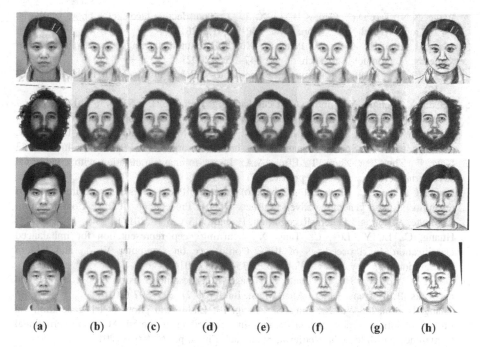

(a)   (b)   (c)   (d)   (e)   (f)   (g)   (h)

Fig. 6. Synthesized sketches on the CUFS database by different methods. (a) Photo; (b) LLE; (c) MWF; (d)FCN (e) GAN; (f) Fast-RSLCR; (g) Proposed; (h) Ground-Truth.

4   Conclusions

In this paper, we present a novel face sketch synthesis method via two regularization terms. By incorporating the local similarity regularization term into the neighbor selection, we can select the most relevant patch samples to reconstruct the face sketches versions of the input photos. The discriminant face sketches are generated with detailed features. Moreover, the global nonlocal similarity regularization term is employed to maintain primitive facial features. The thorough experimental results show the effectiveness of the proposed method.

Acknowledgments. This research was supported in part by the National Natural Science Foundation of China under Grant 61702269, and Grant 61671339, in part by the Natural Science Foundation of Jiangsu Province under Grant BK20171074. The Fundamental Research Funds for the Central Universities at Nanjing Forest Police College under Grant No. LGZD201702.

References

1. Uhl, R.G. da Vitoria Lobo Jr., N.: A framework for recognizing a facial image from a police sketch. In: Proceedings of IEEE Conference on Computer Vision and Pattern Recognition, San Francisco, CA, USA, pp. 586–593 (1996)

2. Wang, N.N., Tao, D.C., Gao, X.B., Li, X., Li, J.: Transductive face sketch photo synthesis. IEEE Trans. Neural Netw. Learn. Syst. **24**(9), 1–13 (2013)
3. Wang, J., Bao, H., Zhou, W., Peng, Q., Xu, Y.Q.: Automatic image-based pencil sketch rendering. J. Comput. Sci. Technol. **17**, 347–355 (2002)
4. Li, X., Cao, X.: A simple framework for face photo-sketch synthesis. Math. Probl. Eng. **2012** (2012)
5. LeCun, Y., Bengio, Y., Hinton, G.: Deep learning. Nature **521**(7553), 436 (2015)
6. Johnson, J., Alahi, A., Fei-Fei, L.: Perceptual losses for real-time style transfer and super-resolution. In: Leibe, B., Matas, J., Sebe, N., Welling, M. (eds.) ECCV 2016. LNCS, vol. 9906, pp. 694–711. Springer, Cham (2016). https://doi.org/10.1007/978-3-319-46475-6_43
7. Isola, P., Zhu, J.Y., Zhou, T., Efros, A.A.: Image-to-image translation with conditional adversarial networks. arXiv preprint (2017)
8. Ledig, C., Theis, L., Huszár, F., Caballero, J., et al.: Photo-realistic single image super-resolution using a generative adversarial network. In: Proceedings of IEEE Conference on Computer Vision and Pattern Recognition, Hawaii, USA, p. 4 (2017)
9. Huang, C., Li, Y., Loy, C., Tang, X.: Learning deep representation for imbalanced classification. In: Proceedings of IEEE Conference on Computer Vision and Pattern Recognition, Las Vegas, NV, USA, pp. 5375–5384 (2016)
10. Dong, C., Loy, C., He, K., Tang, X.: Image super-resolution using deep convolutional networks. IEEE Trans. Pattern Anal. Mach. Intell. **38**(2), 295–307 (2016)
11. Zhang, L., Lin, L., Wu, X., Ding, S., Zhang, L.: End-to-end photo-sketch generation via fully convolutional representation learning. In: Proceedings of the 5th ACM on International Conference on Multimedia Retrieval, Shanghai, China, pp. 627–634 (2015)
12. Goodfellow, I., et al.: Generative adversarial nets. In: Advances in Neural Information Processing Systems, Montreal, Canada, pp. 2672–2680 (2014)
13. Dosovitskiy, A., Brox, T.: Generating images with perceptual similarity metrics based on deep networks. In: Advances in Neural Information Processing Systems, Barcelona, Spain, pp. 658–666 (2016)
14. Wang, N.N., Zha, W., Li, J., Gao, X.B.: Back projection: an effective: postprocessing method for GAN-based face sketch synthesis. Pattern Recognit. Lett. **107**, 59–65 (2018)
15. Tang, X., Wang, X.: Face photo recognition using sketch. In: Proceedings of IEEE International Conference on Image Processing, New York, USA, pp. 257–260 (2002)
16. Liu, Q., Tang, X., Jin, H., Lu, H., Ma, S.: A nonlinear approach for face sketch synthesis and recognition. In: Proceedings of IEEE Conference on Computer Vision and Pattern Recognition, San Diego, USA, pp. 1005–1010 (2005)
17. Roweis, S.T., Saul, L.: Nonlinear dimensionality reduction by locally linear embedding. Science **290**(5500), 2323–2326 (2000)
18. Song, Y., Bao, L., Yang, Q., Yang, M.-H.: Real-time exemplar-based face sketch synthesis. In: Fleet, D., Pajdla, T., Schiele, B., Tuytelaars, T. (eds.) ECCV 2014. LNCS, vol. 8694, pp. 800–813. Springer, Cham (2014). https://doi.org/10.1007/978-3-319-10599-4_51
19. Wang, N.N., Gao, X., Li, J.: Random sampling for fast face sketch synthesis. Pattern Recognit. **76**, 215–227 (2018)
20. Tang, S., Xiao, L., Liu, P., Huang, L., Zhou, N., Xu, Y.: Pansharpening via sparse regression. Opt. Eng. **56**, 093105-1–093105-13 (2017)
21. Wright, J., Ma, Y., Mairal, J., Sapiro, G., Huang, T., Yan, S.: Sparse representation for computer vision pattern recognition. Proc. IEEE **98**(6), 1031–1044 (2010)
22. Chang, L., Zhou, M., Han, Y., Deng, X.: Face sketch synthesis via sparse representation. In: Proceedings of International Conference on Pattern Recognition, Istanbul, Turkey, pp. 2146–2149 (2010)

23. Gao, X.B., Zhong, J., Li, J., Tian, C.: Face sketch synthesis algorithm using E-HMM and selective ensemble. IEEE Trans. Circuits Syst. Video Technol. **18**(4), 487–496 (2008)
24. Zhou, H., Kuang, Z., Wong, K.: Markov weight fields for face sketch synthesis. In: Proceedings of IEEE Conference on Computer Vision and Pattern Recognition, Providence, USA, pp. 1091–1097 (2012)
25. Peng, C.L., Gao, X.B., Wang, N.N., Tao, D.C., Li, X., Li, J.: Multiple representations-based face sketch-photo synthesis. IEEE Trans. Neural Netw. Learn. Syst. **27**(11), 2201–2215 (2016)
26. Li, C., Zhao, S., Xiao, K., Wang, Y.: Face recognition based on the combination of enhanced local texture feature and DBN under complex illumination conditions. J. Inf. Process. Syst. **14**, 191–204 (2018)
27. Muntasa, A.: Homogeneous and non-homogeneous polynomial based eigenspaces to extract the features on facial images. J. Inf. Process. Syst. **12**, 591–611 (2016)
28. Buades, A., Coll, B., Morel, J.M.: A non-local algorithm for image denoising. In: Proceedings of IEEE Conference on Computer Vision and Pattern Recognition, San Diego, USA, pp. 60–65 (2005)
29. Zhang, L., Zhang, L., Mou, X., Zhang, D.: FSIM: a feature similarity index for image quality assessment. IEEE Trans. Image Process. **20**(8), 2378–2386 (2011)

Three-Dimensional Coronary Artery Centerline Extraction and Cross Sectional Lumen Quantification from CT Angiography Images

Hengfei Cui[1,2]([✉]), Yong Xia[1,2], and Yanning Zhang[1]

[1] National Engineering Laboratory for Integrated Aero-Space-Ground-Ocean Big Data Application Technology, School of Computer Science, Northwestern Polytechnical University, Xi'an 710072, China
hfcui@nwpu.edu.cn
[2] Centre for Multidisciplinary Convergence Computing (CMCC), School of Computer Science, Northwestern Polytechnical University, Xi'an 710072, China

Abstract. Automatic centerline extraction based on 3D coronary artery segmentation results is a very important step before quantitative evaluation of intravascular lumen cross-section. In this paper, a method based on the combination of fast marching and gradient vector flow (GVF) is proposed to extract the centerline of the complete coronary artery tree in 3D angiographic images. With the centerline of blood vessel, we propose an automatic method to extract the cross-section of blood vessel lumen. This method calculates the tangent vector based on the two adjacent centerline points before and after the midline point, and then calculates the cross-sectional equation through the centerline point, and then obtains the cross-sectional contour of the cross-section and the surface mesh of blood vessel. The new method is designed to extract the cross-section of 3D intravascular lumen in real physical coordinates, which avoids the traditional interpolation processing in pixel coordinates and improves the accuracy of cross-section extraction. Given the accuracy and efficiency, the proposed coronary artery lumen area measurement algorithm can facilitate quantitative assessment of the anatomic severity of coronary stenosis.

Keywords: Medical imaging · Coronary artery segmentation · Centerline · Cross sectional lumen area · Computer-aided diagnosis

1 Introduction

In the clinical diagnosis of cardiovascular diseases, coronary artery stenosis is one of the main causes of coronary heart diseases. In modern medical image processing, one of the purposes of three-dimensional angiography image processing

© Springer Nature Switzerland AG 2019
Z. Cui et al. (Eds.): IScIDE 2019, LNCS 11935, pp. 238–248, 2019.
https://doi.org/10.1007/978-3-030-36189-1_20

is to automatically and accurately reconstruct the complete coronary artery to qualitatively or quantitatively evaluate coronary artery occlusion [1]. In clinical practice, because of the historical popularity of projective invasive angiography, vessel diameter is usually used to quantify coronary artery stenosis. Generally speaking, the traditional assessment of the severity of coronary artery occlusion is based on the ratio of the smallest diameter of the lesion to the proximal normal diameter of the artery (i.e., diameter stenosis). Unfortunately, even for well-trained imaging experts, accurate measurement of the minimum diameter of blood vessels in the lesion area is a challenging task, because it largely depends on the selection of projection plane and reference points [2].

Compared with the traditional diameter measurement, the measurement of intravascular lumen cross-sectional area is directly related to the hemodynamic characteristics of blood vessels, so it is considered more important and applicable in clinical practice than the measurement of blood vessel diameter [3]. Quantitative analysis of cross-sectional profiles based on three-dimensional centerline of coronary artery will help to assess the severity of coronary artery stenosis intuitively and accurately. Therefore, automatic center line extraction based on three-dimensional coronary artery segmentation results is a very important step before quantitative evaluation of intravascular lumen cross-section, and plays a very important role in clinical practice [4].

Coronary artery centerline extraction from three-dimensional Computed Tomography Angiography (CTA) images has contributed a lot to the literature [5–10]. For example, Van Uitert et al. [11] used the distance transformation of the vessel boundary to travel rapidly from the focal point of the vessel center. The center line is extracted by tracing back from the detected endpoint to the source point. Although this method has many advantages over previous methods, it also has some limitations. Firstly, there is no theoretical basis to guarantee the path centrality for the proposed velocity image of propagation wave. In addition, the experimental verification shows that Van's method can guarantee the tracking process along the gradient descent direction of the propagation time to the graph only when the step size is small (0.01 in this case), which makes the whole center line branch tracking process very time-consuming and is not advisable for large-scale CTA images.

The traditional measurement of the cross-sectional area of intravascular lumen is mostly based on two-dimensional Computed Tomography (CT) images [12]. Some advanced medical image processing software is used to segment the tubular structure first, then extract the midline and cut the cross-sectional along the discrete midline. However, this method is based on the pixel level, so it cannot guarantee that every cross-section can be successfully extracted. For those who cannot extract the complete cross-section, it is necessary to interpolate the pre- and post-processing interface information to get the predicted cross-section. In this step, a large error may occur. The method of extracting 3D cross-section proposed in this paper is completely automated and does not depend on any manual intervention and parameters. The method of extracting centerline is based on our previous work. The innovation of this paper is that the extraction of the cross-section of the intravascular lumen is based on the three-dimensional physical coordinates. Given the coordinates of three discrete midline points, tangent vectors along the midline point are calculated, and then the equation of cross-section through the midline point is calculated. The equation intersects with the grid of the vascular surface. The contour generated is the boundary of the cross-section.

The main purpose of this study is to propose an automatic method based on the combination of fast marching method and gradient vector flow to extract the centerline of blood vessels from three-dimensional CTA images, so as to help medical image processing and visualization. Based on the obtained centerline of coronary artery, we develop an accurate and repeatable algorithm for measuring the cross-sectional lumen area. In addition, we propose an improved percentage of stenosis to measure the severity of coronary artery stenosis and to facilitate the clinical diagnosis and treatment planning of coronary heart disease. The rest of this paper is organized as follows. In Sect. 2, the methodologies about centerline computation and lumen area measurement are presented. In Sect. 3, the computed centerline and the cross sectional lumen area results are given. Finally, conclusions are presented in Sect. 4.

2   Methodologies

2.1   Centerline Extraction

Edge Map. The edge map [13] $f(\mathbf{x})$ of the segmentation $I(\mathbf{x})$ can be calculated by

$$f(\mathbf{x}) = |\nabla[\mathbf{G}_\sigma(\mathbf{x}) * \mathbf{I}(\mathbf{x})]| \tag{1}$$

where $G_\sigma(\mathbf{x})$ is the Gaussian function. ∇f usually point toward the edges and have large magnitudes at the edge locations.

Gradient Vector Flow. The GVF field $V(\mathbf{x}) = [u(\mathbf{x}), v(\mathbf{x}), w(\mathbf{x})]$ is defined as [14]

$$E_{GVF}(V) = \iiint \mu(|\nabla u(\mathbf{x})|^2 + |\nabla v(\mathbf{x})|^2 + |\nabla w(\mathbf{x})|^2)$$
$$+|\nabla f(\mathbf{x})|^2|V(\mathbf{x}) - \nabla f(\mathbf{x})|^2 d\mathbf{x} \tag{2}$$

where μ is the smoothness regularization parameter. The GVF field formulation is usually solved by

$$V_t = \mu\nabla^2 V(t) - (V - \nabla f)|\nabla f|^2 \tag{3}$$

with V_t the partial derivative of V.

Speed Image. We first compute the gradient of GVF field $\nabla V(\mathbf{x})$. The centerlines and the edge points can be discerned by checking $|\nabla V(\mathbf{x})|$. The edge indicator function [15] is defined to remove the edges after separating the smooth area

$$g(\mathbf{x}) = \frac{1}{1 + f(\mathbf{x})} \tag{4}$$

The centerline strength function is therefore defined as

$$k(\mathbf{x}) = g(\mathbf{x}) * |\nabla V(\mathbf{x})| \tag{5}$$

then normalizing it to $[0, 1]$,

$$\bar{k}(\mathbf{x}) = \left(\frac{k(\mathbf{x}) - \min(k)}{\max(k) - \min(k)} \right)^{\gamma} \tag{6}$$

where γ is the field strength in $[0, 1]$. Without loss of generality, γ is chosen as 1 in this work. Therefore, a new speed image based on $\bar{k}(\mathbf{x})$ for the fast marching method [16]

$$F = e^{\beta \bar{k}} \tag{7}$$

where β is a speed parameter defined as $\beta = 1/\tau$ and τ is the parameter.

Tracking. Based on the proposed centerline strength function $\bar{k}(\mathbf{x})$, the source point P_s is chosen as the point with maximal value in $\bar{k}(x)$:

$$P_s = \arg \max_x \bar{k}(\mathbf{x}) \tag{8}$$

Then a wave front is propagated using the new speed image F from the source P_s, resulting in the time arriving map T. The start point m_0 is selected as the point with maximum T value. The first centerline branch is tracked by solving the ordinary differential equation

$$\frac{dS}{dt} = -\frac{\nabla T}{|\nabla T|}, \quad S(0) = m_0 \tag{9}$$

where $S(t)$ is the centerline and m_0 is the furthest geodesic point. The two-stage Gauss-Legendre method (order of 4) [17] is applied to achieve higher accuracy. Once the first branch is determined, it will be used as the new wave front source points and all the remaining branches can be computed following the same tracking process.

2.2   Lumen Measurement

Since the 3D coronary artery surface mesh is obtained in voxel coordinate (space), all the mesh results need to be transformed to world coordinate by multiplying the resolution of the original CTA images. Then the centerline results are automatically obtained in world coordinate. Furthermore, consider one centerline point which is defined as $\mathbf{P_0} = (x_0, y_0, z_0)$. The tangent vector to the centerline at point $\mathbf{P_0}$ can be obtained by central finite difference approximation using forward and backward adjacent centerline points, which is denoted as $\mathbf{N} = (a, b, c)$. Assume that $\mathbf{P} = (x, y, z)$ is any point in the normal plane perpendicular to the centerline. It must satisfy

$$a(x - x_0) + b(y - y_0) + c(z - z_0) = 0 \tag{10}$$

The intersection of the vertical plane and the surface mesh of the coronary artery will be a series of vertices, forming a disc-shaped polygon. However, in the practice of graphics, it is a challenge to obtain accurate and complete cross-sectional lumen contours based on voxel shape representation. In order to improve the geometric accuracy of the subvoxel level, the original three-dimensional meshes are generated. Based on the extracted centerline, the cross section can be cut at each midline point on the vascular surface gridding data, instead of on the voxel data, and the polygon contour composed of a set of boundary vertices can be generated. When cutting a set of vertices, V_i $(i = 1, 2, ..., n)$, at the skeleton point $\mathbf{P_0}$, the geometric centroid of the polygon formed by these vertices can be calculated and represented by C. By connecting each pair of adjacent vertices and each vertex v_i to the center of mass C, a set of n triangles is formed to decompose the polygon. For each two-dimensional triangle, T_i, its area can be calculated according to the original resolution of the original CT image. Therefore, the lumen area of the three-dimensional cross-section of the normal plane can be approximated to

$$\text{Area}(S) = \sum_{i=1}^{n} \text{Area}(T_i) \tag{11}$$

3   Results

The presented centerline extraction and 3D lumen area measurement framework were applied over three different real-patient CTA datasets (ccta 1, 14 and 22) obtained from the National Heart Center Singapore. The details of the dataset size and resolution are summarized in Table 1. All the experiments were implemented in Matlab 2014b on a 16 GB RAM Windows laptop.

Table 1. Dataset size and resolution.

	Size	Resolution (mm^3)
ccta 1	$320 \times 320 \times 460$	$0.439 \times 0.439 \times 0.25$
ccta 14	$400 \times 400 \times 470$	$0.31 \times 0.31 \times 0.25$
ccta 22	$400 \times 400 \times 426$	$0.319 \times 0.319 \times 0.3$

Table 2. Coordinates of the nine polygon vertices and the geometric centroid.

	x	y	z
V1	338.0424	159.8646	185.9575
V2	338.3765	159.8514	185.6234
V3	338.6155	159.4158	185.3844
V4	338.7952	158.3011	185.2047
V5	338.4915	157.7357	185.5084
V6	337.6983	157.5017	186.3016
V7	337.2956	158.0384	186.7043
V8	336.8149	159.2199	187.1850
V9	336.8648	159.2920	187.1351
C	337.8884	158.8023	186.1116

In Fig. 1 we show the comparison of 3D centerline results by using two different algorithms, i.e., topological thinning method and the proposed method. Figure 1(a) gives the 3D segmentation result of Left Coronary Artery (LCA) tree from dataset ccta 1. Figure 1(b) shows the obtained centerline result by using topological thinning method. The centerline obtained by thinning method is found to be centrally located, when overlaying the centerline over 3D rendering of LCA in Fig. 1(c). In Fig. 1(d) we display the centerline result obtained by the proposed algorithm. The resulting centerline tends to be continuous and much smoother, and most of the zigzags and extraneous branches are eliminated.

In this section, cross sections perpendicular to the centerline were cut along a specific branch of the obtained left coronary artery tree at six landmark centerline points L1, L2, L3, L4, L5 and L6 (see Fig. 2(a)). The original three-dimensional binary data set on the surface of coronary artery is pre-meshed, while maintaining the same parameters as the three-dimensional geometry. Figure 2(b) depicts the corresponding boundary profile of each vertical plane intersecting with the surface mesh of the coronary artery. At the landmark L1, the contour obtained is a polygon with 9 vertices, v1 to v9. The geometric centroid can be calculated and expressed as C. Figure 3 shows the obtained polygon. Table 2 gives the three-dimensional coordinates of the nine vertices and the centroid of C in world coordinate. In Table 3, we show the boundary information of the cross-section contour. For each discrete centerline point, we calculated the number of polygonal vertices formed when the normal plane intersected with the surface mesh of the original coronary artery, and then approximated the corresponding cross-sectional lumen area (in mm^2). Then the cross-sectional area of the vessel is plotted according to the position of the centerline. The calculation results of cross-sectional contours perpendicular to the centerline can be represented by a polyline graph.

The corresponding cross-sectional area and average diameter are shown in Fig. 5. In Fig. 5(a), the distance between each landmark and the initial point is

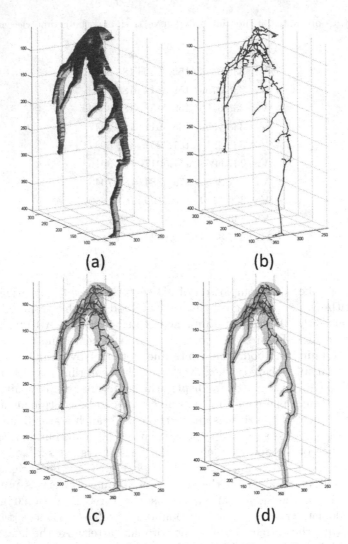

Fig. 1. Left coronary artery tree: (a) 3D rendering. (b) 3D centerline obtained by using thinning method. (c) Overlay of the 3D centerline over the surface. (d) Centerline obtained by the proposed method.

measured horizontally, and the corresponding cross-sectional lumen area at each landmark is measured vertically. From L5, the lumen area of cross section of coronary artery decreased sharply, reaching the minimum at L6. Then the cross-section lumen area begins to rise and reaches the initial value. The decrease in cross-sectional lumen area reflects coronary artery stenosis. By calculating the distance difference between the two landmarks, the narrow length can be obtained. Therefore, changes in coronary artery area and diameter can be quantitatively displayed on the chart, and the length and extent of coronary artery

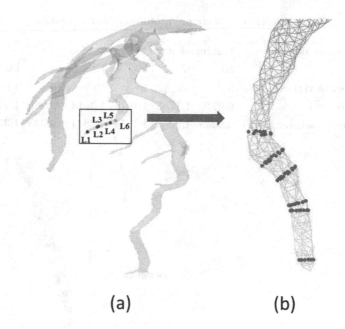

(a)   **(b)**

Fig. 2. Cross sectional lumen planes extraction at 6 landmark centerline points. (a) Six landmarks L1, L2, L3, L4, L5 and L6. (b) Corresponding contours at the vessel boundary.

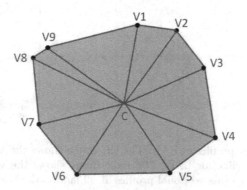

Fig. 3. Obtained polygon when cutting the cross section at landmark L1. We have 9 vertices, $V1 - V9$. Geometric centroid C can be calculated. The polygon can be decomposed into 9 triangles.

stenosis can be determined. In addition, the proposed cross-sectional area measurement algorithm can deal with the whole Left Anterial Descending (LAD). Firstly, the central line is extracted by the proposed method, and then smoothed and pruned at the subvoxel level. The total number of skeleton points in LAD segment is 286. In this paper, only 29 representative skeleton points are selected to obtain a good visual effect, which can clearly see the obtained boundary

Table 3. 3D cross sectional profiles at six landmarks.

Cross sectional	Landmark name					
	L1	L2	L3	L4	L5	L6
Number of vertices	9	12	13	15	18	14
Area (mm^2)	0.6710	1.1785	1.1217	1.4614	2.0966	1.4764
Average Diameter (mm)	0.9245	1.2253	1.1954	1.3644	1.6343	1.3714

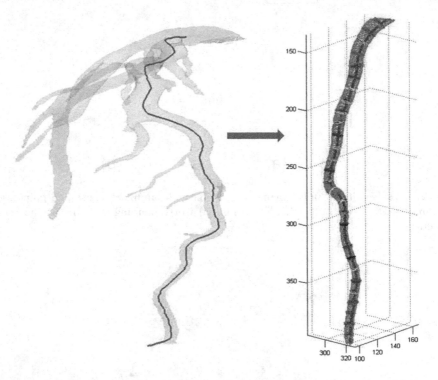

Fig. 4. Cross sectional profiles of LAD. The left frame shows the segmented left coronary artery and centerline for LAD. The right frame shows the vessel surface mesh, the centerline, and the cross sectional profiles at 29 landmark skeleton points.

contour of the section intercepted from the vessel surface mesh and the normal planes perpendicular to the centerline at each centerline point. In Fig. 4, we show the vascular surface meshes, corresponding centerlines and normal planes at 29 landmark skeleton points.

4   Conclusion

In this paper, a method based on the combination of fast marching and gradient vector flow is proposed to extract the centerline of a complete coronary artery

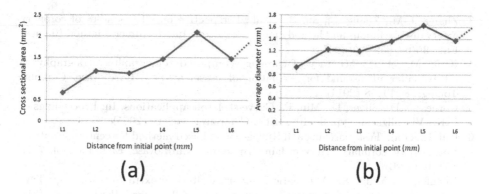

Fig. 5. 3D cross sectional profile at 6 landmark skeleton points. The drop in the cross sectional lumen area reflects the stenosis.

tree in 3D Angiography images. The experimental results show that the obtained vascular centerline is complete and accurate, and highly coincides with the vascular centerline marked by imaging experts. With the centerline of blood vessel, we propose an automatic method to extract the cross-section of blood vessel lumen. This method calculates the tangent vector based on the two adjacent centerline points before and after the centerline point, and calculates the cross-sectional equation through the centerline point, and then obtains the cross-sectional lumen contour along the surface mesh of blood vessels. The new method is to extract the cross-section of 3D intravascular lumen in real physical coordinates, which avoids the traditional interpolation processing in pixel coordinates and improves the accuracy of cross-section extraction. The proposed three-dimensional cross sectional lumen area measurement algorithm, based on centerline extraction and surface mesh generation of coronary artery, will be helpful for quantitative assessment of the severity of coronary artery stenosis and for clinical diagnosis and treatment planning of coronary heart diseases.

Acknowledgment. The study was supported in part by the National Natural Science Foundation of China under Grants 61771397, 61801391, 61801393 and 61801395, in part by the Natural Science Basic Research Project in Shaanxi of China (Program No. 2019JQ-254 and 2019JQ-158), and in part by the Fundamental Research Funds for the Central Universities under Grants 3102018zy031.

References

1. Samuels, O.B., Joseph, G.J., Lynn, M.J., Smith, H.A., Chimowitz, M.I.: A standardized method for measuring intracranial arterial stenosis. Am. J. Neuroradiol. **21**(4), 643–646 (2000)
2. Kirişli, H.A., et al.: Standardized evaluation framework for evaluating coronary artery stenosis detection, stenosis quantification and lumen segmentation algorithms in computed tomography angiography. Med. Image Anal. **17**(8), 859–876 (2013)

3. Zhang, J.-M., Zhong, L., Su, B., et al.: Perspective on CFD studies of coronary artery disease lesions and hemodynamics: a review. Int. J. Numer. Method Biomed. Eng. **30**(6), 659–680 (2014)
4. Kruk, M., et al.: Accuracy of coronary computed tomography angiography vs intravascular ultrasound for evaluation of vessel area. J. Cardiovasc. Comput. Tomogr. **8**, 141–8 (2014)
5. Cornea, N.D., Silver, D., Min, P.: Curve-skeleton applications. In: Proceedings of IEEE Visualization, pp. 95–102. IEEE Computer Society (2005)
6. Cui, H., et al.: Fast marching and Runge-Kutta based method for centreline extraction of right coronary artery in human patients. Cardiovasc. Eng. Technol. **7**(2), 159–169 (2016)
7. Hengfei, C., Yong, X.: Automatic coronary centerline extraction using gradient vector flow field and fast marching method from CT images. IEEE Access **1**(1), 41816–41826 (2018)
8. Cui, H., Xia, Y., Zhang, Y., et al.: Validation of right coronary artery lumen area from cardiac computed tomography against intravascular ultrasound. Mach. Vis. Appl. **29**(8), 1287–1298 (2018)
9. Zheng, Y., Tek, H., Funka-Lea, G.: Robust and accurate coronary artery centerline extraction in CTA by combining model-driven and data-driven approaches. In: Mori, K., Sakuma, I., Sato, Y., Barillot, C., Navab, N. (eds.) MICCAI 2013. LNCS, vol. 8151, pp. 74–81. Springer, Heidelberg (2013). https://doi.org/10.1007/978-3-642-40760-4_10
10. Huang, Z., Zhang, Y., Li, Q., Zhang, T., Sang, N., Hong, H.: Progressive dual-domain filter for enhancing and denoising optical remote-sensing images. IEEE Geosci. Remote Sens. Lett. **15**(5), 759–63 (2018)
11. Uitert, R.V., Bitter, I.: Subvoxel precise skeletons of volumetric data based on fast marching methods. Med. Phys. **34**(2), 627 (2007)
12. Luo, T., Wischgoll, T., Koo, B.K., et al.: IVUS validation of patient coronary artery lumen area obtained from CT images. PLOS ONE **9**, e86949 (2014)
13. Jain, A.K.: Fundamentals of Digital Image Processing. Prentice-Hall Inc., Upper Saddle River (1989)
14. Xu, C., Prince, J.L.: Snakes, shapes, and gradient vector flow. IEEE Trans. Image Process. Publ. IEEE Signal Process. Soc. **7**(3), 359–69 (1998)
15. Zhang, S., Zhou, J.: Centerline extraction for image segmentation using gradient and direction vector flow active contours. J. Signal Inf. Process. **4**(4), 407–413 (2013)
16. Hassouna, M.S., Farag, A.A.: Variational curve skeletons using gradient vector flow. IEEE Trans. Pattern Anal. Mach. Intell. **31**(12), 2257–2274 (2009)
17. Press, W.H., Flannery, B.P., Teukolsky, S.A., Vetterling, W.T.: Numerical recipes in C. Contemp. Phys. **10**(1), 176–177 (1992)

A Robust Facial Landmark Detector
with Mixed Loss

Xian Zhang[1,2], Xinjie Tong[1], Ziyu Li[1,2], and Wankou Yang[1,2(✉)]

[1] School of Automation, Southeast University, Nanjing 210096, China
wkyang@seu.edu.cn
[2] Key Lab of Measurement and Control of Complex Systems of Engineering,
Ministry of Education, Southeast University, Nanjing 210096, China

Abstract. Facial landmark detection is one of the most important tasks in face image and video analysis. Existing algorithms based on deep convolutional neural networks have achieved good performance in public benchmarks and practical applications such as face verification, expression analysis, beauty applications and so on. However, the performance of a facial landmark detector degrades significantly when dealing with challenging facial images in the presence of extreme appearance variations such as pose, expression, occlusion, etc. To mitigate these difficulties, we propose a robust facial landmark detection algorithm based on coordinates regression in an end-to-end training fashion. By using the soft-argmax function, the network weights can be optimised with a mixed loss function. The online pose-based data augmentation technology is used to effectively solve the data imbalance problem and improve the robustness of the proposed method. Experiments conducted on the 300-W and AFLW datasets demonstrate that the performance of the proposed algorithm is competitive to the state-of-the-art heatmap regression algorithms, in terms of accuracy. Besides, our method achieves real-time speed on 300-W with 68 landmarks, which runs at 85 FPS on a Tesla v100 GPU.

Keywords: Facial landmark detection · Mixed loss · Soft-argmax · Pose-based data augmentation

1 Introduction

Facial landmark detection, also known as face alignment [29,43], is a fundamental task in various facial image and video analysis applications [31,32,41,44–46]. During the past decades, the facial landmark detection area has made significant progress. Nevertheless, many existing approaches have difficulties in dealing with in-the-wild faces with extreme appearance variations in pose, expression, illumination, blur and occlusion.

Existing facial landmark detection algorithms can be roughly divided into three categories: global appearance based approaches, constrained local models and regression-based methods. Global appearance based methods detect the key

© Springer Nature Switzerland AG 2019
Z. Cui et al. (Eds.): IScIDE 2019, LNCS 11935, pp. 249–261, 2019.
https://doi.org/10.1007/978-3-030-36189-1_21

points using the whole facial textural information and global shape information [3–5,13,25,30]. Constrained local model [17] is based on global face shape
and independent local textural information around each key point that captures
more robust information for illumination and occlusion variations. Regression-
based methods can be divided into direct regression, cascade regression and
regression with deep neural networks. At present, the most widely used and the
most accurate methods are all based on deep Convolutional Neural Networks
(CNNs) [12,16]. In this paper, the proposed facial landmark detection method
is based on CNNs as well.

The key innovations of the proposed method include:

- For data augmentation, we adopt the online Pose-based Data Balancing
  (PDB) [8] method that balances the original training dataset. To be more specific, we copy the samples of low proportion defined by PDB and randomly
  modify the samples (flip, rotate, blur, etc.) including changing the copied
  samples with different styles since the intrinsic variance of image styles can
  also affects the performance of a trained network [6].
- The baseline of this paper is CPM [34] that generates a heatmap image as
  the final output of a network. In order to apply Wing Loss that is specially
  designed for coordinates regression models in this work, we introduce the soft-
  argmax function [21]. The function converts heatmaps to coordinates thus the
  network is differentiable.
- The original Wing Loss function [8] focuses on small and medium errors, but
  pays less attention to the samples with large errors. To address this issue,
  we design a new loss function, namely mixed loss, that considers the samples
  with errors at various magnitudes.

2   Related Work

2.1   Pose Variation

The aim of data augmentations is to reduce the bias in network training due
to the imbalance of a training dataset. STN [22] applies spatial transformer
network to learn transformation parameters thus to automatically initialise a
training dataset. SAN [6] translates each image to four different styles by a
generative adversarial module. Both of them try to inject diversity to a training
dataset and balance the training samples.

2.2   Regression Model

The regression methods used from facial landmark detection can be divided
into two categories: coordinate regression and heatmap regression. A coordinate
regression network performs well on a dataset with sparse landmarks, but not as
well as heatmap regression on dense landmarks. However, heatmap regression has
been proved that the prediction can be worsen despite MSE improving during the
regression of heatmap matching [26]. Luvizon et al. [21] propose the soft-argmax

function to convert heatmaps to coordinates to make the network differentiable. Nibali et al. [26] use a new regularisation strategy to improve the prediction accuracy of a network.

2.3 Loss Function

For a CNN-based facial landmark detector, a loss function has to be defined to supervise the network training process. Most existing facial landmark detection approaches are based on the L2 loss, which is sensitive to outliers. Feng et al. [8] propose a new loss function, i.e. wing loss, to balance the sensitivity of small errors and big errors for the training of a deep CNN model. Guo et al. [11] introduce a loss that can adjust weights for different samples during the training process according to the tag that describes the pose of each sample. Merget et al. [23] proposes a loss function that judges whether each landmark is labelled and within the image boundary at the beginning, which gives each landmark a specific weight according to the judgement.

3 Methodology

3.1 Data Augmentation

Data imbalance is a common issue in deep learning, which limits the accuracy and robustness of a trained network [11]. From Table 1 and Fig. 1, we can see that most datasets contain a large number of frontal faces, but lack of samples with large poses, expressions, illuminations and occlusions [42]. The imbalance of a dataset in gesture is very significant. If we train a network using an imbalanced dataset, the network may not able to generalise well to practical applications. Besides, the distribution variations among training and test sets can influence the performance of a trained network significantly.

Table 1. Distribution of the 300-W dataset in gesture [29].

Pose	$-30°:-15°$	$-15°:0°$	$0°:15°$	$15°:30°$
Pitch	14.59%	61.05%	24.02%	0.34%
Yaw	19.11%	26.02%	22.00%	32.87%

To address the data imbalance problem, various algorithms have been proposed, including both geometric and textural transformations [9]. The main methods used for geometric transformation are flipping, scaling, translation and rotation. For textural transformation, Gaussian noise and brightness transformation are widely used. Nevertheless, if we randomly apply the above methods to the training samples of a dataset, we don't know how many times a training sample should be augmented/copied.

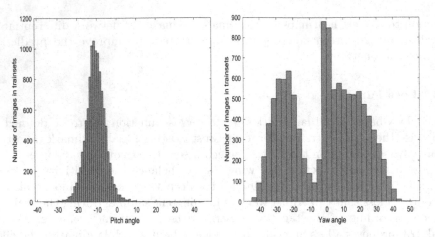

Fig. 1. Distribution of the ICME2019 GC facial landmark datasets in gesture [19]. The X-axis stands for pitch angle in the left figure and yaw angle in the right figure. The Y-axis denotes the number of samples of the training set.

To improve the balance of a dataset, we introduce the Pose-based Data Balancing (PDB) [8] strategy (Algorithm 1) in our work. PDB is a statistical method that aims to analyse the distribution of a face dataset in shape and posture. To adapt PDB to our network, first, we take Procrustes Analysis [10] to align all the faces in a training dataset to the mean face. Procrustes Analysis learns a affine transformation from a shape to another shape with minimum mean square error. By applying PCA to the training set and analysing the distribution of the principle component, we can balance the training set by copying each sample for a fixed number of times which is set to balance the distribution.

Algorithm 1. Pose-based data balancing

Require:
    All the images in a training set, I_n;
    Bounding boxes of the faces, B_n;
    $N * 2$ coordinates of the facial landmarks, $\{(x_n, y_n)\}$;

Ensure:
    Images after copying and random transformation of the training set, E_n;
    Bounding boxes of faces after PDB, B_n;
    Landmark coordinates after PDB (x_n, y_n)

1: Read all the samples in the training set $D_n = \{I_n, B_n, \{(x_n, y_n)\}\}$
2: Calculate the distribution of the training set by Procrustes Analysis;
3: Divide the training set according to the interval of the principal component;
4: Estimate copying times of each image and extend the dataset;
5: Perform random transformation on the samples obtained in step 5;
6: **return** $D'_n = \{I'_n, B'_n, \{(x'_n, y'_n)\}\}$;

In order to minimise the impact of dataset imbalance on facial landmark detection accuracy, the PDB process is applied in each epoch at the beginning. Since the modification of each sample is random, the online PDB process can substantially enhance the variety of samples in different attributes. In each epoch, the data is copied the same number of times, but in different epochs, the data is randomly transformed independently. In this case, dataset is invariably expanded by a period of multiple times and each epoch can be regarded as sampling in a large dataset. According to our experiments, the offline data augmentation has a very good improvement on the performance of a detector. When we convert the offline data augmentation to online data augmentation, the performance of a trained facial landmark detector can be further improved. However, it is worth noting that offline data augmentation does not require many CPU resources. If one does not perform multi-thread data augmentation, each online PDB training process needs to multiply the original running time by several times.

3.2   Network and Mixed Loss

The backbone network of our facial landmark detector is shown in Figs. 2 and 3. The network is based on VGG16 [33] + CPM [7,34], which uses first four convolutions of VGG16 to extract coarse feature maps, followed by three stages of

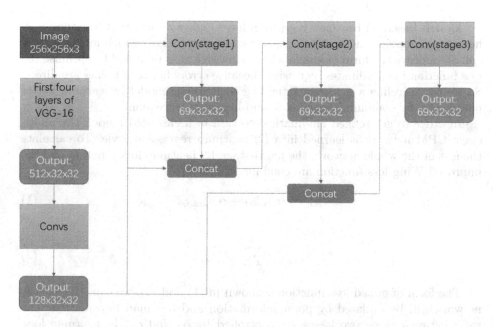

Fig. 2. Backbone network of the proposed facial landmark detector. All the inputs must be resized to 256 × 256. Concat means splicing the feature maps by channels and changing channels by 1 × 1 convolution, the channels 69 in output means 68 landmarks and 1 mask denoting the visibility.

CPM structure. The detailed architecture of conv in Fig. 2 is shown in Fig. 3. We use the convolutional pose machines (CPM) as the main architecture, CPMs combine and concatenate outputs of each stage in the network, in order to hold the geometric constraint and semantic information in feature maps. The ground truth is transformed into heatmap style via taking Gaussian blur on the landmark point. After down-sampling the image with landmark points to the same size and channels with the output of CPM, the error between the predicted and ground truth values is back propagated in each stage of CPM since each stage is intermediately supervised by the L1/L2 loss function.

Fig. 3. Detailed kernel size and channels of convolution layers in Conv in Fig. 2, the first row is Conv(stage1), and the second row is the Conv(stage>1). The L in last layers denote the number off facial landmarks and mask.

Models based on heatmap regression have higher accuracy. Additionally, all heatmap regression methods are supervised by the L2 loss, which makes it difficult to improve the form of loss function. However, it is practical to optimise the loss function in coordinates regression, because errors between points are direct. So we try to refine a detector, intending to help the model in learning better parameters by combining coordinate and heatmap regression.

In order to get refined landmark detection in a cascaded model, the multi-stage CPM network is learned in a L2 heatmap regression style. To calculate the loss of the whole network, the multi-stage L2 heatmap loss function and the improved Wing loss function are combined.

$$l_{mix} = \alpha_1 l_{point} + \alpha_2 l_{stage}, \tag{1}$$

$$l_{stage} = \sum_{i=1}^{3} \beta_i l_{stage}^{(i)}. \tag{2}$$

The form of mixed loss function is shown in (1) and (2), we can see that the network will be updated by point information and heatmap information. The ratio between these two losses are controlled by α_1 and α_2. In heatmap loss, the output of each stage can also contributes to total loss, β_1, β_2 and β_3 are hyper-parameters.

As aforementioned, it is well-known that the proportion of difficult samples is relatively small in a training data set, causing data imbalance issue. Additionally, the simple samples usually dominate the network training. In this case, the widely used L2 loss is not necessarily the best loss function. The L2 loss function amplifies the effects of samples with large errors and neglects small errors. In contrast, the Wing loss function focuses on small and medium errors, but pay less attention to the samples with large errors. In order to design a new loss function that considers the samples with various errors, we formulate the function in Eq. (3) [8]:

$$l_{point} = wing\{x\} = \begin{cases} w\ln\left(1 + \frac{x}{\varepsilon}\right) & if\ |x| < w \\ |x| - C & otherwise \end{cases} \tag{3}$$

3.3   Heatmap to Point Regression

It is easy to convert a heatmap to key point coordinates, by just finding the peak locations in the heatmap throughout the argmax function. However, the process is not trivial because the gradients cannot be back-propagated through argmax. To address this issue, this paper adopts soft-argmax, which can guarantee the differentiation in the training process while searching for the maximum value. We represent the argmax function as a parsed form to explain the expectation of the idea. The expectation on the idea of representing the argmax function as a parsed form.

Assuming that one channel of heatmap can be represented as $I(x, y)$, which has the size of $W \times H \times C$, where W and H are the width and height of heatmap, and C denotes channels. The maximum point can be calculated by [26]:

$$softargmax(I) = \left(\sum_{i,j} W_x(i,j) I(i,j), \sum_{i,j} W_y(i,j) I(i,j) \right), \tag{4}$$

$$W_x(i,j) = \frac{i}{W}, \tag{5}$$

$$W_y(i,j) = \frac{j}{H}. \tag{6}$$

In fact, considering that in our model, each heatmap has the order of 10^{-5}, to avoid truncation errors and insufficient precision, we use the adapted Algorithm 2.

Algorithm 2. Modified soft-argmax in our model

1: Input the heatmap;
2: Set an expansion factor α ;
3: Take an exponential functional $e^{\alpha x}$ on each item x in the heat map;
4: Choose a value x in heatmap, and let all the heatmap divides this;
5: Use the original soft-argmax on the transformed heatmap.

---

4   Experimental Results

4.1   Datasets

In this paper, we conduct experiments on two datasets: the 300-W [29] and AFLW facial landmark datasets [15].

300-W is an open facial landmark dataset, which is composed by LFPW [1], AFW [40], HELEN [18], XM2VTS [24] and IBUG [20] datasets. The whole 300-W dataset contains 3148 training images and 687 test images. Each image in 300-W is labelled with 68 facial landmark (Fig. 4).

AFLW is another classic datasets in face alignment. AFLW consists of more than 25000 images with 21 landmarks. In our experiments, we follow AFLW-Full protocol [15], which contains 24386 images in total, 20000 images for training and others for testing. The images are annotated with 19 landmarks since the landmarks of two ears are ignored in this protocol.

Fig. 4. Partial visualization of the results of our model on 300-W.

4.2   Experimental Settings

We conduct all the experiments on an Intel E5-2650 v4 CPU with two Tesla v100 GPUs. The proposed method was implemented with Pytorch 1.1 [27,28] and Python 3.7. All the input images are resized to $256 \times 256 \times 3$ and the output is N*2 landmark coordinates. The type of heatmap is Gaussian. Our models is updated by Stochastic Gradient Descent (SGD), with the momentum of 0.9 and weight decay of 0.0005. For the 300-W dataset, the learning rate is 0.00005, while for ALFW we set the learning rate to 0.00001. From epoch 30 to 40, the learning

Table 2. Results on 300-W and AFLW datasets. For 300-W, we use inter-pupil distance to compute NME. For AFLW, we use the face size for NME.

Methods	Common set	Challenging set	Full set	AFLW
RCPR [2]	6.18	17.26	8.35	5.43
CFAN [37]	5.50	16.78	7.69	–
TCDCN [38]	4.80	8.60	5.54	–
RAR [35]	4.12	8.35	4.94	–
3DDFA [41]	6.15	10.59	7.01	–
LBF [39]	4.95	11.98	6.32	4.25
ERT [14]	–	–	6.40	4.35
SDM [36]	5.57	15.40	7.50	4.05
Baseline (heatmap + L2 loss)	4.40	9.92	5.49	1.93
Baseline + mixed loss	4.35	9.12	5.43	1.82
Baseline + offline PDB	4.32	8.67	5.38	1.67
Baseline + mixed loss + online PDB	4.26	8.11	5.02	–

rate will decay by a factor of 0.2. After 40 epochs, the learning rate will decay by a factor of 0.1. We train the model for more than 60 epochs. The batch size is set to 64. In the mixed loss function, we try to combine different coefficients with grid search. We get the best result when setting $\alpha_1 = 0.7$ and $\alpha_1 = 0.3$. Meanwhile β_i are set as $\{0.5, 0.5, 1\}$. In the training step, it cost above half a day on 300-W without PDB while about one day with offline PDB. When applying online PDB, it costs 8 days on the same CPU and GPU. For the AFLW dataset we don't do online PDB due to the time limitation.

4.3 Results

We use the backbone network with L2 point regression as the baseline method. Then we try to observe the effects of different methods on 300-W, we use NME as the evaluation metric, which is defined as:

$$NME = \frac{1}{N} \sum_{k=1}^{N} \frac{\|x_k - y_k\|_2}{d}. \tag{7}$$

where x denotes the ground truth landmarks for a given face, y denotes the corresponding prediction and d can be computed as the face size, using the inter-ocular distance or the pupil distance.

4.3.1 Results on 300-W

We apply different innovations to our experiments on 300-W. The performance of different state-of-the-art methods as well as the proposed method in terms of

NME are reported in Table 2. We can see that, in spite of the accuracy loss of point regression our method achieves competitive result. The test batch size is 16 and the proposed method achieves 85 FPS on a Tesla v100 GPU.

4.4 Results on AFLW

As shown in Table 2, we conduct similar experiments on the AFLW dataset. The speed of the proposed method can also achieve more than 80 FPS under the same environment. We also summarise the important parameters, e.g. model size, FLOP and so on. The model size is similar to the model used for 300-W except for the last output layer. The number of parameters is 15.94 M, model size is 127 MB, and the GFLOPs is 2.57 billions.

5   Conclusion

In this paper, we presented a robust facial landmark detector that combines coordinate and heatmap information, thus improving the performance of a trained CNN network in terms of accuracy. Besides, we used the soft-argmax instead of argmax as well as online PDB for training data augmentation. The main purpose of the proposed method is to mitigate the dataset imbalance problem. In addition, we designed a mixed loss function consisting of more information for network training. The experiments obtained on 300-W and AFLW demonstrate the effectiveness of the proposed method compared with the state-of-the-art approaches.

Acknowledgement. This work is partly supported by the National Natural Science Foundation of China (61773117, 61703096 and 61473086), the Jiangsu key R&D plan (BE2017157) and the Natural Science Foundation of Jiangsu Province (BK20170691).

References

1. Belhumeur, P.N., Jacobs, D.W., Kriegman, D.J., Kumar, N.: Localizing parts of faces using a consensus of exemplars. IEEE Trans. Pattern Anal. Mach. Intell. **35**, 2930–2940 (2011)
2. Burgos-Artizzu, X.P., Perona, P., Dollár, P.: Robust face landmark estimation under occlusion. In: 2013 IEEE International Conference on Computer Vision, pp. 1513–1520 (2013)
3. Cootes, T.F., Edwards, G.J., Taylor, C.J.: Active appearance models. In: Burkhardt, H., Neumann, B. (eds.) ECCV 1998. LNCS, vol. 1407, pp. 484–498. Springer, Heidelberg (1998). https://doi.org/10.1007/BFb0054760
4. Cootes, T.F., Taylor, C.J., Cooper, D.H., Graham, J.: Active shape models-their training and application. Comput. Vis. Image Underst. **61**, 38–59 (1995)
5. Cootes, T.F., Walker, K.N., Taylor, C.J.: View-based active appearance models. Image Vision Comput. **20**, 657–664 (2000)
6. Dong, X., Yan, Y., Ouyang, W., Yang, Y.: Style aggregated network for facial landmark detection. In: 2018 IEEE/CVF Conference on Computer Vision and Pattern Recognition, pp. 379–388 (2018)

7. Dong, X., Yu, S.-I., Weng, X., Wei, S.-E., Yang, Y., Sheikh, Y.: Supervision-by-registration: an unsupervised approach to improve the precision of facial landmark detectors. In: 2018 IEEE/CVF Conference on Computer Vision and Pattern Recognition, pp. 360–368 (2018)

8. Feng, Z.-H., Kittler, J., Awais, M., Huber, P., Wu, X.: Wing loss for robust facial landmark localisation with convolutional neural networks. In: 2018 IEEE/CVF Conference on Computer Vision and Pattern Recognition, pp. 2235–2245 (2018)

9. Feng, Z.-H., Kittler, J., Xiaojun, W.: Mining hard augmented samples for robust facial landmark localization with CNNs. IEEE Signal Process. Lett. **26**(3), 450–454 (2019)

10. Gower, J.C.: Generalized procrustes analysis. Psychometrika **40**, 33–51 (1975)

11. Guo, X., et al.: PFLD: a practical facial landmark detector. ArXiv, abs/1902.10859 (2019)

12. He, K., Zhang, X., Ren, S., Sun, J.: Deep residual learning for image recognition. In: 2016 IEEE Conference on Computer Vision and Pattern Recognition (CVPR), pp. 770–778 (2016)

13. Kahraman, F., Gökmen, M., Darkner, S., Larsen, R.: An active illumination and appearance (AIA) model for face alignment. In: 2007 IEEE Conference on Computer Vision and Pattern Recognition, pp. 1–7 (2007)

14. Kazemi, V., Sullivan, J.: One millisecond face alignment with an ensemble of regression trees. In: 2014 IEEE Conference on Computer Vision and Pattern Recognition, pp. 1867–1874 (2014)

15. Köstinger, M., Wohlhart, P., Roth, P.M., Bischof, H.: Annotated facial landmarks in the wild: a large-scale, real-world database for facial landmark localization. In: 2011 IEEE International Conference on Computer Vision Workshops (ICCV Workshops), pp. 2144–2151 (2011)

16. Krizhevsky, A., Sutskever, I., Hinton, G.E.: Imagenet classification with deep convolutional neural networks. Commun. ACM **60**, 84–90 (2012)

17. Kumar, N., Belhumeur, P., Nayar, S.: FaceTracer: a search engine for large collections of images with faces. In: Forsyth, D., Torr, P., Zisserman, A. (eds.) ECCV 2008. LNCS, vol. 5305, pp. 340–353. Springer, Heidelberg (2008). https://doi.org/10.1007/978-3-540-88693-8_25

18. Le, V., Brandt, J., Lin, Z., Bourdev, L., Huang, T.S.: Interactive facial feature localization. In: Fitzgibbon, A., Lazebnik, S., Perona, P., Sato, Y., Schmid, C. (eds.) ECCV 2012. LNCS, vol. 7574, pp. 679–692. Springer, Heidelberg (2012). https://doi.org/10.1007/978-3-642-33712-3_49

19. Liu, Y., et al.: Grand challenge of 106-point facial landmark localization. ArXiv, abs/1905.03469 (2019)

20. Luo, B., Shen, J., Wang, Y., Pantic, M.: The iBUG eye segmentation dataset. In: ICCSW (2018)

21. Luvizon, D.C., Tabia, H., Picard, D.: Human pose regression by combining indirect part detection and contextual information. CoRR, abs/1710.02322 (2017)

22. Lv, J.-J., Shao, X., Xing, J., Cheng, C., Zhou, X.: A deep regression architecture with two-stage re-initialization for high performance facial landmark detection. In: 2017 IEEE Conference on Computer Vision and Pattern Recognition (CVPR), pp. 3691–3700 (2017)

23. Merget, D., Rock, M., Rigoll, G.: Robust facial landmark detection via a fully-convolutional local-global context network. In: 2018 IEEE/CVF Conference on Computer Vision and Pattern Recognition, pp. 781–790 (2018)

24. Messer, K., Matas, J., Kittler, J., Luettin, J., Maître, G.: XM2VTSDB: The extended M2VTS database (1999)

25. Milborrow, S., Nicolls, F.: Locating facial features with an extended active shape model. In: Forsyth, D., Torr, P., Zisserman, A. (eds.) ECCV 2008. LNCS, vol. 5305, pp. 504–513. Springer, Heidelberg (2008). https://doi.org/10.1007/978-3-540-88693-8_37

26. Nibali, A., He, Z., Morgan, S., Prendergast, L.: Numerical coordinate regression with convolutional neural networks. CoRR, abs/1801.07372 (2018)

27. Paszke, A., et al.: Automatic differentiation in PyTorch, Alban Desmaison (2017)

28. Rumelhart, D.E., Hinton, G.E., Williams, R.J.: Learning representations by back-propagating errors. Nature **323**, 533–536 (1986)

29. Sagonas, C., Tzimiropoulos, G., Zafeiriou, S.P., Pantic, M.: 300 faces in-the-wild challenge: the first facial landmark localization challenge. In: 2013 IEEE International Conference on Computer Vision Workshops, pp. 397–403 (2013)

30. Saragih, J.M., Goecke, R.: A nonlinear discriminative approach to AAM fitting. In: 2007 IEEE 11th International Conference on Computer Vision, pp. 1–8 (2007)

31. Benitez-Quiroz, C.F., Srinivasan, R., Martínez, A.M.: EmotioNet: an accurate, real-time algorithm for the automatic annotation of a million facial expressions in the wild. In: 2016 IEEE Conference on Computer Vision and Pattern Recognition (CVPR), pp. 5562–5570 (2016)

32. Taigman, Y., Yang, M.W., Ranzato, M., Wolf, L.: DeepFace: closing the gap to human-level performance in face verification. In: 2014 IEEE Conference on Computer Vision and Pattern Recognition, pp. 1701–1708 (2014)

33. Simonyan, K., Zisserman, A.: Very deep convolutional networks for large-scale image recognition. CoRR, abs/1409.1556 (2015)

34. Wei, S.-E., Ramakrishna, V., Kanade, T., Sheikh, Y.: Convolutional pose machines. In: 2016 IEEE Conference on Computer Vision and Pattern Recognition (CVPR), pp. 4724–4732 (2016)

35. Xiao, S., Feng, J., Xing, J., Lai, H., Yan, S., Kassim, A.: Robust facial landmark detection via recurrent attentive-refinement networks. In: Leibe, B., Matas, J., Sebe, N., Welling, M. (eds.) ECCV 2016. LNCS, vol. 9905, pp. 57–72. Springer, Cham (2016). https://doi.org/10.1007/978-3-319-46448-0_4

36. Xiong, X., De la Torre, F.: Supervised descent method and its applications to face alignment. In: 2013 IEEE Conference on Computer Vision and Pattern Recognition, pp. 532–539 (2013)

37. Zhang, J., Shan, S., Kan, M., Chen, X.: Coarse-to-fine auto-encoder networks (CFAN) for real-time face alignment. In: Fleet, D., Pajdla, T., Schiele, B., Tuytelaars, T. (eds.) ECCV 2014. LNCS, vol. 8690, pp. 1–16. Springer, Cham (2014). https://doi.org/10.1007/978-3-319-10605-2_1

38. Zhang, Z., Luo, P., Loy, C.C., Tang, X.: Facial landmark detection by deep multi-task learning. In: Fleet, D., Pajdla, T., Schiele, B., Tuytelaars, T. (eds.) ECCV 2014. LNCS, vol. 8694, pp. 94–108. Springer, Cham (2014). https://doi.org/10.1007/978-3-319-10599-4_7

39. Zhu, S., Li, C., Loy, C.C., Tang, X.: Unconstrained face alignment via cascaded compositional learning. In: 2016 IEEE Conference on Computer Vision and Pattern Recognition (CVPR), pp. 3409–3417 (2016)

40. Zhu, X., Ramanan, D.: Face detection, pose estimation, and landmark localization in the wild. In: 2012 IEEE Conference on Computer Vision and Pattern Recognition, pp. 2879–2886 (2012)

41. Zhu, X., Lei, Z., Liu, X., Shi, H., Li, S.Z.: Face alignment across large poses: a 3d solution. In: 2016 IEEE Conference on Computer Vision and Pattern Recognition (CVPR), pp. 146–155 (2016)

42. Zhu, X., Lei, Z., Yan, J., Yi, D., Li, S.Z.: High-fidelity pose and expression normalization for face recognition in the wild. In: 2015 IEEE Conference on Computer Vision and Pattern Recognition (CVPR), pp. 787–796 (2015)

43. Liu, F., Zeng, D., Zhao, Q., Liu, X.: Joint face alignment and 3D face reconstruction. In: Leibe, B., Matas, J., Sebe, N., Welling, M. (eds.) ECCV 2016. LNCS, vol. 9909, pp. 545–560. Springer, Cham (2016). https://doi.org/10.1007/978-3-319-46454-1_33

44. Liu, F., Zhao, Q., Liu, X., Zeng, D.: Joint face alignment and 3d face reconstruction with application to face recognition. IEEE Trans. Pattern Anal. Mach. Intell. **37**(6), 1312–1320 (2017)

45. Lu, J., Liong, V.E., Zhou, X., Zhou, J.: Learning compact binary face descriptor for face recognition. IEEE Trans. Pattern Anal. Mach. Intell. **37**, 2041–2056 (2015)

46. Lu, J., Tan, Y.-P., Wang, G.: Discriminative multimanifold analysis for face recognition from a single training sample per person. In: 2011 International Conference on Computer Vision, pp. 1943–1950 (2011)

Object Guided Beam Steering Algorithm for Optical Phased Array (OPA) LIDAR

Zhiqing Wang[✉], Zhiyu Xiang, and Eryun Liu

College of Information Science and Electronic Engineering, Zhejiang University,
Hangzhou, China
{wangzhiqing,xiangzy,eryunliu}@zju.edu.cn

Abstract. As a fundamental sensor for autonomous driving, light detection and ranging (LIDAR) has gained increasing attentions in recent years. Optical phased array (OPA) LIDAR as a solid-state solution with the advantages of durability and low cost has been actively researched in both the academic and industry fields. Beam steering is a critical problem in OPA LIDAR where the beam can be controlled by software instantaneously. In this paper, we propose an object guided beam steering algorithm where the beams are allocated according to the detected objects in current frame of image. A series of rules are designed to assign different weights to different regions in the scene. We evaluated the algorithm in a simulated environment and the experimental results demonstrated the effectiveness of the proposed algorithm.

Keywords: OPA LIDAR · Beam steering · Object detection · Point cloud segmentation

1 Introduction

In recent years, light detection and ranging (LIDAR) has gained increasing attentions in both industrial and academic area. As a 3D perception device, LIDAR uses pulsed lasers to collect measurements which can be used for object detection, localization and mapping. Therefore, LIDAR has broad applications in the fields of autonomous driving, robotics, aerial mapping, and atmospheric measurements.

Beam steering is a critical problem in a LIDAR system. Many LIDAR systems available today utilize a mechanical mechanism for beam steering [11,12,14]. Unfortunately, the mechanical mechanism limits the scan rate, decreases reliability and increases the system cost.

Recently, solid-state beam steering LIDAR system has been emphasized to increase system durability. Optical phased array (OPA) LIDAR is the most promising technology to achieve this goal [5–7,13,27].

Different from the mechanical spinning LIDAR with fixed scanning pattern and resolution, the OPA LIDAR is extremely flexible in controlling the laser beam. The spatial scanning resolution can be instantaneously changed according

© Springer Nature Switzerland AG 2019
Z. Cui et al. (Eds.): IScIDE 2019, LNCS 11935, pp. 262–272, 2019.
https://doi.org/10.1007/978-3-030-36189-1_22

to the environment. This flexibility introduces a new way of 3D perception of environment, i.e., active 3D perception. The beam can be steered actively based on perception results and focused on the area of interests.

In this paper, we propose an object guided beam steering algorithm for OPA LIDAR by fusing object detection results in image. The goal of beam steering of OPA LIDAR is to detect objects, understand the environment, and avoid obstacles. To reach this goal, the beam resources are allocated according to the attributes of objects detected in previous frame of image. The algorithm has been evaluated in a simulated environment with a point cloud segmentation task and experimental results show the superior performance of OPA LIDAR over spinning LIDAR. As far as we know, this is the first attempt of steering OPA LIDAR beams with object guided strategy in simulated environment in the literature.

The paper is organized as follows. In Sect. 2, we introduce our proposed object guided beam steering algorithm for OPA LIDAR. A point cloud semantic segmentation method for cars and pedestrians based on 3D LIDAR point cloud is presented for making comparison between the OPA LIDAR and the spinning LIDAR in Sect. 3. Experimental results and a discussion are presented in Sect. 4 which is followed by a summary in Sect. 5.

2   Beam Steering

In this section, we firstly introduce how we use a simulator to simulate the OPA LIDAR. Then we propose an object guided beam steering algorithm.

2.1   System Overview

Our system uses two sensors, i.e., camera and OPA LIDAR, and fuses them in the data collection stage. The image captured by camera is used as a means of coarse scene understanding and the detection results are used to steer the LIDAR beams. Figure 1 is the flowchart of our proposed system, which consists of two parts. The first part is object detection in the image. The over detected objects are some bounding boxes of all probable objects. In the second part, based on these bounding boxes and corresponding object attributes, the laser beams are allocated for the next frame.

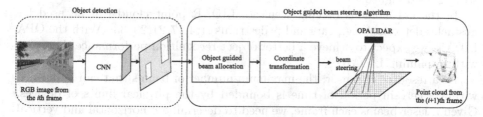

Fig. 1. The flowchart of proposed beam steering algorithm for OPA LIDAR.

2.2 OPA LIDAR Simulator

There is no physical OPA LIDAR device available in the market. However, some famous open source simulators for autonomous vehicles [3,16,26] provide a great convenience for researchers. There are many research works [2,15,22,28] conducted on CARLA simulator [3]. Thus, we verify our proposed beam steering algorithm by using CARLA simulator.

We mount two virtual LIDAR scanners atop an in-game car. One is a mechanical spinning LIDAR for comparison purpose, the other is an OPA LIDAR. The spinning LIDAR is a simulation of Velodyne HDL-32 LIDAR with a 360° horizontal field of view (FOV) and a 40° vertical FOV. It fires approximately 70,000 laser beams per frame. Simultaneously, we simulate an OPA LIDAR by sending arbitrary vertical and horizontal angles of lasers to the simulator and then collecting measurements.

To summary, in our setup, two virtual LIDAR scanners and one camera are placed at the same position for data collection.

2.3 Initial Object Detection

Object detection in image is a deeply researched area in computer vision [9,10, 18–20,23–25]. In our system, the object detection plays a role of coarse scene understanding for OPA LIDAR control. Two issues are emphasized in our system, the first is detection speed, since the detection algorithm runs on the fly. The second is object missing rate. The output (i.e., bounding boxes) of the object detection algorithm is used as a guide to find object in the next frame. So, the object missing rate should be low.

In this paper, we use tiny YOLO v3 [24] for candidate object bounding boxes detection. The confidence level is set to a lower value than the conventional scenario so as to overly detect more objects. For demonstration purpose, only cars and pedestrians are used as targets.

2.4 Beam Allocation

The OPA LIDAR needs horizontal and vertical angles parameters for lasers irradiating. To take the advantage of the OPA LIDAR, an intelligent beam allocation strategy is designed.

In the autonomous driving scenario, LIDAR point cloud can be used for obstacles detection, e.g., cars and pedestrians [1,4,17,21,29–33]. With the OPA LIDAR, we expect to achieve a better object recognition performance than that with a spinning LIDAR.

As a resource of sensor, the maximum number of beams that can be formed within a certain period of time is bounded by the physical limits of devices. Given n laser beams each frame, we need to determine n horizontal and vertical angles. In designing the beam allocation policy, we expect the LIDAR will focus on the more important objects, but still with attention on the background in case some abrupt obstacles appear.

The total n laser beams are divided into two parts. The first part with at most γn laser beams are used to scan the area of interest, where γ is the percentage of the whole laser beams. We set γ to 0.5. The second part is used to scan the area uniformly with fixed resolution. Like Velodyne HDL-32 LIDAR, we use 32 laser beams in the vertical direction to cover a $40°$ vertical FOV.

For simplicity, the virtual LIDAR and camera are placed at the same position. According to the camera model, we can convert pixel coordinate in the image coordinate system to horizontal and vertical angles of laser beams in LIDAR coordinate system.

Supposing the image size is $w \times h$ pixels. Given the pixel coordinate $p = (u, v)$, we can get horizontal and vertical angles from pixel coordinate, which display in formula as θ_h and θ_v, respectively.

$$\theta_h = \arctan((\frac{w}{2} - u) * \frac{-1}{f}) \tag{1}$$

$$\theta_v = \arctan(\frac{(\frac{h}{2} - v)}{\sqrt{(\frac{w}{2} - u)^2 + f^2}}) \tag{2}$$

where f is obtained by multiplying the focal length by the number of pixels per unit length.

Supposing there are L bounding boxes $\{B_1, B_2, ..., B_L\}$, with each bounding box having three attributes, i.e., area in image space, distance from LIDAR and type of object (car or pedestrian). For a candidate object with bounding box B_i, the number of laser beams allocated is $P_i \times \gamma n$, where the ratio P_i is relevant to the position, confidence score and object type of the bounding box prediction and can be computed as follows:

$$P_i = \frac{q_i}{\sum\limits_{j=1}^{L} q_j} \tag{3}$$

where

$$q_i = S_{obj,i} \times S_{dist,i} \times S_{conf,i} \tag{4}$$

$S_{obj,i}$ is an object class based weight. Different objects have different importance in the autonomous driving scenario. For example, the pedestrian on road is expected to be detected earlier and more accurate than the car. In this paper, the weight is manually set by experience.

$S_{dist,i}$ is the weight for distance of object from the OPA LIDAR. The difficulties of object detection at various distances are different. $S_{dist,i}$ increases with increasing distance, since objects that are farther away are more difficult to detect. Here, $S_{dist,i}$ is the sigmoid function:

$$S_{dist,i} = \sigma(d_i; \alpha, d_0) = \frac{1}{1 + e^{\alpha(d_i + d_0)}} \tag{5}$$

where α and d_0 are the hyper parameters.

$S_{conf,i}$ is the weight for confidence output of the object detection algorithm. In general, object with lower confidence needs more LIDAR beam resources. Thus, $S_{conf,i}$ should increase with decreasing of the confidence level of object. Here, the $S_{conf,i}$ is inversely proportional to the sigmoid function $\sigma(conf)$:

$$S_{conf,i} = \sigma(c_i; \beta, c_0) = \frac{1}{1 + e^{\beta(c_i+c_0)}} \tag{6}$$

where β and c_0 are the hyper parameters.

Supposing there are n_{B_i} pixels within the ith bounding boxes, we can get n_{B_i} horizontal and vertical angles of laser beams. If $n_{B_i} > P_i \times \gamma n$, we sample $P_i \times \gamma n$ pixels uniformly for conversion to horizontal and vertical angles of laser beams. If $n_{B_i} < P_i \times \gamma n$, n_{B_i} pixels are all used for conversion and thus P_i is updated to $\frac{n_{B_i}}{\gamma n}$. Consequently, $(1 - \sum_{j=1}^{L} P_j \times \gamma)n$ laser beams are distributed uniformly with fixed resolution like spinning LIDAR.

In the simulator, we use 800×600 RGB image with a FOV of $90°$ and f is 400. The scanning results of the two kinds of LIDAR are shown in Fig. 2.

Fig. 2. Scanning results projected to the image plane. **Left**: the OPA LIDAR. **Right**: the spinning LIDAR.

3   Point Cloud Segmentation of OPA LIDAR

To evaluate effectiveness of our proposed beam steering algorithm, the 3D LIDAR point cloud collected at next frame is used for object segmentation. A LIDAR point is represented as Cartesian coordinate (x, y, z). Since the 3D

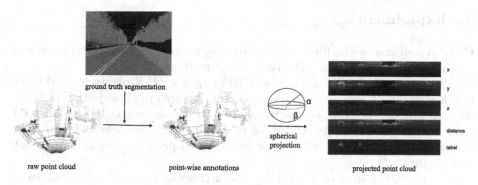

Fig. 3. The spherical projection process from raw point cloud and ground truth semantic segmentation to a dense spherical image with labels.

LIDAR point cloud is usually sparse and huge, spherical projection is used to transform the LIDAR point cloud data to a dense, grid-based representation, as that in SqueezeSeg [30] and PointSeg [29]. The projection grid is defined as follows:

$$\alpha = \arcsin\left(\frac{z}{\sqrt{x^2 + y^2 + z^2}}\right), v = \lfloor \frac{\alpha}{\Delta\alpha} \rfloor \tag{7}$$

$$\beta = \arcsin\left(\frac{y}{\sqrt{x^2 + y^2}}\right), u = \lfloor \frac{\beta}{\Delta\beta} \rfloor \tag{8}$$

where α and β are the azimuth and zenith angles, as shown in Fig. 3. $\Delta\alpha$ and $\Delta\beta$ are the resolutions and (u, v) is index which denotes the position of a point on the 2D spherical image. If more than one point is projected to the same grid (u, v), we only use the point that is closest to the LIDAR.

By applying Eqs. (7) and (8) to each LIDAR point, we obtain a tensor of size $H \times W \times C$, where H is set to 32. Considering that in a real self-driving system most attentions are focusing on the front view, we filter the LIDAR data only to consider the front view area $(-45°, 45°)$ and discretize it into 560 bins, so W is 560. C is the number of channel of input data. For each LIDAR point, we consider 4 feature channels, i.e., coordinate values x, y, z, and distance $d = \sqrt{x^2 + y^2 + z^2}$. Hence, we can obtain the transformed tensor as $32 \times 560 \times 4$. The new representation of 3D point cloud is fed into convolutional layers.

For each point of both spinning LIDAR and OPA LIDAR, we need to assign one of the class labels, i.e., car or pedestrian or background. With the LIDAR-camera transformation matrix T and the camera intrinsic matrix P, we can project each LIDAR point to a pixel in the transformed coordinate system. A LIDAR point and its corresponding pixel are considered to have the same class label. The semantic segmentation result provided by the simulator is used as ground truth. Therefore, we can apply segmentation labels for point-wise annotations.

4   Experiment

CARLA contains two highly realistic urban towns with cars, pedestrians, as well as buildings and vegetation. The two towns are shown in Fig. 4.

We collect raw data that includes RGB images, LIDAR points, ground truth depth images and semantic segmentation results. Spherical projection is firstly used to generate training and testing data and corresponding labels. Since for point cloud segmentation in outdoor scene, PointSeg [29] has achieved impressive results on KITTI dataset [8], we use PointSeg as point segmentation network architecture.

We use 4302 frames of Town 1 for model training and 3045 frames of Town 2 for evaluation as an unseen environment.

Fig. 4. Two CARLA towns. **Left**: Town 1 (for training). **Right**: Town 2 (for testing). Map on top, sample view from camera below.

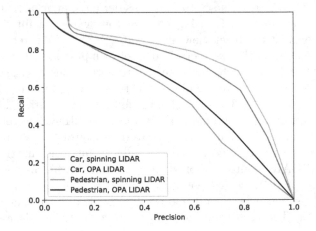

Fig. 5. Recall-Precision curves.

Table 1. Point cloud segmentation performance comparison between OPA LIDAR and spinning LIDAR.

IoU (%)	Car	Pedestrian
Spinning LIDAR	42.4	17.1
OPA LIDAR	45.1	18.7

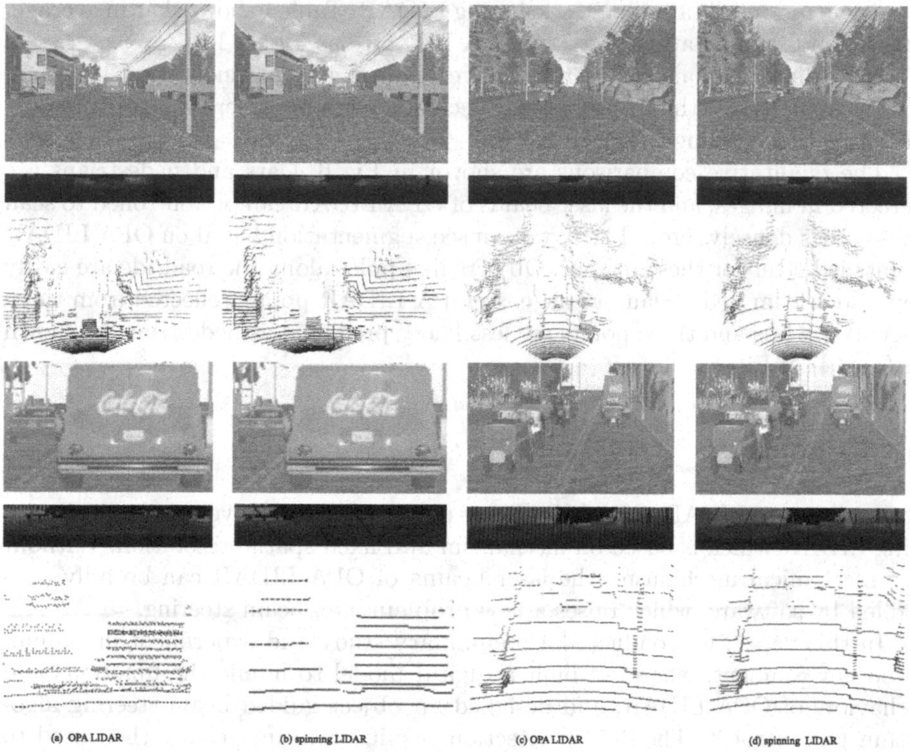

(a) OPA LIDAR   (b) spinning LIDAR   (c) OPA LIDAR   (d) spinning LIDAR

Fig. 6. The first row shows RGB images. The second row shows network predictions. The third row shows results that are projected back to the 3D space. The last three rows are the zoom-in view.

According to our proposed object guided beam steering algorithm, tiny YOLO v3 is firstly used for candidate object detection. In the next frame, OPA LIDAR points reflected from these regions of interest are much more than spinning LIDAR points, while OPA LIDAR points reflected from background areas are less than spinning LIDAR points. Obviously, when segmenting point cloud, the reduction of the number of background points improve the accuracy of car and pedestrian segmentation. What is more, objects that may hardly be detected based on LIDAR points can be detected easily based on image. Therefore, our proposed algorithm can be considered as a new direction of

LIDAR-camera fusion. It's important to detect more objects in image and we set the confidence thresh to 0.05.

The segmentation network outputs three confidence maps that are used to draw recall-precision curves. For evaluation, we also use intersection over union (IoU) metric that quantifies the percent overlap between the object mask and prediction output.

The recall-precision curves are shown in Fig. 5. The OPA LIDAR has higher precision and recall rate than the spinning LIDAR. Table 1 shows that in average, point cloud segmentation based on OPA LIDAR has higher IoU.

According to quantitative results, we confirm that segmentation based on the OPA LIDAR with our proposed object guided beam steering algorithm outperforms the spinning LIDAR.

The qualitative comparisons are shown in Fig. 6. Cars and pedestrians are detected in images, and the laser beams of OPA LIDAR can be controlled to scan these areas densely. From Fig. 6, we can see segmentation based on OPA LIDAR points is better for these regions. Objects like poles along the roadside are easily separated from pedestrian in image, so OPA LIDAR points reflected from these regions are less and these points are less likely predicted as pedestrian, as shown in (c), (d) of Fig. 6.

5   Conclusion

As a solid state LIDAR, OPA LIDAR has many advantages over mechanical spinning LIDAR which is based on mechanism and fixed spatial resolution. Without the mechanical mechanism, the laser beams of OPA LIDAR can be fully controlled by software, which raises a new problem, i.e., beam steering.

In this paper, we conducted a preliminary study and experiment in a simulated environment, where we built a digital model to mimic the functions and behaviors of OPA LIDAR, and designed an object guided beam steering algorithm to control it. The object detection results from images are then used to compute the horizontal and vertical angles of laser beams. Our semantic segmentation of 3D point cloud experimental results show that the OPA LIDAR can have denser resolution over candidate object and the segmentation accuracy is higher than a mechanical spinning LIDAR.

Our work can be further improved in many directions. Currently, the beam allocation policy is designed according to experiences, which can be improved by a learning strategy. Another future work that can be done is the criterion of beam steering algorithm. Finally, how the OPA LIDAR can be used in a fully autonomous driving car is also an interesting problem. The algorithm for a new active perceptual device is completely different from the traditional device.

References

1. Bo, L.: 3d fully convolutional network for vehicle detection in point cloud (2016)
2. Brekke, Å., Vatsendvik, F., Lindseth, F.: Multimodal 3d object detection from simulated pretraining. arXiv preprint arXiv:1905.07754 (2019)
3. Dosovitskiy, A., Ros, G., Codevilla, F., Lopez, A., Koltun, V.: CARLA: an open urban driving simulator. In: Proceedings of the 1st Annual Conference on Robot Learning, pp. 1–16 (2017)
4. Douillard, B., et al.: On the segmentation of 3d lidar point clouds. In: IEEE International Conference on Robotics & Automation (2011)
5. Eldada, L.: Planar beam forming and steering optical phased array chip and method of using same, 5 September 2017. US Patent 9,753,351
6. Eldada, L.: Three-dimensional-mapping two-dimensional-scanning lidar based on one-dimensional-steering optical phased arrays and method of using same, 16 January 2018. US Patent 9,869,753
7. Fujita, J., Eldada, L.: Low cost and compact optical phased array with electro-optic beam steering, 3 May 2018. US Patent App. 15/342,958
8. Geiger, A., Lenz, P., Stiller, C., Urtasun, R.: Vision meets robotics: the kitti dataset. Int. J. Robot. Res. **32**(11), 1231–1237 (2013)
9. Girshick, R.: Fast R-CNN. In: The IEEE International Conference on Computer Vision (ICCV), December 2015
10. Girshick, R., Donahue, J., Darrell, T., Malik, J.: Rich feature hierarchies for accurate object detection and semantic segmentation. In: The IEEE Conference on Computer Vision and Pattern Recognition (CVPR), June 2014
11. Hall, D.S.: High definition lidar system, 28 June 2011. US Patent 7,969,558
12. Hall, D.S.: Color lidar scanner, 18 March 2014. US Patent 8,675,181
13. Heck, M.J.: Highly integrated optical phased arrays: photonic integrated circuits for optical beam shaping and beam steering. Nanophotonics **6**(1), 93 (2017)
14. Jensen, T., Siercks, K.: Laser scanner, 26 April 2011. US Patent 7,933,055
15. Kaneko, M., Iwami, K., Ogawa, T., Yamasaki, T., Aizawa, K.: Mask-slam: robust feature-based monocular slam by masking using semantic segmentation. In: 2018 IEEE/CVF Conference on Computer Vision and Pattern Recognition Workshops (CVPRW) (2018)
16. Kato, S., Takeuchi, E., Ishiguro, Y., Ninomiya, Y., Takeda, K., Hamada, T.: An open approach to autonomous vehicles. IEEE Micro **35**(6), 60–68 (2015)
17. Ku, J., Mozifian, M., Lee, J., Harakeh, A., Waslander, S.L.: Joint 3d proposal generation and object detection from view aggregation. In: 2018 IEEE/RSJ International Conference on Intelligent Robots and Systems (IROS), pp. 1–8. IEEE (2018)
18. Lin, T.Y., Dollár, P., Girshick, R., He, K., Hariharan, B., Belongie, S.: Feature pyramid networks for object detection. In: Proceedings of the IEEE Conference on Computer Vision and Pattern Recognition, pp. 2117–2125 (2017)
19. Lin, T.Y., Goyal, P., Girshick, R., He, K., Dollár, P.: Focal loss for dense object detection. In: Proceedings of the IEEE International Conference on Computer Vision, pp. 2980–2988 (2017)
20. Liu, W., et al.: SSD: Single Shot MultiBox Detector. In: Leibe, B., Matas, J., Sebe, N., Welling, M. (eds.) ECCV 2016. LNCS, vol. 9905, pp. 21–37. Springer, Cham (2016). https://doi.org/10.1007/978-3-319-46448-0_2
21. Qi, C.R., Liu, W., Wu, C., Su, H., Guibas, L.J.: Frustum PointNets for 3D object detection from RGB-D data. In: Proceedings of the IEEE Conference on Computer Vision and Pattern Recognition, pp. 918–927 (2018)

22. Kiran, B.R., et al.: Real-time dynamic object detection for autonomous driving using prior 3D-Maps. In: Leal-Taixé, L., Roth, S. (eds.) ECCV 2018. LNCS, vol. 11133, pp. 567–582. Springer, Cham (2019). https://doi.org/10.1007/978-3-030-11021-5_35

23. Redmon, J., Divvala, S., Girshick, R., Farhadi, A.: You only look once: unified, real-time object detection. In: The IEEE Conference on Computer Vision and Pattern Recognition (CVPR), June 2016

24. Redmon, J., Farhadi, A.: Yolov3: an incremental improvement. arXiv preprint arXiv:1804.02767 (2018)

25. Ren, S., He, K., Girshick, R., Sun, J.: Faster R-CNN: towards real-time object detection with region proposal networks. In: Cortes, C., Lawrence, N.D., Lee, D.D., Sugiyama, M., Garnett, R. (eds.) Advances in Neural Information Processing Systems, vol. 28, pp. 91–99. Curran Associates, Inc. (2015). http://papers.nips.cc/paper/5638-faster-r-cnn-towards-real-time-object-detection-with-region-proposal-networks.pdf

26. Shah, S., Dey, D., Lovett, C., Kapoor, A.: AirSim: high-fidelity visual and physical simulation for autonomous vehicles. In: Hutter, M., Siegwart, R. (eds.) Field and Service Robotics. SPAR, vol. 5, pp. 621–635. Springer, Cham (2018). https://doi.org/10.1007/978-3-319-67361-5_40

27. Skirlo, S., et al.: Methods and systems for optical beam steering, 16 April 2019. US Patent App. 10/261,389

28. Sobh, I., et al.: End-to-end multi-modal sensors fusion system for urban automated driving (2018)

29. Wang, Y., Shi, T., Yun, P., Tai, L., Liu, M.: PointSeg: real-time semantic segmentation based on 3D lidar point cloud (2018)

30. Wu, B., Wan, A., Yue, X., Keutzer, K.: SqueezeSeg: Convolutional neural nets with recurrent CRF for real-time road-object segmentation from 3d lidar point cloud. arXiv preprint arXiv:1710.07368 (2017)

31. Wu, B., Zhou, X., Zhao, S., Yue, X., Keutzer, K.: SqueezeSegV2: improved model structure and unsupervised domain adaptation for road-object segmentation from a lidar point cloud. arXiv preprint arXiv:1809.08495 (2018)

32. Xu, D., Anguelov, D., Jain, A.: PointFusion: deep sensor fusion for 3D bounding box estimation. In: Proceedings of the IEEE Conference on Computer Vision and Pattern Recognition, pp. 244–253 (2018)

33. Zhou, Y., Tuzel, O.: VoxelNet: end-to-end learning for point cloud based 3D object detection. In: Proceedings of the IEEE Conference on Computer Vision and Pattern Recognition, pp. 4490–4499 (2018)

Channel Max Pooling for Image Classification

Lu Cheng[1], Dongliang Chang[1(✉)], Jiyang Xie[1], Rongliang Ma[2], Chunsheng Wu[3], and Zhanyu Ma[1]

[1] Beijing University of Posts and Telecommunications, Beijing 100876, China
{chengl,changdongliang,xiejiyang2013,mazhanyu}@bupt.edu.cn
[2] Institute of Forensic Science, Ministry of Public Security, Beijing 100876, China
marl2013@163.com
[3] Forensic Science Institution of Beijing Public Security Bureau, Beijing 100876, China
wucs@sccas.cn

Abstract. A problem of deep convolutional neural networks is that the channel numbers of the feature maps often increases with the depth of the network. This problem can result in a dramatic increase in the number of parameters and serious over-fitting. The 1×1 convolutional layer whose kernel size is 1×1 is popular for decreasing the channel numbers of the feature maps by offer a channel-wise parametric pooling, often called a feature map pooling or a projection layer. However, the 1×1 convolutional layer has numerous parameters that need to be learned. Inspired by the 1×1 convolutional layer, we proposed a channel max pooling, which reduces the feature space by compressing multiple feature maps into one feature map via selecting the maximum values of the same locations from different feature maps. The advantages of the proposed method are twofold as follows: the first is that it decreases the channel numbers of the feature maps whilst retaining their salient features, and the second is that it non-parametric which has no increase in parameters. The experimental results on three image classification datasets show that the proposed method achieves good performance and significantly reduced the parameters of the neural network.

Keywords: Machine learning · Image classification · Deep neural networks · Channel-wise pooling

1 Introduction

Over the past several years, deep learning has gained great success in academic and industry for their excellent performance on various tasks [3,5,17,20,26,31], especially in computer vision. Convolutional neural networks (CNNs), a major part of deep learning, is one of the most powerful tools applied to solve various visual tasks [16,25,29,30]. In a conventional CNN [8,9,11,16,27], the convolutional layers are formulated to generate resulting activations, or called feature maps containing response to different convolutional kernel filters. Then the

Z. Cui et al. (Eds.): IScIDE 2019, LNCS 11935, pp. 273–284, 2019.
https://doi.org/10.1007/978-3-030-36189-1_23

resulting activations of a convolutional layer are passed to a pooling layer. Pooling layer is used to achieve invariance to image transformations, more compact representations, and better robustness to noise and clutter [2]. There are two conventional choices for the pooling function: average pooling [18,19] and max pooling [1,12]. Max pooling usually works better than average-pooling empirically. In addition, max pooling also provides nonlinearity for the network.

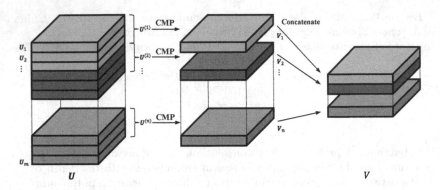

Fig. 1. The structure of the channel max pooling (CMP) layer assuming $k = s = 3$.

It is well known that the deeper a neural network is, the more discriminative features can be extracted [8,11,21,27]. However, an increase in the layer number causes problems such as the gradient disappearance, which leads to a decrease in classification performance. He *et al.* [8] proposed ResNet is a good solution to solve the problem of gradient disappearance through the residual module. DenseNet [11] adopts a structure similar with the ResNet. As a result, the deep neural networks have reached more than 100 layers, and the performance of the models has been greatly improved in many task [7,13,32]. However, the size of CNNs ofter makes overfitting a significant problem, even with a large number of labeled training examples [4,23].

In order to solve the above problems, many researchers have conducted in-depth researches. There are two conventional ways to reducing the parameters in a model: downsampling feature maps and reduce the channel numbers of the feature maps. The pooling layer is a conventional method for downsampling feature maps and includes two classes of methods: the local region pooling [1,12,18,19] and the global region pooling [21]. Especially, the feature maps contain many feature maps of the same size, in which the information in small local are subsampled by the local region pooling. In addition, the local region pooling can also make the network have a certain invariance to image transformations [2]. Therefore, the local region pooling is widely used in models [8,9,11,16,27]. However, even though the local region pooling reduces the parameters of the feature maps, the parameters of the following fully connected layer are still huge [21]. The parameters of the fully connected layer is proportional to the size of the feature maps extracted by the last convolutional layer. Therefore, the global region pooling is often used between the last convolutional layer and the following fully connected layer, which aggregates the information in each feature map into a scalar value [21]. The global region pooling is widely used in some newly proposed networks, such as ResNet [8], DenseNet [11], *etc.* However, the global region pooling has limited benefits for fine-grained image classification, since it usually requires more local information in the feature maps [10]. Moreover, the parameters of fully connected layers in some newly proposed methods are forced to concern channel numbers of the feature maps, for example, bilinear model [22] and its variants. [6]. Therefore, even if the size of the feature maps is reduced, the parameters of the model is still large in those models [6].

In addition to the above methods, the 1×1 convolutional layer is widely used in many deep learning models to decrease the channel numbers of the feature maps by offering a channel-wise pooling, often called feature map pooling or a projection layer [11,28]. Lin *et al.* [21] introduced the 1×1 convolutional layer to generate more discriminative features from the original features, which also can enable the cross-channel information interaction and integration. Furthermore, it can adjust the number of convolution channels. Especially, since the DenseNet needs to concatenate feature maps, the channel numbers of the feature maps are becoming larger and larger; therefore, it uses 1×1 convolutional layer to reduce the channel numbers. Moreover, the 1×1 convolutional layer can significant reduce the parameters of the fully connected layers in some newly proposed methods, which is concerned with the channel numbers. However, the 1×1 convolutional layer has many parameters which should be learned. This means that the 1×1 convolutional layer not only change the channel numbers of original feature maps, but also generates new feature maps. It is irrelevant for the models that trained from scratch, but has a huge influence for the fine-tune with pretrained models, because the 1×1 convolutional layer is randomly initialized which increase the difficulty of fine-tune [29].

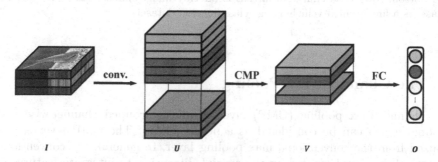

Fig. 2. The CMP-CNN architecture. Note that $I \in R^{3 \times W^I \times H^I}$ and $O \in R^c$ are the input image and the output probability vector for classification respectively, where W^I and H^I are the width and height of I and c is category number; U and V are the input and the output of the CMP layer mentioned above.

Following the aforementioned idea, we propose a new methodology: named as channel max pooling (CMP) to overcome above drawbacks of the aforementioned methods, which reduces the feature space by compressing multiple feature maps into one feature map via selecting the maximum values of the same locations from different feature maps.

The advantages of the proposed method are the following: (1) the number of parameters in a neural network is reduced such that the network becomes more compact, (2) the generalization ability of the compressed neural network is improved, (3) the CMP layer has no parameters that need to be optimized, and (4) it greatly preserves the salient features of the original features.

In this paper, we focus on two types of image classification tasks: conventional image classification and fine-grained image classification with CMP modified CNNs. In fine-grained image classification, the goal is to identify the fineg-rained categories of the given images, rather than to classify the main categories [14]. Experimental results show that, on the Cars-196 [14] dataset and the cifar-100 dataset [15], a network employed CMP obtains better performance than the one without CMP.

Fig. 3. Examples of the cifar-100 dataset (the upper row), and the Cars-196 dataset (the bottom row). The cifar-100 dataset is an tiny image dataset, and the Cars-196 dataset is a fine-grained vehicle image classification dataset.

2   The Channel Max Pooling

The channel max pooling (CMP) layer conducts grouped channel-wise max pooling, which can be considered as a pooling layer. The CMP layer is generalized from the conventional max pooling layer. In general, the conventional max pooling is implemented on the spatial dimension to integrate features on each feature map. Here, the proposed CMP layer turns the implementation of the conventional max pooling from the spatial dimension to the channel dimension. In this case, the CMP layer would conduct 1-D max pooling for elements on the same location of each feature map. As Fig. 1, the input feature maps $U = \{U_1, \cdots, U_i, \cdots, U_m\} \in R^{m \times W \times H}$ where m, W, and H are the feature channel number, the width, and the height, respectively. Meanwhile,

$U_i \in R^{W \times H}$ is the ith feature map is input into the CMP layer. U can be the output feature maps of any convolutional layer. Then, U is grouped into n groups $\{U^{(1)}, \cdots, U^{(i)}, \cdots, U^{(n)}\}$ with k feature maps of each. Assuming an integer $s \leq k$ as the slide stride between neighbouring groups, we can represent the ith group by feature maps as

$$U^{(i)} = \{U_{s \cdot (i-1)+1}, \cdots, U_{s \cdot (i-1)+k}\},$$
$$i = 1, \cdots, n, \tag{1}$$

and m, n, k, and s satisfy the relationship as

$$m = s \cdot (n - 1) + k. \tag{2}$$

When $k = s$, $i.e.$, no overlap between neighbouring groups, $U^{(i)}$ can be degenerated as

$$U^{(i)} = \{U_{k \cdot (i-1)+1}, \cdots, U_{k \cdot i}\}. \tag{3}$$

In the next step, we implement 1-D channel-wise max pooling for each group on channel dimension obtaining outputs $\{V_1, \cdots, V_n\} \in R^{n \times W \times H}$ as

$$V_{i,w,h} = \max_j U^{(i)}_{j,w,h},$$
$$w = 1, \cdots, W; h = 1, \cdots, H, \tag{4}$$

where $U^{(i)}_{j,w,h}$ is the element located at (w, h) on the jth feature map of the ith group and $V_{i,w,h}$ is the element located at (w, h) on the ith feature map. Finally, concatenating the outputs of each group, we can obtain the output V of the CMP layer. The compression rate r between U and V is equal to $\frac{m}{n}$.

In the implementation, the CMP layer can be placed after any convolutional layer in a convolutional neural network (CNN) replacing the normal max pooling layer and is recommended to apply behind the last convolutional layer of the CNN as the integrator of convolutional features in distinct feature channels. Figure 2 shows the CMP-CNN architecture divided into three parts: a feature extractor, the CMP layer, and a classifier. The feature extractor which usually includes multiple convolutional (conv.) layers and the classifier consisting of a fully connected (FC) network are common components of standard CNN architectures for image classification. Meanwhile, the CMP layer is cascaded between the aforementioned parts as Fig. 2.

In addition, due to the use of the back-propagation and the gradient descend algorithms in deep convolutional neural networks (DCNNs) training, the CMP layer should be derivable for gradient propagation. We apply the trick mentioned in [24] for the CMP layer when training. Here, the error signals in the CMP layer are only propagated to the position at the maximum value of each group.

Table 1. Statistics of datasets

Datasets	#Category	#Training	#Testing
Cifar-100	100	50000	10000
Cars-196	196	8144	8041

Table 2. ConvNet configurations (shown in columns). In particular, VGG16, VGG16-A, and VGG16-B denote modified VGG16 for Cars-196, modified VGG16 with the proposed method, and modified VGG16 with the 1×1 convolutional layer (1×1 conv.). The convolutional layer parameters are denoted as "conv(receptive field size)-(number of channels)". The ReLU activation function is not shown for brevity.

ConvNet Configuration		
VGG16	VGG16-A	VGG16-B
input (224×224 RGB image)		
Conv3-64	Conv3-64	Conv3-64
Conv3-64	Conv3-64	Conv3-64
maxpool		
Conv3-128	Conv3-128	Conv3-128
Conv3-128	Conv3-128	Conv3-128
maxpool		
Conv3-256	Conv3-256	Conv3-256
Conv3-256	Conv3-256	Conv3-256
Conv3-256	Conv3-256	Conv3-256
maxpool		
Conv3-512	Conv3-512	Conv3-512
Conv3-512	Conv3-512	Conv3-512
Conv3-512	Conv3-512	Conv3-512
maxpool		
–	CMP	1×1 conv.
FC-512	FC-512	FC-512
FC-196	FC-196	FC-196
softmax	softmax	softmax

3 Experiment Results and Discussions

3.1 Experiment Setup

Datasets. We evaluate the proposed CMP on two widely used image classification datasets, Cifar-100 [15] and Cars-196 [14] datasets. The detail statistics with category numbers and data splits of the two datasets are summarized in Table 1, and images are shown in Fig. 3 as examples. In order to keep consistency with other methods, datasets are divided into training and test set only. Moreover, we only use the category lables in our experiments, rather than bounding box information.

Implementation Details. To evaluate the classification performance of the proposed CMP layer method on the aforementioned two datasets, we compare it with the following methods: (1) the modified VGG16 and (2) the 1×1 *conv.* Detailed structures see Table 2.

Regarding the implementation of the modified VGG16, we replace the classifier of VGG16 by two FC layers with a batch normalization (BN) layer before each layer and an rectified linear unit (ReLU) layer before the latter one. The hidden node numbers and the feature map sizes input are 256 and $512 \times 1 \times 1$ for CIFAR-100 dataset, and 512 and $512 \times 7 \times 7$ for Cars-196 dataset, respectively.

For the implementation of the 1×1 *conv.*, we insert a 1×1 convolutional layer between feature extractor and classifier of the modified VGG16. Except for this difference, other settings of the 1×1 *conv.* modified VGG16 are identical to the modified VGG16. Especially, the range of output channel is in the set $\{256, 128, 64, 32, 16, 8\}$ with the corresponding compression rate $r \in \{2, 4, 8, 16, 32, 64\}$.

With respect to CMP, we insert a CMP layer between the feature extractor and the classifier of the modified VGG16. Except for this difference, other settings of the CMP modified VGG16 are identical to the modified VGG16. Especially, the range of output channel is same with the 1×1 *conv.* modified VGG16.

All the networks are trained using the stochastic gradient descent (SGD) algorithm in 300 epochs with batch size as 128 for the CIFAR-100 dataset and 32 for the Cars-196 dataset, respectively. The input images are randomly cropped to 32×32 and 224×224 with a padding as 4 for the CIFAR-100 and the Cars-196 datasets. Moreover, data augmentation is applied for image preprocessing, horizontally flipping with a probability of 0.5. The learning rate is initially set as 0.1 and decayed by the cosine annealing schedule [24]. Following, we set a weight decay as 0.0005 and momentum as 0.9. For better performance, we initialize models on the Cars-196 dataset by ImageNet pre-trained models.

Table 3. Experimental results (%) on the Cifar-100 datasets using VGG16 as backbone architecture (trained from scratch). The best result is marked in **bold**.

Methods	Measure	1	2	4	8	16	32	64
VGG16	Acc.	74.49	–					
	params(M)	14.88						
CMP	Acc.	74.49	74.46	74.40	**74.96**	74.92	74.74	73.76
	params(M)	14.88	14.82	14.78	14.77	14.76	14.75	14.75
1×1 *conv.*	Acc.	73.63	73.51	73.42	73.32	73.25	73.95	73.46
	params(M)	15.14	14.95	14.85	14.80	14.77	14.76	14.76

3.2   Evaluation on the Cifar-100 Dataset

We investigate the effectiveness of the proposed method and the 1×1 convolutional layer under different compression factor. The compression factor affects the overall performance and the number of model parameters. We set the value of the compression factor in the set $\{2, 4, 8, 16, 32\}$. Table 3 presents comparisons of classification accuracies obtained by various methods with different compression factors. From Table 3, the CMP achieves the best result on the Cifar-100 dataset when the compression factor is equal to 8.

As the compression factor increases, the accuracy of the CMP and the 1×1 conv. decreases, but the performance of the CMP is always better than the 1×1 conv.

Meanwhile, since the feature map size input the FC layers on the CIFAR-100 dataset is equal to $512 \times 1 \times 1$, we can see that the parameters of the models are only slightly dropped with different compression factors. But the parameters of the CMP is always less than 1×1 conv., due to the non-parametric property of the CMP layer.

Table 4. Experimental results (%) on the Cars-196 datasets with pre-trained VGG16. The best result is marked in **bold**.

Methods	Measure	1	2	4	8	16	32	64
VGG16	Acc.	88.02	–					
	params(M)	27.72						
CMP	Acc.	88.02	**88.67**	87.80	87.64	87.24	86.22	83.48
	params(M)	27.72	21.27	18.05	16.44	15.63	15.23	15.03
1×1 conv.	Acc.	87.98	87.93	87.53	87.07	86.83	86.28	83.57
	params(M)	27.98	21.40	18.11	16.47	15.65	15.24	15.03

3.3   Evaluation on the Care-196 Dataset

The Cars-196 dataset contains 16185 images from 196 car models. This dataset provides the ground-truth labels and the annotations with bounding boxes on both the training set and the test set. In our experiments, we evaluated the proposed method on the Cars-196 dataset only with the ground-truth labels on 196 car model categories.

We investigate the effectiveness of the proposed method and the 1×1 conv. layer under different compression factors. Table 4 presents the comparisons of classification accuracies obtained by various methods with different compression factors.

From Table 4, the CMP also achieves the best result on Cars-196 dataset when the compression factor is equal to 2.

Similarly with the experimental results on the CIFAR-100 dataset, as the compression factor increases, the accuracies of the CMP and the 1×1 *conv.* decrease, but the performance of CMP is always better than the 1×1 *conv.* Meanwhile, as compression factor increases, the accuracy decays faster.

Since the feature map size input the FC layers on Cars-196 dataset is equal to $512 \times 7 \times 7$, we can see that the parameters of the models are significantly reduced with different compression factor, and the parameters of the CMP is still always less than 1×1 *conv.*

Table 5. Experimental results (%) on the Cars-196 datasets with pre-trained VGG16 modified by the bilinear pooling. The best result is marked in **bold.**

Methods	Measure	1	2	4	8	16	32	64
VGG16	Acc.	91.57	–					
	params(M)	149.57						
CMP	Acc.	91.57	**91.85**	91.46	91.12	90.20	88.73	87.07
	params(M)	149.57	48.51	23.25	16.93	15.35	14.96	14.86
1×1 *conv.*	Acc.	91.53	91.39	90.98	90.57	90.26	88.96	87.20
	params(M)	149.83	48.64	23.31	16.96	15.37	14.97	14.86

3.4   Experiment on the Bilinear Structure

To further verification the performance of our proposed method on decreasing the channel numbers of the feature maps whilst maintain their salient features, we evaluate the proposed method on the bilinear structure.

The bilinear structure has shown impressive performance on a wide range of visual tasks, including semantic segmentation, fine grained recognition, and face recognition. However, bilinear features are high dimensional, typically on the order of hundreds ofthousands to a few million, which makes them impractical for subsequent analysis.

In this part, we use the modified VGG16 based on the bilinear structure as the base model to evaluate the performance of the proposed method and the 1×1 *conv.* on decreasing the channel numbers of the feature maps whilst retaining the salient features.

We set the value of the compression factors as $\{2, 4, 8, 16, 32\}$. Table 5 presents the comparisons of the classification accuracies obtained by various methods with different compression factors.

From Table 5, we can observe that the CMP has achieves the best result when the compression factor is equal to 2 which outperforms the full bilinear model, and the parameters of CMP is only 32% of it. Meanwhile, the performance of the 1×1 *conv.* is worse than the full bilinear model in all cases. This demonstrates that the proposed method not only reduce the amount of parameters, but even obtained more discriminative features for classification. When the compression rate is equal to 4 and 8, the accuracies of the proposed CMP are around 0.5%

higher than those of the 1×1 *conv.*, respectively, while they obtain similar performance when the compression rate is no smaller than 16. In addition, the CMP obtains similar performance with the full bilinear model when the compression rate is equal to 4, but the parameter number of the CMP is only 16% of that of the full bilinear model. Therefore, the proposed CMP can make the bilinear models apply to more fields by reducing parameter quantity with performance unreduced.

3.5 Discussions

The experimental results on the Cifar-100 and the Cars-196 datasets illustrate that the referred method, the 1×1 convolutional layer, does not show any performance improvement compared with the base network without the 1×1 convolutional layer.

The reason is that the 1×1 convolutional layer has numerous parameters which should be learned. This means that the 1×1 convolutional layer not only changes the original features, but also generates new features. This is irrelevant for the model that trained from scratch, but has a huge influence for fine-tune with the pre-trained models, because the 1×1 convolutional layer is randomly initialized, which increases the difficulty of fine-tune. Due to the characteristics of the 1×1 convolutional layer, the most salient features of the feature maps is difficult to be kept, so that the models with it cannot work well on these image classification datasets and are difficult to be optimized. In contrast, the model introduced the proposed CMP converges more easily than that with the 1×1 *conv.* and do not introduces new parameters that should be optimized.

The proposed method shows higher accuracies, fewer parameters, and easy optimization, which are mainly attributed to it decreases the channel numbers of the feature maps whilst retaining their salient features. The proposed method does not increase difficulty to parameter optimization. In contrast, since the proposed method reduces the parameter space, the optimization speed is increased. Especially, the proposed method can decrease the channel numbers of the feature maps whilst retaining the size of each feature map. Larger feature maps will contain more spatial information about the objects, which is benifit for fine-grained image classification. Furthermore, the proposed method obtains the best performance on the Cifar-100 and the Cars-196 datasets. The experimental results on these two datasets also suggest that the proposed method obtains good performance regardless of the input size.

4   Conclusion

In this paper, we introduce a new pooling layer for CNNs called channel max pooling (CMP). The proposed method aims to force the neural networks to learn more robust discriminative features leading to improving the performance of image classification while reducing the parameters of the model. The effectiveness and robustness of the proposed method is demonstrated by experiments on two general purpose image datasets. Future work includes replacing all of 1×1 convolutional layer in DenseNet with the proposed method and experimenting on different types of networks as well as different kinds of data, such as speech and text, to evaluate the effectiveness of the proposed pooling layer.

Acknowledgements. This work was supported in part by the National Key Research and Development Program of China under Grant 2018YFC0807205, in part by the National Natural Science Foundation of China (NSFC) No. 61773071, 61922015, in part by the Beijing Nova Program No. Z171100001117049, in part by the Beijing Nova Program Interdisciplinary Cooperation Project No. Z181100006218137, in part by the Fundamental Research Funds for the Central University No. 2018XKJC02, in part by the scholarship from China Scholarship Council (CSC) under Grant CSC No. 201906470049, and in part by the BUPT Excellent Ph.D. Students Foundation No. CX2019109, XTCX201804.

References

1. Boureau, Y.l., Cun, Y.L., et al.: Sparse feature learning for deep belief networks. In: Advances in Neural Information Processing Systems, pp. 1185–1192 (2008)
2. Boureau, Y.L., Ponce, J., LeCun, Y.: A theoretical analysis of feature pooling in visual recognition. In: Proceedings of the 27th International Conference on Machine Learning (ICML 2010), pp. 111–118 (2010)
3. Chen, Y.C., Zhu, X., Zheng, W.S., Lai, J.H.: Person re-identification by camera correlation aware feature augmentation. IEEE Trans. Pattern Anal. Mach. Intell. **40**(2), 392–408 (2018)
4. Cogswell, M., Ahmed, F., Girshick, R., Zitnick, L., Batra, D.: Reducing overfitting in deep networks by decorrelating representations. arXiv preprint arXiv:1511.06068 (2015)
5. Ding, X., Li, B., Xiong, W., Guo, W., Hu, W., Wang, B.: Multi-instance multi-label learning combining hierarchical context and its application to image annotation. IEEE Trans. Multimed. **18**(8), 1616–1627 (2016)
6. Gao, Y., Beijbom, O., Zhang, N., Darrell, T.: Compact bilinear pooling. In: Proceedings of the IEEE Conference on Computer Vision and Pattern Recognition, pp. 317–326 (2016)
7. He, K., Gkioxari, G., Dollár, P., Girshick, R.: Mask R-CNN. In: Proceedings of the IEEE International Conference on Computer Vision, pp. 2961–2969 (2017)
8. He, K., Zhang, X., Ren, S., Sun, J.: Deep residual learning for image recognition. In: Proceedings of the IEEE Conference on Computer Vision and Pattern Recognition, pp. 770–778 (2016)
9. Howard, A.G., et al.: MobileNets: efficient convolutional neural networks for mobile vision applications. arXiv preprint arXiv:1704.04861 (2017)
10. Hu, Q., Wang, H., Li, T., Shen, C.: Deep CNNs with spatially weighted pooling for fine-grained car recognition. IEEE Trans. Intell. Transp. Syst. **18**(11), 3147–3156 (2017)
11. Huang, G., Liu, Z., Van Der Maaten, L., Weinberger, K.Q.: Densely connected convolutional networks. In: Proceedings of the IEEE Conference on Computer Vision and Pattern Recognition, pp. 4700–4708 (2017)
12. Jarrett, K., Kavukcuoglu, K., LeCun, Y., et al.: What is the best multi-stage architecture for object recognition? In: 2009 IEEE 12th International Conference on Computer Vision, pp. 2146–2153. IEEE (2009)

13. Jégou, S., Drozdzal, M., Vazquez, D., Romero, A., Bengio, Y.: The one hundred layers tiramisu: fully convolutional densenets for semantic segmentation. In: Proceedings of the IEEE Conference on Computer Vision and Pattern Recognition Workshops, pp. 11–19 (2017)
14. Krause, J., Stark, M., Deng, J., Fei-Fei, L.: 3d object representations for fine-grained categorization. In: Proceedings of the IEEE International Conference on Computer Vision Workshops, pp. 554–561 (2013)
15. Krizhevsky, A., Hinton, G.: Learning multiple layers of features from tiny images. Technical report, Citeseer (2009)
16. Krizhevsky, A., Sutskever, I., Hinton, G.E.: ImageNet classification with deep convolutional neural networks. In: Advances in Neural Information Processing Systems, pp. 1097–1105 (2012)
17. LeCun, Y., Bengio, Y., Hinton, G.: Deep learning. Nature 521(7553), 436 (2015)
18. LeCun, Y., et al.: Handwritten digit recognition with a back-propagation network. In: Advances in Neural Information Processing Systems, pp. 396–404 (1990)
19. LeCun, Y., Bottou, L., Bengio, Y., Haffner, P., et al.: Gradient-based learning applied to document recognition. Proc. IEEE 86(11), 2278–2324 (1998)
20. Li, B., Xiong, W., Hu, W., Funt, B.: Evaluating combinational illumination estimation methods on real-world images. IEEE Trans. Image Process. 23(3), 1194–1209 (2013)
21. Lin, M., Chen, Q., Yan, S.: Network in network. arXiv preprint arXiv:1312.4400 (2013)
22. Lin, T.Y., RoyChowdhury, A., Maji, S.: Bilinear CNN models for fine-grained visual recognition. In: Proceedings of the IEEE International Conference on Computer Vision, pp. 1449–1457 (2015)
23. Liu, R., Gillies, D.F.: Overfitting in linear feature extraction for classification of high-dimensional image data. Pattern Recognit. 53, 73–86 (2016)
24. Loshchilov, I., Hutter, F.: SGDR: stochastic gradient descent with warm restarts. arXiv preprint arXiv:1608.03983 (2016)
25. Ma, Z., Yu, H., Chen, W., Guo, J.: Short utterance based speech language identification in intelligent vehicles with time-scale modifications and deep bottleneck features. IEEE Trans. Veh. Technol. 68(1), 1–13 (2019)
26. Salamon, J., Bello, J.P.: Deep convolutional neural networks and data augmentation for environmental sound classification. IEEE Signal Process. Lett. 24(3), 279–283 (2017)
27. Simonyan, K., Zisserman, A.: Very deep convolutional networks for large-scale image recognition. arXiv preprint arXiv:1409.1556 (2014)
28. Szegedy, C., et al.: Going deeper with convolutions. In: Proceedings of the IEEE Conference on Computer Vision and Pattern Recognition, pp. 1–9 (2015)
29. Wang, Y., Morariu, V.I., Davis, L.S.: Learning a discriminative filter bank within a CNN for fine-grained recognition. In: Proceedings of the IEEE Conference on Computer Vision and Pattern Recognition, pp. 4148–4157 (2018)
30. Zhang, T., et al.: Predicting functional cortical ROIs via DTI-derived fiber shape models. Cerebral cortex 22(4), 854–864 (2011)
31. Zhou, Y., Fadlullah, Z.M., Mao, B., Kato, N.: A deep-learning-based radio resource assignment technique for 5G ultra dense networks. IEEE Netw. 32(6), 28–34 (2018)
32. Zhu, J.Y., Park, T., Isola, P., Efros, A.A.: Unpaired image-to-image translation using cycle-consistent adversarial networks. In: Proceedings of the IEEE International Conference on Computer Vision, pp. 2223–2232 (2017)

A Multi-resolution Coarse-to-Fine Segmentation Framework with Active Learning in 3D Brain MRI

Zhenxi Zhang[1], Jie Li[1], Zhusi Zhong[1], Zhicheng Jiao[2], and Xinbo Gao[1(✉)]

[1] School of Electronic Engineering, Xidian University, Xi'an 710071, China
xbgao@mail.xidian.edu.cn
[2] Perelman School of Medicine at University of Pennsylvania, Philadelphia, USA

Abstract. Precise segmentation of key tissues in medical images is of great significance. Although deep neural networks have achieved promising results in many medical image segmentation tasks, it is still a challenge for volumetric medical image segmentation due to the limited computing resources and annotated datasets. In this paper, we propose a multi-resolution coarse-to-fine segmentation framework to perform accurate segmentation. The proposed framework contains a coarse stage and a fine stage. The coarse stage with low-resolution data provide high semantic cues for the fine stage. Moreover, we embed active learning processes into coarse-to-fine framework for sparse annotation, the proposed multiple query criteria active learning methods can select high-value slices to label. We evaluated the effectiveness of proposed framework on two public brain MRI datasets. Our coarse-to-fine networks outperform other competitive methods under the condition of fully supervised training. In addition, the proposed active learning method only need 30% to 40% slices of one scan to produce relatively better dense prediction results than non-active learning method and one query criteria active learning methods.

Keywords: Deep learning · Coarse-to-fine framework · Active learning · Sparse annotation · Tissue segmentation

1 Introduction

Automatic segmentation of medical images such as computed tomography (CT) and magnetic resonance imaging (MRI) has become an active topic in the field of medical image processing and computer vision. It is of great significance for clinical diagnosis, pathological analysis, and surgical planning. With the rapid development of deep neural networks (DNNs), a variety of deep learning methods have been proposed for medical image processing and analysis [1–9, 39, 40]. However, it is still a challenge for volumetric medical image segmentation due to the limited GPU memory and less available 3D data annotation.

A few methods [3–5] use 2D fully convolutional networks (FCNs) to perform segmentation slice by slice. Although some works [4, 5] design a view aggregation

Z. Cui et al. (Eds.): IScIDE 2019, LNCS 11935, pp. 285–298, 2019.
https://doi.org/10.1007/978-3-030-36189-1_24

strategy to fuse different output from three views (axial, coronal and sagittal plane), they lose some spatial contexts of the third dimension in 3D data. 2D FCNs can use the whole 2D slices of one view, but it needs higher computational cost and memory consumption when 3D FCN meets the whole 3D data. Patch-based method [6–8], 2D patch or 3D patch, lose lots of global contexts of the whole tissue, and the sliding window strategy to predict each pixel is slow and redundant due to the overlapping patches. Thus, several coarse-to-fine segmentation frameworks [3, 14–16, 41] have been proposed. But these methods mostly focus on the problem of segmenting the small target on a high-resolution image. The coarse stage is to find a small target region and crop it for the fine stage. The fine stage supplements the target details. The current coarse-to-fine framework is effective for segmenting small target. However, this framework is not suitable for solving the whole image segmentation problems, such as whole-brain segmentation, different tissues have complex structures and irregular boundaries. So, we propose a multi-resolution coarse-to-fine segmentation network. Specifically, the coarse stage with low resolution input localizes the tissues and the fine stage with original resolution input refines the tissues' boundaries. We further study the feature fusion between two segmentation stages.

Secondly, methods in the DNNs fashions require a large number of annotated datasets to train a segmentation network and gain improved performance. As well known, obtaining manual per-pixel annotation of medical images is quite time-consuming and labor-intensive. Adaptive selecting the most informative to-be-annotated samples from the unlabeled samples has the potential to achieve maximum performance of the segmentation model with minimum labeling efforts [22]. 3D u-net [2] got dense volumetric segmentation from sparse annotation, which can reduce the annotation efforts, but it is lack of in-depth research on how to select informative slices to label. Several previous works [22, 24–28] use active learning (AL) methods to select informative samples to train deep learning models. It has certain significance to combine active learning with our proposed multi-resolution coarse-to-fine segmentation networks, because low-resolution annotation in the coarse stage can reduce a lot of annotation cost compared with direct annotation of original resolution image, and high-resolution annotation in the fine stage can supplement more segmentation details. So, we propose a coarse-to-fine selective annotation framework based on AL which serves for coarse-to-fine segmentation framework (Fig. 1).

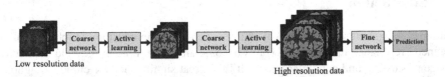

Fig. 1. The proposed multi-resolution coarse-to-fine segmentation framework. The architectures of coarse network, fine network, and the method of proposed active learning method will be shown in detail in the figures below.

The contributions of this paper are as follows: (a) we propose a multi-resolution coarse-to-fine segmentation network for whole-brain segmentation. (b) We study the feature fusion strategies between the coarse stage and the fine stage. (c) We propose a multiple query criteria active learning method for slice selection, and incorporate this method into coarse-to-fine segmentation framework. (d) We evaluate the proposed framework on two public brain MRI datasets, and gain segmentation improvement compared with previous competitive methods. Furthermore, the proposed multiple query criteria active learning method helps the coarse-to-fine framework achieve relatively better results with less labeling cost.

2 Related Work

2.1 U-Net like Models in Medical Image Segmentation

U-net like models [1, 2, 9–11] are widely used in deep learning based medical image segmentation. These models have two paths, the left path encodes semantic contexts by convolutions and down-samplings. The right path combines the high-level semantic contexts with fine local information by skip connections, and gets better segmentation results by up-samplings and convolutions. 3D u-net [2] and V-net [9] are the 3D implementation of u-net. V-net introduces residual connection into a structure with better optimization and convergence. What issues need to be addressed in using this structure? The first one is the depth of down-sampling. The depth of down-sampling may be determined by specific tasks and related to the input size, and it is important to make full use of multi-scale features. The input size in [10] is set as $24 \times 96 \times 96$ voxels. To avoid neglecting information along first axis, low layers use flat kernels, and set down-sampling stride as $1 \times 2 \times 2$. Another concern is the semantic gap between encoder and decoder. Zhang [12] pointed out that simple fusion of low-level and high-level features may be less effective. Unet++ [11] redesigned skip pathways to reduce the semantic gap between encoder and decoder sub networks. The last point is the feature extraction capability of sub networks. Many models use common $3 \times 3 \times 3$ filter kernels in sub networks, which share the same size receptive fields in each layer.

2.2 Coarse-to-Fine Segmentation Networks

There are two main reasons for the popularity of coarse-to-fine framework in medical image segmentation. Firstly, as the limitation of GPU memory, the deep networks can't accept the whole 3D data or a high resolution 2D image as input. Another reason is that fine segmentation results with high resolution are more useful in clinic situations. Many coarse-to-fine segmentation networks [3, 14–16, 41] are proposed to segment some organs which occupy a small region in the input image, such as the pancreas segmentation from abdominal CT scans. Feature fusion between the coarse stage and the fine stage is worth studying. The algorithms in [3, 14] only used the coarse stage to provide localization of target for the fine stage, ignored the multi-stage visual cues. The approaches of [15] addressed this problem by using the probability map produced in the previous stage to generate an updated input image. Zhong et al. proposed a two stage attention-guided deep regression model for landmark detection in [41]. Another concern is that unlike abdominal organ segmentation, the boundaries between different brain tissues are always irregular. Crop operation is not suitable for this situation. So, we design a multi-resolution coarse-to-fine framework inspired by [16, 17]. Han et al. [16] proposed multi-resolution V-net networks to segment thoracic organs in CT images. The submission results rank first in the SegTHOR 2019 challenge. Tokunaga et al. [17] used three different magnification images with different spatial resolutions as input to train three expert CNNs and aggregate these features by estimated weights.

2.3 Active Learning in Deep Learning

Previous works [22, 24–28] use AL to select valuable samples for deep learning models in different computer vision tasks, aiming to achieve maximum performance with minimum annotation [24, 25] select to-be-annotated samples based on uncertainty estimation and similarity estimation. Yang et al. [24] trained a lot of FCNs to calculate the uncertainty of pixels, while [25] used Monte Carlo (MC) dropout [29] to estimate uncertainty. Both of two works estimate similarity based on the feature from network layers. In [26], a sparse annotation strategy based on attention-guided active learning was proposed for 3D medical image segmentation. In [27, 28], the authors compute the entropy of the unlabeled samples to select large entropy ones.

3 Materials and Methods

3.1 Dataset

(1) CANDI dataset [32, 33] has 103 T1-weighted MRI brain scans and labels of 39 brain structures segmented by expert manually. The dataset is provided by Child and Adolescent Neuro Development Initiative (CANDI) at UMass Medical School and publicly available on (https://www.nitrc.org/projects/candi_share). The MRI scans have 256×256 pixels in the coronal plane, and the number of slices ranges from 128 to 158. We assigned 39 structures into GM, WM, CSF, and background referring to [13].

(2) IBSR18 consists of 18 T1-weighted MRI brain scans, which is publicly available from the Internet Brain Segmentation Repository (IBSR) [34], and provides segmentation ground truth with four-class tissues (GM, WM, CSF, and background) for evaluation (https://www.nitrc.org/projects/ibsr). All volumes have $256 \times 128 \times 256$ voxels.

3.2 Coarse-to-Fine Framework

The proposed coarse-to-fine framework is illustrated in Fig. 2. For the coarse stage, we resize the coronal slices from 256×256 to 128×128 by bilinear interpolation, the ground truth by nearest interpolation. The input to the coarse stage is set as $128 \times 32 \times 128$ voxels. We aim to get the global information of the whole coronal slices and as much spatial information as possible from another dimension. We use u-net like models in coarse stage. There are some details to be annotated. To tackle the anisotropic spacing ($128 \times 32 \times 128$), we use flat kernel size ($3 \times 1 \times 3$) for convolution and flat stride $(2 \times 1 \times 2)$ for max pooling. The number of down-sampling is set according to specific task. For example, V-net [9] downsamples 4 times for prostate segmentation. 3D u-net [2] downsamples 3 times for the Xenopus kidney segmentation. In order to preserve more information of details, we only downsample twice. The residual blocks are used to extract features.

SEnet (Squeeze and Excitation block), proposed in [18], won the first place in ILSVRC 2017 classification task. A series of studies [19–21] have been produced based on [18]. Cbam [19] proposed channel attention module and spatial attention module to boost representation power of CNNs. Three variants of SE modules (cSE, sSE, scSE) were proposed in [20] and incorporated in three popular F-CNNs to achieve improvements in different segmentation tasks. In the skip connection between encoder and decoder, we add scSE [20] modules to adaptively rescale the features from encoder and reduce the semantic gap between encoder and decoder. As shown in Fig. 3, scSE has two branches. The first one is channel attention mechanism (CAM). The CAM is the original SEnet [18] in essence. Another one is spatial attention mechanism (SAM).

In the fine segmentation stage, we also use u-net like model. Several details are different from the coarse segmentation stage. The first is the input size which is set as $256 \times 8 \times 256$. The purpose of this set is to produce better segmentation result for

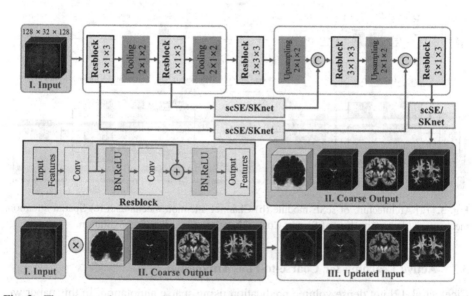

Fig. 2. The proposed coarse-to-fine segmentation framework. In coarse stage, we use scSE modules to rescale features. In the fine stage, we use SKnet to reduce semantic gap between the feature maps of encoder and decoder. The last row shows different input parts of the fine stage.

finer details. Secondly, we use the coarse stage cues in fine stage. The four feature maps generated from the coarse stage provides multi-scale high-level semantic contexts and richer spatial contexts from the third dimension. At the same time, it greatly improves the convergence speed of the fine stage. We try three ways to use the coarse stage cues in the fine training. Firstly, we resize the coarse output to the original resolution. Then we multiply fine input by the resized coarse output and get 4 new updated inputs. We set the fused input to the fine stage as concatenating I and II, I and III, all three parts separately. The results of three fusion ways will be shown in Sect. 4.2.

The last change is that we use the SKnet (Selective Kernel Networks) [21] between encoder and decoder. SKnet was inspired by the receptive field size of visual cortical neurons are adjusted by the stimulus. SKnet combined the operation of SEnet and Inception net [23], and designed a dynamic selection mechanism, in which the features from different kernel branches are fused in an attention-guided way. The SKnet structure is shown in Fig. 3(d), it is divided into two steps generally. First, we use different kernels to generate multi-scale features and aggregate them by simple addition. Then we use global average pooling and fully connected layers to generate different attention tensors and refuse the new feature maps again.

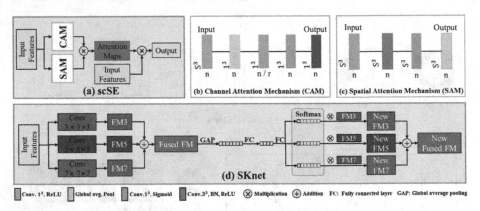

Fig. 3. (a) Architecture of scSE module (b) Channel attention mechanism (c) Spatial attention mechanism (d) Architecture of SKnet. Conv. 3^3 denotes $3 \times 3 \times 3$ convolution.

3.3 Active Learning for Coarse-to-Fine Annotation

Cicek et al. [2] got dense volume predication using sparse annotation. In this paper we propose a deep multiple query criteria active learning framework to select informative slices for sparse annotation to train models, and we integrate the active learning process into coarse-to-fine training process. Followed by [25, 29–31], which use dropout as a Bayesian approximation, we add dropout layers in proposed networks and get 10 Monte Carlo (MC) samples of the predication by using dropout at inference time. Then we compute four uncertainty measures for slice selection.

Predication Variance. In testing time, we get 10 MC samples of predication. We use the variance of 10 MC samples as a measure of prediction uncertainty. Then we cut the predictions into coronal slices, and compute the slice predication variance by adding voxel-wise variance values.

Entropy. The predication is generated from the four-class probability maps. For a given voxel x, the voxel-wise uncertainty U_x can be estimated by entropy over 10 MC probability maps as formula (1).

$$U_x = \sum_{i=1}^{10} \sum_{c=1}^{4} P_i^c(x) log P_i^c(x) \tag{1}$$

Suppose the total number of pixels in one slice is N. The slice-wise uncertainty U_s is measured by formula (2).

$$U_s = \sum_{n=1}^{N} U_x \tag{2}$$

Dice Score. Dice similarity coefficients are usually used to measure the segmentation accuracy between segmentation results and ground truth. We use Dice score between samples as a measure of uncertainty. We divide 10 samples into 5 groups, and compute average Dice score over 5 groups in a slice-wise manner. S_i means the different predication of the slice to-be-computed.

$$Avg\ Dice = \frac{1}{5} \sum_{i=1}^{5} \frac{2|S_i \cap S_{2i}|}{|S_i| + |S_{2i}|} \tag{3}$$

Weighted Rank Aggregation. We first choose the slices which have higher predication variance, higher entropy or lower average Dice as preferred annotations separately in one scan. Then we use a multiple query criteria active learning method [35] based on all these three measures. Compared with traditional one query criteria active learning method, the multiple query criteria [35] can make full use of all three measures and has a higher potential performance. The proposed strategy is divided into 3 steps. First, we get three different rankings by 3 query criterions on the priority to be labeled for every slice. Second, we set up the weights of each criterion according to segmentation performance. Third, we implement weighted rank aggregation of each ranking, then select the most valuable slices to label (Fig. 4).

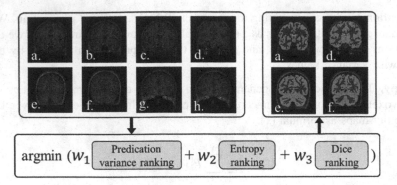

Fig. 4. The proposed multiple query criteria active learning method. The weighted rank aggregation strategy is used to select to-be-labeled slices.

4   Experimental Results

4.1   Training

We evaluated our method on two brain MRI datasets with T1-weighted MRI brain scans. Our model was implemented based on PyTorch deep learning framework. Stochastic gradient descent (SGD) optimization, momentum of 0.9, learning rate of 0.001 were used for network optimization. The number of epochs in different training stages was set to 200 epochs. In the coarse stage, we used a batch size of 4 and adopted cross-entropy loss. In the fine stage, we used a batch size of 2 and adopted Generalized Dice Loss (GDL) [42] as a loss function. GDL can be used as a loss function in case of class imbalanced problem. All experiments were implemented on a NVIDIA Titan XP Pascal card with 12 GB GPU memory.

4.2   Segmentation Results

The performance of proposed coarse-to-fine segmentation framework is evaluated on 121 images. For CANDI dataset, 60 images are used for training, 14 images are used for validation, and 29 images are used for testing. IBSR18 dataset is another independent test set. The segmentation results are shown in Fig. 5.

To validate the effectiveness of the proposed coarse-to-fine framework, we compare our method with four segmentation methods with full supervision, 3D u-net [2], V-net [9], RP-Net [13], VoxResnet [38] with same dataset division. Table 1 shows the segmentation performance of CSF, GM and WM on CANDI dataset. Our method outperforms other three methods and gain 1.11%, 1.80%, 3.70% improvement on three tissues respectively compared with the second best method RP-Net [13]. We also study feature fusion ways between coarse stage and fine stage. The different kinds of input to

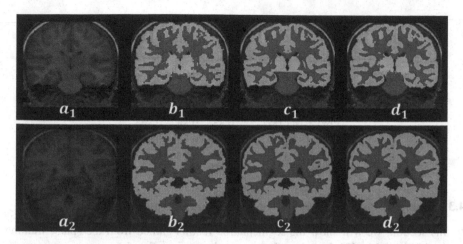

Fig. 5. The first row is one scan from CANDI test set, and the second is one scan from IBSR18 dataset. Column b represents the segmentation result of one coronal slice from coarse stage. Column c represents segmentation result from fine stage. The last column shows ground truth.

the fine stage is shown in Fig. 2. In Table 1, the contents in brackets represent different input to fine stage. Benefitting from the rich semantic information provided by coarse output feature maps, all three fusion ways gain a faster convergence speed (50 epochs) than the original input I (100 epochs). The combination of original input I and updated input I achieves better results. Directly concatenating input I and coarse output II performs worse than two other ways. The reason may be that the feature maps of CSF are not accurate enough (0.792 in coarse stage). The updated III is a tradeoff between fine local information for details and high context information for classification. The reason why the fusion of the three parts has not significantly improved may be the information redundancy between II and III, the updated III includes the same high-level semantic information as II, the addition of II will result in duplication of information. The better segmentation performance on IBSR18 also validates the robustness of our method (Table 2).

Table 1. Dice of CSF, GM, WM for CANDI test set

Method	CSF	GM	WM
3D u-net [2]	0.822	0.925	0.897
V-net [9]	0.810	0.910	0.885
RP-Net [13]	0.875	0.931	0.902
VoxResNet [38]	0.802	0.905	0.882
Coarse stage	0.792	0.917	0.885
Fine stage (I)	0.848	0.940	0.929
Fine stage (I + II)	0.849	0.947	0.933
Fine stage (I + III)	**0.886**	**0.949**	**0.939**
Fine stage (I + II + III)	0.880	0.947	0.932

Table 2. DSC of CSF, GM, WM for IBSR18 dataset

Method	CSF	GM	WM
U-SegNet [36]	0.666	0.903	0.892
Modified U-Net [37]	0.868	0.918	0.907
3D u-net [2]	0.850	0.910	0.899
V-net [9]	0.842	0.905	0.875
VoxResNet [38]	0.833	0.882	0.863
RP-Net [13]	0.882	0.921	0.912
Our method	**0.901**	**0.936**	**0.933**

4.3   Comparison with Different Active Learning Strategies

In this section, we will show the segmentation results gained from the sparse training method followed by [2]. Based on the proposed coarse-to-fine framework, we compare the proposed multiple query criteria active learning method with one non-active learning method and three one query criteria active learning methods, which is guided by predication variance [31], entropy [31], Dice score [30] respectively. The ablation experiments show the efficacy of our proposed weighted rank aggregation strategy. In Fig. 6, we show the change of segmentation accuracy under different annotation ratios. In the initial coarse stage, we use 10% random low resolution annotated slices to pre-train network. In fine stage, we use the high-resolution annotation combined with the low-resolution annotation after up-sampling. According to segmentation performance when deploying only one criterion, the weights of three measures are set as 0.4, 0.4, 0.2 for predication variance, entropy and Dice score respectively. The line charts show the effectiveness of the proposed multiple query criteria active learning method, which achieves comparative results with only 30% to 40% annotated slices, reducing labeling cost to a large extent.

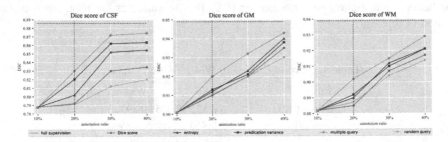

Fig. 6. The comparisons of different strategies. The black dotted line is the dividing line between the coarse stage and the fine stage.

The mean Dice score for CSF, GM, and WM are 0.874, 0.943, and 0.929 respectively when using 40% annotated slices. The segmentation results from different strategies are shown in Fig. 7. The annotation ratio is 40% in both strategies.

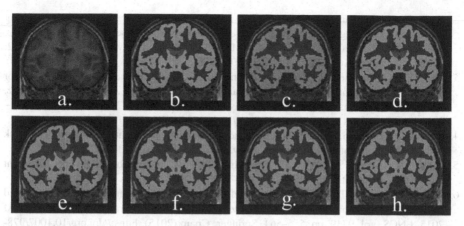

Fig. 7. (a) One coronal slice from a CANDI testing scan (b) Ground truth (c) Pre-training result with 10% random low-resolution annotation (d) Random query (e) Predication variance query (f) Dice score (g) Entropy query (h) The proposed multiple query

5   Conclusion

In this paper, we designed a novel multi-resolution coarse-to-fine segmentation framework, and analyzed its robustness on two public brain MRI datasets in automatically segmenting three key tissues. Our method uses u-net like models in both stages, and reduces the semantic gaps by embedding SEnet or SKnet between encoder and decoder in coarse stage or fine stage respectively. An integration of high-level contextual information from coarse stage and low-level appearance information accelerates the convergence of the fine stage and boosts the segmentation performance. Furthermore, the proposed multiple query criteria active learning method selects valuable slices, which helps the coarse-to-fine framework achieve relatively better results with less labeling cost. The proposed method also offers a powerful framework for other whole 3D medical image segmentation tasks, such as multi-organ segmentation.

Acknowledgements. This work was supported in part by the National Natural Science Foundation of China under Grant 61432014, 61772402, U1605252 and 61671339, and in part by National High-Level Talents Special Support Program of China under Grant CS31117200001.

References

1. Ronneberger, O., Fischer, P., Brox, T.: U-net: convolutional networks for biomedical image segmentation. In: Navab, N., Hornegger, J., Wells, W., Frangi, A. (eds.) MICCAI 2015. LNCS, vol. 9351, pp. 234–241. Springer, Cham (2015). https://doi.org/10.1007/978-3-319-24574-4_28

2. Çiçek, Ö., Abdulkadir, A., Lienkamp, S.S., Brox, T., Ronneberger, O.: 3D U-net: learning dense volumetric segmentation from sparse annotation. In: Ourselin, S., Joskowicz, L., Sabuncu, M., Unal, G., Wells, W. (eds.) MICCAI 2016. LNCS, vol. 9901, pp. 424–432. Springer, Cham (2016). https://doi.org/10.1007/978-3-319-46723-8_49

3. Zhou, Y., Xie, L., Shen, W., Wang, Y., Fishman, E.K., Yuille, A.L.: A fixed-point model for pancreas segmentation in abdominal CT scans. In: Descoteaux, M., Maier-Hein, L., Franz, A., Jannin, P., Collins, D., Duchesne, S. (eds.) MICCAI 2017. LNCS, vol. 10433, pp. 693–701. Springer, Cham (2017). https://doi.org/10.1007/978-3-319-66182-7_79

4. Roy, A.G., Conjeti, S., Navab, N., et al.: QuickNAT: a fully convolutional network for quick and accurate segmentation of neuroanatomy. NeuroImage **186**, 713–727 (2019)

5. Kumar, S., Conjeti, S., Roy, A.G., et al.: InfiNet: fully convolutional networks for infant brain MRI segmentation. In: ISBI, pp. 145–148 (2018)

6. Roth, H.R., et al.: DeepOrgan: multi-level deep convolutional networks for automated pancreas segmentation. In: Navab, N., Hornegger, J., Wells, W., Frangi, A. (eds.) MICCAI 2015. LNCS, vol. 9349, pp. 556–564. Springer, Cham (2015). https://doi.org/10.1007/978-3-319-24553-9_68

7. Moeskops, P., et al.: Deep learning for multi-task medical image segmentation in multiple modalities. In: Ourselin, S., Joskowicz, L., Sabuncu, M., Unal, G., Wells, W. (eds.) MICCAI 2016. LNCS, vol. 9901, pp. 478–486. Springer, Cham (2016). https://doi.org/10.1007/978-3-319-46723-8_55

8. Wachinger, C., Reuter, M., Klein, T.: DeepNAT: deep convolutional neural network for segmenting neuroanatomy. NeuroImage **170**, 434–445 (2018)

9. Milletari, F., et al.: V-net: fully convolutional neural networks for volumetric medical image segmentation. In: 2016 Fourth International Conference on 3D Vision, pp. 565–571. IEEE (2016)

10. Huang, Y.J., et al.: HL-FCN: hybrid loss guided FCN for colorectal cancer segmentation. In: ISBI, pp. 195–198 (2018)

11. Zhou, Z., Rahman Siddiquee, M.M., Tajbakhsh, N., Liang, J.: UNet++: a nested U-Net architecture for medical image segmentation. In: Stoyanov, D., et al. (eds.) DLMIA/ML-CDS -2018. LNCS, vol. 11045, pp. 3–11. Springer, Cham (2018). https://doi.org/10.1007/978-3-030-00889-5_1

12. Zhang, Z., Zhang, X., Peng, C., Xue, X., Sun, J.: ExFuse: enhancing feature fusion for semantic segmentation. In: Ferrari, V., Hebert, M., Sminchisescu, C., Weiss, Y. (eds.) ECCV 2018. LNCS, vol. 11214, pp. 273–288. Springer, Cham (2018). https://doi.org/10.1007/978-3-030-01249-6_17

13. Wang, L., Xie, C., Zeng, N.: RP-Net: a 3D convolutional neural network for brain segmentation from magnetic resonance imaging. IEEE Access **7**, 39670–39679 (2019)

14. Zhu, Z., et al.: A 3D coarse-to-fine framework for volumetric medical image segmentation. In: International Conference on 3D Vision, pp. 682–690 (2018)

15. Yu, Q., et al.: Recurrent saliency transformation network: incorporating multi-stage visual cues for small organ segmentation. In: CVPR, pp. 8280–8289 (2018)

16. Han, M., et al.: Segmentation of CT thoracic organs by multi-resolution VB-nets. SegTHOR@ ISBI (2019)

17. Tokunaga, H., et al.: Adaptive weighting multi-field-of-view CNN for semantic segmentation in pathology. In: CVPR, pp. 12597–12606 (2019)

18. Hu, J., Shen, L., Sun, G.: Squeeze-and-excitation networks. In: CVPR, pp. 7132–7141 (2018)

19. Woo, S., Park, J., Lee, J.-Y., Kweon, I.S.: CBAM: convolutional block attention module. In: Ferrari, V., Hebert, M., Sminchisescu, C., Weiss, Y. (eds.) ECCV 2018. LNCS, vol. 11211, pp. 3–19. Springer, Cham (2018). https://doi.org/10.1007/978-3-030-01234-2_1

20. Roy, A.G., Navab, N., Wachinger, C.: Concurrent spatial and channel 'squeeze and excitation' in fully convolutional networks. In: Frangi, A., Schnabel, J., Davatzikos, C., Alberola-López, C., Fichtinger, G. (eds.) MICCAI 2018. LNCS, vol. 11070, pp. 421–429. Springer, Cham (2018). https://doi.org/10.1007/978-3-030-00928-1_48
21. Li, X., et al.: Selective Kernel Networks. arXiv preprint arXiv:1903.06586 (2019)
22. Wu, J., et al.: Active learning with noise modeling for medical image annotation. In: ISBI, pp. 298–301 (2018)
23. Szegedy, C., et al.: Going deeper with convolutions. In: CVPR, pp. 1–9 (2015)
24. Yang, L., Zhang, Y., Chen, J., Zhang, S., Chen, D.Z.: Suggestive annotation: a deep active learning framework for biomedical image segmentation. In: Descoteaux, M., Maier-Hein, L., Franz, A., Jannin, P., Collins, D., Duchesne, S. (eds.) MICCAI 2017. LNCS, vol. 10435, pp. 399–407. Springer, Cham (2017). https://doi.org/10.1007/978-3-319-66179-7_46
25. Bhalgat, Y., Shah, M., Awate, S.: Annotation-cost minimization for medical image segmentation using suggestive mixed supervision fully convolutional networks. arXiv preprint arXiv:1812.11302 (2018)
26. Zhang, Z., et al.: A sparse annotation strategy based on attention-guided active learning for 3D medical image segmentation. arXiv preprint arXiv:1906.07367 (2019)
27. Shao, W., Sun, L., Zhang, D.: Deep active learning for nucleus classification in pathology images. In: ISBI, pp. 199–202 (2018)
28. Zhou, Z., et al.: Fine-tuning convolutional neural networks for biomedical image analysis: actively and incrementally. In: CVPR, pp. 7340–7351 (2017)
29. Gal, Y., Ghahramani, Z.: Dropout as a Bayesian approximation: representing model uncertainty in deep learning. In: ICML, pp. 1050–1059 (2016)
30. Roy, A.G., Conjeti, S., Navab, N., Wachinger, C.: Inherent brain segmentation quality control from fully convnet Monte Carlo sampling. In: Frangi, A., Schnabel, J., Davatzikos, C., Alberola-López, C., Fichtinger, G. (eds.) MICCAI 2018. LNCS, vol. 11070, pp. 664–672. Springer, Cham (2018). https://doi.org/10.1007/978-3-030-00928-1_75
31. Nair, T., Precup, D., Arnold, D.L., Arbel, T.: Exploring uncertainty measures in deep networks for multiple sclerosis lesion detection and segmentation. In: Frangi, A., Schnabel, J., Davatzikos, C., Alberola-López, C., Fichtinger, G. (eds.) MICCAI 2018. LNCS, vol. 11070, pp. 655–663. Springer, Cham (2018). https://doi.org/10.1007/978-3-030-00928-1_74
32. Kennedy, D.N., Haselgrove, C., Hodge, S.M., Rane, P.S., Makris, N., Frazier, J.A.: CANDIShare: a resource for pediatric neuroimaging data. Neuroinformatics 10(3), 319–322 (2012)
33. Frazier, J.A., et al.: Diagnostic and sex effects on limbic volumes in early-onset bipolar disorder and schizophrenia. Schizophr. Bull. 34(1), 37–46 (2007)
34. Rohlfing, T., et al.: Evaluation of atlas selection strategies for atlas-based image segmentation with application to confocal microscopy images of bee brains. NeuroImage 21(4), 1428–1442 (2004)
35. Zhao, Y., et al.: A novel active learning framework for classification: using weighted rank aggregation to achieve multiple query criteria. Pattern Recogn. 93, 581–602 (2019)
36. Kumar, P., et al.: U-Segnet: fully convolutional neural network based automated brain tissue segmentation tool. In: ICIP, pp. 3503–3507 (2018)
37. Deng, Y., et al.: A strategy of MR brain tissue images' suggestive annotation based on modified U-net. arXiv preprint arXiv:1807.07510 (2018)
38. Chen, H., et al.: VoxResNet: deep voxelwise residual networks for brain segmentation from 3D MR images. NeuroImage 170, 446–455 (2018)
39. Jiao, Z., et al.: A deep feature based framework for breast masses classification. Neurocomputing 197, 221–231 (2016)

40. Jiao, Z., et al.: Deep convolutional neural networks for mental load classification based on EEG data. Pattern Recogn. **76**, 582–595 (2018)
41. Zhong, Z., et al.: An attention-guided deep regression model for landmark detection in cephalograms. arXiv preprint arXiv:1906.07549 (2019)
42. Sudre, C.H., Li, W., Vercauteren, T., Ourselin, S., Jorge Cardoso, M.: Generalised dice overlap as a deep learning loss function for highly unbalanced segmentations. In: Cardoso, M., et al. (eds.) DLMIA/ML-CDS-2017. LNCS, vol. 10553, pp. 240–248. Springer, Cham (2017). https://doi.org/10.1007/978-3-319-67558-9_28

Deep 3D Facial Landmark Detection on Position Maps

Kangkang Gao[1], Shanming Yang[2], Keren Fu[1(✉)], and Peng Cheng[3]

[1] College of Computer Science, Sichuan University, Chengdu, China
fkrsuper@scu.edu.cn
[2] National Key Laboratory of Fundamental Science on Synthetic Vision,
Sichuan University, Chengdu, China
[3] College of Aeronautics and Astronautics, Sichuan University, Chengdu, China

Abstract. 3D facial landmark detection is a crucial step for many computer vision applications, such as 3D facial expression analysis, 3D face recognition, and 3D reconstruction. Pose variations, expression changes and self-occlusion yet make 3D facial landmark detection a very challenging task. In this paper, we propose a novel *3D Face Landmark Localization Network (3DLLN)*, which is robust to the above challenges. Different from existing methods, the proposed 3DLLN utilizes the position maps as an intermediate representation, from which 3DLLN detects 3D landmark coordinates. Further, we demonstrate the usage of a deep regression architecture to improve the accuracy and robustness of a large number of landmarks. The proposed scheme is evaluated on two public datasets FRGCv2 and BU_3DFE and achieves superior results to state-of-the-arts.

Keywords: 3D landmark detection · Deep convolutional neural network · Position map · UV map

1 Introduction

In the fields of biometrics, computer vision and graphics, finding key points has been a long-standing topic because of their rich semantic information. As the key points on faces, 3D facial landmarks are widely used in many face-related applications including recognition, expression analysis, semantic segmentation, rendering and relighting. Therefore, an automatic and effective 3D landmark localization system is very valuable. There have been many literatures on 3D landmark detection/localization so far. Texture-based detection methods were reported to achieve decent results [12,29]. In our work, texture information is not mandatory since we aim at detecting landmarks from pure 3D shape information. This is a more general consideration for the cases where texture information acquired is inaccurate or not available. Here below we summarize three main challenges for 3D landmark detection under unconstrained conditions:

© Springer Nature Switzerland AG 2019
Z. Cui et al. (Eds.): IScIDE 2019, LNCS 11935, pp. 299–311, 2019.
https://doi.org/10.1007/978-3-030-36189-1_25

– **Expression variations**. Expression changes lead to point cloud deformation, which may make the landmarks scattered and hard to locate.
– **Pose variations**. Pose variations may lead to the self-occlusion problem of 3D face data when viewing from an un-calibrated view, making some landmarks invisible. However, determining pose parameters without landmarks is difficult.
– **Detecting a large number of landmarks**. Some existing methods can only capture a few points of distinctive facial areas, such as nose tip, eye corners and mouse corners. This is because they resort to hand-craft features such curvature [20] and Spin Image [25]. Unfortunately, in many face-related applications such as deformation analysis, a lot of landmarks (e.g., 68 landmarks) are needed.

Recently, deep convolutional neural networks have been deployed for 2D landmark detection [5] and shown to obtain impressive performance. Inspired by the work of [3,13,30] and to tackle the above-mentioned challenges, we make the following attempts for 3D landmark detection: (1) utilize pose normalization before detection so that the latter is facilitated; (2) find a better intermediate representation of 3D data for deep convolutional networks; (3) deploy deep convolutional network for locating a large number of 3D landmarks. As a result, we propose a novel *3D Face Landmark Localization Network* (*3DLLN*), which is the first to introduce the UV position map (UVPM) as an effective intermediate representation for 3D landmark discovery. Our contributions are below:

1. We propose a novel *3D Face Landmark Localization Network*(*3DLLN*) that detect 3D landmarks from UV position map. To the best of our knowledge, it is *the first time* that UV position map are jointly used with deep convolutional neural network to locate a large number of 3D landmarks. Besides, unlike existing 2D landmark networks which output heatmaps, 3DLLN leverages a deep regression architecture to obtain landmarks' 3D coordinates directly. Also, unlike [13] which discovers landmarks jointly with the 3D shape by regressing an UV position map from an unconstrained 2D image, their landmarks have fixed known locations on the map (see Fig. 4 in [13]), while we aim to "detect" landmarks "on" the UV position map.
2. We propose an effective pose calibration scheme for 3D facial data, which employs Shape Index and Principle Component Analysis (PCA). This ensures one to perform landmark detection regardless of pose changes.
3. We evaluate 3DLLN on two representative databases FRGCv2 and BU_3DFE, which are widely-used for 3D point cloud research. 3DLLN is show to surpass existing methods and achieve state-of-the-art results.

The reminder of the paper is organized as follows. Section 2 describes related work on 3D facial landmark detection. Section 3 describes the proposed method in details. Experimental results, performance evaluation and comparisons are included in Sect. 4. Finally, conclusion is drawn in Sect. 5.

2   Related Work

3D facial landmark detection has attracted the attention of researchers because of its importance in 3D research. Surface curvatures is firstly used as a 3D model geometric feature for landmark positioning [6,8,10,26,27]. Gaussian and mean curvatures and the combination between surface curvature classification and depth relief curves are effective to detect landmarks with distinct feature. But fail to extreme expression variations. As another representative work, Berretti *et al.* [1] propose a method based on SIFT features and depth map. Yet even with fewer landmarks to detect, the accuracy seems low on the BU_3DFE database. Jahanbin *et al.* [18] use Gabor features computed at similarity maps for 3D landmark detection. But any pose variations will affect the accuracy of landmark detection. Li *et al.* [20] aim to pose variations in 3D landmark detection by using two curvature-based detectors that can repeatedly identify complementary locations with high local curvature, but the method is sensitive to noise and time consuming. Meanwhile, some researchers try to achieve better results in 3D landmark detection research by model matching. Feng *et al.* [14] believe that this method cannot obtain accurate landmarks. Gilani *et al.* [31] propose a shape-based 3D landmark detection algorithm, which automatically extracted effective seed points and implement dense correspondence by using the adaptive evolution level set curve of geometric velocity function, and decent results are achieved.

In recent years, some researchers have done further work on this basis, such as [7,9,12,19,25]. Perakis *et al.* [25] propose the Full Face Statistical Landmark Model(FLM) and a combination of Shape Index and Spin Image to achieve higher accuracy on FRGCv2 for large yaw and expression variations. Fan *et al.* [12] attempt to solve the pose variations problem by mapping the multi-pose 3D face model and corresponding texture to 2D space. But inaccurate texture will bring larger test error. Recently, [19] propose a landmark detection method for expression and pose variations by combining the SIFT feature with the designed

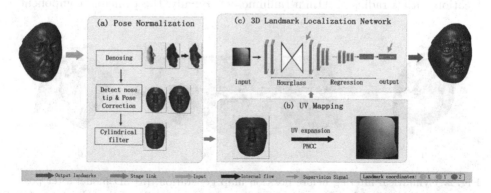

Fig. 1. The processing flow of our scheme. The flow arrows with different colors indicate different meanings as shown in the bottom.

Grid Function, achieving a higher accuracy. However, it can only detect the visible landmarks on 3D scans and the robustness to expression variation is limited. Terada *et al.* [28] put forward the latest results by using the deep convolution neural network on the depth map and return the detected landmarks to the 3D face model. Yet the results are not satisfactory. Paulsen *et al.* [24] use multiview consensus convolutional neural networks to achieve outstanding results in 3D face landmark detection tasks.

Different from all the works mentioned above, our method utilizes the position maps, a new feature map, as an intermediate representation. To the best of our knowledge, it is the first time that UV position map are jointly used with deep convolutional neural network to locate a large number of 3D landmarks.

3 Methodology

The processing flow of our method is given in Fig. 1. It consists of three key steps: (a) Pose normalization, which performs data pre-processing and pose correction; (b) UV position map generation, which expands the 3D model of the pose-normalization output and then generates a UV position map; (c) 3D Landmark Localization Network (3DLLN) with the UV position map (UVPM) as input. The final output of 3DLLN is the vector of landmarks' 3D coordinates.

3.1 Pose Normalization

Given a 3D facial data, we first conduct outlier filtering [11, 21] so that holes and noises are removed. We fill holes by cubic interpolation [21]. As a result, outlier filtering benefits subsequent processing. To perform pose calibration, we first find the nose tip [10, 11, 21] as a key reference. The *Shape Index* measure [16, 25] is employed in this task and we follow [25] to make a rough prediction of nose-tip area, and then the centroid of the area is used as the nose tip position. After the nose tip is obtained, the 3D facial data is cropped by a sphere centered on this location with a radius of 90 mm (millimeter). Finally, the principle component

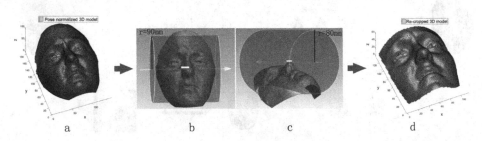

Fig. 2. Cylindrical filtering before position map generation. (a) 3D model after pose correction. (b) Cylindrical filter with a radius of 90 mm in the vertical direction. (c) Cylindrical filter with a radius of 80 mm in the horizontal direction. (d) Cropped 3D model.

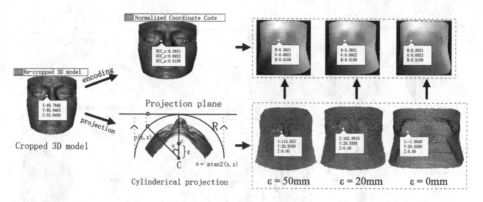

Fig. 3. Position map generation process. The cylinderical projection is employed to associate a 3D point with a 2D coordinates on the position map. The corresponding values on the map are the so-called *Projected Normalized Coordinate Code* (PNCC) [30].

analysis (PCA) is computed to find the three principle axes. Thanks to the intrinsic structure of 3D facial data and also the nose-tip reference, one can easily calibrate the 3D facial data to standard X, Y, Z axis-system with the frontal face pointing to the Z-axis. An example is shown in Fig. 2(a).

3.2   Position Map Generation

After pose normalization, we use two cylindrical filters centering on the nose tip to re-crop out a face area. The other cylindrical filter is along vertical direction (Y-axis) and has a radius 90 mm (Fig. 2(b)). One cylindrical filter is along horizontal direction (X-axis) and has a radius of 80 mm (Fig. 2(c)). Note that both radius values are determined empirically. The face area after cropping is then closer to a square size when viewing onto the X-Y plane. This is beneficial to the UV position map generation. In addition, more face regions can be reserved than the spherical cropping used during the aforementioned PCA.

The UV maps are widely used to represent 3D models [3,12]. One its advantage is that it can fully represent 3D structures in a 2-dimensional space, and meanwhile avoid self-occlusion. Position maps are recent advances [30] which go one step further upon UV maps. Position maps represent 3D coordinates as RGB information via an encoding scheme called *Projected Normalized Coordinate Code* (PNCC) [30]. By encoding, the coordinates of a 3D point are normalized into interval [0, 1] and can be represented as a color value on the UV map. The resultant UV map is called UV position map [13]. Unlike [13], we generate UV position map without the help of the Basel Face Model (BFM) [2], because aligning to the BFM model will change the structure and point number of input 3D facial data. In contrast, we use the cylinderical projection technique mentioned in [3] (Fig. 3), which projects a 3D mesh to a 2D plane as below:

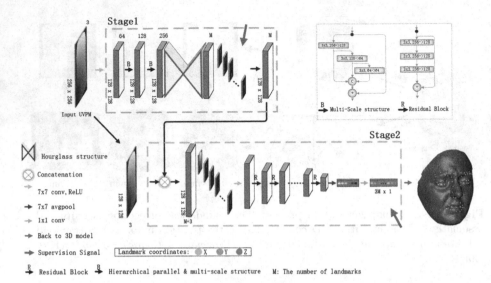

Fig. 4. The network architecture of 3DLLN. The input is the UV position map (UVPM) with 3 channels and the output is the 68 landmark coordinates. (Color figure online)

$$\mathcal{P} = (\mathcal{U}, \mathcal{V}) \tag{1}$$

In (1), $\mathcal{P} = (\mathcal{U}, \mathcal{V})$ denotes the projected 2D coordinates of a 3D point $V = (x, y, z)$ in UV space, where $\mathcal{U} = atan2(\frac{x}{z})$, $\mathcal{V} = y$. Note that here $V = (x, y, z)$ denotes the coordinate with the projection center C rather than the model centroid c as origin (Fig. 3). We locate the projection center on the Z-axis and is ε away from the model centroid. Figure 3 bottom-right shows the effect of ε, where $\varepsilon = 0$ means using the model centroid as projection center. In practice, we find $\varepsilon = 20$ mm is a good choice for projection.

3.3   3D Face Landmark Localization Network

The UV position map (UVPM) inherit the advantages of both UV map and position map, and allows one to directly infer 3D location information. This section elaborates how we locate a large number of face landmarks on UV position map. To the best of our knowledge, it is the first time that the UVPMs are used for 3D landmark detection. Our deep network is called *3D Face Landmark Localization Network* (Fig. 4), or 3DLLN for short.

The entire network architecture is shown in Fig. 4. In stage 1, we adopt the hourglass structure proposed in [22] as our backbone (with input size 256×256) to predict a heatmap for each landmark from the input UV position map. We use a hierarchical, parallel and multi-scale structure in [4] throughout the entire Hourglass network. The position of the 2D landmark is considered as the position of the maximum value in the corresponding heatmap, which is shown by the red highlight of each heatmap in Fig. 4. In stage 2, the input UV position map is concatenated with the obtained heatmaps from hourglass structure and then fed to a regression network (with a smaller input size 128×128 for efficiency). We base our regression architecture on the residual blocks described in [17]. The final output from the regression network is a vector containing 3D coordinates of landmarks. It is worthy noting that since we use a UV position map which contains 3D coordinate information and combine with the landmark location in the heatmap as input. This facilitates the regression of final coordinates because the UV position map itself contains coordinate information. As a comparison, we have attempted to regress final 3D coordinates from a texture UV map but found the regression was likely to fail. This may also be the reason that the training data set was small, and we will try in more detail in the future. Finally, we find the nearest neighbor on the original 3D point cloud of each landmark so that all found landmarks locate on the 3D mesh.

Regarding to the overall loss function $\mathcal{L}$, 3DLLN is jointly supervised by two Euclidean losses, namely the loss of heatmaps and the loss of landmark coordinates:

$$\mathcal{L} = \underbrace{\sum_{i=1}^{M} ||H_i - \hat{H}_i||^2}_{\text{heatmap loss}} + \underbrace{\lambda ||L - \hat{L}||_2^2}_{\text{landmark loss}} \qquad (2)$$

where H_i is the predicted heatmap and $\hat{H}_i$ is the ground truth heatmap with the size of 128×128 associated with the ith landmark. M denotes the total number of landmarks. L is the final coordinate vector of the predicted landmarks and $\hat{L}$ is the corresponding ground truth. λ is the balancing weight between the two losses.

4   Experiment and Results

In this section we validate the effectiveness of our approach for three challenges (as mentioned in Sect. 1) on the BU_3DFE and subset of FRGCv2 databases and compare with the state-of-the-art methods.

4.1   Databases and Evaluation Criteria

The FRGCv2 database consists of 4,007 3D face models of 466 people, accompanied by minor pose variations and extreme expressions variations. There is a subset of FRGCv2, called DB00F, which is filtered by Perakis et al. [25] with 300 training models and 975 testing models. Meanwhile, the DB00F database was manually classified into three subclasses according to expression intensities: "neutral", "mild", "extreme". The detected manual annotated landmarks on 3D model of DB00F is eight (as shown in Fig. 5(b)).

The BU_3DFE database contains 2500 3D facial expression models with a large number of manual annotated facial landmarks. Each 3D facial expression scan is also associated with a raw 3D face mesh, a cropped 3D face mesh, a pair of texture images with two-angles of view (about $\pm 45°$ yaw angle), a set of 83 manually annotated facial landmarks. We choose 1486 models for training, 1000 models for testing and 68 facial landmarks for localization.

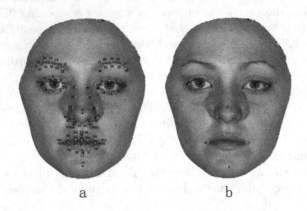

a   b

Fig. 5. (a) 68 landmarks on BU_3DFE database and (b) 8 landmarks on FRGCv2 database.

The evaluation criteria of 3DLLN is presented by computing the following three values:

1. **Mean error**. The Euclidean distance between the predicted landmark and the manually annotated landmark, which is considered ground truth.
2. **Success rate**. The ratio of successful predictions of a landmark over a test database. And successful detection is considered as the predicted landmark with Mean error under a certain threshold (e.g., 10 mm).
3. **Standard deviation**. The dispersion of Euclidean distance of detected landmarks.

In our experiments, the success rate is calculated as the ratio of the mean error less than 10 mm. Note that, as pointed out in [23], the UR3D-S face recognition method can tolerate landmark localization errors up to 10 mm.

4.2   Comparison to State-of-the-art Methods

In this section we compare the evaluation result of 3DLLN on different facial maps (as shown in Fig. 6) with the state-of-the-art (as shown in Tables 1 and 2) on the public databases DB00F and BU_3DFE. The detected landmarks are shown in Fig. 5 and the red dots in Fig. 5(a) are the position of comparison points in Table 2.

The summary comparison with the state-of-the-art is shown in Table 1. Both of the performance on DB00F (8 landmarks each face) and BU_3DFE database (68 landmarks each face) are very good. For the method only based on 3D

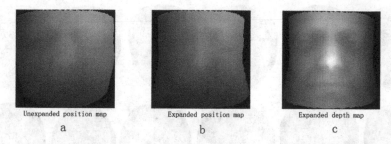

Unexpanded position map	Expanded position map	Expanded depth map
a	b	c

Fig. 6. Different facial maps. (a) Unexpanded position map (PM), with pixel value according to point coordinates. (b) Expanded position map (UVPM), projected by our method with pixel value according to points coordinates. (c) Expanded depth map (UVDM), projected by our method with pixel value according to z axis value.

Table 1. Comparison of landmark detection summary results with state-of-the-arts on BU_3DFE and DB00F databases. The bold is the best.

Method	Images(FRGCv2/BU_3DFE)	Texture	Mean Error ± Std(mm) (Success rate)				
			DB00F-neutral	DB00F-mild	DB00F-extreme	DB00F	BU3DFE
Perakis et al.[25]	975/0	No	4.52±1.51 (99.32%)	4.95±1.46 (99.72%)	6.28±2.60 (90.40%)	5.00±1.85 (97.85%)	*
Križaj J et al.[19]	975/0	No	3.00±1.60 (99.6%)	3.40±1.70 (99.70%)	3.90±1.80 (99.40%)	3.20±1.70 (99.60%)	*
Gilani et al.[31]	975/2500	NO	3.38±0.94	3.81±1.28	4.46±1.77	3.72±1.15	5.85±4.26
Xin Fan et al.[12]	975/1100	**Yes**	2.59 (98.61%)	2.84 (97.95%)	3.70 (92.73%)	2.89 (97.30%)	4.66±2.50 (93.52%)
3DLLN(UVPM)	975/1000	No	2.78±1.79 (99.56%)	2.87±1.87 (99.19%)	3.07±2.05 (98.23%)	**2.87±1.81(99.58%)**	2.66±1.89(99.54%)
3DLLN(UVDM)	975/1000	No	*	*	*	*	2.72±1.93(99.48%)
3DLLN(PM)	975/1000	No	*	*	*	*	2.61±1.90(99.52%)

Table 2. Comparison of certain landmark detection results with the state-of-the-arts on BU_3DFE and DB00F databases. The bold is the best. The abbreviations for certain landmarks are given in Fig. 5

Database	BU_3DFE database							DB00F database			
	Mean Error± Std(mm) (Success rate)							Mean Error ± Std(mm) (Success rate)			
Method	Grewe[15]	Segundo[27]	Gilani[31]	Paulsen[24]	3DLLN(UVPM)	3DLLN(UVDM)	3DLLN(PM)	Passalis[23]	Perakis[25]	Xin Fan[12]	3DLLN(UVPM)
Images	2500	2500	2500	500	1000	1000	1000	975	975	975	975
Texture	NO	NO	NO	Yes	NO	NO	NO	NO	NO	Yes	NO
REIC	3.23 *	* *	3.29±2.67	**1.80±0.89**	2.51±1.79	2.55±1.79	2.39±1.66	5.03±1.66	4.15±2.35	**1.33±1.47**	2.63±1.73
REOC	3.22 *	6.33±5.04	4.35±2.70	2.85±1.50	1.82± 1.46	1.80±1.41	**1.66±1.42**	5.79±3.45	5.58±3.33	**2.53±1.62**	3.24±1.93
LEIC	3.04 *	6.33±4.82	4.75±2.64	**1.89±0.98**	2.58±1.72	2.63±1.77	2.45±1.73	5.48±2.59	4.41±2.49	**2.49±1.67**	2.81±1.81
LEOC	2.95 *	* *	4.43±2.74	2.59±1.53	1.75±1.49	1.67±1.50	**1.48±1.39**	5.62±3.47	5.83±3.42	**1.39±1.84**	3.41±2.13
NT	* *	* *	* *	* *	* *	* *	* *	4.91±2.49	4.09±2.41	4.38±2.90	**1.87±1.30**
CT	* *	* *	* *	* *	* *	* *	* *	6.31±4.43	4.92±3.74	3.85±3.05	**3.02±1.85**
LS	* *	* *	3.20±2.68	2.33±1.31	2.05±1.52	2.01±1.54	**1.92±1.44**	* *	* *	* *	* *
LI	* *	* *	6.90±5.31	2.50±1.41	2.88±1.95	2.86±1.92	**2.79±1.88**	* *	* *	* *	* *
LAC	* *	6.66±3.36	4.30±2.73	2.61±1.41	2.53±1.77	2.51±1.75	**2.27±1.65**	* *	* *	* *	* *
RAC	* *	6.49±3.40	4.28±2.71	2.96±1.56	**2.51±1.80**	2.56±1.78	2.57±1.80	* *	* *	* *	* *
LSBAL	* *	* *	4.86±2.80	* *	2.14±1.71	**2.11±1.74**	2.16±1.67	* *	* *	* *	* *
RSBAL	* *	* *	3.57±2.59	* *	2.26±1.77	2.19±1.76	**2.22±1.70**	* *	* *	* *	* *
MLC	* *	* *	6.00±3.94	2.18±1.44	**2.60±1.80**	2.71±1.81	2.64±1.80	6.47±4.26	5.42±3.84	3.06±2.82	**2.90±1.85**
MRC	* *	* *	5.45±3.12	2.42±1.44	2.65±1.76	2.78±1.89	**2.46±1.71**	5.65±4.34	5.56±3.93	4.07±3.36	**3.06±1.85**
Mean	3.11 *	6.45±4.15	4.61±4.05	2.41±1.34	2.35±1.71	2.36±1.72	**2.25±1.65**	5.65±3.34	5.0(97.85%)	2.89(97.30%)	**2.87±1.81(99.58%)**

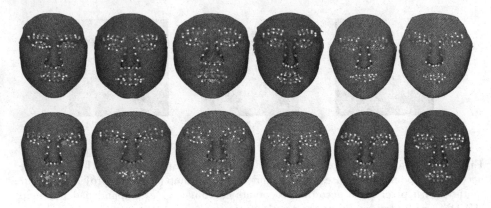

Fig. 7. Example 3D landmark detection results of the proposed method on BU_3DFE database. The white dots are predicted landmarks. The pink dots are manual annotation landmarks. The yellow dots are the coincidence dots between the manual annotation and the predicted landmarks. (Color figure online)

structural information, we get a obvious improvement, even better than the method with texture (Xin Fan's method [12]). For DB00F database, the mean error and standard deviation as low as *2.87* mm and *1.81* mm on the all test data, respectively, with success rate over 99.5% . As expression variations from nature to the extreme, mean error growth is less than 0.2 mm. For BU_3DFE database, we made a more obvious ascension, achieved the state-of-the-art performance.

A more detailed comparison results are shown in Table 2. For BU_3DFE database, the most robust landmark is the eye corner. While the maximum mean error occurs at LI, it is still lower than the state-of-the-art. For DB00F database, the most robust landmark is the nose tip, followed by the eye inner corners. We believe that most methods can give good results for strong robustness landmarks because of its distinct geometric features. However, in the face of a large number of landmark detection, our method can still accurately locate the landmark at the location where the geometric features are not obvious (as shown in Fig. 8). Finally, the performance of 3DLLN over a test database is shown in Fig. 7. The white points are detected by 3DLLN and the pink points are the manually annotated landmarks. When the white point is too close to the pink point, there will be overlap, and we show them with yellow dot.

Table 3. Comparison of landmark detection results before and after regression network on BU_3DFE database.

Method	Images(BU_3DFE)	Texture	BEFORE			AFTER		
			Mean error	Std	Success rate	Mean error	Std	Success rate
3DLLN(UVPM)	1000	No	2.92	1.98	98.94%	**2.66**	1.89	99.54%
3DLLN(UVDM)	1000	No	3.16	2.08	98.01%	**2.72**	1.93	99.48%
3DLLN(PM)	1000	No	3.01	2.03	98.83%	**2.61**	1.90	99.52%

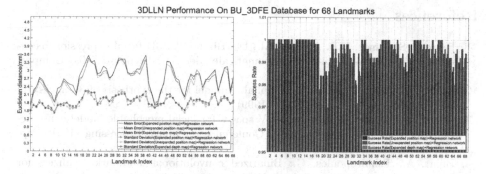

Fig. 8. Comparison results between different facial maps for 68 detected landmarks. (**LEFT**) The mean error and standard deviation curve of Euclidean distance between each predicted landmark and the manually marked landmark. (**RIGHT**) The success rate bar of our method on different facial maps. (Color figure online)

4.3 The Role of Regression Network

In this section, we assess the proposed regression network's role in the 3D landmark detection task. We show the comparison results of 3D landmark detection with and without regression network in Table 3. Obviously, the accuracy of 3D landmark detection gets a whole promotion with the help of regression network and we also show the detailed result on expanded position map in Fig. 8. In the absence of regression network, the difference between UVPM and PM mainly comes from the prediction of nasal saddle point, with 2.88 mm mean error on UVPM and 3.11 mm on PM. But we found that the advantage disappeared with the help of regression network in Fig. 8.

5 Conclusion

We propose a novel method for 3D landmark detection using deep convolutional neural network based on UV position map without texture. It has been evaluated using challenging 3D facial databases which contain 3D scans with extreme expression variations or a large number of landmarks (more than 68) to be detected. Our method achieves state-of-the-art accuracy with high success rate. Besides, we have performed ablation studies and achieved the following observations: (a) The position map retains richer 3D model feature than the depth map. (b) The expanded position map is superior to the unexpanded position map for landmark detection near the saddle. (c) The regression architecture in the second stage of 3DLLN brings significant improvement and is able to compensate for the high error of the unexpanded position map at the nasal saddle.

Acknowledgement. This work was supported by the National Science Foundation of China, No. 61703077, U1833128, the Fundamental Research Funds for the Central Universities, No. YJ201755, and the Sichuan Science and Technology Major Projects (2018GZDZX0029).

References

1. Berretti, S., Amor, B.B., Daoudi, M., Del Bimbo, A.: 3D facial expression recognition using sift descriptors of automatically detected keypoints. Vis. Comput. **27**(11), 1021 (2011)
2. Blanz, V., Vetter, T., et al.: A morphable model for the synthesis of 3D faces. In: SIGGRAPH, vol. 99, pp. 187–194 (1999)
3. Booth, J., Zafeiriou, S.: Optimal UV spaces for facial morphable model construction. In: 2014 IEEE International Conference on Image Processing (ICIP), pp. 4672–4676. IEEE (2014)
4. Bulat, A., Tzimiropoulos, G.: Binarized convolutional landmark localizers for human pose estimation and face alignment with limited resources. In: Proceedings of the IEEE International Conference on Computer Vision, pp. 3706–3714 (2017)
5. Bulat, A., Tzimiropoulos, G.: How far are we from solving the 2D & 3D face alignment problem? (and a dataset of 230,000 3D facial landmarks). In: Proceedings of the IEEE International Conference on Computer Vision, pp. 1021–1030 (2017)
6. Chang, K.I., Bowyer, K.W., Flynn, P.J.: Multiple nose region matching for 3D face recognition under varying facial expression. IEEE Trans. Pattern Anal. Mach. Intell. **28**(10), 1695–1700 (2006)
7. Cihan Camgoz, N., Struc, V., Gokberk, B., Akarun, L., Alp Kindiroglu, A.: Facial landmark localization in depth images using supervised ridge descent. In: Proceedings of the IEEE International Conference on Computer Vision Workshops, pp. 136–141 (2015)
8. Colombo, A., Cusano, C., Schettini, R.: 3D face detection using curvature analysis. Pattern Recogn. **39**(3), 444–455 (2006)
9. Creusot, C., Pears, N., Austin, J.: A machine-learning approach to keypoint detection and landmarking on 3D meshes. Int. J. Comput. Vis. **102**(1–3), 146–179 (2013)
10. Dibeklioğlu, H., Gökberk, B., Akarun, L.: Nasal region-based 3D face recognition under pose and expression variations. In: Tistarelli, M., Nixon, M.S. (eds.) ICB 2009. LNCS, vol. 5558, pp. 309–318. Springer, Heidelberg (2009). https://doi.org/10.1007/978-3-642-01793-3_32
11. Emambakhsh, M., Evans, A.N., Smith, M.: Using nasal curves matching for expression robust 3D nose recognition. In: 2013 IEEE Sixth International Conference on Biometrics: Theory, Applications and Systems (BTAS), pp. 1–8. IEEE (2013)
12. Fan, X., Jia, Q., Huyan, K., Gu, X., Luo, Z.: 3D facial landmark localization using texture regression via conformal mapping. Pattern Recogn. Lett. **83**, 395–402 (2016)
13. Feng, Y., Wu, F., Shao, X., Wang, Y., Zhou, X.: Joint 3D face reconstruction and dense alignment with position map regression network. In: Proceedings of the European Conference on Computer Vision (ECCV), pp. 534–551 (2018)
14. Feng, Z.H., Hu, G., Kittler, J., Christmas, W., Wu, X.J.: Cascaded collaborative regression for robust facial landmark detection trained using a mixture of synthetic and real images with dynamic weighting. IEEE Trans. Image Process. **24**(11), 3425–3440 (2015)
15. Grewe, C.M., Zachow, S.: Fully automated and highly accurate dense correspondence for facial surfaces. In: Hua, G., Jégou, H. (eds.) ECCV 2016. LNCS, vol. 9914, pp. 552–568. Springer, Cham (2016). https://doi.org/10.1007/978-3-319-48881-3_38

16. Harris, C.G., Stephens, M., et al.: A combined corner and edge detector. In: Alvey Vision Conference, vol. 15, pp. 10–5244. Citeseer (1988)
17. He, K., Zhang, X., Ren, S., Sun, J.: Deep residual learning for image recognition. In: Proceedings of the IEEE Conference on Computer Vision and Pattern Recognition, pp. 770–778 (2016)
18. Jahanbin, S., Choi, H., Bovik, A.C.: Passive multimodal 2-D+ 3-D face recognition using gabor features and landmark distances. IEEE Trans. Inf. Forensics Secur. 6(4), 1287–1304 (2011)
19. Križaj, J., Emeršič, Ž., Dobrišek, S., Peer, P., Štruc, V.: Localization of facial landmarks in depth images using gated multiple ridge descent. In: 2018 IEEE International Work Conference on Bioinspired Intelligence (IWOBI), pp. 1–8. IEEE (2018)
20. Li, H., Huang, D., Morvan, J.M., Wang, Y., Chen, L.: Towards 3D face recognition in the real: a registration-free approach using fine-grained matching of 3D keypoint descriptors. Int. J. Comput. Vis. 113(2), 128–142 (2015)
21. Mian, A., Bennamoun, M., Owens, R.: An efficient multimodal 2D–3D hybrid approach to automatic face recognition. IEEE Trans. Pattern Anal. Mach. Intell. 29(11), 1927–1943 (2007)
22. Newell, A., Yang, K., Deng, J.: Stacked hourglass networks for human pose estimation. In: Leibe, B., Matas, J., Sebe, N., Welling, M. (eds.) ECCV 2016. LNCS, vol. 9912, pp. 483–499. Springer, Cham (2016). https://doi.org/10.1007/978-3-319-46484-8_29
23. Passalis, G., Perakis, P., Theoharis, T., Kakadiaris, I.A.: Using facial symmetry to handle pose variations in real-world 3D face recognition. IEEE Trans. Pattern Anal. Mach. Intell. 33(10), 1938–1951 (2011)
24. Paulsen, R.R., Juhl, K.A., Haspang, T.M., Hansen, T., Ganz, M., Einarsson, G.: Multi-view consensus CNN for 3D facial landmark placement. In: Jawahar, C.V., Li, H., Mori, G., Schindler, K. (eds.) ACCV 2018. LNCS, vol. 11361, pp. 706–719. Springer, Cham (2019). https://doi.org/10.1007/978-3-030-20887-5_44
25. Perakis, P., Passalis, G., Theoharis, T., Kakadiaris, I.A.: 3D facial landmark detection under large yaw and expression variations. IEEE Trans. Pattern Anal. Mach. Intell. 35(7), 1552–1564 (2012)
26. Segundo, M.P., Queirolo, C., Bellon, O.R., Silva, L.: Automatic 3D facial segmentation and landmark detection. In: 14th International Conference on Image Analysis and Processing (ICIAP 2007), pp. 431–436. IEEE (2007)
27. Segundo, M.P., Silva, L., Bellon, O.R.P., Queirolo, C.C.: Automatic face segmentation and facial landmark detection in range images. IEEE Trans. Syst. Man Cybern. Part B (Cybern.) 40(5), 1319–1330 (2010)
28. Terada, T., Chen, Y.W., Kimura, R.: 3D facial landmark detection using deep convolutional neural networks. In: 2018 14th International Conference on Natural Computation, Fuzzy Systems and Knowledge Discovery (ICNC-FSKD), pp. 390–393. IEEE (2018)
29. Zhao, X., Dellandrea, E., Chen, L., Kakadiaris, I.A.: Accurate landmarking of three-dimensional facial data in the presence of facial expressions and occlusions using a three-dimensional statistical facial feature model. IEEE Trans. Syst. Man Cybern. Part B (Cybern.) 41(5), 1417–1428 (2011)
30. Zhu, X., Liu, X., Lei, Z., Li, S.Z.: Face alignment in full pose range: a 3D total solution. IEEE Trans. Pattern Anal. Mach. Intell. 41(1), 78–92 (2019)
31. Zulqarnain Gilani, S., Shafait, F., Mian, A.: Shape-based automatic detection of a large number of 3D facial landmarks. In: Proceedings of the IEEE Conference on Computer Vision and Pattern Recognition, pp. 4639–4648 (2015)

Joint Object Detection and Depth Estimation in Multiplexed Image

Changxin Zhou[iD] and Yazhou Liu[✉][iD]

Nanjing University of Science and Technology, Nanjing, China
{cxzhou,yazhouliu}@njust.edu.cn

Abstract. This paper presents an object detection method that can simultaneously estimate the positions and depth of the objects from multiplexed images. Multiplexed image [28] is produced by a new type of imaging device that collects the light from different fields of view using a single image sensor, which is originally designed for stereo, 3D reconstruction and broad view generation using computation imaging. Intuitively, multiplexed image is a blended result of the images of multiple views and both of the appearance and disparities of objects are encoded in a single image implicitly, which provides the possibility for reliable object detection and depth/disparity estimation. Motivated by the recent success of CNN based detector, a multi-anchor detector method is proposed, which detects all the views of the same object as a clique and uses the disparity of different views to estimate the depth of the object. The method is interesting in the following aspects: firstly, both locations and depth of the objects can be simultaneously estimated from a single multiplexed image; secondly, there is almost no computation load increase comparing with the popular object detectors; thirdly, even in the blended multiplexed images, the detection and depth estimation results are very competitive. There is no public multiplexed image dataset yet, therefore the evaluation is based on simulated multiplexed image using the stereo images from KITTI, and very encouraging results have been obtained.

Keywords: Object detection · Depth estimation · Multiplexed image

1 Introduction

Object detection is a fundamental and challenging problem in the computer vision. The goal of object detection is to obtain the location and category information of each object instance in the given image. However, applications like autonomous driving, not only need the positions of objects in image space but also require the actual depth of these detected objects. In the computer vision community, this task can be divided into two sub-tasks: object detection and depth estimation, as shown in Fig. 1. Object detection methods [14, 20, 27, 33] provide the location information (bounding boxes) of detected objects and depth estimation methods [11, 15] predict the disparity of each pixel in the image. Since not all pixels inside the bounding box belong to the object, an object's disparity

© Springer Nature Switzerland AG 2019
Z. Cui et al. (Eds.): IScIDE 2019, LNCS 11935, pp. 312–323, 2019.
https://doi.org/10.1007/978-3-030-36189-1_26

Fig. 1. Combining object detection with depth estimation to get the depth of detected objects.

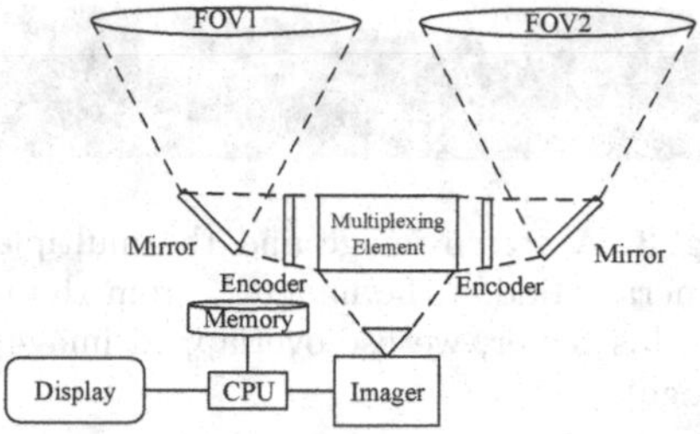

Fig. 2. The architecture of the device for multiplexed imaging [28].

can be roughly estimated by the average disparity or the median disparity of pixels in it. However, this method will fail if the object is occluded unless getting pixel-wise classification information from instance segmentation [12]. What's more, this scheme suffers from the large latency and memory requirement resulting from the combination of multi-tasks.

Another solution is 3D object detection that predicts the 3D location, dimensions (height, width, and length) and orientation of objects. Benefiting from the accurate depth information, methods [5,17] based on LiDAR data achieve the state-of-the-art performance in 3D object detection. But, LiDAR has the disadvantage of high cost, relatively short perception range and sparse information. Methods [4,23] whose input is a monocular image cannot predict the accurate depth of objects, especially for unseen scenes. Stereo R-CNN [16] is a 3D object detection method that utilizes the sparse and dense, semantic and geometry information in the stereo image. But, its inference speed (0.28 s per image) is far from the real-time demand for autonomous driving because of extracting features from two images (a stereo image) and post-processing.

To this end, we propose a simple and fast detector capable of depth estimation based on the multiplexed image. Shepard and Rachlin [28] proposed a new imaging device that collects multiple channels of light by a single sensor, as illustrated in Fig. 2. They also propose methods to disambiguate a captured multiplexed image to produce stereo images. Comparing with multiplexed imaging, the method that uses two or more cameras to create a stereo image suffers from the added cost, power, volume and complexity of using multiple cameras. Notice that the purpose of our work is to estimate depth information of detected objects, the multiplexed imaging in this paper has two horizontal camera lens like stereo imaging but only uses a single imaging sensor. This makes the multiplexed image equal to an overlap of a stereo image pair, as shown in Fig. 3.

Our work is based on the observation that both the appearance and disparity of the object are encoded implicitly in the multiplexed image. Our method, named Disparity Detector, firstly detects all the views of the same object as a clique by the strategy we proposed, then uses the disparity of different views to estimate the depth of each object. There is no public multiplexed image

Fig. 3. A stereo image and the multiplexed image. Left: the image I_l from the left camera; Middle: the image I_r from the right camera; Right: the multiplexed image I (in this paper, we use overlapped image $I = (I_l + I_r)/2$ to simulate the multiplexed image).

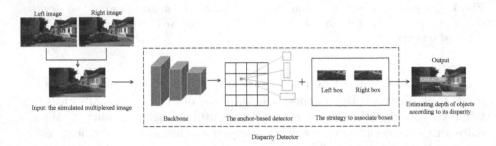

Fig. 4. The whole pipeline of our Disparity Detector framework. Disparity Detector takes the multiplexed image as input and consists of three flexible modules (backbone, anchor-based detector, and the strategy). Disparity Detector estimates the depth of detected objects by their disparities of left and right views.

dataset yet; therefore, the experiments are conducted on the simulated multiplexed image using the stereo images from KITTI [8]. The proposed method is developed based on the VGG16 [30] backbone and SSD detector framework [20], but it can be easily cooperated with other anchor-based CNN detectors (e.g. DSSD [7]) and backbones (e.g. ResNet [13]) for a better performance. The whole pipeline of our work is shown in Fig. 4.

To our best knowledge, the Disparity Detector proposed this paper is the first detector for the multiplexed images. The works in this paper are interesting in the following aspects: (1) The proposed method can simultaneously estimate the 2D positions and the actual depth of objects in the multiplexed image; (2) Comparing with popular object detector, the proposed method has almost no extra computation load or latency time; (3) Even in blended multiplexed images, the proposed method achieves competitive detection and depth estimation results.

2   Related Work

In this section, we are going to briefly review the advances of the related works from two aspects: object detection and disparity estimation framework.

Object Detection. The goal of object detection is to obtain the location and category information of each object instance in a given image. In the recent several years, deep convolution neural networks are widely used for vision tasks.

Different from classic detectors, CNN based object detectors use image features extracted by a base network (e.g. VGG16 [30]) to find objects. Due to the outstanding performance, CNN based object detectors become the main force in the detection field. Usually, CNN based detectors can be roughly divided into two categories, i.e., the two-stage approach and the one-stage approach. The two-stage approach [9,10,27] have two steps, where the first one produces a fixed number of potential object proposals, and the second one predicts the offsets of the spatial location and category labels. The two-stage methods have been achieving top results on several benchmarks [6,19]. Recently plenty of novel techniques are used for better performance, such as iterative bounding box [1] regression and training strategy [29]. The one-stage approach directly predicts class probabilities and bounding box offsets with a convolutional neural network. Therefore, the one-stage approach has a better trade-off of speed and accuracy. SSD [20] and YOLO [24–26] are representative object detectors of the one-stage approach. YOLO [24] directly predicts the object category and the offsets of spatial location with a single convolution network and has a fast inference speed. Liu et al. [20] propose a single shot object detector, named SSD, which predicts objects using feature maps with different receptive fields. DSSD [7] applies deconvolution operation to SSD for additional context and uses a more complex prediction module for better accuracy. RetinaNet [18] rethinks the extreme class imbalance problem in the current one-stage approach and solves it by re-designing the loss function.

Disparity Estimation. The goal of the depth estimation task is to predict the disparity of every pixel in the input image. Benefit from the rapid development of neural networks, several depth estimation methods have made great progress. Zbontar and LeCun [32] calculate patch similarities of a stereo image pair with a Siamese convolutional network. Their method inspired more study on depth estimation using convolution networks. DispNet [21] formulates the depth estimation as a supervised learning problem and predicts disparities directly with a convolutional network. PSMNet [3] uses spatial pyramid pooling to take advantage of the capacity of global context information and achieves state-of-the-art performance. To achieve a better trade-off between accuracy and speed, AnyNet [31] estimates depth in several stages, during which the model can be queried at any time to output its current best estimate. Monodepth [11] is one of the unsupervised methods that attempt to generate a dense disparity map by training the network with an image reconstruction loss. It only requires the stereo image pair for training and enables the network to learn to perform single image depth estimation at a faster speed.

3 Disparity Detector Framework

In this section, we propose a strategy that can be cooperated with any anchor based object detector to form our Disparity Detector. Disparity Detector can simultaneously detect and estimate the depth of objects in multiplexed images.

Firstly, we analyze the characteristics of multiplexed images and explain the reason why current detectors are not suitable for multiplexed images. Then we introduce the Disparity Detector which is composed of a backbone network, an anchor based object detector, and our proposed strategy.

3.1 The Characteristic of Multiplexed Image

Since the multiplexed image this paper focuses on is a mixture of images from two horizontal views, each object in it has two parts that can be located with a pair of horizontal bounding boxes (dashed boxes in Fig. 5). The disparity of an object can be estimated by the horizontal pixel distance between the centers of its two boxes. In stereo vision, the formula of disparity and depth is:

$$depth = \frac{b \times f}{disparity} \qquad (1)$$

where b is the stereo baseline distance and f is the focal length of the camera. Benefit from this characteristic, the multiplexed image provides the possibility of joint object detection and depth estimation.

However, the current CNN based object detectors have limitations when directly used to detect objects in multiplexed images: Left or right box of the same object may be filtered during non-maximum suppression (NMS) due to their large overlap area; From the output of a detector, a set of predicted bounding boxes, the disparity of each object is unavailable because the detector cannot associate left and right boxes of the same object.

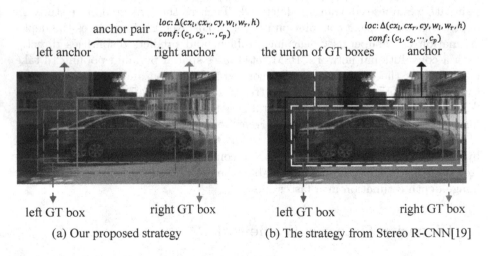

(a) Our proposed strategy   (b) The strategy from Stereo R-CNN[19]

Fig. 5. The strategies to associate boxes of the same object (Sect. 3.2).

3.2   Disparity Detector

Considering the speed-vs-accuracy trade-off, our Disparity Detector is based on SSD [20], which is a representative one-stage detector. SSD is built on top of a backbone network (VGG16 [30]) that ends (or is truncated to end) with some convolutional layers. To detect objects with multiple size, SSD utilizes feature maps with different receptive fields to predict scores and offsets for predefined anchors. These predictions are performed by $3 \times 3 \times \#channels$ dimensional filters, one filter for classification score and one for location offsets of the anchors. Finally non-maximum suppression (NMS) is used to reduce redundancy and obtain detection results. More details can be found in [20].

To enable the detector to detect and associate the left and right boxes from the same object in the multiplexed image, we propose a strategy, named *anchor pair*, and cooperate it with VGG16 [30] backbone and SSD [20] detector to form our Disparity Detector.

Anchor Pair. Inspired by each object having horizontal left and right GT boxes, we propose *anchor pair* which is an extension of *anchor*. Each *anchor pair* consists of horizontal left and right anchors, as shown in Fig. 5(a). For each anchor pair, we calculate its left anchor's IoU (IoU_l) with the left GT box and right anchor's IoU (IoU_r) with the corresponding right GT box. The anchor pair is assigned a positive label if IoU_l and IoU_r are both above 0.5, or a negative label if they are both below 0.5. Each anchor pair predicts a classification score so that its left and right anchor share the classification score. We let the positive anchor pair predict location offsets $[\Delta cx_l, \Delta cx_r, \Delta cy, \Delta w_l, \Delta w_r, \Delta h]$ respecting to the left and right GT boxes, where we use cx, cy to denote the horizontal and vertical coordinates of the box center in image space, w, h for width and height of the box, and the superscript $(\cdot)_l, (\cdot)_r$ for corresponding terms in the left and right box. Note that we use the same cy, h offsets $\Delta cy, \Delta h$ for the left and right boxes because we use rectified stereo images to simulate multiplexed images. Therefore, we have six offsets for each anchor pair instead of four in the origin SSD. Since left and right predicted boxes are generated by the same anchor pair and share the classification score, they are associated as a clique naturally. We use NMS on left and right predicted boxes separately to reduce redundancy and get final detection results. A left-right predicted box pair will be kept if its left and right box are both kept after NMS.

Difference with the Current Strategy. Stereo R-CNN [16] proposed a simple but rough strategy (referred to as *strategy stereo*) to associate boxes of the same object. As shown in Fig. 5(b), *strategy stereo* assigns the union of left and right ground-truth boxes (referred as union GT box) as the target for object classification. And an anchor is assigned a positive label if its IoU with one of the *union GT boxes* is above a threshold T_H, or a negative label if the IoU is below T_L. Each positive anchor predict offsets respecting to the left and right GT boxes contained in the target union GT box. However, the positive anchor having a IoU above T_H with the *union box* cannot guarantee that it also has a high IoU with each box inside the union box. In other word, the anchor with the positive label may have a IoU below T_H (even below T_L) with left or right box.

4   Implementation Details

Anchor Shape. Different from the implementation in [27], the shape of anchor is determined by k-means algorithm proposed by YOLO9000 [25]. We first run Cluster IoU [25] on the training set to automatically choose n_1 ($n_1 = 6$ in this paper) different anchor shapes (w, h). And these n_1 anchor shapes are used by *strategy stereo*. Cluster IoU is a k-mean algorithm with distance metric:

$$d(anchor, centroid) = 1 - IoU(anchor, centroid) \tag{2}$$

Then, in each of n_1 cluster, we use standard k-means with Euclidean distance to get n_2 ($n_2 = 4$ in this paper) different distances d. Therefore, there are $n_1 \times n_2$ shapes of anchor pair for the proposed strategy *anchor pair*.

Network. We have made minor changes when re-implementing SSD [20]: (1) The size of network input is 576×320. (2) We remove layers after Conv_8 in the origin SSD implementation and use three layers' feature map (Con4_3, Conv7 and Conv8_2) for prediction. Other settings, such as data augmentation and hard example mining, are same as the origin SSD. We train the network using SGD with a weight decay of 0.0001. We train 100K iterations (the batch size is 16) in total on a RTX2080 Ti GPU. The learning rate is initially set to 0.001 and reduced by 0.1 at the 60K and 80K iterations.

5   Experiments

In this section, we evaluate the proposed Disparity Detector on the KITTI detection dataset [8]. Firstly, we introduce the preparation of the dataset. Then, we compare our proposed strategy with the strategy from Stereo R-CNN [16] on the performance of object detection and depth estimation respectively.

5.1   Dataset Preparation

The KITTI detection dataset [8] provides 7481 training stereo image pairs and 3D bounding box label (The 2D box label of the left or right image can be calculated by projecting the 3D box to the corresponding image). We simulate multiplexed images using stereo image pairs by $I = (I_l + I_r)/2$. Following Stereo R-CNN [16], this paper only uses *car* category labels for training and evaluation, and uses 50% of the images for training (*training set*), the rest images are used for evaluation (*evaluation set*). Evaluation has three difficulty levels: *easy, moderate,* and *hard*, which are defined in term of the occlusion, size and truncation levels of objects. Checking [8] for a detailed definition of the difficulty levels.

We also use KITTI stereo 2015 [22] to train depth estimation methods [3, 11,32]. It contains 200 training stereo image pairs with sparse ground-truth disparities obtained using LiDAR.

Table 1. Average precision (in %) of detection, evaluated on the KITTI *evaluation set*

Method	Setting	AP_{left}			AP_{stereo}			Time (s)
		Easy	Mode	Hard	Easy	Mode	Hard	
Faster R-CNN [27]	-	99.23	98.40	90.88	-	-	-	0.082
MFFD [2]		91.16	84.01	72.43	-	-	-	0.005
YOLOv3 [26]		95.96	95.51	88.26	-	-	-	0.031
SSD [20]		98.78	96.06	88.49	-	-	-	0.027
Disparity detector	*strategy stereo* [16]	98.37	95.54	85.93	93.17	90.37	81.15	0.027
	anchor pair (ours)	98.55	96.00	88.62	93.69	91.87	83.14	0.027

Fig. 6. Examples of detection results on KITTI *evaluation set* using proposed method. Left: detection results on multiplexed images; Middle: mapping the detection results to left images I_l; Right: mapping the detection results to right images I_r.

5.2   Performance of Object Detection

In this section, we evaluate the proposed Disparity Detector's performance of object detection. We train the Disparity Detector with two strategies on the multiplexed image *training set* and evaluate them on the multiplexed image *evaluation set*. We also train the base detector, SSD and other common detectors (Faster R-CNN [27], MFFD [2], and YOLOv3 [26]) on the left image *training set* and evaluate it on the left image *evaluation set*. For Faster R-CNN, the original image is resized to 600 pixels in the shorter side. For SSD, MFFD and YOLOv3, the input image is resized to 576×320. All detectors share the same anchor shape that introduced in Sect. 4.

Our Disparity Detector aims to simultaneously detect and associate boxes of the same object in the multiplexed image. Besides evaluating the Average Precision (AP) on the left image (mapping the detection results to the left image), we also use the stereo AP metric which defined in Stereo R-CNN [16] to evaluate association performance. In stereo AP, a left-right box pair is considered as the True Positive (TP) if it meets the following conditions:

1. The maximum IoU between the left box and left GT boxes is above the threshold;
2. The maximum IoU between the right box and right GT boxes is above the threshold;

3. The selected left and right GT boxes belong to the same object.

We mark the best method in bold-red. As reported in Table 1, the proposed *anchor pair* outperforms *strategy stereo* [16] by large margins. Specifically, *anchor pair* outperforms *strategy stereo* over 2.69 AP_{left} (*Hard* level) and 1.99 AP_{stereo} (*Hard* level). We show some detection examples of Disparity Detector in Fig. 6.

5.3 Performance of Depth Estimation

In this section, we evaluate the proposed Disparity Detector's performance of object-level depth estimation. We report the results of Disparity Detector with three strategies on *evaluation set*. For evaluation, we use the end-point-error (EPE), which is calculated as the average Euclidean distance between estimated and ground-truth disparity. We also use the percentage of disparities with EPE larger than t pixels (>$t$px). Here an object is considered correct if its disparity EPE is less than t pixels. And the disparity of an object is estimated by the horizontal pixel distance between its centers of the left box and the right box.

Table 2 shows the comparison results of objects with different occlusion levels. We mark the best method in bold-red. In the KITTI label, the objects with *occlusion* = 0 are fully visible, *occlusion* = 0 means the objects are partly occluded, and the objects with *occlusion* = 2 are largely occluded. As expected, our *anchor pair* achieves a better performance than *strategy stereo* in all object occlusion levels. We show some depth estimation results of proposed Disparity Detector in the Fig. 7.

We also conduct an interesting experiment that combines SSD with depth estimation methods [3,11,32] to predict object-level depth. These depth estimation methods are trained with KITTI stereo dataset. We utilize the bounding boxes from SSD to locate objects and get the disparity of each object from the disparity map predicted by depth estimation methods. As introduced in Sect. 1, the disparity of an object can be estimated by the average disparity of its central pixels or the median disparity of pixels inside its bounding box:

$$disp_{mean}^{object} = \frac{1}{0.4w \times 0.4h} \times \sum_{cx-0.2w}^{cx+0.2w} \sum_{cy-0.2h}^{cy+0.2h} disp(i,j) \tag{3}$$

and

$$disp_{median}^{object} = median\{disp(i,j)|(i,j) \in BB\} \tag{4}$$

where $w, h, (cx, cy)$ are the width, height and center of an object's bounding box (BB), respectively. $disp(i,j)$ is the disparity map predicted by a depth estimation method such as MC-CNN [32], PSMNet [3] and Monodepth [11].

Table 3 shows the evaluation results on KITTI detection *evaluation set*. We mark the best method in bold-red. It can be observed that this combination scheme (with the state-of-the-art depth estimation method PSMNet [3]) achieves a comparable performance when the objects are not occluded (*occlusion* = 0). However, their performance drops severely when the object is occluded (*occlusion* = 1 and *occlusion* = 2) because most pixels inside the bounding box do not

Fig. 7. Examples of object-level depth estimation results on KITTI *evaluation set* using Disparity Detector (with *anchor pair*). The number in bold-red is the ground-truth depth, and the number in bold-yellow is the predicted depth from our method. (Color figure online)

belong to the object. By contrast, our proposed Disparity Detector not only outperforms them with a large margin in all occlusion levels but also achieves steady performance in all occlusion levels.

Table 2. Performance of object-level depth estimation on the KITTI *evaluation set*.

Method	Setting	Occlusion = 0		Occlusion = 1		Occlusion = 2	
		EPE	>3px	EPE	>3px	EPE	>3px
Disparity detector	*strategy stereo* [16]	1.23	7.46	1.37	9.12	1.46	10.59
	anchor pair (ours)	1.11	6.58	1.26	8.00	1.38	9.54

Table 3. Performance of the scheme combing object detection and depth estimation.

Method	Setting	Occlusion = 0		Occlusion = 1		Occlusion = 2		Time(s)
		EPE	>3px	EPE	>3px	EPE	>3px	
Disparity detector	*anchor pair*	1.11	6.58	1.26	8.00	1.38	9.54	0.027
SSD [20] + MC-CNN [32]	*median*	2.21	23.05	3.44	34.83	7.62	77.54	0.704
	mean	1.62	14.75	3.55	33.88	7.76	80.44	
SSD [20] + Mono [11]	*median*	1.69	16.44	2.87	32.71	6.21	46.33	0.072
	mean	1.78	16.95	3.29	35.52	8.29	68.35	
SSD [20] + PSMNet [3]	*median*	0.98	7.53	1.72	15.12	5.98	62.74	0.612
	mean	0.93	7.10	2.34	23.03	6.61	73.19	

6 Conclusion

In this paper, we have presented a new method for simultaneous object detection and depth estimation from a single multiplexed image. The multiplexed image can encode the appearance and disparity of the object by blending multiple views. The object detection task is formulated as a clique detection task that can detect and associate all the views of the same object in the image. Then, the actual position on any single view and the disparity/depth of the object can be recovered. The evaluation results show that the proposed method can yield very competitive results compared with the state-of-the-art.

References

1. Cai, Z., Vasconcelos, N.: Cascade R-CNN: delving into high quality object detection. arXiv preprint. arXiv:1712.00726 (2017)
2. Cao, S., Liu, Y., Lasang, P., Shen, S.: Detecting the objects on the road using modular lightweight network. arXiv preprint. arXiv:1811.06641 (2018)
3. Chang, J.R., Chen, Y.S.: Pyramid stereo matching network. In: Proceedings of the IEEE Conference on Computer Vision and Pattern Recognition, pp. 5410–5418 (2018)
4. Chen, X., Kundu, K., Zhang, Z., Ma, H., Fidler, S., Urtasun, R.: Monocular 3D object detection for autonomous driving. In: Proceedings of the IEEE Conference on Computer Vision and Pattern Recognition, pp. 2147–2156 (2016)
5. Chen, X., Ma, H., Wan, J., Li, B., Xia, T.: Multi-view 3D object detection network for autonomous driving. In: Proceedings of the IEEE Conference on Computer Vision and Pattern Recognition, pp. 1907–1915 (2017)
6. Everingham, M., Van Gool, L., Williams, C.K., Winn, J., Zisserman, A.: The pascal visual object classes (voc) challenge. Int. J. Comput. Vis. **88**(2), 303–338 (2010)
7. Fu, C.Y., Liu, W., Ranga, A., Tyagi, A., Berg, A.C.: DSSD: deconvolutional single shot detector. arXiv preprint. arXiv:1701.06659 (2017)
8. Geiger, A., Lenz, P., Urtasun, R.: Are we ready for autonomous driving? The kitti vision benchmark suite. In: Conference on Computer Vision and Pattern Recognition (CVPR) (2012)
9. Girshick, R.: Fast R-CNN. In: Proceedings of the IEEE International Conference on Computer Vision, pp. 1440–1448 (2015)
10. Girshick, R., Donahue, J., Darrell, T., Malik, J.: Rich feature hierarchies for accurate object detection and semantic segmentation. In: Proceedings of the IEEE Conference on Computer Vision and Pattern Recognition, pp. 580–587 (2014)
11. Godard, C., Mac Aodha, O., Brostow, G.J.: Unsupervised monocular depth estimation with left-right consistency. In: 2017 IEEE Conference on Computer Vision and Pattern Recognition (CVPR), pp. 6602–6611. IEEE (2017)
12. He, K., Gkioxari, G., Dollár, P., Girshick, R.: Mask R-CNN. In: Proceedings of the IEEE International Conference on Computer Vision, pp. 2961–2969 (2017)
13. He, K., Zhang, X., Ren, S., Sun, J.: Deep residual learning for image recognition. In: Proceedings of the IEEE Conference on Computer Vision and Pattern Recognition, pp. 770–778 (2016)
14. Hu, H., Gu, J., Zhang, Z., Dai, J., Wei, Y.: Relation networks for object detection. In: Proceedings of the IEEE Conference on Computer Vision and Pattern Recognition, pp. 3588–3597 (2018)
15. Kendall, A., et al.: End-to-end learning of geometry and context for deep stereo regression. In: Proceedings of the IEEE International Conference on Computer Vision, pp. 66–75 (2017)

16. Li, P., Chen, X., Shen, S.: Stereo R-CNN based 3D object detection for autonomous driving. In: CVPR (2019)

17. Liang, M., Yang, B., Wang, S., Urtasun, R.: Deep continuous fusion for multi-sensor 3D object detection. In: Proceedings of the European Conference on Computer Vision (ECCV), pp. 641–656 (2018)

18. Lin, T.Y., Goyal, P., Girshick, R., He, K., Dollár, P.: Focal loss for dense object detection. In: Proceedings of the IEEE International Conference on Computer Vision, pp. 2980–2988 (2017)

19. Lin, T.-Y., et al.: Microsoft COCO: common objects in context. In: Fleet, D., Pajdla, T., Schiele, B., Tuytelaars, T. (eds.) ECCV 2014. LNCS, vol. 8693, pp. 740–755. Springer, Cham (2014). https://doi.org/10.1007/978-3-319-10602-1_48

20. Liu, W., et al.: SSD: single shot multibox detector. In: Leibe, B., Matas, J., Sebe, N., Welling, M. (eds.) ECCV 2016. LNCS, vol. 9905, pp. 21–37. Springer, Cham (2016). https://doi.org/10.1007/978-3-319-46448-0_2

21. Mayer, N., et al.: A large dataset to train convolutional networks for disparity, optical flow, and scene flow estimation. In: Proceedings of the IEEE Conference on Computer Vision and Pattern Recognition, pp. 4040–4048 (2016)

22. Menze, M., Geiger, A.: Object scene flow for autonomous vehicles. In: Proceedings of the IEEE Conference on Computer Vision and Pattern Recognition, pp. 3061–3070 (2015)

23. Mousavian, A., Anguelov, D., Flynn, J., Kosecka, J.: 3D bounding box estimation using deep learning and geometry. In: Proceedings of the IEEE Conference on Computer Vision and Pattern Recognition, pp. 7074–7082 (2017)

24. Redmon, J., Divvala, S., Girshick, R., Farhadi, A.: You only look once: unified, real-time object detection. In: Proceedings of the IEEE Conference on Computer Vision and Pattern Recognition, pp. 779–788 (2016)

25. Redmon, J., Farhadi, A.: YOLO9000: better, faster, stronger. In: Proceedings of the IEEE Conference on Computer Vision and Pattern Recognition, pp. 7263–7271 (2017)

26. Redmon, J., Farhadi, A.: YOLOv3: an incremental improvement. arXiv preprint. arXiv:1804.02767 (2018)

27. Ren, S., He, K., Girshick, R., Sun, J.: Faster R-CNN: towards real-time object detection with region proposal networks. In: Advances in Neural Information Processing Systems, pp. 91–99 (2015)

28. Shepard, R.H., Rachlin, Y.: Devices and methods for optically multiplexed imaging. US Patent App. 14/668,214 (2018)

29. Shrivastava, A., Gupta, A., Girshick, R.: Training region-based object detectors with online hard example mining. In: Proceedings of the IEEE Conference on Computer Vision and Pattern Recognition, pp. 761–769 (2016)

30. Simonyan, K., Zisserman, A.: Very deep convolutional networks for large-scale image recognition. arXiv preprint. arXiv:1409.1556 (2014)

31. Wang, Y., et al.: Anytime stereo image depth estimation on mobile devices. arXiv preprint. arXiv:1810.11408 (2018)

32. Zbontar, J., LeCun, Y.: Stereo matching by training a convolutional neural network to compare image patches. J. Mach. Learn. Res. **17**(1–32), 2 (2016)

33. Zhu, C., He, Y., Savvides, M.: Feature selective anchor-free module for single-shot object detection. arXiv preprint. arXiv:1903.00621 (2019)

Weakly-Supervised Semantic Segmentation with Mean Teacher Learning

Li Tan, WenFeng Luo, and Meng Yang[✉]

School of Data and Computer Science, Sun Yat-Sen University, Guangzhou, China
tanlihn1997@gmail.com, yangm6@mail.sysu.edu.cn

Abstract. Weakly-supervised semantic segmentation with image-level labels is a important task as it directly associates high-level semantic to low-level appearance, which can significantly reduce human efforts. Despite the remarkable progress, it is still not as good as fully supervised segmentation methods. To improve the accuracy, in this paper, we proposed a novel framework of weakly-supervised semantic segmentation with mean teacher (WSSS-MT) learning to advance the class estimation of image pixels. More specifically, our proposed framework includes a student network and a teacher network in the segmentation module, which aims to effectively utilize information of the training process. The student learns the semantic segmentation network with an updated supervision, while the teacher uses the exponential moving average of the student to achieve a more accurate estimation of supervision. WSSS-MT employs the trained teacher as final segmentation network. Experimental results on the PASCAL VOC 2012 dataset show that the performance of our framework is better than the competing methods.

Keywords: Weakly-supervised learning · Image semantic segmentation · Deep learning

1 Introduction

In recent years, deep convolutional neural networks have made great progress in image semantic segmentation [5,21,27]. Unfortunately, it takes a lot of time and money to collect a large number of labeled data for fully-supervision semantic segmentation. Therefore, the development of training segmentation model from unlabeled or weakly-labeled images is a very significant direction in image semantic segmentation. Common examples of weakly-labeled supervisions include bounding boxes [11], points [15], scribble [12], and image-level labels [3,9,14]. In this paper, we focus on image-level labels to tackle the semantic segmentation problem since it is one of the most economical and effective methods.

Simonyan *et al.* [18] proposed the classification activation map (CAM) with classification network to establish the correspondence between image-level labels and pixels. However, these recognition areas only focus on a small portion of the

© Springer Nature Switzerland AG 2019
Z. Cui et al. (Eds.): IScIDE 2019, LNCS 11935, pp. 324–335, 2019.
https://doi.org/10.1007/978-3-030-36189-1_27

objects, without accurately indicating their boundary. Seed, Expand, and Constrain (SEC) [1] employed an image classification network with CAM method to select the most discriminative regions, and used the regions as pixel-level supervision for segmentation network. But the learning strategy was static supervision, which deviated from the requirement of semantic segmentation task that required accurate and complete object regions.

Ahn et al. [2] presented a method of learning pixel-level semantic dependency under the supervision of the initial CAM. It can identify pixels belonging to the same object, and then expend labeling by random walk. But the learning of semantic relevance required additional training, and the results depended on the quality of the initial area. Deep Seeded Region Growing (DSRG) [3] can dynamically expand the discriminative area to cover the entire object in the process of training segmentation network. DSRG classified image pixels using the classical Seeded Region Growth (SRG) [17] method starting from the initial region. However, the model trained by this method was not robust. Once the pixels was incorrectly marked, it can not be modified. And the boundary of the object was heavily dependent on the correction of conditional random field (CRF) [16].

In order to address above issues, we propose a novel framework of weakly-supervised semantic segmentation network with mean teacher (WSSS-MT) learning. We use a classification network to generate the initial labeled area and CRF to decorate the image boundary. More importantly, WSSS-MT defines two roles in semantic segmentation module, named teacher network and student network respectively. As a student, it learns as before; as a teacher, it produces a target for student to learn. During the process of training, the student learns the semantic segmentation network with an updated supervision and the teacher employs the exponential moving average (EMA) of the student weights. It can preserve the valuable information about the training process and produce more accurate segmentation network. In particularly, the consistency error between student and teacher is an important part of supervision. It improves the student to reduce false positives and capture objects more robustly, resulting in a better teacher. Finally, we utilize the trained teacher as our segmentation network.

To sum up, the main contributions of this paper are summarized as follows:

- We propose WSSS-MT for the dense labeling task under only image-level supervision. There are two networks in the segmentation module to effectively utilize information of the training process.
- Mean teacher learning is proposed to average student weights to form a better target-generating teacher. It enables our method to faster convergence during training and achieve optimal segmentation performance with a small number of iterations.
- Experimental results show that our work achieves remarkable segmentation performance on the PASCAL VOC segmentation benchmark dataset.

The rest of this paper is organized as follows. We first review some related works in Sect. 2 and introduce the architecture of our method in Sect. 3. Section 4 illustrates in details of experimental results in representative datasets. Section 5 presents our conclusion.

2   Related Work

Weakly-supervised methods have been widely studied for semantic segmentation. It relies on very slight human labeling, which benefits many computer vision tasks. In various setting of weak labels, the class information is a more natural supervision and requires the least amount of time for annotation. In this paper, we aim at tackling the semantic segmentation task under image-level labels. Early methods attempted to train the segmentation model directly from the image-level labels [9]. But the weak label lacks sufficiently semantic information, which makes this method unsatisfactory. Weakly-supervised semantic segmentation methods of recent introduction employ the erasure and saliency maps based image-level, or the way of seeded region growing.

2.1   Image-Level Processing

Image-level erasure and saliency maps have been proposed [2, 14, 19, 22] as ways of preventing a classifier from focusing exclusively on the discriminative parts of objects. Wei et al. [2] used a method of adversarial erasing to expand the discriminative region by iteratively training multiple classification networks. Although it progressively expanded regions belonging to an object, it relied on the classification network to sequentially produce the most discriminative regions in erased images. This caused error accumulation and the mined object regions had coarse object boundary. The Guided Attention Inference Network (GAIN) [14] had a initial area that is trained to erase regions by deliberately confusing the classifier. However, the classifier mainly responded to high activation, so it was confusing if the only discriminative parts of the object were erased.

Shimoda et al. [22] proposed a method based on CNN-based class-specific saliency maps. Similarly, STC [19] presented the saliency map of the simple image as a starting-point to train the initial segmentation network. However, the network framework of this method was complicated, and the process of training model was time-consuming.

2.2   Seeded Region Growing

A class activation map (CAM) [18] initially identified just the small but the most discriminative parts of an object. It is a good start for classifying pixels from image-level annotations. Seed, Expand, and Constrain (SEC) [1] used the initial seed generated by the CAM. It then employed seeded region growing (SRG) method to extend the localization map and constrained it to the object boundary using the CRF.

Huang et al. [3] proposed the method that refined the initial localization map during the training process of the segmentation network. On each iteration, it took the generated localization map as supervision and borrowed the method of SEC to expand the surrounding pixels. The seed growth was based on a principle that an image of a small uniform area where the pixels should have the same label. However, the process of widely expanding seeds was randomness, which resulted the segmentation effect will heavily depend on CRF.

3   Proposed Method

CAM generates very coarse and sparse object seeds as initial supervision, which is not enough for the dense semantic labeling task. In order to solve the problem, we present a novel weakly-supervised semantic segmentation with mean teacher (WSSS-MT) learning framework. WSSS-MT employs the method of SRG [17] and gradually expands from the initial target seeds to the unmarked region, which is similar to DSRG. What's more, We propose a mean teacher learning method to train the model by using two segmentation networks, named student network and teacher network respectively. It can effectively utilize information of the training process to produce an accurate model.

The rest of this section, we first introduce the architecture of WSSS-MT and the composition of the total loss function. We then detailedly explain the mean teacher learning method and the consistency loss function in the semantic segmentation module. At last, we will denote the seeding loss and boundary loss. The whole process is summarized as Algorithm 1.

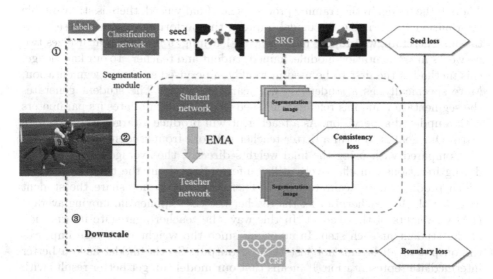

Fig. 1. Overview of the proposed WSSS-MT approach. The segmentation module employs two networks, where we average student weights to form a target-generating teacher and teacher becomes final output network.

3.1 Architecture of the Proposed WSSS-MT

As shown in Fig. 1, WSSS-MT is mainly divided into three parts. The first part is to input images and image-level labels into the classification network to get the initial seeds. The second part is to input images into the segmentation module, in which student network and teacher network generate two kinds of segmentation images. The seeds are then expanded to unlabeled region using the semantic segmentation map that is generated by the student network. And each iteration of the teacher network will be updated by using the EMA parameters of the student network. The third part is to use CRF to process images boundary and get more accurate segmentation images. At the end of model training, we will utilize the teacher network as our segmentation model. In this way, the segmentation features are explored to generate a more robust teacher segmentation network.

Overall, we use the following loss function to train the segmentation network:

$$L = \lambda L_{consistency} + L_{seed} + L_{boundary} \tag{1}$$

where $L_{consistency}$ is the consistency loss between the student network and the teacher network. And L_{seed} and $L_{boundary}$ respectively are the seeding loss and the boundary loss. The parameter λ is the consistency loss weight. In the rest of paper, we will explain each loss term in detail.

3.2 Mean Teacher Learning

Though the model in the training process is not finally used, there is still valuable information which make the model improve the performance of image segmentation. Therefore, we propose mean teacher learning, a method that utilizes two networks in segmentation module, named student and teacher. To our knowledge, this method is the first to be used in weakly-supervised semantic segmentation. More specifically, as a student, it will learn as before. The student generates the segmentation map as reference for seeds expand and updates its parameters with a updated supervision. As a teacher, it will produce a target for student to learn. Our goal is to form a better teacher network from student network.

Compared with using the final weights directly, the average model weights during iterations can effectively utilize information about the training process, which produce more accurate model. Therefore, we do not share the student weight with the teacher, but let the teacher use the exponential moving average (EMA) weights of the student. In this way, the teacher aggregate information of the student at each step. In addition, since the weighted average improves the output of all layers, not just the top output. So the model have a better intermediate representation. It means that our model can get better results with less training times.

The consistency loss between student and teacher is an important part of supervision. It enables the student to reduce false positives and capture targets more robustly, which results in more accurate teacher. We define the consistency loss as the mean square error (MSE) between the softmax output of the student (with weights θ) and the teacher (with weights θ'). Meanwhile, $f(u, c, \theta)$ and $f(u, c, \theta')$ respectively denote the probability of class c at position u of segmentation map in the student and the teacher. The consistency loss $L_{consistency}$ is defined as follows:

$$L_{consistency} = \frac{1}{\sum_{c \in C} |F_c|} \sum_{c \in C} \sum_{u \in F_C} \|f(u, c, \theta) - f(u, c, \theta')\|^2 \qquad (2)$$

in which, F_c is a set of locations that are classified to class c.

Meanwhile, we define teacher weights θ_t' at training step t as the exponential moving average of successive student weights θ:

$$\theta_t' = \alpha \theta'_{t-1} + (1 - \alpha)\theta_t \qquad (3)$$

Where α is a hyperparameter of the smoothing coefficient, and its value determines the speed of the teacher network weight update.

3.3 Seeding and Boundary Loss

In fact, the initial seeds are obtained through the CAM, and only 40% of the pixels have labels. We utilize the seeding algorithm to encourage the prediction of the segmentation network to the seeds, while ignoring the remaining pixels in the image. The seeding loss L_{seed} is defined as follows:

$$L_{seed} = -\frac{1}{\sum_{c \in C} |F_c|} \sum_{c \in C} \sum_{u \in F_c} \log H_{u,c} - \frac{1}{\sum_{c \in \bar{C}} |F_c|} \sum_{c \in \bar{C}} \sum_{u \in F_c} \log H_{u,c} \qquad (4)$$

where C is a set of classes that exist in the image (excluding the background), and $\bar{C}$ is background. F_c is a set of locations that are classified to class c. $H_{u,c}$ denotes the probability of class c at position u of segmentation map H.

The idea of the boundary loss is to penalize the segmentation network for producing semantic maps that are discontinuous with respect to spatial and color information in the input image. Therefore, it encourages the network to generate segmentation masks that conform to the object boundary. We construct a boundary loss $L_{boundary}$ which utilizes the method proposed in [1].

Algorithm 1. Training Procedure of WSSS-MT

Input: Image I, Image-level label c, Initial seed s;
Output: Semantic segmentation *images* and trained *teacher network*;
1: **While** iteration is effective **do**
2:    Input I to student and teacher and get $L_{consistency}$;
3:    $Expand(s)$ with segmentation map generated by student and get L_{seed};
4:    Constrain the seed boundary and get $L_{boundary}$;
5:    $L = L_{seed} + L_{boundary} + \lambda L_{consistency}$;
6:    Update student by total loss L;
7:    Update teacher parameters from student by EMA;
8: **End while**

---

4 Experiments

4.1 Experimental Setup

Dataset and Evaluation Metric. We evaluate experiments on the PASCAL VOC 2012 image segmentation benchmark. It contains 21 object classes, including one background class. Following the common practice, we trained network using image-level augmented dataset which contains 10582 training images. In addition, it also contains validation set (1449 images) and test set (1456 images). In our experiment, we compare WSSS-MT with other methods on the validation and test set. The standard intersection over union (IOU) criterion is adopted for evaluation dataset. The result on the test set is obtained by submitting the predicted results to the official PASCAL VOC evaluation server.

Training Settings. We use a slightly modified version of the 16-layer VGG network in [1] as the classification network, and DeepLab-ASPP in [8] as the segmentation network. They are all initialized by the VGG-16 [13] that pretrained on ImageNet. We use 0.9 momentum and 0.0005 weight decay. The smoothing coefficient of EMA α is set to 0.99. The batch size is 16 and the dropout rate is 0.5 We define the initial learning rate as 1e−3 and it is decreased by 10 times every 4 epochs. For the similarity criteria in our model, we respectively set the foreground and background thresholds to 0.7 and 0.5.

Our approach is implemented based on Tensorflow. All the experiment were performed on an NVIDIA TITAN X GPU. In order to obtain the object-related region based on a dense localization map, the pixels belonging to the top 20% of the unique largest value are selected as object regions. We use the saliency map generated by [26] to provide background clues. We set the normalized pixel with a significance value less than 0.06 as the background.

Other Details. We use the softmax output of the student to get the seeding loss and boundary loss. We believe that the two losses are equally important throughout the training cycle of the model. While the consistency loss between the student and teacher, we set a loss coefficient λ. The value of λ is very small early in the training of the model, but it will gradually increase. This is to learn more semantic information about images as soon as possible.

At the beginning of training, we use 8 epochs to increase the loss coefficient λ from 0 to its maximum value, using the sigmoid-shape $e^{-5(1-x)^2}$, where $x \in [0,1]$. It is worth noting that there are many options for the function for λ. In general, it is a good choice to set the value of 1 in the middle of model training steps. We think that the sigmoid function is a natural change, which improves the performance of the model.

4.2 Compare with Other Methods

In this subsection, we mainly compare coarse pixel-level annotations (such as graffiti, borders, and points) and image-level annotations as methods of supervising information. We provide the result of the PASCAL VOC validation and test dataset with other state-of-the-art weakly-supervised semantic segmentation solutions in Table 1.

Some results can be used as a reference, but they should not be directly compared to our method because they are different training set or different labels. Among the approaches, MIL-seg [6] and AF-MCG [4] implicitly use pixel-level supervision, and TransferNet [20] implicitly use box supervision. While STC [19] use image-level supervision and more images (50K) for training. All the other methods are based on 10K training images and built on top of the VGG-16 model and DSRG method is implemented by ourselves.

Compare with other competing techniques of weakly-supervision, the results show that our proposed WSSS-MT achieves the best performance. For the validation and test images, our mIoU value achieves 59.6% and 60.4%.

Table 1. Comparison of weakly-supervised semantic segmentation methods on PAS-CAL VOC 2012 validation and test image sets. The methods we have listed here use VGG-16 for segmentation.

Methods	Training	Validation	Test
MIL-FCN [24]	10K	25.7	24.9
EM-Adapt [9]	10K	38.2	39.6
MIL-seg [6]	700K	42.0	40.6
DCSM [22]	10K	44.1	45.1
STC [19]	50K	49.8	51.2
SEC [1]	10K	50.7	51.1
TransferNet [20]	70K	52.1	51.2
AF-MCG [4]	10K	54.3	55.5
AE-PSL [2]	10K	55.0	55.7
GAIN [14]	10K	55.3	56.8
MCOF [23]	10K	56.2	57.6
DSRG [3]	10K	58.8	59.7
WSSS-MT (ours)	10K	**59.6**	**60.4**

4.3 Comparison to the Role of CRF

The conditional random field (CRF) is a probability map model based on Markov. In weakly-supervised semantic segmentation, it can capture image boundary information and improve the segmentation ability of the model. However, under the condition of not using CRF, it can better reflect the basic performance of the model. As shown in Table 2, WSSS-MT achieves 59.6% mIoU on validation set when using CRF operation, and the result of not using CRF is 57.1% mIoU. The delta of WSSS-MT is 2.5%. Compared with the DSRG and GAIN methods, the delta value of our method is the smallest.

Note that WSSS-MT is 0.8% and 4.3% higher than the DSRG and GAIN method with CRF and is 1.7% and 6.3% higher than them without CRF. It means that the basic performance of WSSS-MT is superior to other methods and our method is less dependent on CRF.

Table 2. Comparison to with and without CRF on validation set. The delta denotes the difference with and without CRF.

Methods	With CRF	Without CRF	Delta
GAIN [14]	55.3	50.8	4.5
DSRG	58.8	55.4	3.4
WSSS-MT (ours)	**59.6**	**57.1**	**2.5**

4.4   Dynamic Observation over Epoch

We record the dynamic result of the WSSS-MT and DSRG experiment on validation set, as shown in Table 3. Our method is higher mIoU results every epoch than the DSRG method. At the 8th epoch, our method is more than 1.6% mIoU higher than the DSRG model. Not that our method can achieve 59.4% mIoU at the 16th epoch, which is only 0.2% mIoU off the final result. This is very closely to the highest level of results. While the DSRG method differed from the final result by 0.7% at the same epoch. So this indicates that our model can converge faster.

Table 3. The mIoU values of WSSS-MT and DSRG method with epoch changes on the validation set

Methods	Epoch 8	Epoch 16	Epoch 32
DSRG	55.9	58.1	58.8
WSSS-MT (ours)	**57.5**	**59.4**	**59.6**

4.5   Qualitative Results

Figure 2 shows some successful results of predicted segmentation masks in WSSS-MT. It shows our method can produce accurate segmentations even for complicated images and recover fine details of the boundary. The main reason is that we suitably use two networks in segmentation module. With supervision provided by the consistency error, the student can reduce false positives and capture targets more robustly, resulting in more accurate teacher.

Image   Ground Truth Our Method

Fig. 2. Examples of predictive segmentation mask by WSSS-MT for PASCAL VOC 2012 validation set in weakly-supervised manner.

5   Conclusion

In this paper, we propose a new model of weakly-supervised semantic segmentation with mean teacher (WSSS-MT) learning. WSSS-MT sets up two networks in segmentation module, named student and teacher respectively, where we average student weights to form a target-generating teacher. It effectively utilizes information of the model in the training process and improves the performance of the model. The experimental results on the representative database demonstrated the advantages of our proposed model in accuracy and efficiency.

Acknowledgement. This work is partially supported by the National Natural Science Foundation of China (Grant no. 61772568), the Guangzhou Science and Technology Program (Grant no. 201804010288), and the Fundamental Research Funds for the Central Universities (Grant no. 18lgzd15).

References

1. Kolesnikov, A., Lampert, C.H.: Seed, expand and constrain: three principles for weakly-supervised image segmentation. In: Leibe, B., Matas, J., Sebe, N., Welling, M. (eds.) ECCV 2016. LNCS, vol. 9908, pp. 695–711. Springer, Cham (2016). https://doi.org/10.1007/978-3-319-46493-0_42
2. Wei, Y., Feng, J., Liang, X., Cheng, M.-M., Zhao, Y., Yan, S.: Object region mining with adversarial erasing: a simple classification to semantic segmentation approach (2017). arXiv preprint arXiv:1703.08448
3. Huang, Z., Wang, X., Wang, J., Liu, W., Wang, J.: Weakly-supervised semantic segmentation network with deep seeded region growing. In: CVPR, pp. 7014–7023 (2018)
4. Qi, X., Liu, Z., Shi, J., Zhao, H., Jia, J.: Augmented feedback in semantic segmentation under image level supervision. In: Leibe, B., Matas, J., Sebe, N., Welling, M. (eds.) ECCV 2016. LNCS, vol. 9912, pp. 90–105. Springer, Cham (2016). https://doi.org/10.1007/978-3-319-46484-8_6
5. Chen, L.C., Papandreou, G., Kokkinos, I., Murphy, K., Yuille, A.L.: DeepLab: semantic image segmentation with deep convolutional nets, atrous convolution, and fully connected CRFs. IEEE Trans. Pattern Anal. Mach. Intell. **PP**(99), 1 (2017)
6. Everingham, M., Eslami, S.M.A., Van Gool, L., Williams, C.K.I., Winn, J., Zisserman, A.: The Pascal visual object classes challenge: a retrospective. Int. J. Comput. Vis. **111**(1), 98–136 (2015). https://doi.org/10.1007/s11263-014-0733-5
7. Hariharan, B., Arbelaez, P., Bourdev, L., Maji, S., Malik, J.: Semantic contours from inverse detectors. In: Proceedings of ICCV, pp. 991–998. IEEE (2011)
8. Chen, L.-C., Papandreou, G., Kokkinos, I., Murphy, K., Yuille, A.L.: DeepLab: semantic image segmentation with deep convolutional nets, atrous convolution, and fully connected CRFs (2016). arXiv preprint arXiv:1606.00915
9. Papandreou, G., Chen, L.-C., Murphy, K., Yuille, A.L.: Weakly- and semi-supervised learning of a DCNN for semantic image segmentation. In: Proceedings of the IEEE International Conference on Computer Vision (ICCV), pp. 1742–1750 (2015)
10. Ahn, J., Kwak, S.: Learning pixel-level semantic affinity with image-level supervision for weakly supervised semantic segmentation (2018). arXiv preprint arXiv:1803.10464
11. Khoreva, A., Benenson, R., Hosang, J., Hein, M., Schiele, B.: Simple does it: weakly supervised instance and semantic segmentation. In: Proceedings of the IEEE Conference on Computer Vision and Pattern Recognition (CVPR), pp. 876–885 (2017)
12. Vernaza, P., Chandraker, M.: Learning random-walk label propagation for weakly-supervised semantic segmentation. In: Proceedings of the IEEE Conference on Computer Vision and Pattern Recognition (CVPR), July 2017
13. Simonyan, K., Zisserman, A.: Very deep convolutional networks for large-scale image recognition (2014). arXiv preprint arXiv:1409.1556
14. Li, K., Wu, Z., Peng, K.-C., Ernst, J., Fu, Y.: Tell me where to look: guided attention inference network (2018). arXiv preprint arXiv:1802.10171
15. Bearman, A., Russakovsky, O., Ferrari, V., Fei-Fei, L.: What's the point: semantic segmentation with point supervision. In: Leibe, B., Matas, J., Sebe, N., Welling, M. (eds.) ECCV 2016. LNCS, vol. 9911, pp. 549–565. Springer, Cham (2016). https://doi.org/10.1007/978-3-319-46478-7_34

16. Koltun, V.: Efficient inference in fully connected CRFs with Gaussian edge potentials. In: Proceedings of NIPS, vol. 2, no. 3, p. 4 (2011)
17. Adams, R., Bischof, L.: Seeded region growing. IEEE TPAMI **16**(6), 641–647 (1994)
18. Zhou, B., Khosla, A., Lapedriza, A., Oliva, A., Torralba, A.: Learning deep features for discriminative localization. In: Proceedings of CVPR, pp. 2921–2929 (2016)
19. Wei, Y., et al.: STC: a simple to complex framework for weakly-supervised semantic segmentation. IEEE Trans. Pattern Recogn. Mach. Intell. **39**(11), 2314–2320 (2017)
20. Hong, S., Oh, J., Han, B., Lee, H.: Learning transferrable knowledge for semantic segmentation with deep convolutional neural network. In: IEEE CVPR (2016)
21. Zheng, S., et al.: Conditional random fields as recurrent neural networks. In: Proceedings of the IEEE International Conference on Computer Vision (ICCV), pp. 1529–1537 (2015)
22. Shimoda, W., Yanai, K.: Distinct class-specific saliency maps for weakly supervised semantic segmentation. In: Leibe, B., Matas, J., Sebe, N., Welling, M. (eds.) ECCV 2016. LNCS, vol. 9908, pp. 218–234. Springer, Cham (2016). https://doi.org/10.1007/978-3-319-46493-0_14
23. Wang, X., You, S., Li, X., Ma, H.: Weakly-supervised semantic segmentation by iteratively mining common object features. In: Proceedings of the IEEE Conference on Computer Vision and Pattern Recognition, pp. 1354–1362 (2018)
24. Pathak, D., Shelhamer, E., Long, J., Darrell, T.: Fully convolutional multi-class multiple instance learning (2014). arXiv preprint arXiv:1412.7144
25. Bachman, P., Alsharif, O., Precup, D.: Learning with Pseudo-Ensembles, December 2014. arXiv:1412.4864 [cs, stat]
26. Jiang, H., Wang, J., Yuan, Z., Wu, Y., Zheng, N., Li, S.: Salient object detection: a discriminative regional feature integration approach. In: Proceedings of CVPR, pp. 2083–2090 (2013)
27. Long, J., Shelhamer, E., Darrell, T.: Fully convolutional networks for semantic segmentation. In: Proceedings of CVPR, pp. 3431–3440 (2015)
28. Srivastava, N., Hinton, G., Krizhevsky, A., Sutskever, I., Salakhutdinov, R.: Dropout: a simple way to prevent neural networks from overfitting. J. Mach. Learn. Res. **15**(1), 1929–1958 (2014)

APAC-Net: Unsupervised Learning of Depth and Ego-Motion from Monocular Video

Rui Lin, Yao Lu, and Guangming Lu[✉]

Harbin Institute of Technology (ShenZhen), ShenZhen 518055, China
linrui_1995@163.com, yaolu_1992@126.com, luguangm@hit.edu.cn

Abstract. We propose an unsupervised novel method, Attention-Pixel and Attention-Channel Network (APAC-Net), for unsupervised monocular learning of estimating scene depth and ego-motion. Our model only utilizes monocular image sequences and does not need additional sensor information, such as IMU and GPS, for supervising. The attention mechanism is employed in APAC-Net to improve the networks' efficiency. Specifically, three attention modules are proposed to adjust feature weights when training. Moreover, to minimum the effect of noise, which is produced in the reconstruction processing, the Image-reconstruction loss based on $PSNR$ L_{PSNR} is used to evaluation the reconstruction quality. In addition, due to the fail depth estimation of the objects closed to camera, the Temporal-consistency loss L_{Temp} between adjacent frames and the Scale-based loss L_{Scale} among different scales are proposed. Experimental results showed APAC-Net can perform well in both the depth and ego-motion tasks, and it even behaved better in several items on KITTI and Cityscapes.

Keywords: Depth estimation · Ego-motion estimation · Attention mechanism

1 Introduction

Ego-motion estimation can provide effective and accurate position information from the cameras, and have been implemented in many fields, such as the visual odometry (VO) and the vehicle assistance driving system.

The traditional estimation methods can achieve high accuracy and be classified into the feature-based methods [11,15] and the direct methods [5,16]. And the early works based on Markov Random Fields [20] and non-parametric learning [1] also have precise depth estimation. However, these methods are particularly susceptible to the light change, occlusion and motion blur.

G. Lu—The work is supported by the NSFC fund (61332011), Shenzhen Fundamental Research fund (JCYJ20170811155442454, JCYJ20180306172023949), and Medical Biometrics Perception and Analysis Engineering Laboratory, Shenzhen, China.

© Springer Nature Switzerland AG 2019
Z. Cui et al. (Eds.): IScIDE 2019, LNCS 11935, pp. 336–348, 2019.
https://doi.org/10.1007/978-3-030-36189-1_28

Recently, plenty of methods based on deep-learning are proposed to estimate ego-motion. Pillai *et al.* [17] gets density motion estimation from optical flow vectors and predicts ego-motion through a Conditional Variational Autoencoder (CVA). VINet [2] fuses the images and IMU by LSTMs and estimates poses by a core LSTM. DeepVO [22] retrieves effective features from the image sequences by CNNs and models ego-motion through RNNs.

Since those methods with GPS/IMU have high requirements of calibration, some methods [12,19] are proposed, which only utilize the parallax between different viewpoints. However, they require the advanced calibration of cameras. Thus, Zhou *et al.* [24] applies a joint CNNs framework to estimate depth and ego-motion from monocular short sequences and reconstructs views with the predicted results. It assumes that only ego-motion causes frames difference and supervises itself by the difference between the original and the reconstructed views. GeoNet [23] and SfM-Net [21] divides scenes into static landscapes and dynamic objects. GeoNet [23] obtains dynamic objects from the optical flows and utilizes a geometric loss to the occlusion. SfM-Net [21] proposes a geometry-aware structure and a forward-backward consistency of motion as its constrains.

The attention mechanism is employed to guide networks to focus the meaningful information at the pixel level and improve the quality of features. Similarly, SE-Net [9] is proposed to allocate importance at the channel level. Currently, the performance in many tasks has been improved due to the two mechanism.

On this basis, Attention-Pixel and Attention-Channel Network (APAC-Net, see the Fig. 1) is proposed for unsupervised depth and ego-motion estimation from scene sequences and reconstruct images based on the results of the subnets.

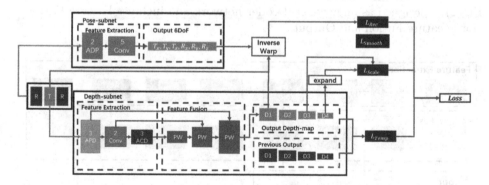

Fig. 1. Overview of APAC-Net. T and R represent the target images and reference images. Depth-subnet and Pose-subnet simultaneously estimate depth and ego-motion. The output and previous output are produced by the current T and the former T. The loss functions L_{Temp} and L_{Scale} are introduced in the Sect. 2.3.

The contributions of APAC-Net are summarized as following:

(1) In the feature extraction of Depth-subnet and Pose-subnet, the pixel-based attention module and the channel-based attention modules are proposed to explicitly predict the feature importance at different levels.
(2) In feature fusion of Depth-subnet, the Predicted-Weight Module (PW Module) is employed to adjust the weights of feature groups.
(3) Three novel loss, the $PSNR$ based Image-reconstruction loss L_{PSNR}, the Temporal-consistency loss L_{Temp} and the Scale-based loss L_{Scale}, are presented. L_{PSNR} can can reduce the influences of reconstruction noise, and L_{Temp} and L_{Scale} can deal with occlusion and objects closed to camera.

APAC-Net is evaluated on the depth and ego-motion tasks on KITTI [7] and Cityscape [3] databases. Compared with the state-of-the-art methods, APAC-Net only used monocular images and simple $2D$ loss functions, and obtained the accurate estimation without any groundtruth of depth and ego-motion.

2   APAC-Net

APAC-Net (see in Fig. 1) is based on Zhou *et al.* [24] and consists of Depth-subnet and Pose-subnet. The depth maps and $6DoF$ are estimated simultaneously and combined to reconstruct new images.

2.1   Depth-Subnet

The depth estimation is regarded as a segmentation task, so Depth-subnet (see in Fig. 2) is designed as an encoder-decoder network and includes Feature Extraction, Feature Fusion and Output.

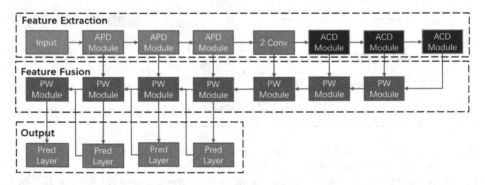

Fig. 2. Overview of Depth-subnet. It can be composed into 7 feature-extraction blocks and 7 feature-fusion blocks. The last 4 blocks produce a group of multi-scale depth.

Feature Extraction. It is composed of 3 modules, including the Attention-Pixel Module of Depth (APD Module), the Attention-Channel Module of Depth (ACD Module) and the 2-layer convolution module.

APD Module. Although low-level features contain lots of visual information, they also include duplicate information. APD Module (see in Fig. 3) is expected to focus on feature-dense areas and filter out the noise and invalid areas directly.

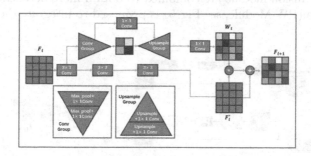

Fig. 3. The structure of APD Module. APD Module extracts obvious features by Conv Group and a 1×1 convolutions layer, and calculate the weight mask W_i by Upsample Group. The features F_i^* are obtained from two convolution layers. Then the final output F_{i+1} are the summation of the $F_i^* * W_i$ and F_i^*.

ACD Module. In the high-level networks, there are many redundant features and the network is easy to over-fitting. Inspired by SE-Net [9], ACD Module (see in Fig. 4) is used to adjust the feature weights. ACD Module attaches more attention to effective features and prevent network from over-fitting.

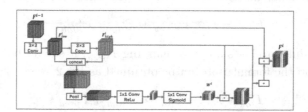

Fig. 4. The structure of ACD Module. In ACD Module, the low-level features F_{low}^i and the high-level features F_{high}^i are extracted by convolution layers. And F_{low}^i and F_{high}^i are concatenated and pooled to obtain global features. Then, the feature weights W^i are predicted by 1×1 convolution layers and multiplied by F_{high}^i. The results will add F_{high}^i and F_{low}^i to obtain the final output F^i.

APD Module pays more interests to the internal relationship of features and enhances the meaningful features areas. It performs better in the low-level features, which contain mass information at the pixel level. ACD Module is sensitive to the relationships among the channels and can effectively select useful features. It can perform better in the high-level network, where lots of features are redundant. Finally, as shown in Fig. 2, Feature Extraction consists of 3 APD Modules, a two-convolution module and 3 ACD Modules.

Feature Fusion. Since each feature extraction module contains a 2-stride layer, the size of output is reduced by 1/2 in each module. Referring to the previous work [24], high-level semantic features F_{high} and low-level visual features F_{low} at each feature fusion module are combined to obtain more detailed. Predicted-Weight Module (PW Module, see Fig. 5) is proposed to integrate these features.

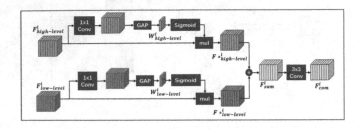

Fig. 5. The structure of PW Module.

PW Module. In previous works, F_{high}^i and F_{low}^i are simply concatenated along the channel extent. However, F_{high}^i and F_{low}^i have different contributions to the depth estimation. In the PW Module, the low-level feature weight w_{low}^i is computed by a 1×1 convolution layer and a global average pooling layer (GAP). And $F*_{low}^i$ is obtained by w_{low}^i and F_{low}^i. So does F_{high}^i. Then, two group features with different weights are added to get the fused features F_{sum}^i.

$$F_{sum}^i = F_{low}^i * w_{low}^i + F_{high}^i * w_{high}^i \tag{1}$$

If this PW module produces a depth mapping F_{pred}^i, it should be expanded as F_{low}^i firstly. And the formulation can be obtained as Eq. 2.

$$F_{sum}^i = F_{low}^i * w_{low}^i + F_{high}^i * w_{high}^i + F_{pred}^i * w_{pred}^i \tag{2}$$

At last, F_{sum}^i is processed to obtain the final output F_{com}^i, which is regarded as the high-level features F_{high}^{i-1} of the $(i-1)th$ layer.

Output. To overcome the trouble of gradient locality and the filter noise, as shown in Fig. 2, 4 1×1 convolution layers are used to correspond to the outputs of the last 4 feature fusion blocks and produce depth mappings F_{pred}. The size of F_{pred} is increased layer by layer.

2.2 Pose-subnet

As shown in Fig. 6, the Pose-subnet consists of 7 blocks to exact features and an output block to compute the corresponding poses between the target views and the reference views.

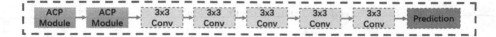

Fig. 6. The overview of Pose-subnet.

Feature Exaction. The differences between the sequence, which is connected by a target view and N reference views, are focused by Pose-subnet. In this work, the Attention-Channel Module of Pose (ACP Module) learns the difference between pixels and the diversity between feature maps.

ACP Module. Similar to the SE-Net [9], which assigns weights according to the validity of features, ACP Module (see Fig. 7) reduces the weights of invalid features and makes full use of the effective features. In ACP Module, groups of features are exacted by a 1×1 convolution layer from F_i and transformed into groups of real values through a GAP layer. Then, two 1×1 convolution layers are equipped to predict the weights W_i. Next, the features F_i are re-weighted by W_i and obtain the outputs F_{i+1} by the 3×3 convolution layer.

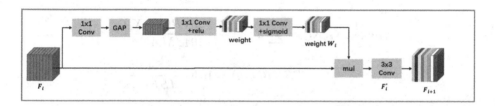

Fig. 7. The structure of ACP Module.

Many structures of Pose-subnet with different numbers of ACP Module and convolution modules are tested to determine the final structure. According to these evaluations, Pose-subnet contains 2 ACP Modules at low-level layers and 5 convolution modules with stride 2. It's important to note that the number of output channels in the last module is $6 \times N$.

Output. As the above illustrations, Pose-subnet produces $6DoF$ poses, which includes the translation information T_x, T_y, T_z and rotation information R_x, R_y, R_z. To output N groups of $6DoF$, the extracted feature are pooled by a global average pooling layer and resized to $N * 6$.

2.3 Loss Function

To constrain APAC-Net without ground-truth, the loss L only includes self-constrain losses: the Image-reconstruction loss L_{Rec}, the Temporal-consistency loss L_{Temp}, the Scale-based loss L_{Scale} and the Depth-smooth loss L_{Smooth}.

$$L = L_{Rec} * w_{Rec} + L_{Temp} * w_{Temp} + L_{Scale} * w_{Scale} + L_{Smooth} * w_{Smooth} \quad (3)$$

Image-Reconstruction Loss. In many works such as [23, 24], the photometric loss L_{Pixel} is used to measure the Image-reconstruction loss L_{Rec}. X_t^s and $\hat{X}_t^s$ represent the input images and the reconstructed images in time t on the last sth block. The reconstruction is rough and cannot well explain the effects of shadow. And this photometric loss is so noise-sensitive that it can cause over-fitting easily.

$$L_{Pixel}(X_t, \hat{X}_t) = \sum_s 2^{-s} * \left| X_t^s - \hat{X}_t^s \right| \quad (4)$$

Similar to [18], $SSIM$ is used to evaluate the reconstructions in this work. $SSIM$ is a measurement for the similarity between two images and can measure image brightness, image contrast and the structure similarity between images.

$$L_{SSIM}(X_t, \hat{X}_t) = \sum_s 2^{-s} * (1 - SSIM(X_t^s, \hat{X}_t^s)) \quad (5)$$

The reconstruction is considered as image restoration, so the peak signal to noise ratio ($PSNR$) can evaluate the reconstructed quality.

$$L_{PSNR}(X_t, \hat{X}_t) = \sum_s 2^{-s}(1 - \frac{PSNR(X_t, \hat{X}_t)}{10 \lg MAX^2})$$

$$PSNR(x, y) = 10 \lg \frac{MAX^2}{MSE(x, y)} \quad (6)$$

$$MSE(x, y) = \frac{1}{H * W} \sum_i^H \sum_j^W (P_x^{ij} - P_y^{ij})^2$$

The Image-reconstruction loss can be expressed by Eq. 7.

$$L_{Rec} = L_{Pixel} * w_{Pixel} + L_{SSIM} * w_{SSIM} + L_{PSNR} * w_{PSNR} \quad (7)$$

Temporal-Consistency Loss. In a very short time, the scenes captured by cameras have very small changes. If the adjacent frames are similar, the depth estimation have high similarity. $SSIM$ can measure the similarity of images from the structure. And in this work, the noise is assumed discrete points and the occlusions are seen as clustered points with high structure. Hence, the Temporal-consistency loss L_{Temp} based on $SSIM$ is proposed. L_{Temp} can quantify the occlusion, light and shadow, and ignore the noise produced in the reconstruction process.

$$L_{Temp}(D_{t-1}, D_t) = \sum_s 2^{-(s+1)} * (1 - SSIM(D_{t-1}, D_t)) \quad (8)$$

Scale-Based Loss. For smaller images, more attention is focused on the global information, and on the larger images, details are more noticeable. The quality of compressed images can be evaluated by $PSNR$. The smaller depth maps can be approximated by the compressed the larger. The Scale-based loss L_{Scale} based on $PSNR$ can judge the prediction quality of multi-scale depth maps. D_s is the larger depth map and UpD_t is enlarged from the small-size one D_t.

$$L_{Scale}(D_s, UpD_t) = \sum_s 2^{-s} \sum_t PSNR(D_s, UpD_t) \qquad (9)$$

Depth-Smooth Loss. In [14], the Depth-smooth loss L_{Smooth} is proposed and it utilizes the gradients of the corresponding input images to regularize the depth estimations D_s and keep sharp details. In order to keep clear changes of different areas on the grayscale map, the edge parts in the $Input_s$ are sharped by the sobel operator.

$$L_{Smooth} = \sum_s 2^{-s} \sum_t \left| D_s * e^{-|\nabla Input_s|} \right| \qquad (10)$$

3   Experiment

APAC-Net is implemented with Tensorflow 1.3 and optimized by Adam optimizer [10], where $\beta_1 = 0.9$, $\beta_2 = 0.999$ and $lr = 0.0002$. We apply the same training/validation/test split as [24] and resize images to 128×416. APAC-Net is trained with monocular resized images and evaluated with the LiDAR $3D$ and IMU messages. To evaluate the APAC-Net's performance, it is compared with the other popular methods on the standard benchmark KITTI [7] and Cityscape [3]. The loss weights in Eqs. 3 and 7 are determined by many times test, where $w_{Rec} = 1.0$, $w_{Temp} = 0.1$, $w_{Scale} = 0.3$, $w_{Smooth} = 0.5$, $w_{Pixel} = 0.5$, $w_{SSIM} = 0.4$ and $w_{PSNR} = 0.1$.

3.1   Depth Estimation

Following the previous works [24], APAC-Net is evaluated with the difference metric and the accuracy metric (see Table 1) on C1 (depth range from 0 to 80 m) and C2 (depth range from 0 to 50 m).

Three versions of APAC-Net are employed on the KITTI dataset (see Table 1): APAC-Net loss, APAC-Net without loss (APAC-Net w/o loss), APAC-Net. Zhou *et al.* [24] without the explainability mask network is the baseline. All versions achieve better performances than the baseline.

To evaluate the performance more intuitively, APAC-Net is compared quantitatively with the previous methods. As shown in Tables 2 and 3, APAC-Net outperforms both Zhou *et al.* [24] and Eigen *et al.* [4]. Because L_{SSIM} and L_{PSNR} in L_{Rec} can process light changes and noise well and L_{Temp} and L_{Scale} can reduce the influence of occlusion and moving objects. Additionally, the relationships between high-level features and low-level features are adjusted by PW Module according to their contribution.

Table 1. Monocular depth C1 and C2 on the KITTI. The three cases of improved are experimented respectively: APAC-Net loss, APAC-Net w/o loss and APAC-Net. The first part and the second part show the result of depth C1 and C2, respectively.

Method	Abs_Rel	Sq_Rel	RMSE	RMSE_log	$\delta < 1.25$	$\delta < 1.25^2$	$\delta < 1.25^3$
Baseline	0.208	2.750	7.170	0.260	0.720	0.893	0.953
APAC-Net loss	0.184	1.909	6.705	0.270	0.705	0.906	0.984
APAC-Net w/o loss	0.195	2.092	6.674	0.272	0.733	0.907	0.960
APAC-Net	**0.158**	**1.281**	**5.982**	**0.242**	**0.782**	**0.925**	**0.970**
Baseline	0.195	1.778	5.368	0.271	0.736	0.905	0.957
APAC-Net loss	0.174	1.367	5.036	0.252	0.768	0.917	0.965
APAC-Net w/o loss	0.184	1.448	5.005	0.255	0.751	0.919	0.967
APAC-Net	**0.151**	**0.964**	**4.488**	**0.226**	**0.789**	**0.936**	**0.974**

Table 2. Monocular depth C1 on the KITTI with the split of Eigen *et al.* [4]. For training, K, CS and All are the KITTI, Cityscapes and Cityscapes+KITTI. Eigen *et al.* [4] and Liu *et al.* [13] are supervised by the depth ground-truth. Godard *et al.* [8] is supervised by the pose ground-truth. Zhou *et al.* [24], Mahjourian *et al.* [14] and ours are unsupervised. The all methods are evaluated on the KITTI dataset.

Method	Data	Abs_Rel	Sq_Rel	RMSE	RMSE_log	$\delta < 1.25$	$\delta < 1.25^2$	$\delta < 1.25^3$
Eigen. [4] Course	K	0.214	1.605	6.563	0.292	0.673	0.884	0.957
Eigen. [4] Fine	K	0.203	1.548	6.307	0.282	0.702	0.890	0.958
Liu. [13]	K	0.202	1.614	6.523	0.275	0.678	0.895	0.965
Godard. [8]	K	**0.148**	1.344	**5.927**	0.247	**0.803**	0.922	0.964
Zhou. [24]	K	0.208	1.768	6.856	0.283	0.678	0.885	0.957
Mahjourian. [14]	K	0.163	**1.240**	6.220	0.250	0.762	0.916	0.968
APAC-Net	K	0.158	1.281	5.982	**0.242**	0.782	**0.925**	**0.970**
Godard. [8]	CS	0.699	10.060	14.445	0.542	0.053	0.326	0.862
Zhou. [24]	CS	0.267	2.686	7.580	0.334	0.577	0.840	0.937
APAC-Net	CS	**0.256**	**2.609**	**7.472**	**0.327**	**0.591**	**0.852**	**0.944**
Mahjourian. [14]	All	**0.159**	**1.231**	**5.912**	0.243	0.784	0.923	**0.970**
Zhou. [24]	All	0.208	1.768	6.856	0.283	0.678	0.885	0.957
APAC-Net	All	0.160	1.438	6.120	**0.241**	**0.787**	**0.927**	**0.970**

Table 3. Monocular depth C2 on the KITTI with the split of Eigen *et al.* [4]. Measures and dataset are same as in Table 2.

Method	Data	Abs_Rel	Sq_Rel	RMSE	RMSE_log	$\delta < 1.25$	$\delta < 1.25^2$	$\delta < 1.25^3$
Gary. [6]	K	0.169	1.080	5.104	0.273	0.740	0.904	0.962
Zhou. [24]	K	0.201	1.391	5.452	0.264	0.696	0.900	0.966
Mahjourian. [14]	K	0.155	**0.927**	4.549	0.231	0.781	0.931	**0.975**
APAC-Net	K	**0.151**	0.964	**4.488**	**0.226**	**0.789**	**0.936**	0.974
Zhou. [24]	CS	0.260	2.232	6.148	0.321	0.590	0.852	0.945
APAC-Net	CS	**0.248**	**2.079**	**6.031**	**0.314**	**0.604**	**0.862**	**0.950**
Zhou. [24]	All	0.190	1.436	4.975	0.258	0.735	0.915	0.968
Mahjourian. [14]	All	**0.151**	**0.949**	**4.383**	0.227	0.802	0.935	0.974
APAC-Net	All	**0.151**	1.040	4.581	**0.225**	**0.805**	**0.938**	**0.975**

As seen in Table 3 and Fig. 8, APAC-Net preforms better at shorter-distances. The distant objects are regarded as static scenes by APAC-Net. From the objects, which are closed to camera, APD Module can extract more high-quality and stable features, and ACD Module can select effective features. Hence, APAC-Net can obtain more effective features to estimate scene depth.

| Input | Sfm-Learnr (K) | Ours (K) | Ours (CS+K) |

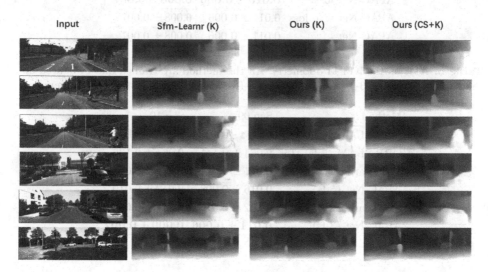

Fig. 8. The depth mappings. The raw images are shown on the first column. The depth images on the second column are estimated by Zhou *et al.* [24], which is trained on the KITTI. The depth estimations by APAC-Net are displayed on the third column and the forth column, which are pre-trained on the KITTI and Cityscape+KITTI.

3.2   Pose Estimation

There are 11 sequences have both raw senor images and accurate ground-truth camera poses in the KITTI odometry. To compare with Zhou *et al.* [24], the 11 sequences are divided into two parts: sequences $00 - 08$ for training and $09 - 10$ for evaluating. The Absolute Trajectory Error (ATE, see Table 4) is a very important criterion in ego-motion estimation. The absolute errors of the estimated trajectory and ground-truth trajectory are an evelution index of the global consistency.

APAC-Net is compared with the deep-learning based methods [24] and the traditional methods. Here are a few things to note: ORB-SLAM (full) and ORB-SLAM (short) run all-frames sequences and the 5-frames sequences. APAC-Net is trained both with 3-frames and 5-frames sequences. And shorter sequences produce better training results. As shown in Table 5, although our method contains only single-view image information without IMU, it performs as well as others.

Due to the IMU in the KITTI odometry, the sampling frequencies of the KITTI odometry and KITTI are different. In addition, some static scenes

Table 4. ATE on 3-frames the KITTI dataset. Baseline is compared with three cases of APAC-Net loss, APAC-Net w/o loss and APAC-Net.

Method	Seq.09	Seq.10
Baseline	0.011 ± 0.007	0.009 ± 0.007
APAC-Net loss	**0.010 ± 0.006**	**0.008 ± 0.007**
APAC-Net w/o loss	0.011 ± 0.006	0.008 ± 0.007
APAC-Net	0.011 ± 0.006	0.008 ± 0.007

Table 5. ATE on the KITTI odometry. All the method are trained and evaluated on the KITTI odometry. 3-fs and 5-fs are the 3-frames and 5-frames training sequences.

Method	Seq.09	Seq.10
ORB-SLAM (full) [15]	**0.014 ± 0.008**	**0.012 ± 0.011**
ORB-SLAM (5-fs) [15]	0.064 ± 0.141	0.064 ± 0.130
Sfm-Learner (5-fs) [24]	0.021 ± 0.017	0.020 ± 0.015
APAC-Net (5-fs)	0.016 ± 0.008	0.013 ± 0.010
Sfm-Learner (3-fs) [24]	0.012 ± 0.012	0.011 ± 0.012
APAC-Net (3-fs)	**0.011 ± 0.006**	**0.009 ± 0.007**

removed from the training set of the KITTI. More importantly, accurate and robust results can be achieved on simple datasets. Thus, APAC-Net is trained on the KITTI raw database first, then the pre-trained model is fine-tuned on the KITTI odometry [7]. To judge quantitatively, Zhou *et al.* [24] is trained and tested on the same datasets. The comparisons are shown in the Table 6.

Table 6. ATE on the KITTI odometry dataset. Zhou *et al.* [24] and APAC-Net are both trained on the KITTI dataset and evaluated on the KITTI odometry.

Method	Seq.09	Seq.10
Sfm-Learner (5-fs) [24]	0.023 ± 0.021	0.018 ± 0.021
APAC-Net (5-fs)	**0.015 ± 0.008**	**0.013 ± 0.010**
Sfm-Learner (3-fs) [24]	0.011 ± 0.007	0.009 ± 0.007
APAC-Net (3-fs)	**0.011 ± 0.006**	**0.008 ± 0.006**

Compared with Zhou *et al.* [24], APAC-Net, only using simple $2D$ loss functions, estimates ego-motion with smaller standard deviation and preforms better. On the one hand, the difference between features are focused by ACP Module can enhances the feature diversity in the low-level network. On the other hand, the noise produced in the reconstructing process is ignored by L_{Temp}. And the occluded object and the light changes can be quantified by L_{Scale}.

4 Conclusion

In this work, APAC-Net is proposed for monocular unsupervised learning of depth and ego-motion without sensor information. To adjust features weights, improve the network's efficiency and avoid over-fitting, three new attention based feature exaction modules, APD Module, ACD Module and ACP Module, are proposed. APD Module pays more attention to the density information areas, ACD Module deals with redundant features in high-level network and ACP Module enhances the difference between adjacent frames. In addition, the PW Module adjusts the weights between the high-level features and low-level features. On this basis, Depth-subnet and Pose-subnet are built. Further, to overcome the trouble of occlusion, noise and lighting changes, three new loss functions, L_{PSNR} about the noise in the processing of reconstruction, L_{Temp} about the relationship between adjacent frames and L_{Scale} about the difference between mutilate scale images depths, are proposed into this work. Besides, to quantitatively measure the performance, APAC-Net is compared with the state-of-art supervised methods and unsupervised methods. APAC-Net, only utilizing the simple loss functions, behaves better than Zhou *et al.* [24] in both depth and ego-motion tasks and performs as well as those methods with extra data and complex $3D$ loss functions. In the future, the semantic information will be employed in APAC-Net to detect positions of the dynamic objects.

References

1. Choi, S., Min, D., Ham, B., Kim, Y., Oh, C., Sohn, K.: Depth analogy: data-driven approach for single image depth estimation using gradient samples. IEEE Trans. Image Process. **24**(12), 5953–5966 (2015)
2. Clark, R., Wang, S., Wen, H., Markham, A., Trigoni, N.: VINet: visual-inertial odometry as a sequence-to-sequence learning problem. National Conference on Artificial Intelligence, pp. 3995–4001 (2017)
3. Cordts, M., et al.: The cityscapes dataset for semantic urban scene understanding. In: Computer Vision and Pattern Recognition, pp. 3213–3223 (2016)
4. Eigen, D., Puhrsch, C., Fergus, R.: Depth map prediction from a single image using a multi-scale deep network. Neural Information Processing Systems, pp. 2366–2374 (2014)
5. Engel, J., Koltun, V., Cremers, D.: Direct sparse odometry. IEEE Trans. Pattern Anal. Mach. Intell. **40**(3), 611–625 (2018)
6. Garg, R., Kumar, B.G.V., Carneiro, G., Reid, I.: Unsupervised CNN for single view depth estimation: geometry to the rescue. In: Leibe, B., Matas, J., Sebe, N., Welling, M. (eds.) ECCV 2016. LNCS, vol. 9912, pp. 740–756. Springer, Cham (2016). https://doi.org/10.1007/978-3-319-46484-8_45
7. Geiger, A., Lenz, P., Stiller, C., Urtasun, R.: Vision meets robotics: the KITTI dataset. Int. J. Robot. Res. **32**(11), 1231–1237 (2013)
8. Godard, C., Aodha, O.M., Brostow, G.J.: Unsupervised monocular depth estimation with left-right consistency. In: Computer Vision and Pattern Recognition, pp. 6602–6611 (2017)
9. Hu, J., Shen, L., Sun, G.: Squeeze-and-excitation networks. In: Computer Vision and Pattern Recognition, pp. 7132–7141 (2018)

10. Kingma, D.P., Ba, J.: Adam: a method for stochastic optimization (2015)

11. Klein, G., Murray, D.W.: Parallel tracking and mapping for small AR workspaces, pp. 1–10 (2007)

12. Li, R., Wang, S., Long, Z., Gu, D.: UnDeepVO: monocular visual odometry through unsupervised deep learning. In: International Conference on Robotics and Automation, pp. 7286–7291 (2018)

13. Liu, B., Gould, S., Koller, D.: Single image depth estimation from predicted semantic labels, pp. 1253–1260 (2010)

14. Mahjourian, R., Wicke, M., Angelova, A.: Unsupervised learning of depth and ego-motion from monocular video using 3D geometric constraints. In: Computer Vision and Pattern Recognition, pp. 5667–5675 (2018)

15. Murartal, R., Montiel, J.M.M., Tardos, J.D.: ORB-SLAM: a versatile and accurate monocular SLAM system. IEEE Trans. Robot. **31**(5), 1147–1163 (2015)

16. Newcombe, R.A., Lovegrove, S., Davison, A.J.: DTAM: dense tracking and mapping in real-time, pp. 2320–2327 (2011)

17. Pillai, S., Leonard, J.J.: Towards visual ego-motion learning in robots, pp. 5533–5540 (2017)

18. Pinard, C., Chevalley, L., Manzanera, A., Filliat, D.: Learning structure-from-motion from motion. Computer Vision and Pattern Recognition, pp. 363–376 (2018). arXiv

19. Repala, V.K., Dubey, S.R.: Dual CNN models for unsupervised monocular depth estimation. Computer Vision and Pattern Recognition (2018). arXiv

20. Saxena, A., Sun, M., Ng, A.Y.: Make3D: learning 3D scene structure from a single still image. IEEE Trans. Pattern Anal. Mach. Intell. **31**(5), 824–840 (2009)

21. Vijayanarasimhan, S., Ricco, S., Schmid, C., Sukthankar, R., Fragkiadaki, K.: SfM-Net: learning of structure and motion from video. Computer Vision and Pattern Recognition (2017). arXiv

22. Wang, S., Clark, R., Wen, H., Trigoni, N.: DeepVO: towards end-to-end visual odometry with deep recurrent convolutional neural networks. In: International Conference on Robotics and Automation, pp. 2043–2050 (2017)

23. Yin, Z., Shi, J.: GeoNet: unsupervised learning of dense depth, optical flow and camera pose. In: Computer Vision and Pattern Recognition, pp. 1983–1992 (2018)

24. Zhou, T., Brown, M., Snavely, N., Lowe, D.G.: Unsupervised learning of depth and ego-motion from video. In: Computer Vision and Pattern Recognition, pp. 6612–6619 (2017)

Robust Image Recovery via Mask Matrix

Mengying Jin and Yunjie Chen$^{(\boxtimes)}$

Nanjing University of Information Science and Technology,
Nanjing 210044, China
jinmengying_maths@163.com, priestcyj@nuist.edu.cn

Abstract. This paper studies the problem of recovering an unknown image matrix from noisy observations. Existed works, such as Robust Principal Component Analysis (RPCA), are under the case where the image component and error component are additive, but in real world applications, the components are often non-additive. Especially an image may consist of a foreground object overlaid on a background, where each pixel either belongs to the foreground or the background. To separate image components robustly in such a situation, this paper employs a binary mask matrix which shows the location of each component, and proposes a novel image recovery model, called Masked Robust Principal Component Analysis (MaskRPCA). On one hand, the image component and error component are measured by rank function and sparse function, separately. On another hand, the non-additive between components is characterized by mask matrix. Then we develop an iterative scheme based on alternating direction method of multipliers. Extensive experiments on face images and videos demonstrate the effectiveness of the proposed algorithm.

Keywords: Non-additive signal · Low rank · Sparse · Robust image recovery

1 Introduction

With ever-increasing huge volume of images data, how to robustly recover an unknown image matrix from noisy observations has being a challenge. Due to the fact that most observation data usually contains various errors or even corruptions, and we recent years have witnessed plenty of efforts on the studies of low-rank recovery and representation. The low-rank models can be roughly divided into two categories, i.e., principal components analysis and low-rank representation.

Among various principal components analysis models, Principal Component Analysis (PCA) [1] is a classic one, whose purpose is to minimize the reconstruction error between the original data and reconstructed data. But PCA is sensitive to noise and outliers, since it employed Frobenius norm for measuring the reconstruction errors [2, 3]. To improve the robustness of PCA, several enhanced models were recently proposed, such as Robust PCA (RPCA) [3], Inductive RPCA (IRPCA) [4], double nuclear norm-based matrix decomposition (DNMD) [5] and Schatten p-norm based matrix completion [6].

RPCA decomposes the corrupted data into the sum of a low-rank matrix and a sparse matrix exactly by minimizing a weighted combination of the nuclear norm and the L_1 norm. Differing from RPCA and its extension works which characterizes the

Z. Cui et al. (Eds.): IScIDE 2019, LNCS 11935, pp. 349–361, 2019.
https://doi.org/10.1007/978-3-030-36189-1_29

error image in the vector form, Zhang et al. [5] proposed the double nuclear norm-based matrix decomposition (DNMD). DNMD represents the error image in the matrix form and measures it via the nuclear norm. In DNMD, the nuclear norm is used to characterize both the real image data and continuous occlusion. DNMD is further extended and applied in [7] and [8].

For low-rank representation models, the most well-known approach is Low-Rank Representation (LRR) [9–11]. LRR sought the lowest rank representation among all the candidates that represent all vectors as the linear combinations of the basis vectors in a dictionary. The LRR can exactly recover the true subspace structures when the data is clean. However, most real-world data are contaminated by various errors and corruptions. This may depress the subspace segmentation recovery performance directly, due to the noise.

All in all, the abovementioned works and their extensions are limited to the assumption of the image components are additive. However, in real world applications, the image components are often non-additive. For example, in background modeling application, an image of surveillance video may consist of a foreground object overlaid on a background, where each pixel belongs to either the foreground or the background. In such a situation, for separating signal components robustly, this paper aims to find a binary mask which shows the location of each component.

Specifically, to characterize the non-additive components, this paper employs a mask matrix in RPCA model. The mask matrix denotes the position of different components. Thus, we propose a novel model, called Masked Robust Principal Component Analysis (MaskRPCA). In MaskRPCA, following RPCA the image component and error component are measured by rank function and sparse function, separately. Furthermore, the non-additive between components is characterized by mask matrix.

The rest of this paper is organized as follows. Section 2 reviews the related work. Section 3 presents our model and corresponding algorithm. Section 4 reports experimental results. Section 5 offers conclusions.

2 Related Work

We first show the important notations used in this paper. For any matrix $S \in R^{m \times q}$, its nuclear norm $\|S\|_*$ is defined as $\|S\|_* = \sum_i \sigma_i(S)$, which is the sum of the singular values of S. Besides, the $\mathbf{1}$ denotes a matrix with all element being 1.

Let $X = [x_1, x_2, \cdots, x_m] \in R^{n \times m}$ be a set of samples, where n is the dimensionality of the original space and m the number of data points, each column vector x_i denotes an n-dimensional sample. In various applications, it is of interest to recover an unknown signal $D \in R^{n \times m}$ from a noisy observation matrix X satisfying the model:

$$X = D + E \tag{1}$$

where E is an error matrix.

This model suffers from a fundamental identifiability issue since the number of unknowns (2 nm) is greater than the number of observations (nm). Therefore, we inevitably have to assume special types of structure between the signal component D and error component E. A kind of powerful approach called RPCA assumes the signal component D has low rank structure and error component E has sparsity structure, whose formulation can be given by (2):

$$\min_{D,E} \ \text{rank}(D) + \lambda\|E\|_0 \ \text{ s.t. } X = D + E \tag{2}$$

with $\lambda > 0$. The low rank assumption on D is often met in practice when there exists a significant correlation between the columns of D, such as in image classification, collaborative filtering or medical imaging, among many others.

However, this minimization is combinatorially hard. To achieve fast computation, the following convex relaxation is often used:

$$\min_{D,E} \ \|D\|_* + \lambda\|E\|_1 \ \text{ s.t. } X = D + E \tag{3}$$

There exist many algorithms for RPCA, such as alternating direction method of multipliers (ADMM) [12] and gradient descent approach [13, 14].

3 Masked Robust Principal Component Analysis

3.1 Motivation and Model

Abovementioned works are under the case where the components are additive, i.e. $X = D + E$, which means they assume there both low rank component and sparse component among each pixel. But in real world applications, the components are often non-additive [15]. An image may consist of a foreground object overlaid on a background, where each pixel belongs to either the foreground or the background.

For example, in Fig. 1, the left image has been corrupted by a block occlusion. In such a case, RPCA assumes that the image component and error component are additive and obtains the decomposition result (a). One can see that the error part contains more details which belong to the image component.

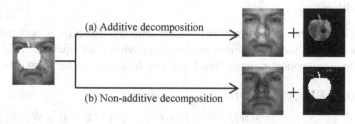

Fig. 1. The decomposition result of an observation image under (a) additive assumption by RPCA and (b) non-additive assumption by MaskRPCA.

However, if we assume the components are non-additive, one can obtain the decomposition result (b) by our MaskRPCA. Compared with additive decomposition, the interaction between low rank component and sparse component is cut down. The key of MaskRPCA is to find a binary mask which shows the location of each component.

Now, we present the formulation of Masked Robust Principal Component Analysis (MaskRPCA). MaskRPCA is formulated based on the idea of RPCA, but improves it explicitly. Given a set of training samples $X = [x_1, x_2, \cdots, x_m] \in R^{n \times m}$, the purpose of MaskRPCA is to recover the clean low-rank component D and the sparse error component E, and at the same time obtaining the error position component W. Thus, we present the following objective function:

$$
\begin{aligned}
&\min_{D,E,W} \|D\|_* + \lambda \|E\|_1 + \gamma \|W\|_1 \\
&s.t. \ X = (1 - W) \circ D + W \circ E, \ W \in \{0, 1\}^{n \times m}
\end{aligned}
\tag{4}
$$

where the L_1-norm is still regularized on the error term E as RPCA does. Furthermore, since the mask term W denotes the position of the error component, it should be as sparse as the error term E. So the mask term W is also regularized by L_1-norm.

In formulation (4), the first constraint $X = (1 - W) \circ D + W \circ E$ decomposes given data X into a low-rank part D fitting recovered data (clean data) and a sparse error part E fitting random corruptions and noise. The symbol $\circ$ denotes the element-wise product, and W is the binary mask matrix denoting the support of each component.

MaskRPCA investigates a class of different signal decompositions, where signal components are superimposed on top of each other, rather than simply added. In other words, only one of all signal components can be active (non-zero) at the same position. Furthermore, we employ an auxiliary variable F for convenience of solving. So, one can rewrite MaskRPCA as (5):

$$
\begin{aligned}
&\min_{D,E,W,F} \|D\|_* + \lambda \|E\|_1 + \gamma \|W\|_1 \\
&s.t. \ X = D - W \circ F, F = D - E, \ W \in \{0, 1\}^{n \times m}
\end{aligned}
\tag{5}
$$

Specifically, the F can separate the variables W and D such that the optimization on W and D can be conducted alternately.

3.2 Algorithm

Since MaskRPCA contains discrete variable, its objective function is non-convex. We solve it based on the inexact Augmented Lagrange Multiplier (inexact-ALM) method [12, 16]. The corresponding augmented Lagrange function can be formulated as:

$$
\begin{aligned}
L(D, E, F, W, Y, Z) = \ &\|D\|_* + \lambda \|E\|_1 + \gamma \|W\|_1 \\
&+ tr(Z^T(F - D + E)) + tr(Y^T(X - D + W \circ F)) \\
&+ \frac{\mu}{2} \left(\|X - D + W \circ F\|_F^2 + \|F - D + E\|_F^2 \right)
\end{aligned}
\tag{6}
$$

where Y, Z are the Lagrange multiplier matrices and μ is a penalty parameter.

Next, we employ alternating direction scheme to update each variable while fixing the other variables. That is, the minimization of the objective function can be alternated between the following five steps.

(i) Updating D. In (6), we fix the other variables and remove terms that are independent of D, then one can obtain

$$
\begin{aligned}
D^{k+1} &= \underset{D}{\operatorname{argmin}} \|D\|_* + tr\left(-(Z+Y)^T D\right) + \frac{\mu}{2}\left(\|X - D + W \circ F\|_F^2 + \|F - D + E\|_F^2\right) \\
&= \underset{D}{\operatorname{argmin}} \|D\|_* + tr\left(-(Z+Y)^T D\right) + \mu \left\|\frac{1}{2}(X + W \circ F + E + F) - D\right\|_F^2 \\
&= \underset{D}{\operatorname{argmin}} \|D\|_* + \mu \left\|\frac{1}{2}\left(X + W \circ F + E + F + \frac{1}{\mu}(Z+Y)\right) - D\right\|_F^2 \\
&= \underset{D}{\operatorname{argmin}} \frac{1}{2\mu}\|D\|_* + \frac{1}{2}\left\|\frac{1}{2}\left(X + W \circ F + E + F + \frac{1}{\mu}(Z+Y)\right) - D\right\|_F^2 \\
&= D_{\frac{1}{2\mu}}\left(\frac{1}{2}\left(X + W^k \circ F^k + E^k + F^k + \frac{1}{\mu}(Z^k + Y^k)\right)\right)
\end{aligned}
$$

$$(7)$$

The symbol $D_\tau(\cdot)$ is a singular value shrinkage operator, given any matrix A and its singular value decomposition $A = U\Sigma V^T$, defined by $D_\tau(A) \triangleq U S_\tau(\Sigma) V^T$ [17].

(ii) Updating E. In (6), we fix the other variables and remove terms that are independent of E, then one can obtain

$$
\begin{aligned}
E^{k+1} &= \underset{E}{\operatorname{argmin}} \lambda\|E\|_1 + tr\left(Z^T E\right) + \frac{\mu}{2}\left(\|F - D + E\|_F^2\right) \\
&= \underset{E}{\operatorname{argmin}} \ \lambda\|E\|_1 + \frac{\mu}{2}\left(\left\|F - D + \frac{1}{\mu}Z + E\right\|_F^2\right) \\
&= \underset{E}{\operatorname{argmin}} \ \frac{\lambda}{\mu}\|E\|_1 + \frac{1}{2}\left(\left\|F - D + \frac{1}{\mu}Z + E\right\|_F^2\right) \\
&= S_{\frac{\lambda}{\mu}}\left(D^{k+1} - F^k - \frac{1}{\mu}Z^k\right)
\end{aligned}
$$

$$(8)$$

The symbol $S_\tau(\cdot)$ is an element-wise soft-thresholding operator, for any number x, defined as [16]:

$$
S_\tau(x) \triangleq \begin{cases} x - \tau, & \text{if } x > \tau \\ x + \tau, & \text{if } x < -\tau \\ 0 & \text{otherwise} \end{cases}
$$

$$(9)$$

(iii) Updating F. In (6), we fix the other variables and remove terms that are independent of F, then one can obtain

$$
\begin{aligned}
F^{k+1} &= \underset{F}{\arg\min}\ L(F) \triangleq tr(Z^T F) + tr(Y^T(W \circ F)) + \frac{\mu}{2}\left(\|X - D + W \circ F\|_F^2\right) \\
&= \underset{F}{\arg\min}\ \frac{\mu}{2}\left\|X - D + \frac{1}{\mu}Y + W \circ F\right\|_F^2 + \frac{\mu}{2}\left\|F - D + E + \frac{1}{\mu}Z\right\|_F^2
\end{aligned}
\tag{10}
$$

By taking derivative $\frac{\partial L}{\partial F}$ and setting it to zero, the closed-form solution of F can be inferred as

$$
F^{k+1} = \left\{D^{k+1} - E^{k+1} - \frac{1}{\mu}Z^k - W^k \circ \left(X - D^{k+1} + \frac{1}{\mu}Y^k\right)\right\} \circ/\ (1 + W^k \circ W^k)
\tag{11}
$$

where the symbol $\circ/$ denotes the element-wise quotient.

(iv) Updating W. We fix the other variables and obtain (12):

$$
\begin{aligned}
W^{k+1} &= \underset{W \in \{0,1\}^{m \times n}}{\arg\min}\ \left\{\gamma\|W\|_1 + tr(Y^T(W \circ F)) + \frac{\mu}{2}\|X - D + W \circ F\|_F^2\right\} \\
&= \underset{W \in \{0,1\}^{m \times n}}{\arg\min}\ \left\{\gamma\|W\|_1 + \frac{\mu}{2}\left\|\frac{1}{\mu}Y + X - D + W \circ F\right\|_F^2\right\} \\
&= \underset{W \in \{0,1\}^{m \times n}}{\arg\min}\ \sum_{i,j} \min\left\{\gamma + \frac{\mu}{2}\left(\frac{1}{\mu}Y_{ij} + X_{ij} - D_{ij} + F_{ij}\right)^2, \frac{\mu}{2}\left(\frac{1}{\mu}Y_{ij} + X_{ij} - D_{ij}\right)^2\right\}
\end{aligned}
\tag{12}
$$

(v) At last, the multipliers Y and Z can be updated by (13):

$$
\begin{cases}
Y^{k+1} = Y^k + \mu(X - D^{k+1} + W^{k+1} \circ F^{k+1}) \\
Z^{k+1} = Z^k + \mu(F - D^{k+1} + E^{k+1})
\end{cases}
\tag{13}
$$

For complete presentation of the algorithm, we summarize the procedures in Algorithm 1.

Algorithm 1. Solving MaskRPCA by inexact-ALM
1. Input: The data matrix X, parameters λ, γ. Let
$W^0 = 1, F^0 = X, \ E^0 = Y^0 = Z^0 = 0, \mu = 1, \mu_{max} = 10^6, \rho = 1.2, k = 0.$
Do
2: Update D by (7);
3: Update E by (8);
4: Update F by (11);
5: Update W by (12);
6: Update Y and Z by (13);
7: Update $\mu = \min(\rho\mu, \mu_{\max})$.
Until Converge.
8: **Output:** $(D^{k+1}, E^{k+1}, F^{k+1})$.

4 Performance Evaluation

We apply the MaskRPCA into data recovery problem from face images and video surveillance. The state-of-the-art signal recovery methods are selected for comparison, including RPCA[1], and RPCA-OM[2] [18], NC-RPCA[3][19], OR-TPCA[4] [20].

4.1 Face Image Recovery Under Lighting Changes

In the Extended Yale B database[5], there are 38 subjects. For each subject, 16 images under various lighting are used and each sample has resolution 48×42. Some samples are shown in Fig. 2. We stack each image as a column of matrix and obtain the observation data X, which is a matrix of 2016×608. Then one can obtain the recovery component D and error component E by kinds of approaches mentioned above, shown in Fig. 3. On the purpose of computation efficient, we can relax the constraints of D and E in practice, alternatively. At meantime, the setting of parameter is similar to RPCA, $\lambda = \frac{1}{\sqrt{\max(n,m)}}$.

Evidently, MaskRPCA outperforms other approaches in removing specularities in the eyes, shadows around the nose region, or brightness saturations on the face, especially for 4th and 5th samples. Notice in the third row of Fig. 3(e) that the mask

---

[1] The code: http://perception.csl.illinois.edu/matrix-rank/sample_code.html.

[2] The code: http://www.escience.cn/people/fpnie/papers.html.

[3] The code: http://www.ece.uci.edu/anandkumar/.

[4] The code: https://panzhous.github.io/.

[5] http://vision.ucsd.edu/~leekc/ExtYaleDatabase/ExtYaleB.html.

images indicate corruption position in images. These results may be useful for conditioning the training data for face recognition, as well as face alignment and tracking under illumination variations.

Fig. 2. Observation samples in X.

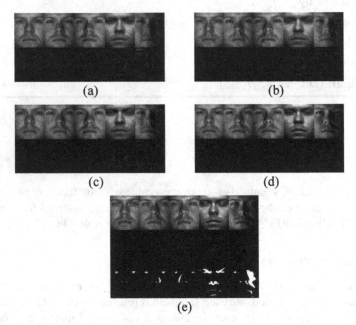

Fig. 3. Recovered images (first row) and error images (second row) by RPCA (a), RPCA-OM (b), OR-TPCA (c), NC-RPCA (d), and MaskRPCA (e). In (e) the images in the 3th row represent the corresponding mask value.

Furthermore, Table 1 gives the recognition rates of nearest neighbor classifier on the component D, where first half samples are for training and the other half for testing. For nearest neighbor classifier, three distance metrics, L_2, L_1 and Cosine, are employed. The proposed algorithm significantly outperforms others.

Table 1. The recognition rates (%) of nearest neighbor classifier under L_2, L_1, and Cosine distances

Distances	RPCA	RPCA-OM	OR-TPCA	NC-RPCA	MaskRPCA
L_2	52.0	64.0	58.0	82.0	88.0
L_1	50.0	58.0	56.0	74.0	74.0
Cosine	80.0	84.0	80.0	98.0	100.0

4.2   Face Image Recovery Under Pixel Occlusions

In this subsection, we evaluate the effectiveness of MaskRPCA for recovering images from occlusions. The dataset is the same X as used in Sect. 4.1. We impose block occlusions on two samples of each subject and obtain the observation data X^{obser}. Then many abovementioned algorithms can be used on X^{obser} and obtain corresponding recovery component D and error component E. We quantify the result of each model by the recovery error which is defined as

$$\text{Recovery Error} = \frac{1}{t}\sum_{i=1}^{t} \|D_i - X_i\|_F$$

where t means the total number of occlusion samples.

The recovery errors under different percentages of pixel occlusions are shown in Fig. 4. The images at the bottom of Fig. 4 illustrate samples with different occlusion rates from 10% to 25%. We can see that MaskRPCA achieves the lowest recovery errors in all cases. In addition, some recovered images are shown in Fig. 5. As it can be seen, MaskRPCA can recover the images and capture occlusion blocks simultaneously.

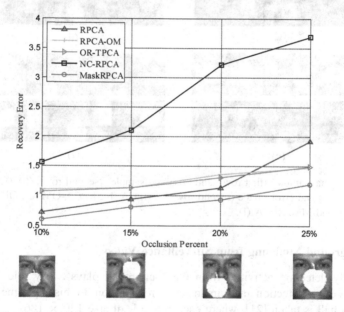

Fig. 4. The recovery error for testing set with different occlusion rates.

Many current methods usually be constrained by $X = D + E$. The flaw of such assumption is quite obvious that their recovery results will partly remove information from expected clean images. As Fig. 5 shows, this situation has become particularly obvious among observation data with white block occlusion. That is to say, recovered error of current methods contained more or less details information while the expected error should be just white. At the same time, the resident of the white occlusion can still be seen at the low rank component. However, the proposed method, MaskRPCA, can obtain the expected clean image and error image. As Fig. 5 shows, benefitting from the particular constraint of non-additive decomposition $X = (1 - W) \circ D + W \circ E$, the non-additivity reduced the interaction between different components well.

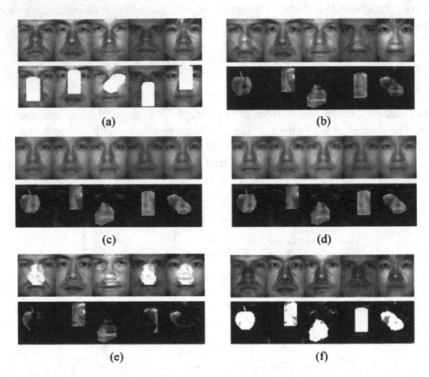

Fig. 5. (a) Original samples (first row), and observed samples (second row); (b)–(f) Recovered samples (first row) and error image (second row) by RPCA (b), RPCA-OM (c), OR-TPCA (d), NC-RPCA (e), and MaskRPCA (f).

4.3   Background Modeling from Surveillance Video

Automatically detecting activities from the background plays a key role in modern time, such as event detection and human action recognition. In this experiment, a video in an airport hall is taken [21], where each frame is of size 120×186.

We randomly select 100 frames and stack each frame into the matrix X. Then, we decompose it via different methods. Table 2 compares the time costs of different methods. We can see that MaskRPCA, RPCA and NC-RPCA are all faster than OR-TPCA and RPCA-OM.

Figure 6 shows the original frames (a) and the recovered backgrounds and the identified activities, respectively. From these results, we can see that MaskRPCA and

OR-TPCA can identify the activities more clearly than others. However, MaskRPCA are faster than OR-TPCA and RPCA-OM.

Table 2. Running time (s) of different methods

Methods	RPCA	RPCA-OM	OR-TPCA	NC-RPCA	MaskRPCA
Times(s)	29.8655	90.3246	105.8416	27.4789	31.3894

(a) Original frame

(b) RPCA

(c) RPCA-OM

(d) OR-TPCA

(e) NC-RPCA

(f) MaskRPCA

Fig. 6. Background modeling from airport hall video. In (b)–(f), the left images are recovered background and right ones are activities.

5 Conclusion

This paper proposes an image recovery model from a signal decomposition perspective. We consider error as an overlaying component on top of a natural scene image, where the pixel values at each point comes from one and only one of the components. This happens in many signal decomposition applications, such as image recovery from corrupted data and background extraction from surveillance videos. An algorithm based on the alternating direction method of multipliers is proposed. Experimental results show the proposed algorithm can provide significantly better performance in image recovery. However, due to the proposed model having non-convex objective function and also 0–1 constraint of mask matrix, the solving algorithm is hard to achieve. How to update the mask matrix more effectively is the direction of our efforts. Meanwhile, we also consider further application of mask matrix in other decomposition models to achieve better results.

Acknowledgments. The authors would like to thank the anonymous reviewers for their valuable comments and suggestions to improve the quality of the paper. This work was supported in part by the National Nature Science Foundation of China 61672291 and Six talent peaks project in Jiangsu Province SWYY-034.

References

1. Jolliffe, I.T.: Principal Component Analysis, 2nd edn. Springer-Verlag, New York (2002). https://doi.org/10.1007/b98835
2. Candès, E., Li, X., Ma, Y., Wright, J.: Robust principal component analysis. J. ACM **58**(3), 1–37 (2011)
3. Wright, J., Ganesh, A., Rao, S., Ma, Y.: Robust principal component analysis: exact recovery of corrupted low-rank matrices by convex optimization. In: Proceedings of the 22nd International Conference on Neural Information Processing Systems, pp. 2080–2088. MIT Press, Vancouver (2009)
4. Bao, B.K., Liu, G., Xu, C., Yan, S.: Inductive robust principal component analysis. IEEE Trans. Image Process. **21**(8), 3794–3800 (2012)
5. Zhang, F., Yang, J., Tai, Y., Tang, J.: Double nuclear norm-based matrix decomposition for occluded image recovery and background modeling. IEEE Trans. Image Process. **24**(6), 1956–1966 (2015)
6. Nie, F., Wang, H., Huang, H., Ding, C.: Joint schatten p-norm and lp-norm robust matrix completion for missing value recovery. Knowl. Inf. Syst. **42**(3), 525–544 (2015)
7. Bouwmans, T., Sobral, A., Javed, S., Jung, S., Zahzah, E.: Decomposition into low-rank plus additive matrices for background/foreground separation: a review for a comparative evaluation with a large-scale dataset. Comput. Sci. Rev. **23**, 1–71 (2017)
8. Zhou, Z., Jin, Z.: Double nuclear norm-based robust principal component analysis for image disocclusion and object detection. Neurocomputing **205**, 481–489 (2016)
9. Liu, G., Lin, Z., Yan, S., Sun, J., Yu, Y., Ma, Y.: Robust recovery of subspace structures by low-rank representation. IEEE Trans. Pattern Anal. Mach. Intell. **35**(1), 171–184 (2013)
10. Gao, G., Jing, X.-Y., Huang, P., Zhou, Q., Wu, S., Yue, D.: Locality-constrained double low-rank representation for effective face hallucination. IEEE Access **4**, 8775–8786 (2016)
11. Li, J., Kong, Y., Zhao, H., Yang, J., Fu, Y.: Learning fast low-rank projection for image classification. IEEE Trans. Image Process. **25**(10), 4803–4814 (2016)
12. Goldstein, T., O'Donoghue, B., Setzer, S., Baraniuk, R.: Fast alternating direction optimization methods. SIAM J. Imaging Sci. **7**(3), 1588–1623 (2014)
13. Yi, X., Park, D., Chen, Y., Caramanis, C.: Fast algorithms for robust PCA via gradient descent. In: International Conference on Neural Information Processing Systems, pp. 4152–4160. MIT Press, Barcelona (2016)

14. Tan, B., Liu, B.: Acceleration for proximal stochastic dual coordinate ascent algorithm in solving regularised loss minimisation with l2 norm. Electron. Lett. **54**(5), 315–317 (2018)
15. Mohammadreza, S., Hegde, C.: Fast algorithms for demixing sparse signals from nonlinear observations. IEEE Trans. Signal Process. **65**(16), 4209–4222 (2017)
16. Lin, Z.C., Chen, M.M., Ma, Y.: The augmented Lagrange multiplier method for exact recovery of corrupted low-rank matrix. Technical Report UILU-ENG-09–2215, UIUC, October 2009
17. Cai, J., Candès, E., Shen, Z.: A singular value thresholding algorithm for matrix completion. SIAM J. Optim. **20**(4), 1956–1982 (2010)
18. Nie, F., Yuan, J., Huang, H.: Optimal mean robust principal component analysis. In: 31st International Conference on Machine Learning, pp. 1062–1070. MIT Press, Beijing (2014)
19. Netrapalli, P., Niranjan, U.N., Sanghavi, S.: Provable non-convex robust PCA. In: International Conference on Neural Information Processing Systems, pp. 1107–1115. MIT Press, Montreal (2014)
20. Zhou, P., Feng, J.: Outlier-robust tensor PCA. In: IEEE Conference on Computer Vision and Pattern Recognition, pp. 3938–3946. IEEE, Honolulu (2017)
21. Li, L., Huang, W., Gu, I., Tian, Q.: Statistical modeling of complex backgrounds for foreground objects detection. IEEE Trans. Image Process. **13**(11), 1459–1472 (2004)

Multiple Objects Tracking Based Vehicle Speed Analysis with Gaussian Filter from Drone Video

Yue Liu, Zhichao Lian[(⊠)], Junjie Ding, and Tangyi Guo

Nanjing University of Science and Technology, Nanjing, China
1294384195@qq.com, lzcts@163.com

Abstract. Vehicle speed analysis based on the video is a challenging task in the field of traffic safety, which has high requirements for accuracy and computational burden. The drone's video is taken from a top-down perspective, providing more complete view comparing to the common surveillance cameras in poles. In this paper, we introduce a Gaussian Filter to deal with the estimated speed data which are extracted by a multiple objects tracking method composed of You Only Look Once (YOLOv3) and Kalman Filter. We exploit the capability of Gaussian Filter to suppress data noise appearing in the process of tracking vehicles from drone videos, and thus use the filter to solve the case where the estimated vehicle speed is fluctuated along the ongoing direction. On the other hand, we built a vehicle dataset from the drone's videos we mentioned above which additionally contains vehicle's real speed information. Experimental results showed that our method is effective to improve the accuracy of vehicle speed estimated by our tracking module. It can improve Mean Squared error (MSE) accuracy 80.5% on experimental data.

Keywords: Gaussian Filter · Data analysis · Multiple object tracking · Traffic safety

1 Introduction

Quickly and accurately extracting vehicle related data from video is a fundamental topic in unmanned driving [11] and road monitoring. In the transportation field, the vehicle speed and density are two important measurements in the vehicle travel. Speeding and other illegal acts will be detected, which has a great impact on traffic safety. This topic is of broad interest for potential applications of traffic supervision, data analysis and so on.

There are many ways to obtain relevant data on the vehicle's travel, which mainly be categorized in two classes. The first one is to analyze the complex signal generated by electronic devices such as radar [8] and vehicle's own attached sensors [9, 10].

The other direction is based on cameras or the like to capture video for object detection and tracking [4]. Road monitoring can cover a certain range of vehicles and the vehicle's data can be measured by the objection detection and tracking method in computer vision [13]. However, there will be a large deviation in the speed measurement due to different monitoring camera's pitch angles to the near and far objects.

© Springer Nature Switzerland AG 2019
Z. Cui et al. (Eds.): IScIDE 2019, LNCS 11935, pp. 362–373, 2019.
https://doi.org/10.1007/978-3-030-36189-1_30

We show this phenomenon in Fig. 1. Since the position of the surveillance camera is generally low, the vehicle will have occlusion and overlap in the road, which raises many problems for analysis algorithms.

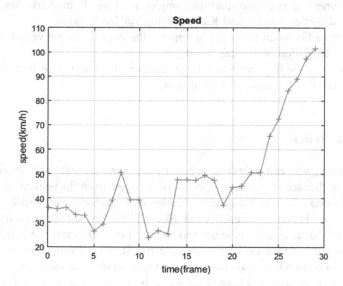

Fig. 1. When some devices measure that the vehicle is approaching a uniform speed, the phenomenon shown in the figure will appear: the speed in the first half is fluctuating up and down, and the speed in the second half is suddenly rising.

Nowadays, it is possible to extract motion information from video sequences thanks to the advancement of multiple object tracking. The pioneer work is to improve the accuracy of object detection like YOLOv3 trained on PASCAL VOC and Microsoft COCO datasets include more labels, such as person, boat and vehicle, etc. On the other hand, more excellent matching mechanism make multiple object tracking better.

State-of-the-art multiple object tracking are based on public datasets which contains multi-classes include oblique perspective vehicle. The model perform not well in vehicle tracking tasks from a special top-town perspective. These problems are due to less information of the vehicle in this view, like the vehicle in drone's video.

The development of drone equipment makes it possible to capture highways at a fixes altitude. This greatly ensures that the scene in the video are more realistically restored and then accurate vehicle data through certain technologies can be obtained. The drone has a panoramic view, which makes the vehicle in the video scene more complete and has no larger scale changes. High-altitude overhead allows for no occlusion between vehicles, simplifying the scene and simultaneously monitoring all vehicles in the scene.

Using video captured by drones, we design a vehicle speed analysis framework that tracks the vehicles in real time, calculates the speed by traces and corrects the speed of vehicles. Through experimental result, it has been proven that the framework can effectively calculate the running speed of vehicle and has a good performance in suppressing the noise occurring in the measurement process. Our main contributions are threefold.

- We implement a real-time multiple object tracking framework based on the YOLOv3 detection system and Kalman Filter for UAV video.
- We propose a Gaussian Filter [5] to remove the noise from the vehicle's data and refine the calculation of vehicle speed.
- We build a vehicle dataset from a large number of drone videos that contains the actual speed of the vehicle at the moment.

2   Related Work

In the following, we review known public traffic scene datasets and drone datasets. Almost all traffic scene datasets consist of images from in-vehicle devices and surveillance videos. In most samples, the scales of the vehicle vary widely, and many vehicles have occlusion from each other like KITTI. For some drone datasets, the proportion of vehicle samples is small and the most critical is that the actual speed of the vehicle is not included in the datasets like the Stanford Drone Dataset.

In order to obtain information such as speed and density in a video sequence, a real-time, stable and accurate tracking framework is often required. This framework is usually combined with the detector and motion prediction model. Driven by YOLO which relies on darknet-53 network, higher accuracy vehicle predictions and real-time detection results are achieved. Prediction of objects in the next frame which utilizing time and space information of vehicles in a video sequence by add Kalman Filter to the tracking framework. To correct the vehicle speed and other data generated by the tracking framework, we proposed a kind of Filter.

Filter selection is roughly divided into two categories. One line is based on some transformation. The data is transformed from the spatial domain to the frequency domain by Fourier Transform [14], processed in the frequency domain, and then processed to the spatial domain by inverse transform [15].

The other method is spatial domain filter. Since they process data or signals directly without any transformation, like operating directly on the pixel in the image. This approach takes advantage of the distribution of data. In the experiment, we notice that the vehicle's data approximates to a certain distribution, and spatial domain filtering is more intuitive and simpler than frequency domain filtering with less computational complexity.

We start with our observation and analysis of the large flutter of the bounding boxes which occurs in object tracking process. The large flutter mentioned above means the coordinates of the bounding boxes are not accurate enough. This is because the tracking framework is not stable enough [1], but actually this is inevitable. So we perform a statistical analysis of the results of the tracking and try to fix it.

To make a good performance in correcting vehicle data, we design a Gaussian filter based on video characteristics and vehicle data distribution rules. This module takes advantage of the fact that the vehicle data closes to the Gaussian distribution [7] and then remove noise by the filter.

Our experiments show that the Gaussian filter has a good effect when dealing with the speed data of moving vehicles in drone video. We use MSE to evaluate our effect on vehicle speed correction. In the testing dataset, data fluctuations in vehicle speed were significantly reduced and closer to truth values. Yet there is still much room to exploit in data analysis and data processing.

3 Our Proposed System

With above analysis, we design a real-time detection system of vehicle based on the YOLO model from the UAV video [16]. In addition, a Kalman Filter has been combined in the system to implement the tracking module [17, 18]. Finally, a Gaussian Filter has been involved to suppress the noise from the result obtained by the tracking module, and more accurate vehicle speed estimations are achieved by the proposed system in Fig. 2.

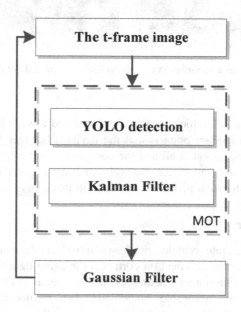

Fig. 2. The proposed system has a Gaussian Filter and a multiple object tracking module consists of YOLO detection and Kalman Filter

3.1 YOLO Detection System

The YOLOv3 detection system [1] is a state-of-the-art framework in real-time object detection, here we use YOLO to refer to it. YOLO creatively uses anchor boxes in network design to get direct location prediction. The image is divided into a 7×7 grid and each grid cell has 9 anchor boxes [2, 6] which are obtained by clustering and have fixed dimension to predict the possible bounding boxes and locations of the object. Although this makes the model predicts more than thousand, the objects are more detectable. In this improvement, the 9 anchor boxes have less coverage on a small scale and 7 × 7 cells to predict the objects which YOLO used make the (x, y) locations of bounding boxes has some instability.

3.2 Kalman Filter

Kalman Filter [3] is a linear filter which can be descrambled by linear stochastic difference equation. It keeps track of the estimated state of the system and the variance or uncertainty of the estimate.

Fig. 3. Since the drone has a panoramic view, the vehicle is crowded but there is no occlusion in the UAV video.

In our drone video, all vehicles are in the lanes and always heading in the same direction. The most important point is that the division between the vehicles is very clear and there is no occlusion at all and the contour of the vehicle remains the same. Some examples are shown in Fig. 3. So Kalman Filter is satisfactory for removing noise and predict the location of vehicle changes in that case.

3.3 Gaussian Filter

To calculate more accurate vehicle speed, we introduce the Gaussian filter, which proves that it has a significant effect on correcting tracking data [19, 20].

In any object detection framework and scientific calculation, errors are inevitable. If the error is within the allowable range, the data results are also acceptable. Although YOLO has taken some strategies to improve the accuracy of location, the prediction of the locations still has some instability [2]. In particular, anchor boxes which YOLO used to help CNN [22] locate targets make an unstable error between the coordinates of bounding boxes and the ground truth [21]. We address this issue by using a suitable Gaussian filter.

The Gaussian filter is a signal processing method as a filter whose impulse response is a Gaussian function, which is commonly used in eliminating Gaussian noise. The tracking data points recorded by the vehicle just tracked or tracked at the end tend to fluctuate greatly, while more stable during the tracking. Sudden acceleration or deceleration of the vehicle can also cause large deviations in the data calculated by our tracking module. This means that tracking sometimes lags behind the actual displacement of the vehicle. Gaussian filter is helpful in this regard to modify those mutation data.

Taking vehicle speed as an example, actually, the speed of the vehicle can be regarded as a constant value in one second. In the actual experiment, we use "frame" as the speed measurement unit instead of seconds for more accurate calculation data.

When measured the speed of the vehicle which can be calculated by $\frac{|x_p - x_n|}{1/fps}$ in every frame, where fps, x_p, x_n denote frames per second, x coordinate of current frame, x coordinate of previous frame respectively. We find that the calculation speed per frame is quite different but regular. We call this phenomenon "data flutter" in Fig. 4, the tracking data float around the true value and satisfy a certain distribution law. According to the characteristics of the data distribution, the correction of vehicle speed can be achieved by utilizing suitable Gaussian filter.

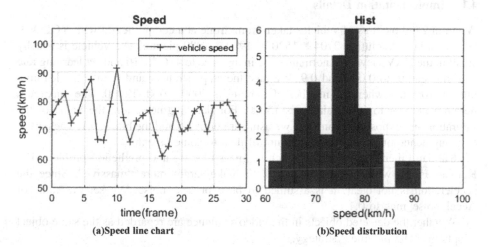

(a)Speed line chart   (b)Speed distribution

Fig. 4. Illustration of data flutter. (a) In one second (assuming the video frame rate is 30), the phenomenon that the forward speed of the uniform moving vehicle in each frame of the image is fluctuated greatly. (b) The vehicle speed calculated by tracking framework is close to the normal distribution.

To analyze these observations, errors are partially related to the precision of the multi-target detection algorithm framework and the video itself has a higher resolution. Since the data has a certain distribution law, a filter with suitable parameters can greatly improve the tracking data results.

We propose a Gaussian Filter to apply to vehicle speed data. We perform confirmatory experiments to show the difference in Sect. 4.

4 Experiments

Our improvement works on vehicle tracking data in drone video. The method we used effectively corrects vehicle tracking data and improves detection accuracy. We notice that although there are many public tracking datasets, they didn't provide motion information for targets in the real world such as speed. So we carry our experiments on our drone video dataset, which contains 45 car tracking data groups, real speed of the vehicle, and 4834 usable data records for training and testing.

For a tracking system, components selected in the framework play an extremely important role in terms of real-time and accuracy. These often determine the efficiency of the tracking and the accuracy of the detection. Our implementation is based on excellent algorithm YOLO which has a good performance in real-time and accuracy as well as a classic prediction method Kalman Filter, then the most important one is that we add a suitable Gaussian filter to smooth the tracking data about vehicle speed.

4.1 Implementation Details

We train 2750 pictures of vehicles taken from drone at a certain height with YOLOv3. We use a high resolution 2704 * 1520 or 1920 * 1080 picture. The vehicle is usually small in the UAV view. We normalize the image size to 416 * 416 and set learning rate and momentum to 0.001 and 0.9. We take the steps strategy and reduce the learning rate by 10 times when the number of training is 5000, 8000, 12000. The maximum iteration number to stop training is 15000. For data augmentation, the value of the saturation and exposure parameters we using are both 1.5 and the hue is 0.1. The part of data augmentation increases the amount of data to some extent.

Kalman Filter is a classic approach to linear filtering and prediction problem. For Kalman Filter, we set delatime to 0.2 to make target more "massive". Since the acceleration is not clear, it is assumed to be a process noise. We set the value of Accel_noise_mag to 0.1.

Whether the detected objects in the video sequence are regarded as the same object adopts the IOU judgment strategy.

The observation of bounding box within a small range flutter made we think about how to stable the detection boxes. By analyzed the tracking data of every car, the error between predicted value and truth value is close to the certain normal distribution in Fig. 4. A Gaussian filter with suitable parameters can yield great performance in the tracking data from analyzing the video.

Datasets. Existing available public tracking datasets lack object motion information while this is part of our system's main contribution. Therefore we built our own vehicle dataset from the drone video which contains its speed information recorded by speedometer. Some examples in Fig. 3. We collect the data points with N (e.g., 30) frames and group them in each second of each video. The data in the Table 1 comes from the video with a frame rate of 30 fps, so every second has 30 data points for analysis. The well grouped data is divided into 810/270 data points for training and testing.

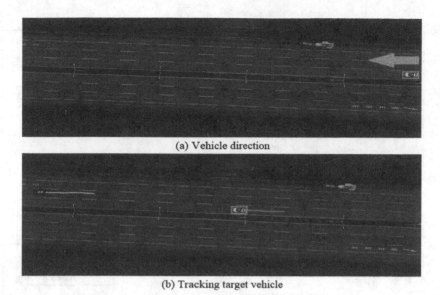

(a) Vehicle direction

(b) Tracking target vehicle

Fig. 5. The vehicle selected by the red rectangle is the target vehicle. (a) The red arrow indicates the direction of motion moves from the right to the left and (b) the red thin line is the trajectory tracked by the vehicle. (Color figure online)

Performance Measure. The process of data analysis is considered as a regression task, so mean square error (MSE) is used as the main performance measure.

We choose the same car in the same section of the road just for justice, and we pick some videos both in the upstream and downstream directions to test.

Upstream Vehicle Video. In the video, "upstream" means that the car drives from right to left in the scene of the video. As shown in Fig. 5, our tracking framework completes the tracking well. We use the tracking data of the previous time that the target vehicle appears in the video scene as the training data to calculate the parameter estimation of the corresponding Gaussian filter, and apply it to the estimation and correction of the vehicle speed for a later period of time.

For this video, we use about 270 data points about the speed of the target vehicle to analysis μ and σ which makes us get a suitable Gaussian filter. Then using the filter to correct the vehicle speed data for the last three seconds of the video.

As shown in the first row and the first column of the second row in Fig. 6, our work is excellent in tracking data. The red lines with diamonds are drawn from unprocessed vehicle speeds. The blue lines is composed of vehicle speed data processed by a Gaussian filter. After the last three seconds of the tracking data of the video is processed by a Gaussian filter, it can be clearly seen that the speed of the vehicle with blue lines is largely stabilized and the jitter is significantly reduced and is closer to real vehicle speed as shown with green line. Our performance measure MSE improves about 71.73%, and another set of data improved about 98.6%. The result as shown in the first three rows of the Table 1.

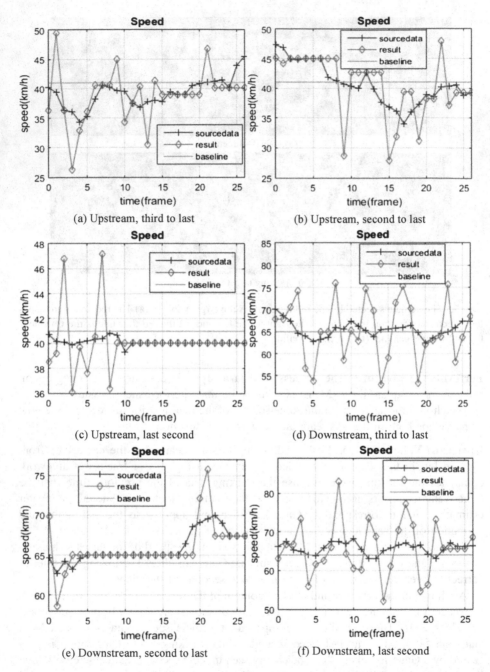

Fig. 6. Overview of our data result graph. Speed line chart of the upstream vehicle obtained in the last third second of the video with (a). The chart of second and last second as shown in (b) and (c). Speed line chart of the downstream vehicle obtained in the last third second of the video with (d). The chart of second and last second as shown in (e) and (f). (Color figure online)

(a) Vehicle direction

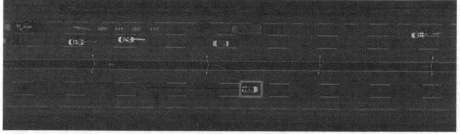

(b) Tracking target vehicle

Fig. 7. The vehicle selected by the red rectangle is the target vehicle. (a) The red arrow indicates the direction of motion move from left to right and (b) the blue thin line with the target vehicle is the trajectory tracked by the vehicle.

Table 1. Evaluate vehicle speed with MSE on test dataset. We choose three cars called "Car 1, Car 2, Car 3" in the video count down three seconds denoted by "Time 1, Time 2, Time 3". The vehicle's label is from the chosen cars that appears in different videos. The data in bold is the better group. Their MSE performance is better. Their mean and the output error is smaller.

Vehicle	Raw data MSE	Result MSE	Improve (%)	AVERAGE
Car1_Time1	**20.20234**	**5.715011**	**71.711143**	**2.813135**
Car1_Time2	27.54991	11.71515	57.476631	2.923800
Car1_Time3	**4.999145**	**0.077600**	**98.447735**	**1.091221**
Car2_Time1	47.59822	3.123054	93.438717	5.439575
Car2_Time2	8.936934	3.644574	59.218967	1.169905
Car2_Time3	47.67669	1.981989	95.842855	5.487424
Car3_Time1	44.39706	13.19164	70.287132	4.698314
Car3_Time2	**42.83019**	**3.381573**	**92.104698**	**4.874536**
Car3_Time3	**6.792341**	**1.213208**	**82.138588**	**1.567593**

Downstream Vehicle Video. In the video, "downstream" means that the car drives from left to right in the scene of the video, as shown in Fig. 7. This set of data contains 210 tracking data points about the speed of the same vehicle which appearing in the "downstream vehicle video" for training and 90 tracking data points for testing. The testing data points are divided into 3 groups. They are the last three seconds of the

Fig. 8. Illustration of vehicle tracking in drone video. The line following the vehicle is the trajectory of the vehicle.

vehicle's speed data. Methods trained the 210 tracking data to get an available Gaussian filter for testing. The result as shown in the middle three rows of the Table 1.

As shown in Table 1, with the suitable Gaussian filter, the three sets of tracking data grouped by vehicle after processing are well corrected, so that the data is distributed near the real data baseline with less jitter and error. The method can improve the MSE of the data about 59.21%–95.84%. We also use the "AVERAGE" about the errors to measure the noise suppression. The MSE from the table is significantly smaller which proved that vehicle speed at different times is closer to the true value by using our method. The "AVERAGE" indicates system output error is less than before. Finally, the system will output these processed data like vehicle speed which can be applied in the analysis process of traffic.

Statistics in Table 1 show that the Gaussian Filter has notable advantage in improving accuracy of the vehicle speed and the like obtained by tracking framework. The above experiments are all in vehicle dataset which contains the real speed of vehicles.

5   Conclusion

After constructing a vehicle dataset containing motion information of vehicle, we introduce YOLO and Kalman Filter to create a tracking module which for motion information such as vehicle speed. This module can better track the vehicle appearing in the drone video in real-time in Fig. 8. By statistically analyzing on the vehicle speed extracted by the tracking module, we propose a kind of Filter to refine the vehicle speed which gets an excellent improvement with MSE evaluation on vehicle dataset we built. The Gaussian Filter we used provides a nice feature in removing noise or present in vehicle speed. In the last part of the system, we obtain the corrected more accurate vehicle speed.

We hope the implementation details publicly available can help the community adopt these useful strategies for dealing with the tracking data and advance related techniques.

References

1. Redmon, J., Farhadi, A.: YOLOv3: an incremental improvement (2018)
2. Redmon, J., Farhadi, A.: YOLO9000: better, faster, stronger. In: IEEE Conference on Computer Vision & Pattern Recognition (2017)
3. Kalman, R.E.: A new approach to linear filtering and prediction problems. J. Basic Eng. Trans. **82**, 35–45 (1960)
4. Wu, Y., Lim, J., Yang, M.H.: Online object tracking: a benchmark. In: Computer Vision & Pattern Recognition (2013)
5. Ito, K.: Gaussian filter for nonlinear filtering problems. In: IEEE Conference on Decision & Control (2002)
6. Ren, S., He, K., Girshick, R., Sun, J.: Faster R-CNN: towards real-time object detection with region proposal networks. arXiv preprint arXiv:1506.01497 (2015)
7. Perret-Gentil, C.: Gaussian distribution of short sums of trace functions over finite fields. Math. Proc. Camb. Philos. Soc. **163**(3), 38 (2017)
8. Eaves, J.L., Reedy, E.K.: Principles of Modern Radar. SciTech Publishing, Chennai (2013)
9. Yun, D.S., et al.: The system integration of unmanned vehicle and driving simulator with sensor fusion system. In: International Conference on Multisensor Fusion & Integration for Intelligent Systems (2002)
10. Im, D.Y., et al.: Development of magnetic position sensor for unmanned driving of robotic vehicle. In: Sensors (2009)
11. Zhang, X., Gao, H., Mu, G., et al.: A study on key technologies of unmanned driving. CAAI Trans. Intell. Technol. **1**(1), 4–13 (2016)
12. Setchell, C., Dagless, E.L.: Vision-based road-traffic monitoring sensor. IEE Proc. – Vis. Image Signal Process. **148**(1), 78–84 (2002)
13. Li, C., Dai, B., Wang, R., et al.: Multi-lane detection based on omnidirectional camera using anisotropic steerable filters. IET Intell. Transp. Syst. **10**(5), 298–307 (2016)
14. Zhe, Y., et al.: Filter design for linear frequency modulation signal based on fractional Fourier transform. In: IEEE International Conference on Signal Processing (2010)
15. Soo, J.S., Pang, K.K.: Multidelay block frequency domain adaptive filter. IEEE Trans. Acoust. Speech Signal Process. **38**(2), 373–376 (1990)
16. Alt, N., Claus, C., Stechele, W.: Hardware/software architecture of an algorithm for vision-based real-time vehicle detection in dark environments. In: Design, Automation & Test in Europe (2008)
17. Zhao, Z., Ping, F., Guo, J., et al.: A hybrid tracking framework based on kernel correlation filtering and particle filtering. Neurocomputing **297**, 40–49 (2018)
18. Wei, C., Zhang, K., Liu, Q.: Robust visual tracking via patch based kernel correlation filters with adaptive multiple feature ensemble. Neurocomputing **214**, 607–617 (2016)
19. Strait, J.C., Jenkins, W.: Filter architectures and adaptive algorithms for 2-D adaptive digital signal processing. In: International Conference on Acoustics (1989)
20. Reid, D.B., Bryson, R.G.: A non-Gaussian filter for tracking targets moving over terrain. In: Asilomar Conference on Circuits (1979)
21. Liu, W., Anguelov, D., Erhan, D., Szegedy, C., Reed, S.E.: SSD: single shot multibox detector. CoRR, abs/1512.02325 (2015)
22. Szegedy, C., et al.: Going deeper with convolutions. CoRR, abs/1409.4842 (2014)

A Novel Small Vehicle Detection Method Based on UAV Using Scale Adaptive Gradient Adjustment

Changju Feng and Zhichao Lian[✉]

Nanjing University of Science and Technology, Nanjing, China
1813071755@qq.com, lzcts@163.com

Abstract. Vehicle detection based on UAV video is a typical small object detection task. In recent years, multi-scale prediction framework has become one of key steps for small object detection. However, the performances of existing methods are still not satisfactory for small object detection. In this paper, inspired by that the scale of object has an impact on gradient descent in the deep learning process, we choose the intersection over union (IOU) as the evaluation metric to analyze the relationship between scale of objects and gradient. We have shown that the gradient adjustment methods should satisfy some rules and thus we propose a new gradient adjustment formula based on our analysis. In addition, we built a mixed small vehicle dataset based on UAV videos for better evaluation of small vehicle detection. In the comparison with existing methods, our proposed method has achieved better results. The performance of our method reveals the potential of scale adaptive gradient descent method.

Keywords: Small vehicle detection · Gradient adjustment · IOU

1 Introduction

As a part of multi-scale prediction. Small object detection has always been the focus of research. The task that we studied–vehicle detection task based on UAV video is a typical task of small object detection. Meanwhile this task has many unique characteristics compared with general object detection, like small scale, environmentally sensitive, background occlusion which are very complicated in object detection and so on. Even through the problem of small scale is many details lost. But it is different from general small object detection, the semantic information of vehicle in drone perspective is being blurred to some extent under the influence of the environment. This makes the object detection more difficult. At the same time, there are more background occlusion appeared in image impacted by the shooting angle and position. It is a rather tricky problem. The background information without any meaning of detection influence the integrity of details and semantic information of object. This makes similar occlusion problem more difficult to solve. So, even though as single category object detection

---

The author of this paper is a student.

© Springer Nature Switzerland AG 2019
Z. Cui et al. (Eds.): IScIDE 2019, LNCS 11935, pp. 374–385, 2019.
https://doi.org/10.1007/978-3-030-36189-1_31

task, vehicle detection in drone perspective has many characteristics and difficulties. And it is a major application with the promotion of drones.

Our main contributions are as follows:

1. We build a dataset which mixed with public vehicle dataset and self-labeling vehicle dataset. Due to the single dataset can not represent a common distribution of UAV vehicle images. We can make the dataset more diverse in perspective, scenes, weather and the classes of car by mixing all these public datasets. So the distribution of scale of vehicles is more universal. In addition, we labeled amounts of UAV vehicle images (more than 1K images with 10K vehicles) and added these images into mixed dataset. For verifying the performance of improved model, we collected and label other images (more than 600 images with more than 24K vehicles) which are different from mixed dataset.
2. We claimed the imbalance of objects with different scales can be removed by gradient scale factor. And we proposed an equal (Eq. (4)) of gradient scale factor which are more suitable to the IOU change of objects in different scales and better suppress the impact of imbalance of scale to gradient.
3. We studied the relationship between IOU change and the scale of objects and proved that our approach is more suitable than basic approach in mathematics. The IOU change represent that the impact of prediction error on evaluation, so the gradient scale factor should fit the relationship of IOU change and scale.

2   Related Work

There are many solutions to the problem of scale change. SPP [1] proposed a spatial pyramid to scale images to different scales to obtain the scale change from large to small; SSD [2] constructs prediction layers with different receptive fields and feature information of VGG [3] to predict objects of different scales respectively. Shallow networks with smaller receptive fields and more detailed information are used to predict small-scale objects. FPN [4] proposed to reconstructed the feature pyramid, and combine detailed information of shallow network with semantic information of deep network to enrich the information contained in shallow prediction layer and improve the detection effect on the small-scale objects. Inspired by FPN, Many methods use different ways to fuse feature. M2Det [5] uses an encoder-decoder to build each layer in the multi-scale prediction structure and make the information of each layer more rich. FCOS [6] increases the size of input to build more predictive layers with different scales to cover more objects finely. Libra-RCNN [7] use the idea of feature balance to improve the utilization efficiency of features, and reconstruct the prediction layer to obtain the improvement of multi-scale prediction.

In addition to FPN, DetNet [8] is a new design for the target detection backbone, introducing more stages to maintain the resolution while expanding the receptive field, in order to solve the problem of insufficient object resolution. Hsieh et al. [9] integrates the spatial characteristics of a specific scene into the RPN to construct candidate proposal with regional information and improves the drone vehicle detection of dense scenes. Li et al. [10] uses PGAN for super-resolution of small objects from another angle to obtain high-resolution objects to improve the detection for small objects.

3 Dataset

The UAV view vehicle dataset refers to the image dataset generated by capturing vehicles in high-altitude view from a drone perspective. But it also includes some images from perspectives similar to a drone perspective. It is mainly used for object detection tasks. The dataset has the characteristics of small object scale, large resolution change, variable scene and relatively dense object distribution. The mixed dataset mainly consists of two types of datasets as shown on Table 1: public datasets, such as CAPRK [9], PUCPR [11] etc.; Self-calibrated data sets.

CAPRK is the dataset of the UAV perspective established by Hsieh et al. [9]. The main research scene is parking lot, and the research target is counting the number of vehicles in the parking lot. The shooting angle is purely overlooked. The dataset contains 1448 images with a total of 89774 objects. Due to the dense distribution of targets and repeated acquisitions in scene, there are a large number of vehicles recurring, so the effective objects is relatively limited. Resolution of the dataset is 1280*720, so the dataset has great learning value as a high resolution dataset.

PUCPR is a set of overhead view datasets established by Razakarivony et al. [11], which mainly studies the influence of the visual effects of illumination and shadow generated by environmental factors such as weather on object detection. This can enrich the diversity of the dataset in the environment, and the resolution of the dataset is 1280*720, so it also has great research value as a high-resolution dataset.

KITTI [12] is a set of vehicle datasets. It is a dataset that mainly studies traffic scenarios, and contains data sets of various traffic scenarios such as parking lots, highways, and elevated roads, which can be used to enrich the dataset. Scene diversity, KIT dataset resolution range (672*377–1038*896) mainly contains low-resolution data samples.

UCAS-AOD [13] is the image dataset obtained by using the satellite perspective image taken by Google Earth. This dataset is similar to other dataset taken by drone's pure overhead view. Due to the perspective of satellite, the dataset contains more diverse scene. The vehicle is small and the distribution of it is variable. So do some research on the dataset helps us to get rid of the constraints of scene.

Other data sets formed by intercepting and labeling the UAV perspective vehicle dataset published on the online website.

Self-labeling dataset formed by autonomously using drones to shoot video along urban road and intercepting video frame to label object. It contains 1117 images with 7885 vehicles, mainly focusing on the vehicles of urban traffic scenes with a resolution of 1920*1080. Compared with most public datasets, it can improve the overall resolution of the mixed dataset. The main research scenes is road, construction sites and other scene. So it can enrich the diversity of dataset and increase object such as engineering vehicles and freight vehicles (Fig. 1).

Table 1. The distribution of mixed vehicle dataset in UAV perspective. The images in Public dataset come from CAPAK, PUCPR, KIT, UCAS-AOD and Other dataset sets.

Dataset	Class	Images	Objects	Resolution
Public	Car	2509	125877	672*377–1280*720
Self-labeling	Car	1275	10390	1920*1080

Fig. 1. Self-labeling dataset and Public dataset, left image is self-labeling image and the right one is a image of CAPRK.

But these public dataset all have their limitation, such as CAPRK's diversity of scenes is restricted to parking lot; PUCPR is also restricted in scene of parking lot and the perspective of image is stable; The images of KITTI almost are low-resolution which are not enough for high-resolution object detection. UCAS-AOD has the same limitation with KITTI.

So we try to mix all these public dataset with our self-labeling dataset to build a mixed dataset that covers more scenes, more resolutions, more environments.

4 Gradient Adjustment Based on Scale

4.1 Scale Imbalance

Similar to You only look once [14], which takes the effects of different scales on gradient into account. In this work, we focus on the scale-imbalance among object with different scales. In order to study the relationship we train a converged object detector and count the error of prediction scale of all objects in training set.

Figure 2 shows the distribution of error of scale in a converged object detector model, which is trained from a large number of vehicles (more than 70K) with different scales. We can note that large scale objects produced larger error than those smaller objects, which will generate a scale-imbalance problem that inhibits model performance. When the dataset comes from many different datasets which have different distributions in scale, the problem mentioned above will be typical.

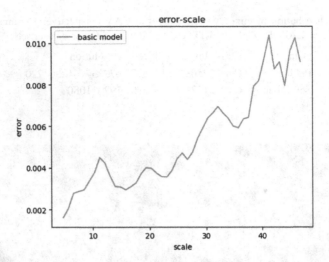

Fig. 2. The distribution of gradient with scale of large amounts of vehicles (more than 70K) from a converged object detector trained on the vehicle dataset.

4.2   Approaches to Suppress the Problem

So when we build a mixed UAV dataset of vehicles, the important problem is how to solve the problem of scale-imbalance. In order to remove the impact of scale on the gradient.

YOLO assigns larger loss and gradient to small-scale objects by squaring root of width and height of prediction box and ground truth in Eq. (1)

$$Loss_{wh} = \left(\sqrt{w_g} - \sqrt{w_p}\right)^2 + \left(\sqrt{h_g} - \sqrt{h_p}\right)^2 \tag{1}$$

Here, $\left(w_p, h_p\right)$ is the width and height of prediction box, $\left(w_g, h_g\right)$ is the width and height of ground truth.

In YOLOv3 [15], it adaptively adjusts the gradient generated by the regression according to scale of ground truth, as Eqs. (2) and (3)

$$Gradient_{wh} = scale_{wh} * \frac{\partial Loss_{wh}}{\partial x} x \in \left\{w_p, h_p\right\} \tag{2}$$

$$scale_{wh} = \left(2 - \frac{w_g * h_g}{W * H}\right) \tag{3}$$

Here $scale_{wh}$ represents gradient scaling coefficient based on the width of object w and the height of object h.

We propose a different function whose second derivative in the interval is greater than zero as the gradient scaling coefficient of the regression of different scales, as shown in Eq. (4),

$$scale_{wh} = e^{\overline{\left(\frac{\ln\left(\frac{w_g}{W} + \frac{h_g}{H}\right)}{2}\right)^{\frac{1}{2}}}} + 1 \tag{4}$$

(w_g, h_g) denotes the width and height of ground truth, W denotes the width of image, and H denotes the height. The First and second derivatives of coefficient and scale of our method on scale is shown on Table 2.

As seen in Table 2, our approach is more suitable than YOLOv3 in function characteristics. In the next work, we will make a proof for this conclusion.

Table 2. The relationships between two methods of gradient adjustment, $f_{\text{change\_IOU}}$ and scale (the width and height of object) of object in mathematics.

	$\frac{\partial f}{\partial w}$ and $\frac{\partial f}{\partial h}$	$\frac{\partial^2 f}{\partial w^2}$ and $\frac{\partial^2 f}{\partial h^2}$
$f_{\text{scaleyolov3}}$	<0	0
$f_{\text{scaleours}}$	<0	<0
$f_{\text{change\_IOU}}$	<0	<0

4.3 The Relationship Between IOU Change and Scale

The IOU recall rate is the most commonly used evaluation metric in object detection, and the calculation formula of IOU is Eq. (5)

$$IOU(P, G) = \frac{area(P) \cap area(G)}{area(P) \cup area(G)} \tag{5}$$

where P is prediction box, and G is ground truth. Thus, Eq. (6) can be explained as the ratio of the intersection to union of the prediction box and the ground truth.

The same errors of different scale of object can represent in the change of IOU which can represent as a loss function when the IOU is used as an evaluation metric. In this paper we use a metric shown as Eq. (6) to study the relationship between scale and gradient.

$$f_{\text{change\_IOU}} = 1 - IOU(P, G) \tag{6}$$

The $f_{\text{change\_IOU}}$ represents the difference between the ground truth and prediction in IOU, P denotes the prediction box, and G represent ground truth. So it can measure the quality of prediction beside Mean Square Error (MSE loss).

In this paper, and the ground truth represents IOU = 1, and the focus of this paper is the scale of object. So we assume that (x_p, y_p) denoting the coordinate of the upper left corner of prediction box is stable. As a result, $x_p = x_g, y_p = y_g$ where (x_g, y_g) is the coordinate of the upper left corner of ground truth. So there are four cases can be discussed in Fig. 3.

The reason that coordinates of the upper left corner of prediction box been set as the same with ground truth is that the regression of the coordinates and scale of prediction

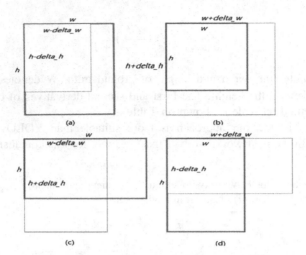

Fig. 3. The green box is ground truth and the red box is prediction box. The errors between prediction box and ground truth are *delta_w* (Δ_w) and *delta_h* (Δ_h) respectively and $\Delta_w > 0, \Delta_h > 0$ (Color figure online)

box are separate and independent. So we assumed the situation mentioned above to make the f_{change_IOU} unaffected by coordinate of prediction.

Here, (w, h) is the width and height of prediction box before regression. The errors between prediction box and ground truth are *delta_w* (Δ_w) and *delta_h* (Δ_h) respectively. In case 1 as shown in Fig. 2(a), the first order partial derivative and the second order partial derivative of IOU with respect to independent variable w is Eq. (7)

$$\begin{cases} IOU = \frac{(w-\Delta_w)(h-\Delta_h)}{wh} \\ \frac{\partial IOU}{\partial w} = \frac{\Delta_w(h-\Delta_h)}{w^2 h} \\ \frac{\partial^2 IOU}{\partial w^2} = -\frac{2\Delta_w(h-\Delta_h)}{w^3 h} \end{cases} \tag{7}$$

In case 2 as shown in Fig. 2(b), the first order partial derivative and the second order partial derivative of IOU with respect to independent variable w is Eq. (8)

$$\begin{cases} IOU = \frac{wh}{(w+\Delta_w)(h+\Delta_h)} \\ \frac{\partial IOU}{\partial w} = \frac{h\Delta_h}{(w+\Delta_w)^2(h+\Delta_h)} \\ \frac{\partial^2 IOU}{\partial w^2} = -\frac{2h\Delta_h}{(w+\Delta_w)^3(h+\Delta_h)} \end{cases} \tag{8}$$

In case 3 as shown in Fig. 2(c), the first order partial derivative and the second order partial derivative of IOU with respect to independent variable w is Eq. (9)

$$\begin{cases} IOU = \dfrac{(w-\Delta_w)h}{(h+\Delta_h)w - \Delta_w\Delta_h} \\[2mm] \dfrac{\partial IOU}{\partial w} = \dfrac{h^2\Delta_w}{((h+\Delta_h)w - \Delta_w\Delta_h)^2} \\[2mm] \dfrac{\partial^2 IOU}{\partial w^2} = -\dfrac{2h^2\Delta_w(h+\Delta_h)}{((h+\Delta_h)w - \Delta_w\Delta_h)^3} \end{cases} \tag{9}$$

In case 4 as shown in Fig. 2(d), the first order partial derivative and the second order partial derivative of IOU with respect to independent variable w is Eq. (10)

$$\begin{cases} IOU = \dfrac{(h-\Delta_h)w}{(w+\Delta_w)h - \Delta_w\Delta_h} \\[2mm] \dfrac{\partial IOU}{\partial w} = \dfrac{(h-\Delta_h)^2\Delta_w}{((w+\Delta_w)h - \Delta_w\Delta_h)^2} \\[2mm] \dfrac{\partial^2 IOU}{\partial w^2} = -\dfrac{2(h-\Delta_h)^2 h\Delta_h}{((w+\Delta_w)h - \Delta_w\Delta_h)^3} \end{cases} \tag{10}$$

By the premise of Eq. (11) we can find that all the cases can draw the same conclusion of $\dfrac{\partial IOU}{\partial w} > 0$ and $\dfrac{\partial^2 IOU}{\partial w^2} < 0$.

$$\begin{cases} w - \Delta_w > 0 \\ h - \Delta_h > 0 \\ (w+\Delta_w)h - \Delta_w\Delta_h > 0 \\ (h+\Delta_h)w - \Delta_w\Delta_h > 0 \end{cases} \tag{11}$$

At the same time, from the perspective of the scale we are discussing, variable w and variable h have the same cases.

Therefore, In the interval $w, h \in (0,1]$, the relationship between change of IOU f_{change_IOU} and scale is $\dfrac{\partial f_{change\_IOU}}{\partial w} < 0$ and $\dfrac{\partial^2 f_{change\_IOU}}{\partial w^2} > 0$, And the change of IOU can represent as a loss function when IOU is used as an evaluation metric.

So we can get a conclusion: With the increasing of scale, the gradient scale factor $scale_{wh}$ should satisfy the following relationship as Eq. (12):

$$\dfrac{\partial scale_{wh}}{\partial w} < 0 \text{ and } \dfrac{\partial^2 scale_{wh}}{\partial w^2} > 0 \tag{12}$$

4.4 The Disadvantage of YOLOv3

YOLOv3 proposed Eq. (3) to solve the imbalance of different scale. We transformed Eq. (3) into $scale_{wh} = 2 - w * h$ and $w, h \in (0, 1]$, Considering variable w separately, we can calculate the partial derivative as Eq. (13)

$$\dfrac{\partial scale_{wh}}{\partial w} = -h, \dfrac{\partial^2 scale_{wh}}{\partial w^2} = 0 \tag{13}$$

In the conditions of $w, h \in (0, 1]$, so $\dfrac{\partial scale_{wh}}{\partial w} < 0$ and $\dfrac{\partial^2 scale_{wh}}{\partial w^2} = 0$. At the same time, from the perspective of the scale we are discussing, variable w and variable h have the same meaning. So the relationship of variable h and the gradient scaling coefficient $scale_{wh}$ is the same with w.

So the approach of YOLOv3 can not satisfy Eq. (12) mentioned above.

4.5 Our Approach

We proposed Eq. (4) to inhibit the impact of scale-imbalance. Considering variable w separately, we can calculate the partial derivative of Eq. (4) as Eqs. (14) and (15)

$$\frac{\partial scale_{wh}}{\partial w} = -\frac{e^{\ln\left(\frac{w+h}{2}\right)^{-\frac{1}{}}}}{w+h} * \frac{1}{\left(\ln\left(\frac{w+h}{2}\right)\right)^2} \tag{14}$$

$$\frac{\partial^2 scale_{wh}}{\partial w^2} = e^{\ln\left(\frac{w+h}{2}\right)^{-\frac{1}{}}} * \frac{1}{\left(\ln\left(\frac{w+h}{2}\right)\right)^2} * \frac{1}{(w+h)^2} * \left(\left(\frac{1}{\ln\left(\frac{w+h}{2}\right)}+1\right)^2 - \frac{1}{\ln\left(\frac{w+h}{2}\right)}\right) \tag{15}$$

In the conditions of $w, h \in (0, 1]$, so $\frac{\partial scale_{wh}}{\partial w} < 0$ and $\frac{\partial^2 scale_{wh}}{\partial w^2} > 0$. And the relationship of variable h and the gradient scaling coefficient $scale_{wh}$ is the same with w.

Our approach can satisfy Eq. (12) mentioned above. It is obviously that the approach which we proposed is more appropriate than YOLOv3 in some condition in gradient adjustment.

5 Experiments

Our experiments are implemented on the mixed UAV dataset. The train set includes 2750 images randomly extracted from mixed UAV dataset. And the test data includes 381 images consists of 12724 vehicles. First test set is also randomly extracted from UAV dataset and without any image in train set.

As shown in Table 3. When the input size is set to 608, our method is equal with YOLOv3 in AP50, but brings 0.90 points higher in AP75.

Table 3. The result of YOLOv3 and ours testing in mixed UAV dataset

Method	AP50	AP75
YOLOv3	89.88%	53.24%
Ours	89.96%	54.14%

In addition to compare our approach with YOLOv3 on mixed dataset. We also create a dataset call HD-UAV dataset about vehicles in drone videos, and the distribution of the new dataset is different from mixed dataset it is used for test. This test set is consist of 650 images intercepted from several HD drone videos, which includes

Fig. 4. The comparison of results between ours (left one) and YOLO 3 (right one) in a image comes from test set of mixed UAV dataset.

Table 4. The result of YOLOv3 and ours testing in HD-UAV dataset

Method	AP50	AP75
YOLOv3	50.46%	11.43%
Ours	50.51%	11.97%

24974 vehicles, and the resolution of images in second test set is 2704*1520. We call this dataset HD-UAV dataset.

As shown in Table 4. The AP50 and AP75 decreased a lot compared with Table 3. But our method still brings 0.54 points higher AP75. The improvement of ours show that well-set gradient scaling coefficient can improve the performance of detection to varying degree. For example, as shown in Fig. 4, this image comes from the HD-UAV test set. Even though there are many objects are not been detected in the upper part of

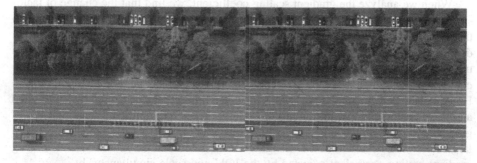

Fig. 5. The comparison of results between ours (left one) and YOLOV3 (right one) in a image comes from test set of HD-UAV dataset.

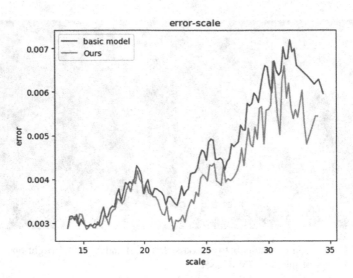

Fig. 6. The statistical results of ours and basic model on prediction scale error

the image (It may be caused by the higher resolution of 2704*1520), our method still detected more vehicles than YOLOv3 (Fig. 5).

We also count the error of our approach and the basic model on training set, and we compared the statistical results between two approaches as shown in Fig. 6. The result proved that our approach can better suppress scale imbalance than basic approach.

6   Conclusion

In this paper, we construct a mix dataset of small vehicles for UAV video for better evaluation of the performance of detection methods. After analyzing the relationship between gradient decent and the scale, we propose a novel loss function which can adjust the gradient based on the scale. Experimental result demonstrate that our proposed method achieved better performance compared to the YOLOv3, a very popular object detection method.

When we analyze the gradient scaling coefficient, we find that the change of factor should be controlled within a certain range. Although it may not be consistent with theory. For example, for an object with infinitely small scale, the influence of a certain degree of error on its prediction can be close to infinitely large(although the situation is unlikely to occur). So we try to use an adaptive weight to change the gradient scaling coefficient, which will make the gradient of small-scale object larger than our approach. However, the performance is not satisfactory. We think it may be caused by a unlimited amplified coefficient, and in training those large gradient amplified by coefficient will guide the training direction to smaller object instead of covering it, so most object with large-scale will be influenced even ignored. So in future we will focus on making gradient scaling coefficient adjusted by the IOU and scale simultaneously.

References

1. He, K., Zhang, X., Ren, S., et al.: Spatial pyramid pooling in deep convolutional networks for visual recognition. IEEE Trans. Pattern Anal. Mach. Intell. **37**(9), 1904–1916 (2014)
2. Liu, W., et al.: SSD: single shot multibox detector. In: Leibe, B., Matas, J., Sebe, N., Welling, M. (eds.) ECCV 2016. LNCS, vol. 9905, pp. 21–37. Springer, Cham (2016). https://doi.org/10.1007/978-3-319-46448-0_2
3. Simonyan, K., Zisserman, A.: Very deep convolutional networks for large-scale image recognition. In: NIPS (2015)
4. Lin, T.Y., Dollár, P., Girshick, R., et al.: Feature pyramid networks for object detection. In: CVPR (2017)
5. Zhao, Q., Sheng, T., Wang, Y., et al.: M2Det: a single-shot object detector based on multi-level feature pyramid network. In: AAAI (2019)
6. Tian, Z., Shen, C., Chen, H., et al.: FCOS: fully convolutional one-stage object detection. https://arxiv.org/abs/1904.01355. Accessed 14 Apr 2019
7. Pang, J., Chen, K., Shi, J., et al.: Libra R-CNN: towards balanced learning for object detection. In: CVPR (2019)
8. Li, Z., Peng, C., Yu, G., Zhang, X., Deng, Y., Sun, J.: DetNet: design backbone for object detection. In: Ferrari, V., Hebert, M., Sminchisescu, C., Weiss, Y. (eds.) ECCV 2018. LNCS, vol. 11213, pp. 339–354. Springer, Cham (2018). https://doi.org/10.1007/978-3-030-01240-3_21
9. Hsieh, M.R., Lin, Y.L., Hsu, W.H.: Drone-based object counting by spatially regularized regional proposal network. In: CVPR (2017)
10. Li, J., Liang, X., Wei, Y., et al.: Perceptual generative adversarial networks for small object detection. In: CVPR (2017)
11. Razakarivony, S., Jurie, F.: Vehicle detection in aerial imagery: a small target detection benchmark. J. Vis. Commun. Image Represent. **34**, 187–203 (2016)
12. Geiger, A., Lenz, P., Stiller, C., et al.: Vision meets robotics: the KITTI dataset. Int. J. Robot. Res. **32**(11), 1231–1237 (2013)
13. Zhu, H., Chen, X., Dai, W., et al.: Orientation robust object detection in aerial images using deep convolutional neural network. In: ICIP (2015)
14. Redmon, J., Divvala, S., Girshick, R., et al.: You only look once: unified, real-time object detection. In: CVPR (2016)
15. Redmon, J., Farhadi, A.: YOLOv3: an incremental improvement. In: CVPR (2018)

A Level Set Method for Natural Image Segmentation by Texture and High Order Edge-Detector

Yutao Yao, Ziguan Cui[(✉)], and Feng Liu

Image Processing and Image Communication Lab,
Nanjing University of Posts and Telecommunications, Nanjing 210003, China
`cuizg@njupt.edu.cn`

Abstract. Active contour model has been a widely used methodology in image segmentation. However, due to the texture complexity of natural images, it unavoidably faces many difficulties. In this paper, we propose a novel method to accurately segment natural images by texture and high order edge-detector. Firstly, we calculate local covariance matrix which is estimated from image gradient information within the local window, and use the eigenvalues of matrix to describe local texture feature of the image. Then, in order to suppress the effect of perplexing background, we introduce a high order edge-detector which can eliminate the background as much as possible while it can save the object boundary. Finally, the intensity term, texture term and edge-detector term are incorporated into the level set method to segment natural images. The proposed method has been tested on many natural images, and experimental results show the segmentation performance of the proposed method is better than prior similar state-of-the-art methods.

Keywords: Active contour model · Level set · Natural image segmentation · Texture feature extraction · High order edge-detector

1 Introduction

Image segmentation plays a key role in computer vision and image analysis fields. In recent decades, active contour models (ACM) have been extensively used for image segmentation. For solving the minimization of ACM, the level set method is widely applied. Generally, existing active contour models can be divided into two categories, i.e., edge-based models and region-based models. Edge-based models exploit the image gradient information to drive the contour toward the object boundary, however, these models are sensitive to noise and unable to identify weak object boundary. Contrary to the edge-based models, region-based models use the region statistical information to drive the contour, they often show better segmentation results than the former especially when deal with intensity inhomogeneity images. In this paper, we focus on region-based models.

The most classic region-based model is the Chan-Vese (CV) [1] model. It approximates the image to two regions with constant intensity. Thus, it has an ideal segmentation result on images with intensity homogeneity on object and background

© Springer Nature Switzerland AG 2019
Z. Cui et al. (Eds.): IScIDE 2019, LNCS 11935, pp. 386–398, 2019.
https://doi.org/10.1007/978-3-030-36189-1_32

areas. However, in practice, the intensity inhomogeneity often occurs in many real-world images, such as medical images and natural images. It is difficult to solve the segmentation problem of these images for the classic CV model.

To address the problem of intensity inhomogeneity in image segmentation, local region-based models have been proposed, which tend to utilize the local region information to better identify local boundary. Li et al. proposed the region-scalable fitting (RSF) [2] energy model based on the Gaussian kernel with a controllable scale, which has desirable performance than the CV model when handles intensity inhomogeneous images. Zhang et al. proposed a local image fitting (LIF) [3] model, which has similar capability of processing intensity inhomogeneity. Later, local Gaussian distribution (LGD) [4] model was proposed, which makes use of Gaussian distribution with different means and variances to better represent local image intensity. Besides, some models are proposed by encoding the intensity inhomogeneity of images. Specifically, the intensity inhomogeneity is estimated by minimizing the level set formulation, and the intensity inhomogeneity is further revised. For example, Li et al. proposed the local intensity clustering (LIC) [5] model, which can be considered as a local intensity clustering method, but this model does not consider the clustering variance. Compared with LIC, locally statistical active contour model (LSACM) [6] considers Gaussian distributions with different means and variances to encode images with intensity inhomogeneity. Recently, Min et al. proposed a method named local approximation of Taylor expansion (LATE) [7], which uses a nonlinear approximation method to solve the intensity inhomogeneity. Liu et al. proposed a level set algorithm based on probabilistic statistics [8]. It behaves well in segmenting MR images. These models have indeed improved the segmentation performance for images with intensity inhomogeneity to some extent. However, natural images often have complex texture feature, which makes the above mentioned models still face with crucial challenges and tend to yield inaccurate segmentation results, and thus further improvements are still required to be made for these images.

In order to improve the capability of level set methods for natural images, several texture-embedded level set methods have been proposed. Most of them use structure tensor or Gabor filter to obtain texture feature of natural images. The local Chan-Vese model (LCV) [9] uses the differences of the averaging convolution image and original image as local information. And the extended structure tensor was proposed and incorporated into the energy functional in LCV model. Wang et al. proposed a unified tensor level set model (UTLSM) [10], which involves the Gabor texture into the unified tensor representation. In [11], Dai et al. proposed an inhomogeneity-embedded active contour (InH-ACM). And the inhomogeneity of pixel is described by its nearby pixels which named pixel inhomogeneity factor (PIF) in InH-ACM. Recently, Min et al. also proposed a local salient fitting (LSF) [12] model. In this model, the original image is converted to a new domain that highlights the contrast of object and background. Based on different texture features, these models can segment natural images with slight texture features. However, level set methods with the above mentioned texture descriptors have still been seen to yield unsatisfactory results in most natural images and needed to be improved further.

Through the above analysis, we can see that existing texture descriptors such as structure tensor and Gabor filter seem not powerful and robust enough to represent local texture feature for image segmentation. In this paper, we propose a new level set method to segment natural images more accurately by exploring local texture feature more effectively and high order edge-detector. Firstly, a covariance matrix is estimated from image gradient information within a local window. It can robustly obtain the local structure of the image. We extract the local texture feature by calculating eigenvalues from this matrix. Secondly, for driving contour towards the object boundary accurately even in the complicated background, we introduce a high order edge-detector inspired by [13]. It can find significant boundary relevant to the object. Finally, we incorporate intensity term, texture term and edge-detector term into an integral level set formulation. In addition, for solving the problem of minimization of our level set formulation efficiently, we perform the numerical solution process by using a semi-implicit gradient descent method like the method in [14] and describe it in detail, which makes the curve evolution more stable. At last, to evaluate the proposed method, we conduct comprehensive experiments on natural images with complex texture. Experimental results demonstrate the superiority of the proposed method.

2 Proposed Method

The flowchart of the proposed method is illustrated in Fig. 1.

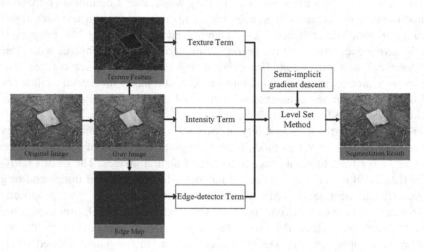

Fig. 1. The flowchart of the proposed method.

In this section, we will first introduce how we extract the local texture feature from local covariance matrix. And then we will introduce the high order edge-detector. At last, we utilize them to construct the integral level set formulation. Besides, we deduce the numerical solution of the formulation using a semi-implicit gradient descent method.

2.1   The Proposed Texture Term

Texture feature plays a key role for image analysis especially for natural images. Structure tensor and Gabor filter are usually used to extract texture feature. There is an extended structure tensor (EST) which has been proposed in [9]. The formulation of EST is written as:

$$J_{EST} = \begin{pmatrix} K * I_x^2 & K * I_x I_y & K * I_x I_i \\ K * I_x I_y & K * I_y^2 & K * I_y I_i \\ K * I_x I_i & K * I_y I_i & K * I_i^2 \end{pmatrix} \tag{1}$$

where K is a Gaussian kernel, I_i denotes the intensity of image, and I_x and I_y denote the gradient related with x-axes and y-axes respectively. The EST describes image texture by a mixture of image intensity and gradient.

Gabor filter can be also considered to extract texture feature. The 2D Gabor function can be defined as follows:

$$G = g_\sigma(x, y) \exp[2\pi j \omega (x \cos \theta + y \sin \theta)] \tag{2}$$

$$g_\sigma(x, y) = \frac{1}{2\pi\sigma^2} \exp\left[-\frac{x^2 + y^2}{2\sigma^2}\right] \tag{3}$$

the σ is the scale of Gabor filter, θ and ω represent the orientation and frequency respectively. Gabor filter varies σ, θ, ω to obtain texture feature with multiple channels. However, these methods seem all not powerful and robust enough to describe natural image texture feature for image segmentation.

We in this section introduce a novel way to extract texture feature from natural image. Firstly, we compute the local covariance matrix from image gradient information within the local window, the selection method of local window is similar in [15]. Specifically, the covariance matrix C can be estimated as follows:

$$C = \begin{bmatrix} I_x(z_1), & I_y(z_1) \\ \cdots & \cdots \\ I_x(z_L), & I_y(z_L) \end{bmatrix}^T \begin{bmatrix} I_x(z_1), & I_y(z_1) \\ \cdots & \cdots \\ I_x(z_L), & I_y(z_L) \end{bmatrix} \in R^{2\times2} \tag{4}$$

where T is the transpose symbol, $I_x(\cdot)$ and $I_y(\cdot)$ denote the gradient related with x-axes and y-axes respectively of each pixel z_i within local window. L is the number of pixels in local window, and we set $L = 9$ in all experiments. Then we calculate the eigenvalues λ_1 and λ_2 of the covariance matrix C. Our texture descriptor for each pixel is written as:

$$I_t(x) = \sqrt{\lambda_1 + \lambda_2} \tag{5}$$

where x denotes the center pixel of each local window. Through the movement of the local window, we can obtain the texture feature of each pixel of the image. Finally, we utilize this texture feature to construct the texture term. Under the framework of CV model, the texture term can be defined as follows:

$$E_t(\phi) = \sum_{i=1,2} \int (I_t(x) - m_i)^2 M_i(\phi(x)) dx \tag{6}$$

where ϕ is a level set function defined on an image domain Ω. The zero level set $\{x: \phi(x) = 0\}$ divides the image into two regions, i.e., inside region $\Omega_1 = \{x: \phi(x) > 0\}$ and outside region $\Omega_2 = \{x: \phi(x) < 0\}$. Meanwhile, the texture constants m_1 and m_2 of regions inside and outside can be computed as follows:

$$m_i(\phi) = \frac{\int I_t(x) M_i(\phi(x)) dx}{\int M_i(\phi(x)) dx} \quad i = 1, 2 \tag{7}$$

where $M_1(\phi) = H(\phi)$ and $M_2(\phi) = 1 - H(\phi)$ where $H(\cdot)$ denotes the smoothed Heaviside function as:

$$H(\phi) = \frac{1}{2}\left(1 + \frac{2}{\pi}\arctan(\phi)\right) \tag{8}$$

In a word, the key covariance matrix C can robustly describe the local structure of natural images. We use the eigenvalues of C to extract the texture features to better reflect the local structural information of the image. Thus, it facilitates to segment objects in natural images greatly.

2.2 The High Order Edge-Detector Term

Due to the cluttered background of natural images, there are still many cases which tend to yield inaccurate segmentation if we only utilize local texture feature of complex natural images. Therefore, we introduce the high order edge-detector to constrain the contour further. A general edge-detector can be defined as:

$$g = \frac{1}{1 + |\nabla K * I|^P} \tag{9}$$

where ∇ is the gradient operator, K is a Gaussian kernel. $P \geq 1$ is a positive integer and is often set to 2 in most cases. However, it is not suitable for natural images. Because it also preserves most background apart from object boundary. Based on the observation, we found this problem can be solved by introducing high order edge-detector by increasing P as shown in Fig. 2. In addition, we use an adaptive method to set the value of P. The formulation of the method can be expressed as:

$$\frac{1}{M}\sum_x |g_{P+1} - g_P| \leq \tau \tag{10}$$

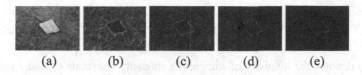

Fig. 2. Edge map with different P, (a) input image, (b) $P = 1$, (c) $P = 2$, (d) $P = 3$, (e) $P = 4$.

where M is the number of pixels in the image, τ denotes a threshold, and τ is set to 0.1 in our experiments. As shown in Fig. 3, the edge map of traditional method has a lot of background boundaries, while our method can extract object boundary from complicated natural images. Besides, our method can adaptively select the P value, which is better than the fixed P value. Therefore, our high order edge-detector term is defined with a level set formulation as follows:

$$E_e(\phi) = \int g(x)|\nabla H(\phi(x))|dx \tag{11}$$

Fig. 3. Edge map, (a) input image, (b) traditional edge map ($P = 2$), (c) adaptive edge map.

It can suppress the contour evolution toward the disturbing boundaries in the background and impel the contour adhere to salient object boundary, and thus obtain better segmentation result.

2.3 Level Set Formulation and Numerical Implementation

Based on the above discussion, our level set formulation can be finally defined as follows:

$$E(\phi) = \lambda_1 E_i(\phi) + \lambda_2 E_t(\phi) + \alpha E_e(\phi) \tag{12}$$

where $\lambda_1 \geq 0, \lambda_2 \geq 0$ and $\alpha \geq 0$ are fixed parameters. In almost all our experiments, we fix $\lambda_1 \geq \lambda_2 \geq 1$ and $\alpha \geq 5000$. $E_i(\cdot)$ is the intensity term which is directly introduced from the classic CV model, defined as:

$$E_i(\phi) = \sum_{i=1,2} \int (I_i(x) - c_i)^2 M_i(\phi(x))dx \tag{13}$$

where $I_i(\cdot)$ denotes the intensity of image, the intensity constants c_1 and c_2 of regions inside and outside can be computed as follows:

$$c_i(\phi) = \frac{\int I_i(x)M_i(\phi(x))dx}{\int M_i(\phi(x))dx} \quad i = 1, 2 \tag{14}$$

To minimize the formulation (12), we can apply the gradient flow as follows:

$$
\begin{aligned}
\frac{\partial \phi}{\partial t} &= -\frac{\partial E}{\partial \phi} \\
&= \delta(\phi(x))[-\lambda_1(I_i(x) - c_1)^2 + \lambda_1(I_i(x) - c_2)^2 - \lambda_2(I_t(x) - m_1)^2 \\
&\quad + \lambda_2(I_t(x) - m_2)^2] + \alpha\delta(\phi(x))\left[g(x)div\left(\frac{\nabla\phi(x)}{|\nabla\phi(x)|}\right) + \nabla g(x)\cdot\frac{\nabla\phi(x)}{|\nabla\phi(x)|}\right]
\end{aligned}
\tag{15}
$$

where $\delta(\cdot)$ is Dirac delta function and it is the derivative of $H(\cdot)$. Here we use method named semi-implicit gradient descent for solving formulation minimization. We recall the notations: let Δt be the time step, and suppose the image I is sampled on a grid $\Omega = \{0, \ldots, M\} \times \{0, \ldots, N\}$. Let n denotes the nth iteration. The central finite differences are:

$$
\begin{cases}
\nabla_x^+ \phi_{i,j} = \phi_{i+1,j} - \phi_{i,j},\ \nabla_x^- \phi_{i,j} = \phi_{i,j} - \phi_{i-1,j},\ \nabla_x^0 \phi_{i,j} = \left(\nabla_x^+ \phi_{i,j} + \nabla_x^- \phi_{i,j}\right)/2 \\
\nabla_y^+ \phi_{i,j} = \phi_{i,j+1} - \phi_{i,j},\ \nabla_y^- \phi_{i,j} = \phi_{i,j} - \phi_{i,j-1},\ \nabla_y^0 \phi_{i,j} = \left(\nabla_y^+ \phi_{i,j} + \nabla_y^- \phi_{i,j}\right)/2
\end{cases}
\tag{16}
$$

where x and y denote alone x-axes and y-axes respectively. Then, we compute ϕ^{n+1} by following discretization of formulation (15) in ϕ:

$$
\begin{aligned}
\frac{\phi_{i,j}^{n+1} - \phi_{i,j}^n}{\Delta t} &= \delta\left(\phi_{i,j}^n\right)[g_{i,j}\left(X_{i,j}\phi_{i+1,j}^n + X_{i-1,j}\phi_{i-1,j}^{n+1} + Y_{i,j}\phi_{i,j+1}^n + Y_{i,j-1}\phi_{i,j-1}^{n+1}\right) \\
&\quad - g_{i,j}\left(X_{i,j} + X_{i-1,j} + Y_{i,j} + Y_{i,j-1}\right)\phi_{i,j}^{n+1} + X_{i,j}\phi_{i+1,j}^n g_{i+1,j} \\
&\quad + X_{i,j}\phi_{i,j}^{n+1}g_{i,j} + Y_{i,j}\phi_{i,j+1}^n g_{i,j+1} + Y_{i,j}\phi_{i,j}^{n+1}g_{i,j} - X_{i,j}\phi_{i,j}^{n+1}g_{i+1,j} \\
&\quad - X_{i,j}\phi_{i+1,j}^n g_{i,j} - Y_{i,j}\phi_{i,j}^{n+1}g_{i,j+1} - Y_{i,j}\phi_{i,j+1}^n g_{i,j} - D]
\end{aligned}
\tag{17}
$$

$$D = \lambda_1\left[-(I_i(x) - c_1)^2 + (I_i(x) - c_2)^2\right] + \lambda_2\left[-(I_t(x) - m_1)^2 + (I_t(x) - m_2)^2\right] \tag{18}$$

where the coefficients X and Y are:

$$X_{i,j} = \alpha \Big/ \sqrt{\left(\nabla_x^+ \phi_{i,j}\right)^2 + \left(\nabla_y^0 \phi_{i,j}\right)^2}, Y_{i,j} = \alpha \Big/ \sqrt{\left(\nabla_x^0 \phi_{i,j}\right)^2 + \left(\nabla_y^+ \phi_{i,j}\right)^2} \qquad (19)$$

The $\phi_{i,j}$, $\phi_{i-1,j}$, $\phi_{i,j-1}$ are evaluated at time step $n + 1$ and all others at time step n. The ϕ at time step $n + 1$ to be solved as follows:

$$
\begin{aligned}
\phi_{i,j}^{n+1} \leftarrow &[\phi_{i,j}^n + \Delta t \delta\left(\phi_{i,j}^n\right)(X_{i-1,j}\phi_{i-1,j}^{n+1}g_{i,j} + X_{i,j}\phi_{i+1,j}^n g_{i+1,j} \\
&+ Y_{i,j-1}\phi_{i,j-1}^{n+1}g_{i,j} + Y_{i,j}\phi_{i,j+1}^n g_{i,j+1} - D)] \\
&/[1 + \Delta t \delta\left(\phi_{i,j}^n\right)(X_{i-1,j}g_{i,j} + X_{i,j}g_{i+1,j} + Y_{i,j-1}g_{i,j} + Y_{i,j}g_{i,j+1})]
\end{aligned}
\qquad (20)
$$

Finally, the concrete implementation process of the proposed method can be outlined by the following steps:

- Step 1: Compute texture feature by Eqs. (4), (5) and compute edge-detector by Eqs. (9) and (10).
- Step 2: Initialize the level set function ϕ^0.
- Step 3: Compute m_1, m_2, c_1 and c_2 by Eqs. (7) and (14).
- Step 4: Update the level set function Eq. (15) by semi-implicit solution Eq. (20).
- Step 5: Check whether the solution is stationary. If not, return to Step3.

3 Experiments

In order to evaluate our proposed method, we conduct comprehensive experiments on natural images with complex texture. We first compare the quality of our texture feature in Sect. 3.1. Then we verify the segmentation performance of the proposed method in MSRA dataset [16] and BSD dataset [17] in Sect. 3.2. Finally, in Sect. 3.3 the Jaccard similarity coefficient (JSC) is used as a quantitative measure for evaluating the segmentation results.

3.1 Texture Feature Comparison

For evaluating the superiority of our texture feature extraction method, we compare our approach with EST [9], Gabor filter with 12 channels and PIF [11]. Figure 4 provides the results of the proposed method and other methods. Here we only utilize the texture feature to construct energy function.

From Fig. 4, it is clear to see that our texture feature extraction method is better than others. The results of EST and Gabor filter lose some significant texture information. Although PIF extracts the object texture very well, it also enhances the background texture. These disadvantages will drive the contour towards background and cause the over-segmentation or less-segmentation phenomena. While our method can robustly obtain the local significant texture around the object in images. Therefore, it is more beneficial to segmentation.

3.2   Segmentation Result Visual Comparison

There are various of comparison experiments which are conducted to show the performance of our method. Firstly, we make the comparison of our method with CV model [1] and several representative local region-based models, such as the RSF model [2], the LIF model [3] and LSACM model [6]. On the other hand, several representative models for natural image segmentation including LCV model [9], InH-ACM model [11] and LSF model [12] are compared. In all experiments, several parameters are fixed in our method: $\lambda_1 = \lambda_2 = 1$, $\alpha = 5000$ and $\Delta t = 0.1$. In Figs. 5 and 6 we show some segmentation results of our method and these models respectively.

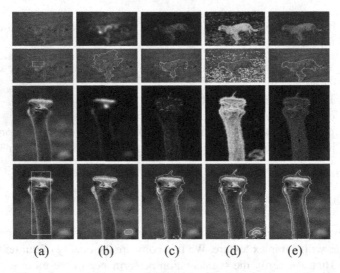

(a)   (b)   (c)   (d)   (e)

Fig. 4. Comparison of extracted texture of different methods and segmentation results. (a) input image and initial contour, (b) EST, (c) Gabor filter, (d) PIF, (e) the proposed method.

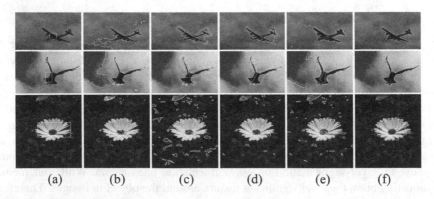

(a)   (b)   (c)   (d)   (e)   (f)

Fig. 5. Comparison of segmentation results, (a) input image and initial contour, (b) CV, (c) RSF, (d) LIF, (e) LSACM, (f) ours.

From Fig. 5, we can see that CV model uses global intensity mean to divide the images directly into two regions, whose results usually contain the background regions. And the other three local region-based models have wrong segmentation in varying degrees, because they do not consider more features of the natural images and add more constraints. Since precise texture feature and high order edge-detector are considered, the proposed method can segment these natural images accurately.

From Fig. 6, the CV model only utilize the image intensity information, result in failing to obtain the desirable segmentation. In LCV model, the wrong segmentation result still occurs because of inaccurate texture feature. In InH-ACM model, pixel inhomogeneity factor (PIF) is proposed to obtain object texture well, but it is still affected by the cluttered background. In the LSF, the new domain of original image enhances the contrast of object and background. However, the segmentation accuracy of this model is slightly worse than our method. For highlighting the robustness of the proposed method, we give more experimental results in Fig. 7. Obviously, our method achieves reliably satisfactory segmentation.

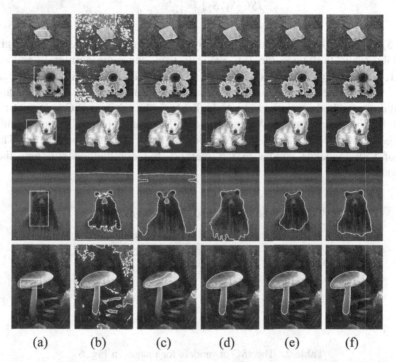

    (a)   (b)   (c)   (d)   (e)   (f)

Fig. 6. Comparison of segmentation results. (a) input image and initial contour, (b) CV, (c) LCV, (d) InH-ACM, (e) LSF, (f) ours.

Fig. 7. Natural image segmentation results of the proposed method.

3.3 Quantitative Comparison

In this section, we use JSC to quantitatively evaluate the performance of different methods. The JSC between two regions S_1 and S_2 is calculated as:

$$JSC(S_1, S_2) = \frac{|S_1 \cap S_2|}{|S_1 \cup S_2|} \tag{21}$$

where S_1 is the segmented object region by different methods and S_2 is the ground-truth. Obviously, the closer the JSC value is to 1, the more similar S_1 is to S_2. In Table 1, the relevant JSC values of the five models for Fig. 5 are shown. It is clear that the JSC value of the CV model and the other three local region-based models are very low. And our method is relatively high. In Table 2, the relevant JSC values of the five models for Fig. 6 are shown. We can see the JSC value of the CV model and LCV model is relatively low. The InH-ACM model is better than the former two. The JSC value of LSF model is similar to our method, but our method still gets the best performance.

Table 1. The JSC of models for images in Fig. 5.

Model	Image1	Image2	Image3
CV	0.3912	0.1775	**0.8368**
RSF	0.2376	0.2253	0.4857
LIF	0.2897	0.1647	0.7368
LSACM	**0.6108**	**0.3818**	0.7570
Ours	**0.9189**	**0.8570**	**0.9218**

Table 2. The JSC of models for images in Fig. 6.

Model	Image1	Image2	Image3	Image4	Image5
CV	0.2420	0.7027	0.9311	0.3561	0.4433
LCV	0.8231	0.5863	0.7346	0.3517	0.3833
InH-ACM	0.9028	0.7003	0.9138	**0.8124**	0.9050
LSF	**0.9417**	**0.7551**	**0.9714**	0.6790	**0.9067**
Ours	**0.9426**	**0.9543**	**0.9760**	**0.8593**	**0.9066**

4 Conclusion

In this paper, we proposed a novel level set method to segment natural image. On one hand, the texture term based on local covariance matrix to capture texture feature has been constructed. On the other hand, the high order edge-detector term can suppress interference form cluttered background as much as possible. Finally, the intensity term, texture term and edge-detector term are combined into level set formulation and we use semi-implicit gradient descent to obtain more stable evolution curve. Experimental results have testified the superiority and effectiveness of this method.

Acknowledgements. This work is supported by National Natural Science Foundation of China (NSFC) (61501260, 61471201, 61471203), Jiangsu Province Higher Education Institutions Natural Science Research Key Grant Project (13KJA510004), The peak of six talents in Jiangsu (RLD201402), and "1311 Talent Program" of NJUPT.

References

1. Chan, T., Vese, L.: Active contours without edges. IEEE Trans. Image Process. **10**(2), 266–277 (2001)
2. Li, C., Kao, C., John, C., Ding, Z.: Minimization of region-scalable fitting energy for image segmentation. IEEE Trans. Image Process. **17**(10), 1940–1949 (2008)
3. Zhang, K., Song, H., Zhang, L.: Active contours driven by local image fitting energy. Pattern Recogn. **43**(4), 1199–1206 (2010)
4. Wang, L., He, L., Mishra, A., Li, C.: Active contours driven by local Gaussian distribution fitting energy. Sign. Process. **89**(12), 2435–2447 (2009)
5. Li, C., Huang, R., Ding, Z., Gatenby, J., Metaxas, D., Gore, J.: A level set method for image segmentation in the presence of intensity inhomogeneities with application to MRI. IEEE Trans. Image Process. **20**(7), 2007–2016 (2011)
6. Zhang, K., Zhang, L., Lam, K., Zhang, D.: A level set approach to image segmentation with intensity inhomogeneity. IEEE Trans. Cybern. **46**(2), 546–557 (2016)
7. Min, H., Zhao, Y., Zuo, W., Ling, H., Luo, Y.: LATE: a level-set method based on local approximation of Taylor expansion for segmenting intensity inhomogeneous image. IEEE Trans. Image Process. **27**(10), 5016–5031 (2018)
8. Liu, J., Wei, X., Li, Q., Li, L.: A level set algorithm based on probabilistic statistics for MR image segmentation. In: International Conference on Intelligence Science and Big Data Engineering (IScIDE), pp. 577–586 (2018)
9. Wang, X., Huang, D., Xu, H.: An efficient local Chan-Vese model for image segmentation. Pattern Recogn. **43**(3), 603–618 (2010)
10. Wang, B., Gao, X., Tao, D., Li, X.: A unified tensor level set for image segmentation. IEEE Trans. Syst. Man Cybern.-Part B: Cybern. **40**(3), 857–867 (2010)
11. Dai, L., Ding, J., Yang, J.: Inhomogeneity-embedded active contour for natural image segmentation. Pattern Recogn. **48**(8), 2513–2529 (2015)
12. Min, H., Lu, J., Jia, W., Zhao, Y., Luo, Y.: An effective local regional model based on salient fitting for image segmentation. Neurocomputing **311**(15), 245–259 (2018)
13. Kim, W., Kim, C.: Active contours driven by the salient edge energy model. IEEE Trans. Image Process. **22**(4), 1667–1673 (2013)
14. Getreuer, P.: Chan-Vese segmentation. Image Process. Line **2**, 214–224 (2012)

15. Seo, H., Milanfar, P.: Static and space-time visual saliency detection by self-resemblance. J. Vis. **9**(12), 1–27 (2009)
16. Achanta, R., Hemami, S., Estrada, F.: Frequency-tuned saliency region detection. In: Proceedings of IEEE Conference on Computer Vision and Pattern Recognition, pp. 1597–1604 (2009)
17. Berkeley Segmentation Dataset 500. https://www2.eecs.berkeley.edu/Research/Projects/CS/vision/grouping/resources.html

An Attention Bi-box Regression Network for Traffic Light Detection

Juncai Ma[1,2], Yao Zhao[1,2(✉)], Ming Luo[1,2], Xiang Jiang[1,2], Ting Liu[1,2], and Shikui Wei[1,2]

[1] School of Computer and Information Technology, Beijing Jiaotong University, Beijing, China
{17120398,yzhao,14112058,16112055,shkwei}@bjtu.edu.cn, phoebe.luo@163.com
[2] The National Engineering Laboratory of Urban Rail Transit Communication and Operation Control, Beijing 100070, China

Abstract. Recently, object detection has made significant progress due to the development of deep learning. Since the traffic lights are extremely small objects, it leads to unsatisfactory performance when directly applying the off-the-shelf methods based on deep convolutional neural networks. To deal with this problem, we propose an improved detection network based on Faster R-CNN framework. By introducing an attention module on the top of the network, the network can focus better on the small object regions. At the same time, the features from shallow layers are leveraged for classification and bounding box regression, in which the features of small objects can be captured better. In addition, we design a two-branch network for detecting the traffic light box and the bulb box at the same time. In this manner, the performance of traffic light detection is improved obviously. Compared with other detection algorithms, our model achieves competitive results on VIVA traffic light challenge dataset.

Keywords: Traffic light detection · Two-branch structure · Attention · Convolutional neural networks

1 Introduction

Recently, autonomous driving has become more and more popular in the field of intelligent vehicles [8,12]. The current autonomous driving technology still relies mainly on the lidar that is cost prohibitive, so the vision-based method is still an important solution. In vision-based ways, traffic light detection plays an crucial rules for autonomous driving in real-world situations [13]. The traffic light detection mission aims to provide drivers with safer and smarter driving modes to deal with complicated scenarios. As we all know, traffic lights generally have low resolution and small size, so the detection of traffic lights are still a challenging problem.

© Springer Nature Switzerland AG 2019
Z. Cui et al. (Eds.): IScIDE 2019, LNCS 11935, pp. 399–410, 2019.
https://doi.org/10.1007/978-3-030-36189-1_33

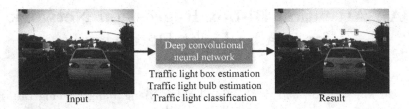

Fig. 1. Our overall framework of traffic detection. Input traffic scene pictures through our network estimated respectively to get the box and bulbs.

In existing works, Lindner et al. [15] proposed a method to detect traffic lights by manually extracting color and shape features for verification and classification. However, the key bottleneck of this method is that both the pixel color and shape feature are not robust to represent the traffic lights, which makes them less possible to handle the wild scenarios, especially on large scale dataset. To involving location information into traffic light detection, the researchers proposed to use GPS data to obtain traffic lights position and combine it into detection procedure [1,6]. Similar to [15], this method is difficult to be applied on a large scale application. In addition to these methods, with the development of deep learning, new methods of object detection have emerged in recent years. Karsten et al. [9] used YOLO [23] to detect traffic lights. In this framework, only the last convolution output layer is employed for detection, so the image is heavily down-sampled. Since the pixels of small objects are few, the information of traffic lights cannot easily captured by the layers of convolutions, which makes the detection of small traffic lights difficult. Recently, [14] proposed an improved method based on ACF [5] and got a good result on VIVA dataset. However, this method is only tested on day-time data, and its effectiveness on data acquiesced at night needs to be further verified. In addition, the detection procedure is not performed at an end-to-end manner, which makes it difficult to be applied in real-world scenarios.

As we all know, deep CNNs have made significant progress in the field of object detection, but in the detection of small objects such as traffic lights, the current detection algorithms still face substantial challenges. This motivates us to design better deep CNN for detecting traffic lights. Therefore, in this paper, we are committed to building an end-to-end deep network model for traffic lights detection. Since object detection is heavily dependent on the positioning of the target, an accurate positioning algorithm will often make the following classification easily. Therefore, inspired by [32], we proposed a method for traffic light detection by regressing bounding boxes for the traffic light box and the traffic light bulb respectively, which help to improve localizing precision of traffic lights. In addition, we also introduced the attention mechanism as the subsequent branches into the object detection network, which is used to select better features for classification. In particular, we used the framework of Faster R-CNN [25] as our main detection framework and VGG-16 as our feature extraction network. Our deep convolutional neural network consists of two parts, one is used to detect traffic light box, and the other is used to detect traffic light bulb. Each branch has both classification and bounding box regression operations. We validated the effectiveness of our approach on the VIVA traffic light dataset.

The main contributions of this paper are as follows:

- We introduced the attention mechanism into the traffic light detection model to improve the detecting capability of small objects.
- We introduced a two-branch bounding box regression at the top of the network, which is used to improve detection precision by detecting both the box and the bulb, and merging the outputs of the two branches.

In the rest of the paper, our structure is as follows: in the Sect. 2, we introduce some state-of-the-art methods; in the Sect. 3, we introduce the detection model we proposed; in the Sect. 4, we discuss the results on the test dataset; finally, we draw a conclusion in the Sect. 5.

2   Related Work

2.1   Traditional Traffic Light Detection

Traditional traffic light detection is usually through a color threshold technology to obtain a possible candidate area, where the threshold is usually set manually, and the color space selection is usually HSV space [7,18,29] and RGB space [4,19,20]. In addition to color information, shape information is often used when detecting traffic light bulb [20], such as hough transform. In [26], the author proposed an algorithm to model hue and saturation based on gaussian distribution, which use lots of images to train the parameters. However, the processing speed of the method is too slow. In [28], the author detected the traffic lights by detecting the characteristics of the black border area and the lighting area of the traffic lights.

2.2   Traffic Light Detection Based on Deep Learning

The first method for detecting traffic lights by deep learning is [30]. In [30], DeepTLR network is used to classify each fine-grained pixel area and conduct bounding box regression for each class. In [2], a complete system of detecting, tracking and classifying traffic lights is proposed. With the development of deep learning in general object detection tasks, more and more solutions have been proposed. In [9], YOLO [23] is used for traffic light detection. The author tested the networks of YOLOv1 [22] and YOLOv2 [23] on the VIVA traffic light challenge dataset.

In the study of the traffic light detection, there is a lot of research on the VIVA challenge dataset. [10] and [21] provide the evaluation results of Area-Under-Curve(AUC), where AUC represents the area under the Precision-Recall curve. In this paper, AP is used to approximate AUC. Due to the small scale of traffic lights and low pixels, traffic light detection is still a great challenge. The

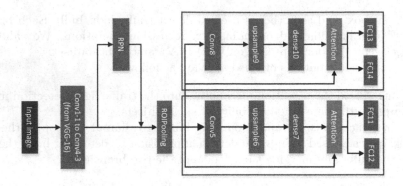

Fig. 2. Detailed network structure. The upper part is the branch of the traffic light box; the lower side is the branch of the traffic light bulb. FC11 and FC13 are both a classify task for the traffic light. FC12 is for traffic light bulb estimation and FC14 is for traffic light box estimation.

simple image processing method cannot achieve effective results, so extracting more robust features by convolutional neural network becomes a better solution. In addition, we found that adding an attention branch to the network can improve detection results and the time it takes is also acceptable. Since the VIVA dataset presents two annotation labels of traffic light box and traffic light bulb respectively, we design a two-branch bounding box regression network, which is called bi-box regression in this paper. In this paper, we compare the state-of-the-art methods on object detection. For the detection of small targets such as traffic lights, our method can achieve better results.

3   Approach

For the traffic light detection, we introduce a modified Faster R-CNN model. The principle of Faster R-CNN is presented in the first part. The second part introduces the bi-box regression network. The third part introduces the attention mechanism and why we use it. Figure 1 shows the overall structure of our method. The input image is estimated by the deep convolutional neural network to locate the traffic light box and the traffic light bulb separately, and the information of the two-branch lights is combined to obtain a better result, and the lights are classified at the same time. The two category labels are the same, but the coordinate information is different. In addition, we apply the attention module at the top of the network, and the detection effect has been well improved.

3.1   Basic Principle

Our proposed approach is based on the Faster R-CNN, which is use VGG-16 as backbone [27]. Figure 2 shows the detailed structure of our network. The whole network consists of three parts, a feature extraction module, a region proposal

(a) traffic light box   (b) traffic light bulb

Fig. 3. Traffic light annotation in VIVA dataset. (a) represents the traffic light box. (b) represents the traffic light bulb.

network, and a classifier and regression module. The feature extraction module that is built on VGG-16 is utilized for capturing features with high resolution. In the region proposal network, we use the original region proposal network to generate the region proposals, and get the reconstructed feature map after ROI Pooling. Since the traffic light belongs to the category of small objects, the scale of the objects will be very small on the reconstructed feature map, which is not conducive to classification and bounding box regression. Therefore, we first perform a one-step upsampling operation on the obtained feature map to increases the resolution of the image. Most importantly, in the third part, we add a branch to detect traffic lights after ROI Pooling. The two-branch network separately estimates the position of the traffic light box and the traffic light bulb and then classifies it, making full use of the information of the traffic light. Besides, to improve the classification effect of the network, an attention branch is added at the top of the network. Here, the loss we used for training is still the loss of the original Faster R-CNN. The loss function is as follows:

$$L(p_i, t_i) = \frac{1}{N_{cls}} \sum_i L_{cls}(p_i, p_i^*) + \lambda \frac{1}{N_{reg}} \sum_i p_i^* L_{loc}(t_i, t_i^*), \tag{1}$$

where p_i and t_i are the classification and regression results, p_i^* and t_i^* are the classification and regression ground truth. The ground truth label p_i^* is 1 if the anchor is positive. Here, the positive is the anchor that has an intersection-over-union (IOU) greater than 0.7 with ground truth bounding box. The negative is the anchor that has an IOU less than 0.3. The classification loss L_{cls} uses the cross entropy loss; the regression loss L_{loc} uses the smooth L1 loss. And the two parts of the loss are normalized by N_{cls} and N_{reg} respectively. N_{cls} represents the mini-batch size, and N_{reg} represents the number of anchor locations. λ is a hyperparameter used to balance classification and regression loss.

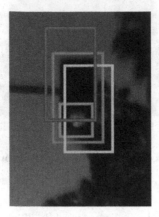

Fig. 4. Proposal determination process after using bi-box regression. The yellow bounding box is a good proposal. The red bounding box is a bad proposal. (Color figure online)

3.2 Bi-box Regression for Traffic Light

For normal detection, an image and a series of generated proposals are feed into the network for classification and traffic light estimation. Figure 3 shows that the VIVA traffic light challenge dataset has both the position label of the box and the position label of the bulb, we have adopted a two-branch detection network to improve the effectiveness of the detection. Here, we let $S = (S^x, S^y, S^w, S^h)$ represent a traffic light box proposal, where S^x and S^y represent the central coordinate of the candidate region S, and S^w and S^h represent the width and height of proposal S. For the proposal S, when it passes the ROI Pooling layer, the traffic light box branch outputs three probabilities and four offsets $f = (f^x, f^y, f^w, f^h)$. The traffic light bulb branch outputs three probabilities and four offsets $g = (g^x, g^y, g^w, g^h)$. Three probabilities indicate the probability that this proposal belongs to a certain type of light. There are three categories in total(green, red, yellow). f represents the offset of the coordinates and the length and width. Since we have two types of labels, we assume $Q = (F, G)$, where F represents the coordinate information of the traffic light box and G represents the coordinate information of the traffic light bulb. A proposal S matches Q when it meets

$$IOU(S, F) > \alpha, O(S, G) > \beta \tag{2}$$

where

$$IOU(S, F) = \frac{S \cap F}{S \cup F}, O(S, G) = \frac{S \cap G}{G} \tag{3}$$

In this paper, we set the values of the super parameters $\alpha = 0.7$ and $\beta = 0.7$. For the two-branch network after ROI Pooling, the final loss is the sum of the loss of the two branches:

$$L = L_{box} + L_{bulb} \tag{4}$$

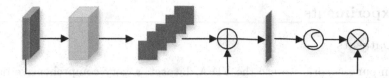

Fig. 5. The attention structure in our network. The input feature and the convolved feature are added to generate the green attention map. The green attention map is multiplied by our previous feature after softmax to get the final result. (Color figure online)

where L_{box} and L_{bulb} are defined by Eq. (1). Figure 4 shows that the traffic light proposals are matched to the ground truth. In Fig. 4, both green box and blue box are ground truth box. Red box and yellow box are traffic light proposals. Yellow box satisfies the condition of formula (2), so the yellow box is a positive bounding box. However, the red box does not meet the condition of formula (2), so the red box is a negative bounding box.

3.3 Attention for Small Object

For the detection of small objects, convolutional neural networks are often designed as multi-scale structures. Classification or regression is performed by using low-level features or by fusing multi-scale features. The original vgg16-based Faster R-CNN features directly into the FC layer after ROI Pooling. In our paper, in order to improve the network's ability to detect small objects, we added an attention branch [11] to the second stage of the network. As we all know, the core goal of the attention is to use attention maps to enable us to have enough visual information when classification. This method is to identify significant areas of the image and amplify its effects while suppressing irrelevant and potentially confusing information in other areas. It can be described by

$$g_a = \langle a, l \rangle, a = l \oplus g \tag{5}$$

where l represents the feature map after ROI Pooling. Here we define g represents the features before the network enters the classification layer, and a represents the feature after adding and convolution of l and g and then passing the vector after softmax. Figure 5 is the detailed structure of attention module [11]. As shown in Fig. 5, the input feature and the convolved feature are added to generate an attention map, which is the green feature in Fig. 5. The obtained feature is multiplied by the input feature to get the output. The specific structure is shown in Fig. 2, after the feature of the RPN, the feature map is obtained after an upsampling, and the feature map is added to the features before the classification and regression layer and convolved to obtain our attention map. Finally, the attention map is multiplied by the feature map and sent to the final classification regression branch. Here we want to make the detection effect more significant, we first perform an upsampling on the feature map.

4 Experiments

4.1 Datasets

To train our network, we use the VIVA dataset, a very comprehensive dataset of traffic light that contains videos of many complex street scenes.

Training Set: The training datasets use the complete data from dayclip1 to dayclip13, nightclip1 to nightclip5. In the training data, the traffic light categories are divided into six categories: go, stop, warning, go left, stop left, warning left. Here we use three categories: go, stop, warning. All models use the SGD optimization method, the momentum is set to 0.9, the backbone network uses the VGG-16 model pre-trained on the imagenet, and we feed the network image size is 300*300. We use SGD to optimize the full training loss. We train the networks for 8 epochs with a learning rate of 0.001, and we train the networks for an extra 8 epochs with a learning rate of 0.0001, and reduce the learning rate to 0.00001 for the last 4 epochs.

Testing Set: Here the test datasets are daysequence1, daysequence2, nightsequence1, nightsequence2. Since there are only six classes in the training set, the goforward class in the test set is deleted. We test our model on a 1080Ti GPU.

4.2 Metrics

The evaluation index AP approximates the area under the Precision-Recall curves, where precision and recall are calculate by the formula:

$$Precision = \frac{TP}{TP + FP} \tag{6}$$

$$Recall = \frac{TP}{TP + FN} \tag{7}$$

where TP represents the number of True Positive samples, and the True positive refers to the candidate box with the overlap indicator IOU greater than 0.7. FP correspond to false positives that are counted for prediction's IOU not overlapping the defined threshold. FN correspond to false negative samples. The larger the area under the Precision-Recall curve indicates the better the effect, that is, the larger the AP value, the better the effect. mAP is the mean of AP of each category.

4.3 Comparison to State-of-the-Art General Object Detection Algorithm

The current general object detection algorithm has been greatly developed, and there are good improvements in PASCAL VOC and COCO datasets. We first analyze and compare one-stage object detection methods. Table 1 shows that the one-stage method is superior in speed in SSD [17] and RFB [16] detection based

Table 1. Comparison of our method and novel architectures on the VIVA datasets.

Method	Go	Stop	Warning	mAP
SSD	34.18%	21.33%	4.8%	20.10%
RFBnet	14.11%	5.32%	0.19%	6.54%
YOLOv3	53.66%	49.89%	7.63%	37.06%
RefineDet	41.72%	37.96%	8.65%	29.44%
R-FCN	51%	53.15%	39.47%	47.87%
Faster R-CNN	49.67%	45.75%	43.01%	46.14%
Ours	49.61%	56.56%	47.24%	51.14%

on SSD framework, but there are still some problems in the small target of the traffic light. YOLOv3 [24] only performed well between the red and green light categories. RefineDet [31] refines the two-stage method's regression idea of box from coarse to fine on the basis of SSD framework to improve the detection effect of small targets. In addition, some new methods of two-stage are compared, and the overall performance of two-stage method is better than that of one-stage method. R-FCN [3] uses the same backbone as Faster R-CNN. Our approach is a two-stage method based on Faster R-CNN. Since the warning and the stop are closer in color on the dataset, the R-FCN algorithm does not distinguish between the two classes well. It is easy to misjudge the warning. Here we have a significant improvement in the detection of traffic lights.

Table 2. Comparison of different values of β.

β	Go	Stop	Warning	mAP
$\beta = 0.5$	45.32%	49.08%	34.77%	43.06%
$\beta = 0.6$	43.17%	49.22%	34.33%	42.24%
$\beta = 0.7$	49.61%	56.56%	47.24%	51.14%
$\beta = 0.8$	41.99%	47.99%	31.04%	40.34%

4.4   Influence of Threshold

As described in Sect. 3.2, the positive proposals we send to the two-branch module have an impact on our results. Therefore, we conduct an experiment to analyze the effects of hyperparameters α, β on our experimental results. In the implementation we set the values of β in Eq. 2 to be 0.5, 0.6, 0.7, 0.8. And α is set to 0.7. The results corresponding to different thresholds are reported in Table 2. We find that the mAP is best when the threshold β is 0.7, which has 8.08 percentage points higher than setting to 0.5. When α is fixed, the value of β affects the quantity and quality of the positive proposal. When the value of β

Table 3. Comparison of the effectiveness of the improved method. (Faster: Faster R-CNN; Att.: Attention; bi-reg: bi-box regression.)

Method	Go	Stop	Warning	mAP
Faster	49.67%	45.75%	43.01%	46.14%
Faster+Att.	49.14%	50.09%	42.54%	47.26%
Faster+Att.+bi-reg	49.61%	56.56%	47.24%	51.14%

is large, the positive proposals are close to the groundtruth box, but the training positive proposals are reduced. When the value of β is small, the positive proposals contain a lot of poor allocations. Therefore, it is important to choose a suitable threshold. We can see from the experiment in Table 2 that the best result can be achieved when α and β are set at 0.7.

4.5 Validity Analysis

Table 3 shows the results obtained by different methods on the VIVA dataset. Faster R-CNN is a classic object detection algorithm. We improve on the basis of Faster R-CNN, and we use VGG-16 for the backbone network. After adding the attention module, the overall mAP of our improved model based on Faster R-CNN improves by 1.12% points. Because the subsequent network has stronger classification ability after adding the attention module, it can pay attention to the information that is more conducive to classification. Although the effect of some categories decreased slightly, the overall effect is improved. In addition, on the basis of adding the attention module, we add the bi-box regression at the top of the network. It can be found from Table 3 that the experimental effect is improved by 3.88% point after two-branch regression is added.

5 Conclusion

In this paper, we proposed a traffic light detection method, and the model we proposed introduced the attention module, which learned a weight for features before classification so that features with higher importance could get higher weight. Furthermore, the characteristics of the output of RPN network were more conducive to the task of small object detection after an upsampling. In addition, we designed a bi-box regression network after ROI Pooling, incorporating the information of traffic light boxes and traffic light bulbs to get more accurately positioned proposals. Experiment showed that our proposed method achieves better results in the VIVA traffic light test set.

Acknowledgements. This work is supported by Fundamental Research Funds for the Central Universities (No. 2018JBZ001).

References

1. Barnes, D., Maddern, W., Posner, I.: Exploiting 3D semantic scene priors for online traffic light interpretation. In: 2015 IEEE Intelligent Vehicles Symposium (IV), pp. 573–578. IEEE (2015)
2. Behrendt, K., Novak, L., Botros, R.: A deep learning approach to traffic lights: detection, tracking, and classification. In: 2017 IEEE International Conference on Robotics and Automation (ICRA), pp. 1370–1377. IEEE (2017)
3. Dai, J., Li, Y., He, K., Sun, J.: R-fcn: object detection via region-based fully convolutional networks. In: Advances in Neural Information Processing Systems, pp. 379–387 (2016)
4. Diaz-Cabrera, M., Cerri, P., Sanchez-Medina, J.: Suspended traffic lights detection and distance estimation using color features. In: 2012 15th International IEEE Conference on Intelligent Transportation Systems, pp. 1315–1320. IEEE (2012)
5. Dollár, P., Appel, R., Belongie, S., Perona, P.: Fast feature pyramids for object detection. IEEE Trans. Pattern Anal. Mach. Intell. 36(8), 1532–1545 (2014)
6. Gomez, A.E., Alencar, F.A., Prado, P.V., Osório, F.S., Wolf, D.F.: Traffic lights detection and state estimation using hidden Markov models. In: 2014 IEEE Intelligent Vehicles Symposium Proceedings, pp. 750–755. IEEE (2014)
7. Gong, J., Jiang, Y., Xiong, G., Guan, C., Tao, G., Chen, H.: The recognition and tracking of traffic lights based on color segmentation and camshift for intelligent vehicles. In: 2010 IEEE Intelligent Vehicles Symposium, pp. 431–435. IEEE (2010)
8. Huang, J., Gao, Y., Lu, S., Zhao, X., Deng, Y., Gu, M.: Energy-efficient automatic train driving by learning driving patterns. In: Thirty-Second AAAI Conference on Artificial Intelligence (2018)
9. Jensen, M.B., Nasrollahi, K., Moeslund, T.B.: Evaluating state-of-the-art object detector on challenging traffic light data. In: Proceedings of the IEEE Conference on Computer Vision and Pattern Recognition Workshops, pp. 9–15 (2017)
10. Jensen, M.B., Philipsen, M.P., Møgelmose, A., Moeslund, T.B., Trivedi, M.M.: Vision for looking at traffic lights: issues, survey, and perspectives. IEEE Trans. Intell. Transp. Syst. 17(7), 1800–1815 (2016)
11. Jetley, S., Lord, N.A., Lee, N., Torr, P.H.: Learn to pay attention. arXiv preprint arXiv:1804.02391 (2018)
12. Korchev, D., Jammalamadaka, A., Bhattacharyya, R.: Automatic rule learning for autonomous driving using semantic memory. arXiv preprint arXiv:1809.07904 (2018)
13. Levinson, J., Askeland, J., Dolson, J., Thrun, S.: Traffic light mapping, localization, and state detection for autonomous vehicles. In: 2011 IEEE International Conference on Robotics and Automation, pp. 5784–5791. IEEE (2011)
14. Li, X., Ma, H., Wang, X., Zhang, X.: Traffic light recognition for complex scene with fusion detections. IEEE Trans. Intell. Transp. Syst. 19(1), 199–208 (2018)
15. Lindner, F., Kressel, U., Kaelberer, S.: Robust recognition of traffic signals. In: 2004 IEEE Intelligent Vehicles Symposium, pp. 49–53. IEEE (2004)
16. Liu, S., Huang, D., Wang, Y.: Receptive field block net for accurate and fast object detection. In: Ferrari, V., Hebert, M., Sminchisescu, C., Weiss, Y. (eds.) ECCV 2018. LNCS, vol. 11215, pp. 404–419. Springer, Cham (2018). https://doi.org/10.1007/978-3-030-01252-6_24
17. Liu, W., et al.: SSD: single shot multibox detector. In: Leibe, B., Matas, J., Sebe, N., Welling, M. (eds.) ECCV 2016. LNCS, vol. 9905, pp. 21–37. Springer, Cham (2016). https://doi.org/10.1007/978-3-319-46448-0_2

18. Lu, K.H., Wang, C.M., Chen, S.Y.: Traffic light recognition. J. Chin. Inst. Eng. **31**(6), 1069–1075 (2008)
19. Omachi, M., Omachi, S.: Traffic light detection with color and edge information. In: 2009 2nd IEEE International Conference on Computer Science and Information Technology, pp. 284–287. IEEE (2009)
20. Omachi, M., Omachi, S.: Detection of traffic light using structural information. In: IEEE 10th International Conference on Signal Processing Proceedings, pp. 809–812. IEEE (2010)
21. Philipsen, M.P., Jensen, M.B., Møgelmose, A., Moeslund, T.B., Trivedi, M.M.: Traffic light detection: a learning algorithm and evaluations on challenging dataset. In: 2015 IEEE 18th International Conference on Intelligent Transportation Systems, pp. 2341–2345. IEEE (2015)
22. Redmon, J., Divvala, S., Girshick, R., Farhadi, A.: You only look once: unified, real-time object detection. In: Proceedings of the IEEE Conference on Computer Vision and Pattern Recognition, pp. 779–788 (2016)
23. Redmon, J., Farhadi, A.: YOLO9000: better, faster, stronger. In: Proceedings of the IEEE Conference on Computer Vision and Pattern Recognition, pp. 7263–7271 (2017)
24. Redmon, J., Farhadi, A.: YOLOv3: an incremental improvement. arXiv preprint arXiv:1804.02767 (2018)
25. Ren, S., He, K., Girshick, R., Sun, J.: Faster R-CNN: towards real-time object detection with region proposal networks. In: Advances in Neural Information Processing Systems, pp. 91–99 (2015)
26. Shen, Y., Ozguner, U., Redmill, K., Liu, J.: A robust video based traffic light detection algorithm for intelligent vehicles. In: 2009 IEEE Intelligent Vehicles Symposium, pp. 521–526. IEEE (2009)
27. Simonyan, K., Zisserman, A.: Very deep convolutional networks for large-scale image recognition. arXiv preprint arXiv:1409.1556 (2014)
28. Siogkas, G., Skodras, E., Dermatas, E.: Traffic lights detection in adverse conditions using color, symmetry and spatiotemporal information. In: VISAPP (1), pp. 620–627 (2012)
29. Wang, C., Jin, T., Yang, M., Wang, B.: Robust and real-time traffic lights recognition in complex urban environments. Int. J. Comput. Intell. Syst. **4**(6), 1383–1390 (2011)
30. Weber, M., Wolf, P., Zöllner, J.M.: DeepTLR: a single deep convolutional network for detection and classification of traffic lights. In: 2016 IEEE Intelligent Vehicles Symposium (IV), pp. 342–348. IEEE (2016)
31. Zhang, S., Wen, L., Bian, X., Lei, Z., Li, S.Z.: Single-shot refinement neural network for object detection. In: Proceedings of the IEEE Conference on Computer Vision and Pattern Recognition, pp. 4203–4212 (2018)
32. Zhou, C., Yuan, J.: Bi-box regression for pedestrian detection and occlusion estimation. In: Ferrari, V., Hebert, M., Sminchisescu, C., Weiss, Y. (eds.) ECCV 2018. LNCS, vol. 11205, pp. 138–154. Springer, Cham (2018). https://doi.org/10.1007/978-3-030-01246-5_9

MGD: Mask Guided De-occlusion Framework for Occluded Person Re-identification

Peixi Zhang[1], Jianhuang Lai[2,3(✉)], Quan Zhang[2], and Xiaohua Xie[2,3]

[1] School of Electronic and Information Technology, Sun Yat-sen University,
Guangzhou, China
zhangpx5@mail2.sysu.edu.cn
[2] School of Data and Computer Science, Sun Yat-sen University, Guangzhou, China
{stsljh,xiexiaoh6}@mail.sysu.edu.cn, zhangq48@mail2.sysu.edu.cn
[3] Guangdong Key Laboratory of Information Security Technology,
Sun Yat-sen University, Guangzhou, China

Abstract. Person re-identification (ReID) is a challenging task in computer vision area due to the dramatic changes across different non-overlapping camera views, e.g., lighting, view angle, and pose, among which occlusion is one of the hardest challenges. Recently, occluded person re-identification (Occluded-ReID) is proposed to address this problem. Nevertheless, current occluded-ReID methods focus on how to learn a matching function between partial-body images and full-body images while ignore the structural information of the full body. To handle this problem, we propose a novel framework called Mask Guided De-occlusion (MGD) for occluded Person Re-identification. The MGD mainly consists of three components, i.e., a Coarse-to-Fine Mask Generation (CFMG) module, a Person De-Occlusion (PDO) module and a Person Feature Extractor (PFE). The key module CFMG aims to locate the occlusion areas by manipulating the instance segmentation masks through a two-stage process. The proposed PDO module is to reconstruct the occluded pedestrian. After that, all the images are fed into the PFE module to obtain their feature vectors. With PDO and CFMG modules, the proposed method MGD reduces the impact of occlusions and thus improves the performance of Occluded-ReID. The extensive experiments conducted on several public occluded ReID datasets show that our method is effective and outperforms the state-of-the-art methods.

Keywords: Occluded person re-identification · Occlusion location · Image reconstruction

1 Introduction

Person re-identification (ReID) has been widely studied in recent years. The mainstream models have achieved a satisfactory performance on the public datasets. However, in many real-world scenarios, pedestrians are inevitably

© Springer Nature Switzerland AG 2019
Z. Cui et al. (Eds.): IScIDE 2019, LNCS 11935, pp. 411–423, 2019.
https://doi.org/10.1007/978-3-030-36189-1_34

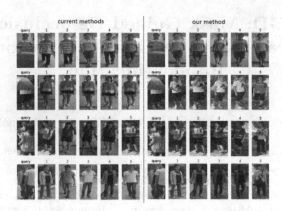

Fig. 1. Some hard samples that current methods cannot solve. Each row represents one identity. The images on the left side represent the retrieval results returned by a traditional model PCB [15] while the images on the right side are our method's results. (Color figure online)

obscured by trees, walls, and other people. As a result, the full-body images are not always available and the existing methods fail to handle it. Traditional models often mistake the occlusion areas as a part of the specified pedestrian. As shown in Fig. 1, the pedestrians on the first two rows are occluded by a desk with stripe and a red umbrella respectively, and the retrieved images of traditional method [15] share some similar low-level patterns with the occlusions.

Occluded-ReID and partial-ReID are proposed to address this problem. As shown in Fig. 2, partial-ReID is a simplified version of occluded-ReID, because the occluded parts of body are erased manually. In real-world application, occluded-ReID is more practicable so we only focus on occluded-ReID task in this work. Recently, some difficulties still have not been explored in occluded-ReID. Most of the existing methods [8,26] attempt to find a robust "partial-to-full" matching function via attention mechanism. Unfortunately, occluded images often lack the overall structural information. Even worse, as shown in the third and fourth row in Fig. 1, there are two pedestrians in the query images, the current approaches cannot distinguish the target person from occlusions and fail to extract the high-level semantics of the target person, leading to poor retrieval results.

In this paper, we propose a novel framework called Mask Guided De-occlusion (MGD) for Occluded ReID. The proposed framework consists of three modules, i.e., a Coarse-to-Fine Mask Generation (CFMG) module, a Person De-Occlusion (PDO) module and a Person Feature Extractor (PFE)

CFMG is to predict precise occlusion masks for occlusion images, which has not been fully investigated before. The concept of occlusion in person ReID is the object that covers the target person. However, common segmentation methods cannot distinguish which of objects are obstructions. Our framework predicts the occlusion areas based on the object masks computed by an instance segmentation model, and then further refines them according to the morphological properties.

Query Gallery   Query Gallery   Query Gallery

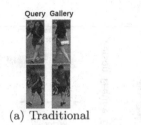  

(a) Traditional   (b) Partial   (c) Occluded

Fig. 2. Three different types of ReID.

Extensive experiments show that the CFMG module can produce promising occlusion masks.

Second, inspired by image inpainting methods [6,10,20,23], we propose a Person De-Occlusion (PDO) module which resorts to an inpainting GAN to recover the occlusion areas of an image. With the PDO module, the network can learn the transformation between the full-body domain and the partial-body domain. The PDO module can also be regarded as a data augmentation operator to provide more training data for learning a robust feature representation.

The **contributions** of this paper are summarized as follows:

- We propose a novel framework called Mask Guided De-occlusion (MGD) for Occluded-ReID, which can transfer the partial-to-full person matching problem to a full-to-full person matching problem.
- We propose a Coarse-to-Fine Mask Generation module (CFMG) for generating the accurate occlusion masks. With the CFMG module, the proposed framework removes noisy occlusions and thus assists the inference of the PFE module.
- Extensive experiments conducted on several occlusion datasets show that the proposed method is quite effective and achieves the state-of-the-arts.

2 Related Work

Traditional ReID. Traditional ReID performs pedestrian search in the full-body person domain. Mainstream traditional ReID methods mainly start from designing the network structure to obtain more discriminative information, including the deep feature of a person [1,11,22], spatial-temporal information [17], view angles [3], and so on. Some studies use pre-defined locations [18,21] and semantic regions [7] to improve the ReID, including stripes, grid patches, person parts and attention regions. For example, Sun et al. [15] proposed a network which outputs a convolutional descriptor consisting of several part-levels features. Wang et al. [18] proposed the multiple granularity network (MGN) with one branch for global features and two branches for local features. Some researchers introduced additional information. For example, Liu et al. [8] proposed a HydraPlus-Net (HP-Net) which uses pedestrian attribute labels.

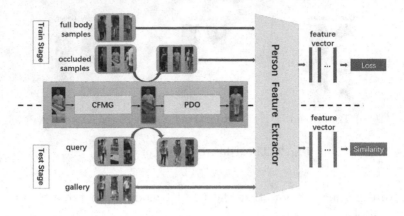

Fig. 3. Our occluded-ReID framework. It consists of three components, a Person Feature Extractor(PFE), a Coarse-to-Fine Mask Generation (CFMG) module and a Person De-occlusion (PDO) module.

Su et al. [12] applied pose estimation to cope with the pose variations. However, the performance of above methods will greatly drop when dealing with occlusions.

Occlusion Methods. Current researches about the occlusion problem attempt to seek a matching pattern between local features and global features. Zheng et al. [24] proposed a local patch level matching model called Ambiguity-sensitive Matching Classifier (AMC), and introduced a global part-based matching model called Sliding Window Matching (SWM) that can provide complementary spatial layout information. However, the computation cost is quite high. He et al. [4] proposed a Deep Spatial feature Reconstruction (DSR) model, which combines fully convolutional network and sparse representation to address the partial-ReID. Sun et al. [13] introduced a Visibility-aware Part Model (VPM), which learns to perceive the visibility of regions through self-supervision. In addition, Fan et al. AFPB [26] introduced an attention framework to concentrate on the non-occluded part of a person. Huang et al. [5] claimed to augment the variation of training data by introducing manually occluded samples to help the model learn robust feature. However, all of the above methods aim to seek a "local-to-global" similarity function, and ignore the impact of full-body pedestrian structure information on person matching.

3   Approach

3.1   Framework

In this section, we introduce a novel Mask Guided De-occlusion (MGD) framework for occluded-ReID, as shown in Fig. 3. Overall, the MGD framework contains three components, i.e., a Person De-Occlusion (PDO) module, a Coarse-to-Fine Mask Generation (CFMG) module and a Person Feature Extractor (PFE).

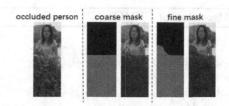

Fig. 4. Visualization for coarse to fine process of CFMG algorithm.

We simply adopt a widely used part-based model PCB [15] as the Person Feature Extractor. And we only focus on the CFMG and PDO modules in this paper. Given an occluded person image, the goal of the CFMG module is to locate occlusions with an instance segmentation network and conduct a coarse-to-fine location scheme. After locating the occlusions, the PDO module is then used to repair the occluded regions by using an inpainting GAN. Finally, both non-occluded person images and the repaired images are fed into the Person Feature Extractor to learn a robust feature descriptor with softmax loss and LSR loss [16]. During the test phase, the PDO module is used to recover each occluded image in the query set. De-occluded images and their original images are fed into the PFE module to obtain two feature vectors for fusion.

3.2 Coarse-to-Fine Mask Generation Module

Directly adopting a segmentation model can only locate the pre-defined objects including the target person in an image but we cannot distinguish which of them are obstructions. Our proposed CFMG module handles it by performing a coarse-to-fine location process. Above all, we make an assumption that the target person should be at the center of an image, while the rest of items including other pedestrians are potential obstructions. This is valid in most of the public ReID datasets.

The first step of CFMG is to feed the occlusion images into an instance segmentation model. We adopt the network MM-detection [2] in this work but note that the other instance segmentation models can also be employed in our framework. We regard the person at the center as the target and denote its mask as M_t. If the segmentation network detects other objects (not handbags and backpacks), we denote their masks as M_{seg}, and regard them as occlusions of the target person.

Objects in appropriate scales can be detected by the segmentation network, while large objects like trees may be out of scope and the segmentation network fails. Such obstructions cover a large area, so finding the maximum inner rectangle outside the target person can coarsely locate the obstructions. But obstructions can be everywhere in the image. So we normalize the image by rotating it so that the obstruction lies at the bottom or bottom-left. And we define the maximum inner rectangle at the bottom or bottom-left as the **maximum bottom rectangle** in Definition 2. Then, in the second step, we formalize

---

Algorithm 1. The Algorithm of CFMG.

Require:
    Input image I.
    A pre-trained instance segmentation network Θ.
Ensure: The occlusion mask M.
  1: Feed I to $\Theta(\cdot)$ and obtain the M_t, M_{seg}.
  2: Calculate I_c by $I_c = I \circ M_t$.
  3: Find maximum inner rectangle A^* of I_c.
  4: Rotate I until A^* satisfies the conditions of maximum bottom rectangle
  5: Calculate downside boundary line C of closed boundary $f(x,y)$ around M_t.
  6: Calculate the region M_f by Eq. 1.
  7: Calculate the total mask M_i by Eq. 2.

---

the original problem as follow. Let a rectangle A stand for the boundary of an image, and f(x,y) denote a closed contour curve of the a target person inside A, we define a function $R(\cdot)$ to represent the region enclosed by the closed curve. Our goal is to find the occlusion areas M of the target person in the region R_p, where $R_p = R(A) - R(f)$. In order to make a clearer description, we clarify the following two definitions.

Definition 1. *For a closed curve $f(x,y) = 0$, the downside boundary line C of $f(x,y) = 0$ can be defined as,*

$$y = C(x) = \min\{y | f(x,y) = 0\}, x_{min} \leq x \leq x_{max},$$

where x_{max} and x_{min} represent the edges of $f(x,y)$ along x-axis.

Definition 2. *Given a rectangle A and a closed curve $f(x,y)$ inside A, We define maximum bottom rectangle A_c^* in the region $R_p = R(A) - R(f)$, which satisfies:*

- *A_c^* is the maximum inner rectangle of R_p.*
- *A_c^* must contain the lower left corner of R_p.*
- *Contact point between A_c^** and curve $f(x,y)$ must be on the upper edge of A_c^*.*

As shown in Fig. 4, we first construct a maximum inner rectangle A^* of the region R_p, and then rotate it to satisfy the conditions of A_c^*. We use this maximum bottom rectangle A_c^* of I_c to estimate a coarse occlusion mask. Further, the obstruction must be connected to the target pedestrian instead of being separated, so the regions between the coarse mask and the target person are also regarded as occluded regions, which can be formally defined as

$$M_f = \{(x,y) | y \leq C(x), x_{min} \leq x \leq x_{max}\}, \tag{1}$$

where x_{max} and x_{min} represent the edges of $f(x,y)$ along x-axis. Finally, the total occlusion mask can be combined as:

$$M = M_{seg} \cup A_c^* \cup M_f \tag{2}$$

The entire algorithm is shown in Algorithm 1.

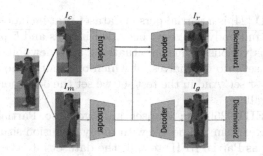

Fig. 5. Overview of PDO module.

3.3   Person De-occlusion Module

Person de-occlusion module (PDO) performs a transformation from a occluded-body person domain to a full-body person domain. Specifically, the PICNet [23] that is originally used to reconstruct occluded human faces are adopted for the PDO module. Differently, we transfer this model to human body reconstruction. Given an image I and its corresponding mask M^1, we compute its occlusion areas I_m and non-occlusion areas I_c by the Hadamard product as Eq. 3.

$$\begin{cases} I_m = I \circ M \\ I_c = I \circ (1 - M) \end{cases} \tag{3}$$

We feed I_c and I_m into the PDO module, as shown in Fig. 5. The PDO module consists of two paths. Each branch contains an encoder and a decoder that share the same weights. Both I_m and I_c are encoded to the latent space. The features in latent space are then used to generate the missing parts by decoder and output I_r and I_g. Note that the upper branch is allowed to utilize the ground-truth information from lower branch to help the learning of decoder.

During the training phase, we conduct a adversarial learning scheme on two branches by using similar strategy as PICNet. The training samples are full-body images and we manually occlude the images with some random masks. We force the PDO module to repair the whole-person images based on person structure semantics. In this way, the PDO module learns how to de-occlude the occlusions. During the test phase, we only the retain the lower branch to reconstruct the occlusions. Specifically, we generate the occlusion masks with the CFMG module and then feed I_m into the PDO module. The network will generate the full-body de-occluded image I_g.

4   Experiment

4.1   Datasets

We conducted our experiments on three public occluded ReID datasets.

---

[1] The mask M is a binary matrix of the same size as the original image. $M(i,j) = 1$ represents that the pixel (i,j) is occluded, and $M(i,j) = 0$ represents not occlusion.

Partial-REID [24] is a partial person dataset that includes 600 images from 60 people, with 5 full-body images, 5 occluded images and 5 partial images per person. It contains various occlusion situations and viewpoints. We follow [24] and split the dataset into two halves, of which one half is the train set and the other half is the test set. And in the test set we set the occlusion images as query and the rest are gallery.

Occluded-REID [26] is a dataset analogous to Partial REID dataset, includes 2000 images from 200 people with many occlusion situations. We adopt the same strategy as Partial REID to split the dataset into the train set and the test set.

P-DukeMTMC [26] is a subset of DukeMTMC [25]. Following [26], we use 24143 images of 1299 identities including both full-body images and occluded images. The train set contains 665 identities while the test set contains 634 identities. Similarly, we adopt the strategy in Partial-REID dataset to split it.

4.2   Experiment Settings

Network Architecture. We utilize the Part-based Convolutional Baseline model (PCB) as our Person Feature Extractor, which is commonly adopted due to its effectiveness in feature extraction. In addition, we adopt PICNet in PDO module to repair the occluded region and employ MM-detection to integrate the instance segmentation model into the proposed CFMG module.

Optimization. We train the person feature extractor (PFE) with SGD optimizer. The weight decay and momentum is set to 5×10^{-4} and 0.9. The initial learning rate of the backbone layer is set to be 0.01. Besides, we resize the images to 240×240 and randomly crop to 224×224 for data augmentation. As for the PICNet model we maintain the author's settings [23] when training. Last but not least, we do not train instance segmentation model by ourselves and directly use the pre-trained model.

Implementation Details. During the training phase, we first train the PDO module with occluded images which are manually generated by calculate the Hadamard product between the random masks and the non-occluded images in target train set. Second, we apply the PDO module to repair the occluded images in the train set. Then the de-occluded images are augmented into the train set for training the PFE (Fig. 3). We do not use any samples from other person re-identification dataset when training the PFE, so we can ensure a fair comparison with other ReID methods.

When testing, we use the CFMG module to estimate the occlusion mask for each image in the query set and then feed the masked person image to the PDO module to generate a reconstructed image. Each generated image, along with the original image are used to extract two feature vectors and finally averaged.

4.3   Comparison with the State-of-the-Art

In this Section, we compare our proposed method with the mainstream occluded and partial ReID models on Partial-REID dataset. As is shown in the Table 1,

Table 1. Comparison with the public occluded ReID method on Partial Dataset

Method	Rank-1	Rank-3
SWM [24] (ICCV2015)	24.33	45
AMC [24] (ICCV2015)	33	46
SWM+AWC [24] (ICCV2015)	36	51
DSR(single-scale) [4] (CVPR2018)	39.3	55.7
DSR(multi-scale) [4] (CVPR2018)	43	60.3
VPM [13] (CVPR2019)	67.7	81.9
AFPB [26] (ICME2018)	78.5	–
Ours	84.7	91.3

Table 2. Comparison with some mainstream ReID methods

Dataset		Partial REID		Occluded REID		P-DukeMTMC	
Method	Ref.	Rank-1	Rank-5	Rank-1	Rank-5	Rank-1	Rank-5
GOG [9]	CVPR2016	41.9	74	40.5	63.2	17.1	29.3
DGD [19]	CVPR2016	56.8	77.7	41.4	65.7	41.4	60.1
SVDNET [14]	ICCV2017	56.1	87.1	63.1	85.1	43.4	63.4
MLFN [1]	CVPR2018	64.3	87.3	64.7	87.8	51.0	70.3
AFPB [26]	ICME2018	78.5	94.9	68.1	88.3	46.2	63.5
Ours(w/o PDO)		72.7	88.1	73.2	85.3	69.6	80.4
Ours(w/o CFMG)		77.3	91.3	78.0	88.1	71.6	81.6
Ours		84.3	94.0	79.2	89.8	74.9	84.8

our method outperforms the state-of-the-art method AFPB with rank-1 accuracy 84.7% vs. 78.5%. And we present the comparison between our method and the mainstream ReID methods in Table 2. Our approach achieves the best performance of rank-1 and rank-5 among all the methods mentioned in Table 2, which indicates the superiority of our method. Besides, we remove the CFMG module and PDO module for contrast. When the CFMG module is removed, we only employ the initial masks generated by MM-detection as occlusion masks and do not perform the coarse-to-fine process. When the PDO module is removed, we can still get rid of the obstructions by the CFMG module but we do not repair the images. The results in Table 2 validate the effectiveness of these two modules.

4.4   Ablation Study

Effects of De-occlusion Data. We further investigate the effects of the de-occluded data during the training phase and the testing phase of feature extractor in the Partial-REID dataset. From Table 3, we notice that employing the de-occluded data in both training and testing phases can boost the retrieval

Table 3. Effects of the de-occlusion images. The first column indicates that if we use the de-occlusion generated data for data augmentation in the training phase. The second column indicates whether we repair the occluded areas of test images.

Train	Test	Rank-1	Rank-5
✗	✗	76.5	89.6
✗	✓	77.1	87.15
✓	✗	72.7	85.3
✓	✓	84.7	94.0

Table 4. Performance of CFMG.

Segmentation speed	Location speed	Average IoU
9 FPS	7 FPS	80.7%

accuracy. When the model is trained without de-occluded data and tested with de-occluded data, we cannot witness a promotion. Analogous result happens when we train with de-occluded data and test without them. One possible reason is that the domain gap between the real samples and the recovered outputs of PDO module cannot be ignored. So we need to augment some generated images for training. Second, we can regard the occluded images as hard examples that are beneficial to train the deep network [5]. Training with too many de-occluded images is harmful to the model's discrimination, which can be validated in the third row of Table 3. If the test images are de-occluded too, the model achieves satisfactory results due to its capability to distinguish person identity on the full-body pedestrian domain. Therefore, de-occluded images can actually help inference and thus our proposed CFMG and PDO modules are beneficial to occluded-ReID problem.

4.5 Mask Evaluation

For further research, we evaluate the accuracy of occlusion masks located by our proposed method. We manually label the occluded masks of 150 images in Partial-REID dataset as ground-truth and then calculate the difference between the ground-truth and our location results using Intersection over Union(IoU) metric. Table 4 presents the accuracy and the speed of our proposed algorithm. The segmentation speed only relies on the MM-detection model while location speed represents the efficiency of our coarse-to-fine approach based on the segmentation output of MM-detection. Table 4 demonstrates that our method achieves a promising performance.

4.6 Visualization

In this section, we show some visualization results of occlusion location and de-occlusion to visually prove the effectiveness of our method. Figure 6 shows 12

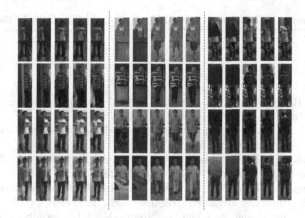

Fig. 6. Examples of occlusion detection and reconstruction.

samples with different identity collected from Partial-REID dataset. We display each person's occluded image together with the corresponding occlusion detection result, which is covered by blue color. Three possible de-occluded images are presented at the following. Samples at the left are occluded by some obstructions on one side of body. The location result is quite precise and we are confident that the de-occluded images are persuasive, because only one side of the body is occluded and the recovery regions are symmetric with the non-occluded regions. The lower body of the samples at the middle are invisible and their recovery regions vary significantly. If the visible parts of body do not provide enough information for image reconstruction, even human cannot judge what is behind the occlusion. Although the color and texture may not be recover, the PDO module in our framework truly reconstruct a normal human body, which somehow gets rid of the disturbance for person feature extractor. Samples on the right show the situation that the target person is occluded on the upper body. The visible parts of these samples can provide some information, such as the color of shirt. Consequently, the possible de-occluded images are more plausible.

5   Conclusion

We propose a novel framework, named Mask Guided De-occlusion (MGD), to address the occluded-ReID problem. The proposed MGD explicitly locates the occlusion and repairs the occluded pedestrian, and thus transfers the partial-to-full person matching problem into a full-to-full matching problem. Given an occluded person image, we use a CFMG module to search for occluded regions. After that, the PDO module is to reconstruct the occluded pedestrian and obtain a full-body image. Finally, the full-body pedestrian images are fed into the PFE module to extract features. Several experiments have been conducted on three public occluded ReID datasets and the experimental results show that our method is effective and outperforms the state-of-the-art methods.

Acknowledgments. This project was supported by the NSFC (61573387, U1611461, 61672544).

References

1. Chang, X., Hospedales, T.M., Xiang, T.: Multi-level factorisation net for person re-identification. In: CVPR, pp. 2109–2118 (2018)
2. Chen, K., et al.: MMDetection: Open MMLab detection toolbox and benchmark. arXiv preprint arXiv:1906.07155 (2019)
3. Feng, Z., Lai, J., Xie, X.: Learning view-specific deep networks for person re-identification. IEEE TIP **27**(7), 3472–3483 (2018)
4. He, L., Liang, J., Li, H., Sun, Z.: Deep spatial feature reconstruction for partial person re-identification: alignment-free approach. In: CVPR, pp. 7073–7082 (2018)
5. Huang, H., Li, D., Zhang, Z., Chen, X., Huang, K.: Adversarially occluded samples for person re-identification. In: CVPR, pp. 5098–5107 (2018)
6. Iizuka, S., Simo-Serra, E., Ishikawa, H.: Globally and locally consistent image completion. ACM Trans. Graph. (ToG) **36**(4), 107 (2017)
7. Li, W., Zhu, X., Gong, S.: Harmonious attention network for person re-identification. In: CVPR, pp. 2285–2294 (2018)
8. Liu, X., et al.: HydraPlus-net: attentive deep features for pedestrian analysis. In: ICCV, pp. 350–359 (2017)
9. Matsukawa, T., Okabe, T., Suzuki, E., Sato, Y.: Hierarchical Gaussian descriptor for person re-identification. In: CVPR, pp. 1363–1372 (2016)
10. Pathak, D., Krähenbühl, P., Donahue, J., Darrell, T., Efros, A.: Context encoders: feature learning by inpainting. In: CVPR (2016)
11. Shen, Y., Lin, W., Yan, J., Xu, M., Wu, J., Wang, J.: Person re-identification with correspondence structure learning. In: ICCV, pp. 3200–3208 (2015)
12. Su, C., Li, J., Zhang, S., Xing, J., Gao, W., Tian, Q.: Pose-driven deep convolutional model for person re-identification. In: ICCV, pp. 3960–3969 (2017)
13. Sun, Y., et al.: Perceive where to focus: learning visibility-aware part-level features for partial person re-identification. In: CVPR, June 2019
14. Sun, Y., Zheng, L., Deng, W., Wang, S.: SVDNet for pedestrian retrieval. In: ICCV, pp. 3800–3808 (2017)
15. Sun, Y., Zheng, L., Yang, Y., Tian, Q., Wang, S.: Beyond part models: person retrieval with refined part pooling (and a strong convolutional baseline). In: Ferrari, V., Hebert, M., Sminchisescu, C., Weiss, Y. (eds.) ECCV 2018. LNCS, vol. 11208, pp. 501–518. Springer, Cham (2018). https://doi.org/10.1007/978-3-030-01225-0_30
16. Szegedy, C., Vanhoucke, V., Ioffe, S., Shlens, J., Wojna, Z.: Rethinking the inception architecture for computer vision. In: CVPR, June 2016
17. Wang, G., Lai, J., Huang, P., Xie, X.: Spatial-temporal person re-identification, pp. 8933–8940 (2019)
18. Wang, G., Yuan, Y., Chen, X., Li, J., Zhou, X.: Learning discriminative features with multiple granularities for person re-identification. In: ACM MM, pp. 274–282. ACM (2018)
19. Xiao, T., Li, H., Ouyang, W., Wang, X.: Learning deep feature representations with domain guided dropout for person re-identification. In: CVPR, pp. 1249–1258 (2016)
20. Yang, C., Lu, X., Lin, Z., Shechtman, E., Wang, O., Li, H.: High-resolution image inpainting using multi-scale neural patch synthesis. In: CVPR, July 2017

21. Zhang, X., et al.: AlignedReID: surpassing human-level performance in person re-identification. arXiv preprint arXiv:1711.08184 (2017)
22. Zhao, R., Ouyang, W., Wang, X.: Unsupervised salience learning for person re-identification. In: CVPR, pp. 3586–3593 (2013)
23. Zheng, C., Cham, T.J., Cai, J.: Pluralistic image completion. arXiv preprint arXiv:1903.04227 (2019)
24. Zheng, W.S., Li, X., Xiang, T., Liao, S., Lai, J., Gong, S.: Partial person re-identification. In: ICCV, pp. 4678–4686 (2015)
25. Zheng, Z., Zheng, L., Yang, Y.: Unlabeled samples generated by GAN improve the person re-identification baseline in vitro. In: ICCV (2017)
26. Zhuo, J., Chen, Z., Lai, J., Wang, G.: Occluded person re-identification. In: ICME, pp. 1–6. IEEE (2018)

Multi-scale Residual Dense Block for Video Super-Resolution

Hetao Cui and Quansen Sun[✉]

Department of Computer Science and Engineering,
Nanjing University of Science and Technology, Nanjing, China
emailcht@qq.com, sunquansen@njust.edu.cn

Abstract. Recent studies on video super-resolution (SR) has shown that convolutional neural network (CNN) combined with motion compensation (MC) is able to merge information from multiple low-resolution (LR) frames to generate high quality images. However, Most SR and MC modules based on deep CNN simply increase the depth of the network, which cannot make full use of hierarchical and multi-scale features, thereby achieving relatively low performance. To address the above problem, a novel multi-scale residual dense Network (MSRDN) is proposed in this paper. A multi-scale residual density block (MSRDB) is first designed to extract abundant local features through dense convolution layer, which helps to adaptively detect image features of different scales with convolution kernels of different scales. Then, we redesign SR module and MC module with MSRDB, which adaptively learns more effective features from local features and uses global feature fusion to jointly and adaptively learn global hierarchical features. Comparative results on Vid4 dataset demonstrate that MSRDB can make more full use of feature information, which helps to effectively improve the reconstruction performance of video SR.

Keywords: Video super-resolution · Optical flow estimation · Convolutional neural network · Multi-scale residual dense network

1 Introduction

Image and Video Super Resolution (SR) aims to restore high resolution (HR) images from their low resolution (LR) counterparts. As one of the long-standing challenges in the field of image processing and computer vision, image and video SR have been extensively investigated for decades [1–3]. Recently, the prevalence of high-definition display further advances the development of SR. For single image SR, image details are restored by intra-frame spatial correlation, and inter-frame temporal correlation can be further used in video SR.

Spatial-temporal correlation is crucial to video SR, How to correctly extract and fuse multi-frame details is the key to this problem. Previous methods [4–6] focus on utilizing optical flow to estimate inter-frame motions. After flow estimation, HR output is reconstructed according to various CNN models. Therefore, these methods integrate SR and MC in a unified framework. Most methods simply increase the depth of the network in order to further improve the performance, while ignoring the full use of hierarchical and multi-scale features in LR images. With the increase of network depth,

© Springer Nature Switzerland AG 2019
Z. Cui et al. (Eds.): IScIDE 2019, LNCS 11935, pp. 424–434, 2019.
https://doi.org/10.1007/978-3-030-36189-1_35

the features gradually disappear in the transmission process. How to make full use of these features is very important for more accurate optical flow estimation and high quality image reconstruction.

In order to make full use of these features from the original LR frames, we propose a novel multi-scale residual dense block (MSRDB) (shown in Fig. 2) for VSR. Firstly, we use the MSRDB to extract image features from different scales with multi-scale convolution, and fuse local feature with dense connected layers. A convolution layer with 1×1 kernel is also introduced to reduce the feature channels. We further design SR and MC modules with MSRDB based on the framework of VESPCN [5] (shown in Fig. 1). Contiguous MSRDBs are utilized to extract the multi-scale features and dense local features, the outputs of each MSRDB are combined for global feature fusion. The combination of local and global features with 1×1 kernel convolution can make fully use of LR frames features.

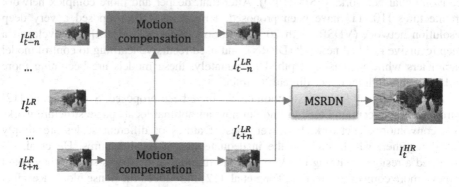

Fig. 1. Framework for video SR Proposed by VESPCN. Our main work is to construct the MC and SR modules with MSRDB.

We construct SR and MC modules with MSRDB based on the framework of VEPCN (shown in Fig. 1), and train the models on the CDVL dataset[1]. The proposed base-model shows superior performance compared with previous methods. Our MSRDB module can be also migrated with other restoration models for feature extraction. Contributions of this paper can be described as follows:

(1) A novel multi-scale residual dense block (MSRDB) is proposed, which helps to achieve feature fusion at different scales, and can also adaptively detect image features. The accumulated features are then adaptively preserved by local feature fusion.

(2) The proposed work is extended to video SR and a new model is designed based on the framework of VESPCN with MSRDB. The performance of proposed model exceed previous methods. Furthermore, MSRDB can be directly extended to other restoration tasks for feature extraction, which can also show promising performance.

---

[1] https://www.cdvl.org/.

2 Related Work

2.1 Single-Image Super-Resolution

Early methods utilized interpolation techniques based on sampling theory, such as linear or bicubic interpolation. These methods run very fast, but they cannot reconstruct detailed and real textures. Recent studies have tended to build an end-to-end CNN model to learn mapping functions from LR to HR images using large training datasets. Dong et al. [7] proposed the pioneering work of using deep learning in single image SR. They used three-layer convolution neural network (CNN) to approximate the non-linear mapping from LR image to HR image. In order to reduce computational complexity caused by pre-processed LR image as input, which was upscaled to HR space via an upsampling operator as bicubic, new methods are proposed, such as Fast Super-Resolution Convolutional Neural Networks (FSRCNN [8]) and Efficient Sub-pixel Convolutional Networks (ESPCN [9]). After that, deeper and more complex network architectures [10, 11] have been proposed. Kim et al. [10] proposed a very deep resolution network (VDSR) with 20 convolution layers. Tai et al. [11] developed a deep recursive residual network (DRRN), and used recursive learning to control model parameters while increasing depth. Unfortunately, these models are becoming more and more in-depth and very difficult to train.

At present, many feature extraction blocks have been proposed. Szegedy et al. [12] introduced the inception block to find out how an optimal local sparse structure works in a convolutional network. However, these features of different scales are simply linked together, which leads to the inadequate use of local features. He et al. [9] proposed a residual learning block to ease the training of networks so that they could achieve more competitive results. Tong et al. [12] introduced the dense block. Residual block and dense block use a single size of convolutional kernel. In order to make full use of image features, we propose a multi-scale residual dense block.

2.2 Video Super-Resolution

In order to exploit pixel-wise correspondences, traditional methods [4, 6] use optical flow to compensate inter-frame motion and iterative framework to estimate HR images. Recently, deep learning has been investigated for video SR. Liao et al. [4] performed motion compensation to generate an ensemble of SR-drafts, and then employed a CNN to recover high-frequency details from the ensemble. Kappeler et al. [16] also performed image alignment through optical flow estimation, and then passed the concatenation of compensated LR inputs to a CNN to reconstruct each HR frame. In these methods, motion compensation is separated from SR model. Therefore, it is difficult for them to obtain the overall optimal solution.

Nowadays, various works aim to establish Deep video SR models with integrated motion compensation. Caballero et al. [5] proposed the first end-to-end CNN framework (VESPCN) for video SR. It comprises a motion compensation module and a sub-pixel convolutional layer used in ESPCN [9]. Since that, end-to-end framework with motion compensation dominates the research of video SR. Tao et al. [15] used the motion estimation module in VESPCN, and proposed an encode-decoder network based on LSTM. This architecture facilitates the extraction of temporal context. Liu et al. [17] customized ESPCN [8] to simultaneously process different numbers of LR frames. They then introduced a temporal adaptive network to aggregate multiple HR estimates with learned dynamic weights. Sajjadi et al. [18] proposed a frame recurrent architecture to use previously inferred HR estimates for the SR of subsequent frames. The recurrent architecture can assimilate previous inferred HR frames without increase in computational demands. It is already demonstrated by traditional video SR methods [4, 6] that simultaneous SR of images and optical flows produces better result. However, more multi-level, multi-scale features are not extracted from the original LR image. we redesign the SR module and MC module with MSRDB under the framework of VESPCN. It is demonstrated that the multi-scale features and feature fusion facilitates our network to achieve the better performance.

3 Proposed Method

The proposed framework is shown in Fig. 1. Our model take $2n + 1$ consecutive LR frames as inputs and can obtain the HR image of central frame. Firstly, $2n$ adjacent frames are compensated the inter-frame motion with the center frame by MC module. Then, the center frame and $2n$ compensated adjacent frames after motion compensation are fed into SR module to infer the HR image of the center frame. In this section, we first introduce the feature extraction module of multi-scale residual dense block (MSRDB), then present the proposed MC and SR module.

3.1 Multi-scale Residual Dense Block (MSRDB)

In order to detect the image features at different scales, we propose multi-scale residual dense block (MSRDB). As shown in Fig. 2, our MSRDB contains three parts: multi-scale features fusion, dense features fusion and local residual learning.

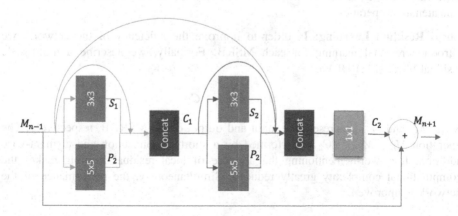

Fig. 2. The structure of multi-scale residual dense block (MSRDB).

Multi-scale Features Fusion: we construct a two-bypass network and each bypass use different convolutional kernel. In this way, the information between those bypass can be shared with each other, which helps to detect the image features at different scales. The operation can be defined as:

$$S_1 = H_{3 \times 3}^{ReLU}(M_{n-1}) \tag{1}$$

$$P_1 = H_{5 \times 5}^{ReLU}(M_{n-1}) \tag{2}$$

$$S_2 = H_{3 \times 3}^{ReLU}(C_1) \tag{3}$$

$$P_2 = H_{5 \times 5}^{ReLU}(C_1) \tag{4}$$

where M_{n-1} represent the input of the MSRDB. $H_{3 \times 3}^{ReLU}(\cdot)$ denotes convolution operation, and the superscripts ReLU represent the activation function Rectified Linear Unit, while the subscripts 3×3 and 5×5 represent the size of the convolutional kernel used in the layer. S_i and P_i denotes the output of 3×3 and 5×5 convolution, and the subscripts i represent the ith convolution layer. $\sigma(x) = max(0, x)$ stands for the ReLU function. Unless otherwise specified, we use ReLU functions as the default activation function in this paper.

Dense Features Fusion: To preserves the feed-forward nature and extracts local dense feature, Dense blocks is applied to adaptively fuse the states from the whole convolutional layers in MSRDB. The operation can be defined as:

$$C_1 = [M_{n-1}, C_1, S_2, P_2] \tag{5}$$

$$C_2 = H_{5 \times 5}^{none}([M_{n-1}, C_1, S_2, P_2]) \tag{6}$$

where $[M_{n-1}, C_1, S_2, P_2]$ and $[M_{n-1}, C_1, S_2, P_2]$ denote the concatenation operation, and *none* mean this layer has no activation function. C_1 and C_2 represent the result of concatenation operation.

Local Residual Learning: In order to improve the efficiency of the network, we introduce residual learning for each MSRB. Formally, we describe a multi-scale residual block (MSRDB) as:

$$M_n = M_{n-1} + C_2 \tag{7}$$

where M_n and M_{n-1} represent the input and output of the MSRDB, respectively. The operation $M_n = M_{n-1} + C_2$ is performed by a shortcut connection and element-wise addition. It is worth mentioning that the use of local residual learning makes the computational complexity greatly reduced. Simultaneously, the performance of the network is improved.

3.2   Motion Compensation Module

For input of $2n + 1$ consecutive LR frames,MC module need compensate the inter-frame motion of each adjacent frame and center frame. Let I_{t-i} and I_t be each adjacent frame and center frame, where $i = [-n, \ldots, -1, 1, \ldots, n]$. The task of optical flow Network is to find the best optical flow representation relating I_{t-i} with I_t. Optical flow is represented with two feature maps corresponding to displacements for the x and y dimensions. Let $\Delta = (\Delta^x, \Delta^y)$ denote the Optical flow. Then we use bilinear interpolation $\Gamma\{\cdot\}$ for motion compensation, a compensated image can be expressed as $I'_{t-i}(x, y) = \Gamma\{I_t(x + \Delta^x, y + \Delta^y)\}$, or more concisely:

$$I'_{t-i} = \Gamma\{I'_{t-i}(\Delta)\} \tag{8}$$

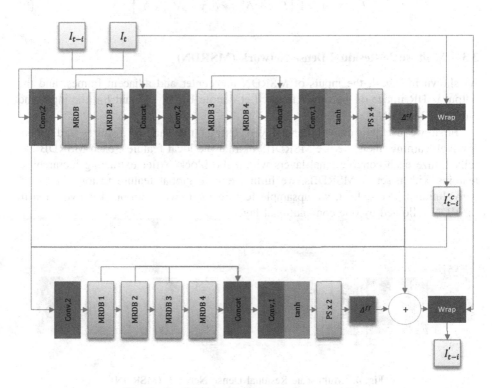

Fig. 3. Spatial transformer motion compensation.

A coarse-fine scale strategy is designed to represent the flow, which has been shown to be effective in classical methods [19, 20] and more recently proposed spatial transformer techniques [21, 22]. A schematic of the design is shown in Fig. 3. First, coarse optical flow Δ^{cf} obtained by from the coarse flow estimation module, and Δ^{cf} is applied to warp the target frame producing I'^c_{t-i}. The warped image is then processed together with the coarse flow and the original images through a fine flow estimation module. The result is finer flow map Δ^{ff}. The final motion compensated frame is obtained by warping the target frame with the total flow $I'_{t-i} = \Gamma\{I'_{t-i}(\Delta^{cf} + \Delta^{ff})\}$. The coarse and fine flow estimation module are designed with MSRDB, these use strided convolution with stride 2, obtain the ×4 coarse estimate and ×2 fine estimate by fusing the local features via MSRDB. The estimated flows are then upscaled with sub-pixel convolution and results are Δ^{cf} and Δ^{ff}.

Output activations use tanh to represent pixel displacement in normalised space, where a displacement of ±1 means the maximum displacement from the center to the border of the image. However, if maximum displacement is too large, optical flow matrix fluctuates greatly, so it is difficult to converge. Therefore, we limit maximum displacement to $[-\delta, \delta]$ with $\delta = 2$:

$$I'_{t-i}(x,y) = \Gamma\{I_t(x + \Delta^x_{t-i} * \delta, y + \Delta^y_{t-i} * \delta)\} \tag{9}$$

3.3 Multi-scale Residual Dense Network (MSRDN)

As shown in Fig. 4, the inputs of MSRDN are center and adjacent frames, and the output is HR image of the center frame. Let denote $I_{LR\{-n,...,n\}}$ and I_{HR} the input and output of MSRDN.

First, we use two convolutional layers extract the shallow features and global residual learning. then, we use MSRDBS extract the local features, each MSRDB can fully utilize each convolutional layers within the block. After extracting hierarchical features with a set of MSRDBs, we further excute global feature fusion via 1×1 convolutional layers. Last, we upsample feature maps with sub-pixel convolution in MSRDN followed by one convolutional layer.

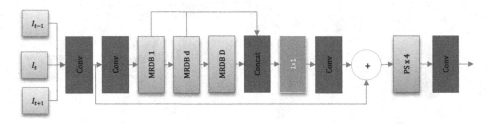

Fig. 4. Multi-scale Residual Dense Network (MSRDN)

3.4 Loss Function

Similar to [5], the loss function $\mathcal{L}$ consists of three terms, described as:

$$L = L_{sr} + \beta L_{mc} + \lambda L_{Huber} \tag{10}$$

where $\mathcal{L}_{sr}$ is the mean square error (MSE) loss between the output HR images of MSRDN and the ground-truth, $\mathcal{L}_{mc}$ is the mean square error (MSE) loss for between the center LR frame and each Adjacent frame, and $\mathcal{L}_{Huber}$ is the Huber loss of the flow map gradients, namely:

$$L_{sr} = \|I_{GT} - I_{HR}\|_2^2 \tag{11}$$

$$L_{mc} = \sum_{i=\pm n} \|I_t - I'_{t-i}\|_2^2 \tag{12}$$

$$L_{Huber} = \sum_{i=\pm n} \sqrt{\epsilon + \sum_{i=x,y} (\partial_x \Delta i^2 + \partial_x \Delta i^2)} \tag{13}$$

where I_{GT} is the ground-truth of HR image, I_{HR} is the output of MSRDN. I_t is the center LR frame, I'_{t-i} is each Adjacent frame, Δi is the flow map and β, λ, ϵ are the parameters to balance the three loss terms.

4 Experiments

4.1 Datasets

We collected 108 1080P HD video clips from the CDVL database. The collected videos cover diverse natural and urban scenes. We used 100 videos as the training set, and other 8 videos as the validation set. We down-sample the video clips to the size of 960×540 as the HR ground-truth using open source software FFmpeg. For fair comparison with state-of-the-arts, we chose the widely used Vid4 bench-mark for further comparison.

4.2 Implementation Details

Following [7], we converted input LR frames into YCbCR color space and only fed the luminance channel to our network. All metrics in this section are computed in the luminance channel. During the training phase, we randomly extracted $2n + 1(n = 1)$ consecutive frames from an HR video clip. The LR images were obtained by down-sampling the HR images using bicubic kernel with a scale factor of 3x or 4×, and randomly cropped a 32×32 patch as the input. Meanwhile, its corresponding patch in HR video clip was cropped as ground-truth.

We implemented our framework in PyTorch, and applied the Adam solver with $\beta_1 = 0.9$, $\beta_2 = 0.99$ and batch size of 32. The parameters of loss functions $\beta = 1$, $\lambda = 0.01$ and $\epsilon = 0.01$. The learning rate is initialized to 0.0001 for all layers and decreases half for every 20K iterations. We trained our network from scratch with 100K iterations. All experiments were conducted on a PC with an Nvidia GTX 1080Ti GPU.

4.3 Experimental Analyses

Comparisons to Previous Methods: We compared our mode with Bicubic [23], SRCNN [7], VDSR [10], VSRnet [16] and VESCPN [5] on the Vid4 dataset. Taking the equality of comparison into account, we evaluate the SR images with two commonly-used image quality metrics: peak signal-to-noise ratio (PSNR) and structural similarity index (SSIM [24]).

Table 1. Comparison with state-of-the-art approaches on Vid4 videos.

Scale		Bicubic [23]	SRCNN [7]	VSRnet [16]	VESPCN [5]	Ours
X3	PSNR	25.39	26.54	26.62	27.22	**27.64**
	SSIM	0.7621	0.8192	0.8228	0.8441	**0.8514**
X4	PSNR	23.78	24.65	24.41	25.31	**25.69**
	SSIM	0.6551	0.7162	0.7374	0.7523	**0.7638**

Quantitative results obtained on the Vid4 dataset are shown in Table 1. It can be observed that our method achieve the best performance for the BI degradation model in terms of all metrics. Specifically, the PSNR and SSIM values achieved by our model are higher than other methods with over 0.4 dB and 0.0115 dB. Compared with the VSEPCN method, the proposed method obtains significant improvement in PSNR and SSIM values.

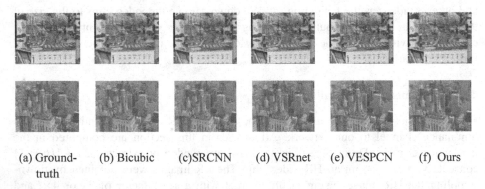

(a) Ground-   (b) Bicubic   (c)SRCNN   (d) VSRnet   (e) VESPCN   (f) Ours
truth

Fig. 5. Visual comparison of ×4 SR results on Calendar and City. Zoom-in regions is in the lower left corner. From left to right: Bicubic [23], SRCNN [7], VDSR [10], VSRnet [16], and VESCPN [5] and Our methods.

Figure 5 illustrates the qualitative results on two scenarios in the Vid4 dataset. From the zoom-in regions, it can be seen that our method can recover finer and more reliable details, such as the lines of the calendar and the stripes of the building, which demonstrates the superior performance of our method.

Convergence Speed of Different Maximum Displacements: Since maximum displacement is from the center to the border of the image, optical flow matrix fluctuates greatly, so it is difficult to converge. As shown in Fig. 6, when $\delta = 2$, the convergence rate of loss function is much faster than when $\delta = H/W$. where H and W is the height and width of LR images.

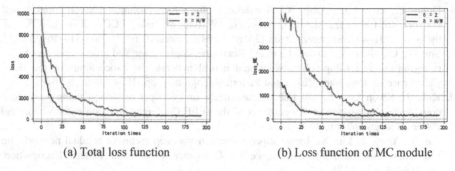

(a) Total loss function   (b) Loss function of MC module

Fig. 6. Convergence curves of different maximum displacements. The green line is $\delta = H/W$, and the red lines is $\delta = 2$. (a) is the total loss function of proposed model, and (b) is the loss function of MC module. (Color figure online)

5 Conclusions

In this paper, we proposed an efficient multi-scale residual Dense block (MSRDB), which is used to adaptively detect the image features at different scales and feature fusion. Based on MSRDB, we redesign the SR module and Optical Flow Estimation module in VESPCN framework. It is a simple and efficient VSR model so that we can fully utilize the local multi-scale features and the hierarchical features to obtain accurate SR image.

References

1. Fattal, R.: Image upsampling via imposed edge statistics. ACM Trans. Graph. (TOG) **26**(3), 95 (2007)
2. Freedman, G., Fattal, R.: Image and video upscaling from local self-examples. ACM Trans. Graph. (TOG) **30**(2), 12 (2011)
3. Nguyen, N., Milanfar, P., Golub, G.: A computationally efficient superresolution image reconstruction algorithm. IEEE Trans. Image Process. **10**(4), 573–583 (2001)
4. Liao, R., Tao, X., Li, R., et al.: Video super-resolution via deep draft-ensemble learning. In: Proceedings of the IEEE International Conference on Computer Vision, pp. 531–539 (2015)
5. Caballero, J., Ledig, C., Aitken, A., et al.: Real-time video super-resolution with spatio-temporal networks and motion compensation. In: Proceedings of the IEEE Conference on Computer Vision and Pattern Recognition, pp. 4778–4787 (2017)
6. Ma, Z., Liao, R., Tao, X., et al.: Handling motion blur in multi-frame super-resolution. In: Proceedings of the IEEE Conference on Computer Vision and Pattern Recognition, pp. 5224–5232 (2015)
7. Dong, C., Loy, C.C., He, K., Tang, X.: Learning a deep convolutional network for image super-resolution. In: Fleet, D., Pajdla, T., Schiele, B., Tuytelaars, T. (eds.) ECCV 2014. LNCS, vol. 8692, pp. 184–199. Springer, Cham (2014). https://doi.org/10.1007/978-3-319-10593-2_13

8. Dong, C., Loy, C.C., Tang, X.: Accelerating the super-resolution convolutional neural network. In: Leibe, B., Matas, J., Sebe, N., Welling, M. (eds.) ECCV 2016. LNCS, vol. 9906, pp. 391–407. Springer, Cham (2016). https://doi.org/10.1007/978-3-319-46475-6_25
9. Shi, W., Caballero, J., Huszár, F., et al.: Real-time single image and video super-resolution using an efficient sub-pixel convolutional neural network. In: Proceedings of the IEEE Conference on Computer Vision and Pattern Recognition, pp. 1874–1883 (2016)
10. Kim, J., Kwon Lee, J., Mu Lee, K.: Accurate image super-resolution using very deep convolutional networks. In: Proceedings of the IEEE Conference on Computer Vision and Pattern Recognition, pp. 1646–1654 (2016)
11. Tai, Y., Yang, J., Liu, X.: Image super-resolution via deep recursive residual network. In: Proceedings of the IEEE Conference on Computer Vision and Pattern Recognition, pp. 3147–3155 (2017)
12. Szegedy, C., Liu, W., Jia, Y., et al.: Going deeper with convolutions. In: Proceedings of the IEEE Conference on Computer Vision and Pattern Recognition, pp. 1–9 (2015)
13. He, K., Zhang, X., Ren, S., et al.: Deep residual learning for image recognition. In: Proceedings of the IEEE Conference on Computer Vision and Pattern Recognition, pp. 770–778 (2016)
14. Tong, T., Li, G., Liu, X., et al.: Image super-resolution using dense skip connections. In: Proceedings of the IEEE International Conference on Computer Vision, pp. 4799–4807 (2017)
15. Tao, X., Gao, H., Liao, R., et al.: Detail-revealing deep video super-resolution. In: Proceedings of the IEEE International Conference on Computer Vision, pp. 4472–4480 (2017)
16. Kappeler, A., Yoo, S., Dai, Q., et al.: Video super-resolution with convolutional neural networks. IEEE Trans. Comput. Imaging 2(2), 109–122 (2016)
17. Liu, D., Wang, Z., Fan, Y., et al.: Robust video super-resolution with learned temporal dynamics. In: Proceedings of the IEEE International Conference on Computer Vision, pp. 2507–2515 (2017)
18. Sajjadi, M.S.M., Vemulapalli, R., Brown, M.: Frame-recurrent video super-resolution. In: Proceedings of the IEEE Conference on Computer Vision and Pattern Recognition, pp. 6626–6634 (2018)
19. Brox, T., Bruhn, A., Papenberg, N., Weickert, J.: High accuracy optical flow estimation based on a theory for warping. In: Pajdla, T., Matas, J. (eds.) ECCV 2004. LNCS, vol. 3024, pp. 25–36. Springer, Heidelberg (2004). https://doi.org/10.1007/978-3-540-24673-2_3
20. Farnebäck, G.: Two-frame motion estimation based on polynomial expansion. In: Bigun, J., Gustavsson, T. (eds.) SCIA 2003. LNCS, vol. 2749, pp. 363–370. Springer, Heidelberg (2003). https://doi.org/10.1007/3-540-45103-X_50
21. Dosovitskiy, A., Fischer, P., Ilg, E., et al.: FlowNet: learning optical flow with convolutional networks. In: Proceedings of the IEEE International Conference on Computer Vision, pp. 2758–2766 (2015)
22. Ahmadi, A., Patras, I.: Unsupervised convolutional neural networks for motion estimation. In: 2016 IEEE International Conference on Image Processing (ICIP), pp. 1629–1633. IEEE (2016)
23. De Boor, C.: Bicubic spline interpolation. J. Math. Phys. 41(1–4), 212–218 (1962)
24. Wang, Z., Bovik, A.C., Sheikh, H.R., et al.: Image quality assessment: from error visibility to structural similarity. IEEE Trans. Image Process. 13(4), 600–612 (2004)

Visual Saliency Guided Deep Fabric Defect Classification

Yonggui He[1], Yaoye Song[1], Jifeng Shen[1(✉)], and Wankou Yang[2]

[1] School of Electronic and Informatics Engineering, Jiangsu University,
Zhenjiang 212013, Jiangsu, China
shenjifeng@ujs.edu.cn
[2] School of Automation, Southeast University, Nanjing 210096, Jiangsu, China

Abstract. Fabric defects have an important influence on the quality of the
fabric product. Automatic fabric defect detection is a crucial part for quality
control in the textile industry. The primary challenge of fabric defects identification is not only to find the existing defects, but also to classify them into
different types. In this paper, we propose a novel fabric defect detection and
classification method consists of three main steps. Firstly, the fabric image is
cropped into a set of image patches and each patch is labeled with specified
defect type. Secondly, the visual saliency map is generated from the patch to
localize defects with specified visual attention. Then, the combination of visual
salience map with raw image input into a convolutional neural network for
robust feature representation, and finally output its predicted defect type. During
the testing section, defect inspection runs in a sliding window schemes using the
trained model, and both the type and position of each defect are obtained
simultaneously. Our method tries to investigate the combination of visual saliency and one-stage object detector with feature pyramid, which fully makes use
of information from multi-resolution guided with visual attention. Besides, soft-
cutoff loss is employed to further improve the performance of the method, and
our network can be learnt in an end-to-end manner. Experiments based on our
fabric defect image datasets, the proposed method can achieve a 98.52%
accuracy of classification. This method is comparable to the usual two-stage
detector with more compact model parameters, makes it valuable in the
industrial application.

Keywords: Fabric defect detection · Visual saliency map · Convolutional
neural networks

1 Introduction

Fabric defect detection is one of the most important problems of the textile industries.
The defects produced by the loom directly determine the grade of the fabric. There are
approximately 40 identified categories of fabric defects defined by the textile industries,
such as barre, bad place, broken end, bias. Surface defect detection based on computer
vision has been widely used in defect detection systems. Traditional methods for
automatic fabric defect detection can be divided into statistical-based, spectroscopy-
based, model-based, learning-based, and structure-based methods. In recent years,

© Springer Nature Switzerland AG 2019
Z. Cui et al. (Eds.): IScIDE 2019, LNCS 11935, pp. 435–446, 2019.
https://doi.org/10.1007/978-3-030-36189-1_36

convolutional neural networks based deep learning has been widely used for target detection [4–6] which makes a deep learning based fabric defect detection more attractive. Compared with the methods using hand-crafted feature set, deep learning based detection algorithms are able to automatically generate distinct features from the training set, free the users from manually identify rules for classification, and in general are able to achieve better detection accuracy [15–17]. In recent years, as a mainstream of research in machine learning, Convolutional Neural Networks (CNNs) based deep learning architectures have achieved tremendous success in the field of object recognition and have proven its suitability for automated detection of fabric defects.

Three main factors make the task of fabric defect detection more difficult, the first is the wide variety of fabrics, whether it is material or pattern, complex structures and fine patterns make the detection of small defects extremely challenging [14]. Algorithms are needed to find representative features on different scales. The second difficulty is that the actual production environment is complex, and the algorithm needs to be extremely robust in the environment such as illumination. Finally, for industrial needs, fabric defect detection must be performed in real time.

Inspired by visual saliency [28], combining the features of multi-scale deep convolutional neural networks [7], we have proposed a lightweight neural network (VMNet) architecture that focuses on subtle and hard samples and improves classification accuracy. Further, to better distinguish similar categories, apart from minimising the cross-entropy loss, we introduce the Focalloss [8] to detect hard defects.

Our main contributions can be summarized as follows:

(1) We propose a novel VMNet to address the problem of hard samples in the fabric defect detection task, achieving multi-scale features.
(2) We generate visual saliency map related to the original image as priors for the classification task, and make visual saliency map an explicit input component of the end-to-end training for the first time, aiming to force the network to focus attention on the most discriminative parts.
(3) To enhance the discriminative power, we introduce a new loss function to reduce the impact of interclass similarity.

The rest of this paper is organized as follows. Section 2 gives a brief review of related works. Section 3 describes the proposed method in detail. Section 4 introduces the experimental setup and presents results. The paper concludes in Sect. 5 with a summary and an outlook.

2   Related Work

In this section, we will briefly review the development of fabric defect detection, and describe the differences between existing methods and the methods we proposed in detail.

2.1   Regular Method

The early works mainly focus on handcrafted features and emerge a series of different types of approaches, including Gabor filtering [11], wavelet transform [12] and so on. These methods have been proven successful in a wide variety of computer vision tasks. However, the limitation of these methods is that they rely on low-level features and involve abundant engineering skills. To make up this deficiency of the above methods, Karlekar et al. [10] proposed a fabric texture detection method combining wavelet transform and morphological algorithm, which achieved good results. The commonly used learning-based method is dictionary learning, using sparse coding method to reconstruct the dictionary set of ordinary fabric images, and then make a difference with the test image, and finally get the defective area [20, 21]. In view of the extensive application of artificial neural network in the field of image target detection, in-depth learning has attracted extensive attention in image texture analysis. Yapi et al. [23] introduced the supervised learning method into fabric defect detection, and trained Bayesian classifier to learn fabric defect features using normal fabric image and fabric defect image. Seker et al. [24] proposed a deep learning model based on automatic coding for fabric defect detection.

All the methods mentioned above, whether using handcrafted features or automatically learning feature via supervised and unsupervised method, not aware that only discriminative regions are essential for identify subtle defects in the entire image area.

2.2   Visual Saliency

Inspired by the neural structure and behavior of the early visual system of early primates, Itti et al. [28] designed a visual attention model. Recently, some research have attempted to extract visual saliency map in detection tasks. The visual attention map is an intensity map aims to highlight salient areas or objects in the image. High intensity represents a region or object that causes human visual attention. In the field of object detection, visual saliency detection is one of the important topics, and there have been many improvements [28–30]. Li et al. [32] proposed a saliency detection method based on frequency domain scale spatial analysis, suppress the repetitive part to highlight the salient area, which easily leads to the loss of details. Yan et al. [25] proposed a multi-layer approach to analyze high contrast regions. Hou et al. [33] extracted the amplitude and phase spectra of the image, preserve the phase spectrum, and change the amplitude spectrum. Attention to existing approaches, our method also takes advantage of the visual saliency map. However, it is completely different from the above method, we only use visual saliency map as supplement of the input.

2.3   Convolutional Neural Network

There are several architecture types in the field of Convolutional Networks. Such as AlexNet [9], ZF Net [13], GoogLeNet [18], VGGNet [19], and MobileNetV2 [22]. They are winners of the ImageNet Large Scale Visual Recognition Challenge (ILSVRC), which is the most famous competition in computer vision. Much progress has been made in recent years on object detection thanks to the use of convolutional neural networks. Popular modern object detectors are Faster R-CNN [4], Mask RCNN [5], SSD [6]. Faster R-CNN and R-FCN are two-stage, proposal-driven detectors that are slower but much more accurate, while the SSD is one-stage detector that is much faster but not as accurate as two-stage detectors. To match the real-time requirement of this project, we decide to use one-stage detector. Except SSD in one-stage detector, there is a new architecture called FPN (Feature Pyramid Network) [7]. We tried both SSD and FPN. After experiments, we decide to use FPN, backbone on top of a feed forward MobileNet model to generate a rich, multi-scale convolutional feature pyramid.

3   Method

As shown in Fig. 1, the proposed VMNet consists of two parts: a visual saliency map generation and a convolutional neural network framework. In this section, we will elaborate each of them in detail.

3.1   Overall Architecture

Figure 1 shows our structure. The left part demonstrates the process of generating visual saliency map. Then we combine the original image with the visual saliency map by point multiplication, using the result as a new input of CNN model. The right half of our framework (Fig. 1) shows our CNN model structure.

3.2   Visual Saliency Map

Most of the saliency map in the current literature are based on scale-space or superpixel [31, 32], they have high computational complexity and hard to retain information of tiny objects. In this paper, we resort to random Region of Interest (ROI) saliency map [26] to generate saliency map of both training and test datasets. In complex lighting conditions, it is necessary to ensure the robustness of color and illumination. According to the characteristics of the tiny defect, sampling the image into random regions of interest (ROI) is more stable to calculate the saliency map.

In order to obtain the grey-scale image, which one-to-one correspondence with the original image, the specific implementation process is as follows. We first convert RGB images to the Lab color space. Then randomly generate n windows for each channel, calculate the ratio of area to grayscale sum in each channel:

$$mean_i = sum_i / Area_i \tag{1}$$

In Eq. 1 $mean_i$ represents the ratio of the ith ROI in current channel, sum_i and $Area_i$ represent the sum of gray value and area of the current ROI.

Then calculate the significance mapping of each pixel $I_{j,k}$ in the window according to the following formula:

$$SM_{j,k} = SM_{j,k} + |I_{j,k} - mean_i| \tag{2}$$

Where (j, k) denotes the current pixel of the *ith* ROI. The final saliency map uses the euclidean distance to fuse the significant values of the color space.

$$FM_{m,n} = \sqrt{LSM^2_{m,n} + ASM^2_{m,n} + BSM^2_{m,n}} \tag{3}$$

(m, n) is the size of the image. LSM, ASM, BSM are saliency maps of three channels.

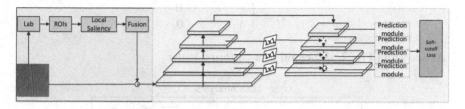

Fig. 1. Overview of our framework, which consists of (1) extracting visual saliency map, (2) CNN model, blending the original image with the saliency map and combining the new loss function.

3.3 Loss Function

Generally speaking, in classification tasks, cross entropy is the most frequently used loss function which can measure the difference in probability distribution, the cross entropy function is presented as follows.

$$L_{ce} = - \sum_{i=1}^{n} y_i \log y' \tag{4}$$

In Eq. 4, y_i is expected output and y' is predictive value.

Although softmax loss can directly solve classification problems, the features of deep learning tend to be separable but not discriminative. Therefore, it is not suitable to directly use these features to identify. In order to enhance the discriminative power of deep learning in neural networks and deal with hard samples of one-stage detector, Lin et al. [8] proposed focalloss to improve the accuracy. Focal loss is mainly to solve the simple-difficult sample imbalance in one-stage target detection, which is based on the cross entropy loss function. For the two classification tasks, as formulated in Eq. 5

$$L_{fl} = \begin{cases} -\alpha(1 - y')^\gamma \log y', & y = 1 \\ -(1 - \alpha)y'^\gamma \log(1 - y'), & y = 0 \end{cases} \tag{5}$$

In Eq. 5, α *and* γ are hyperparameters, y' is the probability of ground truth class.

Although focal loss can focus training on hard samples in fabric defect detection, simply applying focal loss will result in a decrease in classification performance. In order to improve inter-class discriminability, we proposed soft-cutoff loss function.

This helps to reduce confusion between hard and simple samples when they are correctly detected.

$$L_{sc} = -\sum_{i=1}^{m} [\theta(0.9 - \frac{e^{W_{y_i}x_i + b_{y_i}}}{\sum_{j=1}^{n} e^{W_j x_i} + b_j})]^{\gamma} \log \frac{e^{W_{y_i}x_i + b_{y_i}}}{\sum_{j=1}^{n} e^{W_j x_i} + b_j} \tag{6}$$

In Eq. 6, $x_i \in R^d$ represents the *ith* feature, and y_i reprents the *ith* class. $W_j \in R^d$ donates the *jth* row of the weights calculated in the last fully connected layer and b is the bias term. m is the size of mini-batch and is the number of class.

$$\theta(x) = \begin{cases} x & x > 0 \\ 1 & x \leq 0 \end{cases} \tag{7}$$

As shown in Fig. 2, soft-cutoff loss performs better.

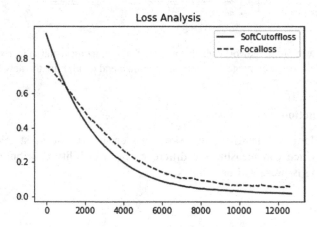

Fig. 2. Analysis of focalloss and soft-cutoff loss. (x-axis denotes the Iteration number and y-axis is the training loss)

4   Experiments

4.1   Implementation Details

For part 1, we generate random ROI to produce visual saliency map consistent with the size of the raw image. For part 2, we experiment the MobilenetV2+FPN architecture. The original image and visual saliency map are point multiplicated as the input of CNN model. We use the ADAM optimizer [27] by setting β1 = 0.9, β2 = 0.999 to train the networks, and terminate after 40 epochs. We implement the proposed models via Keras framework and using NVIDIA Titan V GPUs for training acceleration.

4.2   Data Set Description

The dataset we collected and labeled from the factories includes three kinds of illumination (Green, Yellow and White) of 12 classes, shown in Table 1. The size of the image, a pixel resolution of the industrial camera in the RGB color space is 256×256 pixels. The high within class diversity and the subtle make the data set more challenging. Some examples are shown in Fig. 3. The Green Dataset includes 160000, each class consists of 12000 images of 256×256 pixels. The Yellow Dataset consists of 9,4000 images and the White Dataset includes about 23000 images.

4.3 Experimental Setting

(1) Dataset setting and Evaluation Metrics: We randomly select 80% of the data for training and the rest for testing. Data augmentation is performed during training including setting contrast, saturation, standardization and rerandomly rotated.

Table 1. 12 classes of fabric defects

Barre	Bad place	Bias	Broken end
Broken pick	Coarse end	End out	Fine end
Hole	Knot	Blot	Missing yarn

The most commonly used quantitative evaluation indicators for evaluating classification performance are overall accuracy and confusion matrix. The overall accuracy represents a sample that is correctly detected, regardless of the specific category. The confusion matrix is a situation analysis table that summarizes the prediction results of the classification model in machine learning. The records in the data set are summarized in a matrix form according to the real category and the classification criteria predicted by the classification model. Where the rows of the matrix represent the true values and the columns of the matrix represent the predicted results. In order to obtain reliable experimental results, we repeat the experiment 10 times for each training-test ratio and show mean and standard deviation.

(2) Parameter Setting: Our proposed VMNet method is based on MobilenetV2 and FPN. Learning rate is an important parameter in the end-to-end learning process. For the CNN model, the learning rate is 0.0001, and for every ten epoch, the learning rate is reduced by 10%. For the loss function, the scalar gamma = 2 follows the setting in [8].

4.4 Experiments and Analysis

(1) Generate Visual Saliency Map: Fig. 4 illustrates the saliency map results by [26]. From the RGB-saliency mask, we can see that the VMnet can learn features well from the target images. And the visual saliency map shows the weights of different regions that have different effects on network output decisions.

(2) Ablation study: We analyzed the effects of different components on the frame by experimenting with different settings on the data sets of the three sources and report the results in Table 2. We use VM to represent a model that uses visual saliency, and SC to represent soft-cutoff loss. By comparing the performance of

MobilenetV2, VMNet, VMNet+SC, we found that both improvements have contributed greatly to the system. As shown in Table 2, our method is the best on all three data sets.

(3) Confusion Matrices: In addition to comparing the overall accuracy of the various above algorithms, to further verify distinction performance of our algorithm in different categories, we also compute the ecorresponding confusion matrix, as shown in Fig. 5, the entry in the *ith* row, the *jth* column represents the rate of test

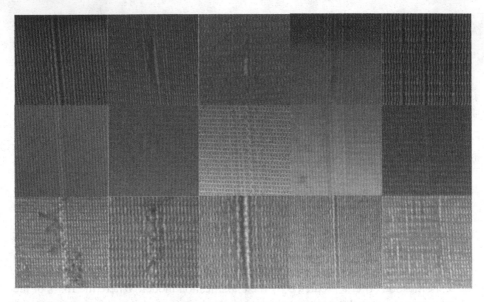

Fig. 3. Partial defects in the green, yellow, and white light data sets. The types of defects are complex and subtle defects are difficult to find. (Color figure online)

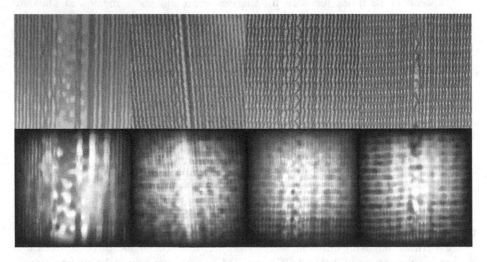

Fig. 4. Visual Saliency Mask results. The first line is the ground truth of the input image, and the second line is the visual saliency map.

images from the *ith* class that are classified into the *jth* class. By analyzing the confusion matrix on our method, we can find that the number of misclassified categories is relatively reduce.

(4) Comparisons with other methods: Since the VMnet+SC use visual saliency map to enhance the representation ability of the CNN features. We mainly compare our method with CNN feature-based methods including Yolo, Faster RCNN, Mask RCNN. We report the mean classification accuracy (AC) and standard deviation (STD) of the proposed method and above mentioned state of the art methods in Table 3. Our approach provides superior performance compared to existing methods.

Table 2. Overall accuracies (%) of three different setting datasets on different components

Dataset	Components	Accuracy
GreenLight	MobileNetV2	91.22 ± 0.16
	VMNet	94.18 ± 0.13
	VMNet+SC	98.52 ± 0.08
YellowLight	MobileNetV2	90.11 ± 0.14
	VMNet	92.05 ± 0.13
	VMNet+SC	96.24 ± 0.09
WhiteLight	MobileNetV2	90.65 ± 0.07
	VMNet	91.17 ± 0.08
	VMNet+SC	92.54 ± 0.08

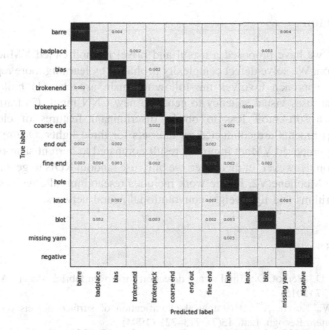

Fig. 5. Confusion matrices of the Green data set under the training ratio of 80% using VMnet+SC.

5 Discussion

In this paper, for the three major challenges of fabric detection, namely, the complexity of the fabric, the diversity of the defects, and the tiny flaws, we proposed a deep learning discrimination method using visual saliency map (VMnet). From the above experimental results, our method can learn better salient features, reduce visual confusion, and improve the performance of defect classification. Our results confirm that visual saliency map using random ROI extraction has great guiding significance for defect classification. Although our research shows important discoveries, there are also limitations. Identifying flaws on patterned fabrics is extremely challenging. And light also affects the detection. We will research on more robust algorithms for complex cloth and try other existing algorithms to generate a lighter framework in the future.

Table 3. Fabric defect classification results (ac%±std%) using the Green data set.

Method	Accuracy
SSD	89.53 ± 0.87
YoloV2	89.17 ± 1.12
Tiny-Yolo	88.69 ± 1.95
Faster RCNN	93.42 ± 0.79
Mask RCNN	94.53 ± 0.92
VMNet+SC (Ours)	98.52 ± 0.08

6 Conclusion

In this paper, we have proposed a useful and novel method called VMnet for fabric defect detection. We solve defect complexity problems by learning more representative features. Our approach involves the following main aspects: (1) building a new framework that uses visual saliency to generate new CNN input. (2) training VMnet combined with soft-cutoff loss to obtain discriminant features of cloth defects. Extensive experiments are conducted on datasets for three lights. Our results clearly show that our proposed VMnet improves result compared to current state-of-the-art for defect detection. However, in this paper, we only use random ROIs to generate saliency maps and use MobilenetV2. Future work includes researching alternatives to existing saliency algorithms and lightweight convolutional neural networks.

References

1. Schneider, D., Merhof, D.: Blind weave detection for woven fabrics. Pattern Anal. Appl. **18**(3), 725–737 (2015)
2. Kim, C.-W., Koivo, A.J.: Hierarchical classification of surface defects on dusty wood boards. Pattern Recogn. Lett. **15**(7), 713–721 (1994)

3. Wen, W., Xia, A.: Verifying edges for visual inspection purposes1. Pattern Recogn. Lett. **20** (3), 315–328 (1999)
4. Ren, S., He, K., Girshick, R., et al.: Faster R-CNN: towards real-time object detection with region proposal networks. IEEE Trans. Pattern Anal. Mach. Intell. **39**(6), 1137–1149 (2015)
5. He, K., Gkioxari, G. Dollar, P., Girshick, R.: Mask R-CNN. In: IEEE International Conference on Computer Vision (ICCV), pp. 2961–2969. IEEE, Venice (2017)
6. Liu, W., et al.: SSD: single shot MultiBox detector. In: Leibe, B., Matas, J., Sebe, N., Welling, M. (eds.) ECCV 2016. LNCS, vol. 9905, pp. 21–37. Springer, Cham (2016). https://doi.org/10.1007/978-3-319-46448-0_2
7. Lin, T.Y., Dollár, P., Girshick, R., et al.: Feature pyramid networks for object detection. In: The IEEE Conference on Computer Vision and Pattern Recognition (CVPR), pp. 2117–2125. IEEE, Hawaii (2017)
8. Lin, T.Y., Goyal, P., Girshick, R., He, K., Dollár, P.: Focal loss for dense object detection. IEEE Trans. Pattern Anal. Mach. Intell. **PP**(99), 2999–3007 (2017)
9. Krizhevsky, A., Sutskever, I., Hinton, G.: ImageNet classification with deep convolutional neural networks. Neural Inf. Process. Syst. Conference **25**(2), 1–9 (2012)
10. Karlekar, V.V., Biradar, M.S., Bhangale K B. Fabric defect detection using wavelet filter. In: International Conference on Computing Communication Control & Automation. IEEE, Pune (2015)
11. Jing, J.: Automatic defect detection of patterned fabric via combining the optimal gabor filter and golden image subtraction. J. Fib. Bioeng. Inform. **8**(2), 229–239 (2015)
12. Wen, Z., Cao, J., Liu, X., et al.: Fabric defects detection using adaptive wavelets. Int. J. Clothing Sci. Technol. **26**(3), 202–211 (2014)
13. Zeiler, M.D., Fergus, R.: Visualizing and understanding convolutional networks. In: Fleet, D., Pajdla, T., Schiele, B., Tuytelaars, T. (eds.) Computer Vision – ECCV 2014. Lecture Notes in Computer Science, vol. 8689, pp. 818–833. Springer, Cham (2014). https://doi.org/10.1007/978-3-319-10590-1_53
14. Tong, L., Wong, W.K., Kwong, C.K.: Fabric defect detection for apparel industry: a nonlocal sparse representation approach. IEEE Access **5**(99), 5947–5964 (2017)
15. Wei, L., Gang, H., Smith, J.R.: Unsupervised one-class learning for automatic outlier removal. In: IEEE Conference on Computer Vision & Pattern Recognition, pp. 3826–3833. IEEE, Columbus (2014)
16. Qu, T., Zou, L., Zhang, Q., et al.: Defect detection on the fabric with complex texture via dual-scale over-complete dictionary. J. Text. Inst. **107**(6), 1–14 (2015)
17. Susan, S., Sharma, M.: Automatic texture defect detection using Gaussian mixture entropy modeling. Neurocomputing **239**, 232–237 (2017)
18. Szegedy, C., Vanhoucke, V., Ioffe, S.: Rethinking the inception architecture for computer vision. In: Proceedings of the IEEE Conference on Computer Vision and Pattern Recognition, pp. 2818–2826. IEEE, Las Vegas (2016)
19. Simonyan, K., Zisserman, A.: Very deep convolutional networks for large-scale image recognition. Comput. Sci. (2014)
20. Zhou, J., Wang, J.: Fabric defect detection using adaptive dictionaries. Textile Res. J. **83**(17), 1846–1859 (2013)
21. Qu, T., Zou, L., Zhang, Q., et al.: Defect detection on the fabric with complex texture via dual-scale over-complete dictionary. J. Text. Inst. **107**(6), 1–14 (2015)
22. Sandler, M., Howard, A., Zhu, M., et al.: MobileNetV2: inverted residuals and linear bottlenecks. In: The IEEE Conference on Computer Vision and Pattern Recognition, pp. 4510–4520. IEEE, Salt Lake City (2018)
23. Yapi, D., Mejri, M., Allili, M.S., et al.: A learning-based approach for automatic defect detection in textile images. IFAC Papersonline **48**(3), 2423–2428 (2015)

24. Şeker, A., Peker, K.A., Yüksek, A.G., et al.: Fabric defect detection using deep learning. In: 24th Signal Processing and Communication Application Conference, pp. 1437–1440. IEEE, Zonguldak (2016)
25. Yan, Q., Xu, L., Shi, J., et al.: Hierarchical Saliency Detection. In: Computer Vision and Pattern Recognition (CVPR), pp. 569–582. IEEE, Portland (2013)
26. Vikram, T.N., Tscherepanow, M., Wrede, B.: A saliency map based on sampling an image into random rectangular regions of interest. Pattern Recogn. 45(9), 3114–3124 (2012)
27. Kingma, D., Ba, J.: Adam: a method for stochastic optimization. Comput. Sci. (2014)
28. Itti, L., Koch, C., Niebur, E.: A model of saliency-based visual attention for rapid scene analysis. IEEE Trans. Pattern Anal. Mach. Intell. 20(11), 1254–1259 (1998)
29. Lu, S., Mahadevan, V., Vasconcelos, N.: Learning optimal seeds for diffusion-based salient object detection. In: The IEEE Conference on Computer Vision and Pattern Recognition (CVPR), pp. 2790–2797. IEEE, Columbus (2014)
30. Cheng, M.M., Zhang, G.X., Mitra, N.J., et al.: Global contrast based salient region detection. In: IEEE Computer Society Conference on Computer Vision and Pattern Recognition, vol. 37, no. 3, pp. 409–416 (2011)
31. Yang, C., Zhang, L., Lu, H., et al.: Saliency detection via graph-based manifold ranking. In: Computer Vision and Pattern Recognition, pp. 3166–3173. IEEE, Portland (2013)
32. Li, J., Levine, M.D., An, X., et al.: Visual saliency based on scale-space analysis in the frequency domain. IEEE Trans. Pattern Anal. Mach. Intell. 35(4), 996–1010 (2013)
33. Hou, X., Zhang, L.: Saliency detection: a spectral residual approach. In: 2007 IEEE Conference on Computer Vision and Pattern Recognition, pp. 1–8. IEEE, Minneapolis (2007)

Locality and Sparsity Preserving Embedding Convolutional Neural Network for Image Classification

Yu Xia and Yongzhao Zhan[(✉)]

School of Computer Science and Communication Engineering,
Jiangsu University, Zhenjiang, China
yzzhan@ujs.edu.cn

Abstract. Convolutional neural networks (CNN) combined with manifold structures have attracted much attention, because they preserve local manifolds that are crucial and effective to reflect the intrinsic structure of data. In view of the excellent performance in image classification of CNN and the success in dimensionality reduction of manifold learning. This paper proposes a new deep learning framework based on deep CNN for image classification, which is called Locality and Sparsity Preserving Embedding Convolutional Neural Network (LSPE-CNN), that simultaneously considers the local information and natural sparsity of data into deep CNN. Compared to existing models, our proposed framework not only better preserve the associated features in the dataset, but also tries to seek the reconstruction relationship among samples by combining different manifold learning methods embedded in the CNN model, which further enhances the feature representation ability of the network. Experiments on CIFAR-10 and CIFAR-100 for image classification indicate that the proposed framework is superior to the other methods proposed in the deep learning literature.

Keywords: Convolutional neural networks · Feature representation ·
Manifold learning · Image classification

1 Introduction

With the rapid development of computer technology and networks, multimedia data is growing fast, and image databases are increasing constantly. At the same time, image classification shows extensive application prospects in object detection [17], image retrieval [22] and so on. How to accomplish the classification, organization and management of massive images on the network has become one of the important research issues for obtaining effective image information.

At present, image classification technology relies on the representation of images and the selection of classification structure. In general, the image classification algorithm globally describes the entire image by manual hand-crafted

© Springer Nature Switzerland AG 2019
Z. Cui et al. (Eds.): IScIDE 2019, LNCS 11935, pp. 447–458, 2019.
https://doi.org/10.1007/978-3-030-36189-1_37

feature or other feature learning methods, and then uses the classifier to determine whether there is a certain type of object. The biggest difference between deep learning and traditional pattern recognition methods is that it automatically learns features from big data rather than using hand-crafted features such as SIFT [14], HOG [5], LBP [1]. Good features can greatly improve the performance of the pattern recognition system. The features of the hand-crafted feature obtain the underlying features of the image through shallow learning, and there is still a large "semantic gap" between the advanced themes of the image. Deep learning was proposed by Hinton [10] in 2006 and has achieved good development in image classification and action recognition. Deep learning, using the already configured network structure, completely learns the hierarchical structural features of the image from the training data, and can extract the abstract features closer to the high-level semantics of the image. So the performance in image classification far exceeds the traditional method.

Deep learning technology represented by CNN has made excellent breakthroughs in many fields in recent years. AlexNet [12] won the championship in the 2012 ImageNet Large Scale Visual Recognition Competition (ILSVRC-2012), after which various improved networks were continuously proposed. VGG [19] increased the depth of the convolution neural network to 19-layer and achieved amazing results on ILSVRC-2014. ResNet [8] introduced the residual network structure and trained a 152 layers neural network and won the championship in the ILSVRC-2015 competition. These network frameworks tend to develop in a deeper network layer. As the depth of the network deepens, the convergence speed becomes slower due to the internal covariate shift. It means that the input distribution of the hidden layer will change as the parameters of each layer change during the training process. These layers need to constantly adapt to the new distribution. This slows down the training by requiring lower learning rates and careful parameter initialization, and makes it very difficult to train models with saturated nonlinearities. To alleviate these problem, Ioffe and Szegedy [11] proposed batch normalization (BN), which normalizes the output to the standard positive distribution by standardizing the internals of each mini-batch. However, although the BN algorithm improves the training speed of the model and reduces the dependence on parameter initialization, it does not take into account the structural feature of a small batch of samples, and the introduction of two additional parameters increases the risk of over-fitting.

It is natural to speculate that if the sample structure features are used during the training of each layer, the data distribution of each layer will become more stable and discriminative. Therefore, the internal covariate shift phenomenon will be further reduced, and the risk of over-fitting will drop sharply. Yang [23] proposed the concept of hidden layer, which can learn image features and hash functions at the same time, and achieved good performance in image retrieval. Yuan [24] took into account the structural features of the image, applied the manifold regularization to the depth model, and achieved good results in scene recognition. Bai [2] introduced maxout activation function in the fully connected layer and achieved good results. By combining multiple graph manifolds, Fang [6] proposed a multi-graph regularization framework based on auto-encoder for image classification, which better proved the effectiveness of manifold regularization. In fact, many manifold learning algorithms have been proposed, such as Locally Linear Embedding (LLE) [18], Locality Preserving Projection (LPP) [9], Locality Sensitive Discriminant Analysis (LSDA) [3], Sparsity Preserving Projections (SPP) [16], just to name a few. Thus how to combine different manifold learning methods embedded in the CNN to model feature representation is an important issue.

Based on above discussions, considering that the manifold is embedded into the CNN model, the model can learn the structural features between different samples independently, which improves the generalization ability of the model. And unlike BN, manifold embedding does not introduce additional parameters, which also reduces the risk of overfitting. We propose a framework that embeds locality and sparsity preserving with CNN, which is called locality and sparsity preserving embedding CNN, motivated by the advances in CNN and manifold learning. In a word, the main contributions of our paper are emphasized as follows:

- By using the manifold learning method to construct multiple graphs whose neighborhood relationship from samples are then embedded into CNN, each layer of the CNN maintains the upper manifold structure. The LSPE-CNN framework can make the distribution of each layer more stable and discriminative.
- Our LSPE-CNN framework provides a guidance and constraints for deep network training, and enables the filter to produce different outputs for the same input in different batches, which serves as a data augmentation technique to improve the generalization ability of the model and reduces the risk of overfitting.
- Our proposed framework demonstrates its superiority through experiments on two datasets, CIFAR-10 and CIFAR-100.

2   Related Work

In this section, a concise review of literature in terms of CNN model will be given. To understand how to construct locality and sparsity graphs, we will introduce some related manifold learning dimensionality reduction methods.

2.1   CNN

CNN is a kind of neural network with convolutional computation and deep structure. It is one of the representative algorithms of deep learning.

This paper designs the network based on the commonly used classic AlexNet structure, because it can be trained much faster. Although only the CNN is considered for illustration in this paper, the framework developed here can also be applied in other deep networks. The classic AlexNet consists of 5 convolution processing operations and 3 fully connected layers. This paper draws on the method proposed by Bai [2] as our CNN framework. He used the maxout activation function instead of the original activation function in Alexnet's fully connected layer, and added the Dropout function after maxout, which achieved good performance. The training process of the network is divided into two stages: forward propagation and backward propagation. The process of forward propagation can be expressed as:

$$a^l = \sigma \left(w^l a^{l-1} + b^l \right) \tag{1}$$

where a^l indicates the output of the layer l, a^{l-1} is the input, w^l is the weight of the convolution kernel, b^l is the bias of layer l, $\sigma(x)$ is a nonlinear activation function. In the process of backward propagation, the objective function is:

$$J = \arg\min \sum_{i=1}^{M} L(z_i) \tag{2}$$

where z_i is the backpropagation input and $L(z_i)$ is the loss function. Generally, the network weights are learned using the gradient descent algorithm.

2.2   Graph Construction Methods

Sparsity Preserving Projections (SPP). Sparsity Preserving Projections [16] is an unsupervised dimensionality reduction method. Through l_1 regularized objective function, it achieves the goal of maintaining the sparse structural relationship of data, with rotation, scale and other invariance. In the absence of a label, it still contains natural discriminant information and can automatically select neighborhood relationship. Given a set of training samples $X = [x_1, x_2, \cdots x_i, \cdots, x_n] \in R^{M \times n}$, where $x_i \in R^{M \times 1}$. We expect to reconstruct each sample x_i with as few samples as possible. This means that we use the l_1 minimization problem to calculate the sparse representation coefficient α_i for each sample x_i:

$$\begin{cases} \min_{\alpha_i} \|\alpha_i\|_1 \quad s.t. \quad x_i = X\alpha_i \\ \qquad 1^T \alpha_i = 1 \end{cases} \tag{3}$$

where $\tilde{\alpha}_i = [\alpha_{i1}, \alpha_{i2}, \cdots, \alpha_{i-1}, 0, \alpha_{i+1}, \cdots, \alpha_{in}]^T$, by using a number of adjacent samples to effectively represent a given sample, the neighborhood is set to a hollow neighborhood, i.e. the sample itself to be represented is not included. Then perform the Eq. (3) on all the samples X in turn, and construct a sparse representation reconstruction matrix M:

$$M = [\tilde{\alpha}_1, \tilde{\alpha}_2, \cdots, \tilde{\alpha}_n] \tag{4}$$

where M is the weight matrix.

Locality Sensitive Discriminant Analysis (LSDA). Locality Sensitive Discriminant Analysis [3] is a classic supervised dimensionality reduction algorithm which respects both discriminant and geometrical structure in the data. This method can well preserve the original local structure of the data manifold, and can retain the original category label of the data, which has better discriminability. Given a set of points $\{(x_1, l_1), (x_2, l_2), \cdots, (x_i, l_i), \cdots, (x_N, l_N)\}$, $x_i \in R^D$ represents a D-dimensional vector, l_i is the category label of the samples and $N(x_i)$ is the k nearest neighbors of x_i, construct an within-class graph and between-class graph G_w and G_b for each data point in order to discover geometrical and discriminant structure of the data manifold. W_w is the weight matrix of G_w and a is a projection vector. The objective function is defined as follow:

$$\arg\max X\left(\alpha L_b + (1-\alpha)W_w\right)X^T a \tag{5}$$

where α is a suitable constant and L_b is the Laplacian matrix of G_b.

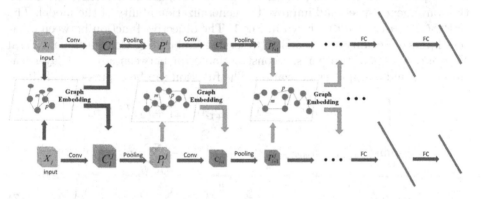

Fig. 1. Framework of our proposed LSPE-CNN

3 Locality and Sparsity Preserving Embedding CNN

3.1 Motivation to Introduce Manifold Regularization into CNN

We find that the introduction of two graph construction method changes the expression of the original convolution function by adding the reconstruction relationship between the samples. SPP is an unsupervised dimensionality reduction algorithm. Different from the traditional graph construction method (such as k-nearest neighbor), SPP obtains the sparse reconstruction relationship between data through sparse representation, which not only effectively utilizes the natural discriminative ability of sparse representation, but also, it alleviates the difficulty of neighboring parameter selection in a way. LSDA is a supervised dimensionality reduction algorithm. This method optimally preserves the local neighbor information and discriminant information by constructing graphs. The introduction of the two graph construction methods not only considers the sparsity and locality of the data, but also obtains the excellent feature representation. Moreover, the manifold regularization will produce different outputs for the same input in different batches, which serves as a data augmentation technique to reduce the overfitting risk.

3.2   The Framework and Manifold Regularization

Locality and Sparsity Preserving Embedding CNN (LSPE-CNN) is a framework that combines locality and sparsity graph constructions. In our framework, to take advantage of the structural information between images, we expect to obtain the manifolds of the previous layer (the previous can be the input layer or pooling layer) by constructing locality and sparsity graphs, and use the manifolds to redesign the mapping relationship between adjacent layers. These graph construction methods make the learned feature more stable and discriminative in the process of deepening the depth of the network, and further accelerate the convergence speed and improve the generalization ability of the model. The LSPE-CNN framework is shown in Fig. 1. The objective function between adjacent layers of Locality and Sparsity Preserving Embedding Convolutional Neural Network consists of two parts: reconstruction error between adjacent layer feature maps and a graph regularization. The function can be expressed as follows:

$$\min_{f_{l+1}} \left\{ \left\| f_{l+1} - h_{w,b}(f_l) \right\|^2 + \lambda \left(r_1 f_{l+1}^T M_l f_{l+1} + r_2 f_{l+1}^T P_l f_{l+1} \right) \right\} \tag{6}$$

$$\Rightarrow \min_{f_{l+1}^i} \left\{ \left\| f_{l+1}^i - h_{w,b}(f_l^i) \right\|^2 + \lambda \left(r_1 \left\| f_{l+1}^i - \sum_{j \in N_i} m_{i,j}^l f_{l+1}^j \right\|^2 \right.\right.$$

$$\left.\left. + r_2 \left\| f_{l+1}^i - \sum_{k \in N_i} p_{i,k}^l f_{l+1}^k \right\|^2 \right) \right\} \tag{7}$$

where $f_{l+1} = \left(f_{l+1}^1, f_{l+1}^2, \cdots, f_{l+1}^N \right)$ represents the feature maps of all training samples, M_l and P_l represent manifolds of all training samples learned by SPP and LSDA of layer l, $h_{w,b}()$ represents convolution mapping function, N_i represents the neighbor of f_l^i, λ is the balance factor, $m_{i,j}^l$ represent edge weight between f_l^i and f_l^j in the lth layer manifold which is learned by SPP, $p_{i,k}^l$ represent edge weight between f_l^i and f_l^k in the lth layer manifold which is learned by LSDA. r_1 and r_2 are constraint terms that control the trade-off among different graph constructions, and the sum of them is one. Then the r_k is calculated as follows:

$$r_1 = \frac{\left(\frac{1}{tr(f^T M f)} \right)}{\left(\frac{1}{tr(f^T M f)} + \frac{1}{tr(f^T P f)} \right)} \tag{8}$$

$$r_2 = \frac{\left(\frac{1}{tr(f^T P f)} \right)}{\left(\frac{1}{tr(f^T M f)} + \frac{1}{tr(f^T P f)} \right)} \tag{9}$$

The gradient of f_{l+1}^i in the function can be calculated as follows:

$$\partial \left\| f_{l+1}^i - h_{w,b}(f_l^i) \right\|^2 / \partial f_{l+1}^i + \lambda \partial \left(\left\| f_{l+1}^i - \sum_{j \in N_i} r_1 m_{i,j}^l f_{l+1}^j \right\|^2 \right.$$

$$\left. + \left\| f_{l+1}^i - \sum_{k \in N_i} r_2 p_{i,k}^l f_{l+1}^k \right\|^2 \right) / \partial f_{l+1}^i$$

$$= 2 \left(f_{l+1}^i - h_{w,b}(f_l^i) \right) + 2\lambda \left(f_{l+1}^i - \sum_{j \in N_i} r_1 m_{i,j}^l f_{l+1}^j + f_{l+1}^i - \sum_{k \in N_i} r_2 p_{i,k}^l f_{l+1}^k \right)$$

$$(10)$$

Then, we can get the expression of f_{l+1}^i as follows:

$$f_{l+1}^i = \frac{h_{w,b}(f_l^i) + \lambda \left(\sum_{j \in N_i} r_1 m_{i,j}^l f_{l+1}^j + \sum_{k \in N_i} r_2 p_{i,k}^l f_{l+1}^k \right)}{1 + 2\lambda} \tag{11}$$

3.3 Deep Network Training

In the process of using backpropagation algorithm to train the network, we replace the original convolution function $h_{w,b}()$ with a new mapping function by manifold regularization constraints in Eq. (11). At the same time, since the feature maps of adjacent samples are unknown, we first initialize them during the training process, and then we update them with the manifold learned by the previous layer. The details of the training process are shown in Table 1.

4 Experimental Results

In this section, we first introduce the specific details of the two datasets. Then, we performed a classification task to verify the performance of Locality and Sparsity Preserving Embedding Convolutional Neural Network. In order to verify the effectiveness of the proposed method, we use tensorflow to implement the optimized deep learning framework. We fine-tune the network parameters on the target dataset by backpropagation algorithms. The image classification performance evaluation index is the accuracy rate. We perform image classification tasks on CIFAR-10 and CIFAR-100 to evaluate the proposed Locality and Sparsity Preserving Embedding Convolutional Neural Network.

4.1 Evaluation Datasets

CIFAR-10 and CIFAR-100 are computer vision dataset for universal object recognition collected by Hinton's students. The CIFAR-10 dataset consists of 60,000 32×32 color images of 10 classes, each with 6000 images. There are 50,000 training images and 10,000 test images. The datasets is divided into five

Table 1. The process of deep network training

Algorithm 1 Manifold Regularization Embedding CNN
1   **Input**: Input dataset X;randomly initialized weights and offsets W,b;batch size B; learing rate ε ;manifold balance factor λ
2   **Output**: Learned network patameters W,b
3   Calculate the manifold parameters(SPP and LSDA) of layer l of each batch of samples
4   Initialize layer $l+1$ by ReLU [7]
5   **for** sample $n=1$ to B on $l+1$ **do**
6   Updadate all sameples by Eq.(11)
7   **End for**
8   Backward propagation
9   Return W,b

training batches and one test batch, each with 10,000 images. The test batch contains exactly 1000 randomly selected images from each category. Training batches contain the remaining images in random order, but some training batches may contain more images from one category than the other. Overall, the sum of the five training sets contains exactly 5,000 images from each class. The image size in the CIFAR-100 dataset is the same as CIFAR-10, which is pixel distribution, but the database has 100 types of images, each of which includes 600 color images, 500 training images and 100 test images.

4.2   Analysis of Experimental Results of CIFAR-10

The cifar-10 dataset has ten categories, and the final output of the network is set to 10 channels to predict the classification of 10 categories of cifar-10 data images. Meanwhile, we set the manifold parameter as 0.2. The learning rate is initially set as 10^{-3}. We use softmax as a classifier and calculate the loss with cross entropy, and train the data using stochastic gradient descent. In order to verify the effect of the size of batchsize on the image classification results, we used different epochs to test the image classification accuracy rate of different batchsize on cifar-10 dataset. The results are shown in Table 2, from which we can see that the model performs best when the batchsize is 128 and the number of epoch is 80. It also proves that batchsize is not "the bigger the better", as bigger batchsize may also lead to lower accuracy.

The results of experiments on CIFAR-10 by the following comparative methods: MGDAE [6], CNN+Spearmint [20], MCDNN [4], NIN [13], AlexNet+ Finetuning [23], AlexNet+FC.maxout [2], AlexNet+FC.maxout+Batch Normalization [2], are as shown in Table 3.

It can be seen from the experimental results that LSPE-CNN achieves best performance among all methods on cifar-10. We note that the result of MGDAE are the worst because although it uses multi-graph regularization, in some cases the feature extraction ability of auto-encoder will be weaker than most CNN models. Some methods based on the CNN model have achieved high performance. AlexNet+FC.maxout [2] introduced maxout activation in the AlexNet

Table 2. Classification accuracy (%) of different Batchsize on CIFAR-10 dataset

Epochs number	Batchsize			
	32	64	128	256
20	75.95%	77.85%	79.60%	78.95%
40	82.37%	84.52%	85.87%	84.32%
80	88.72%	89.42%	**89.89%**	88.65%
100	87.46%	88.76%	89.25%	87.97%

Table 3. Classification accuracy (%) of comparison methods on CIFAR-10 dataset

Method	Accuracy (%)
MGDAE [6]	53.58
CNN+Spearmint [20]	85.02
MCDNN [4]	88.79
NIN+Dropout [13]	89.59
AlexNet+Fine-tuning [23]	89.40
AlexNet+FC.maxout [2]	88.90
AlexNet+FC.maxout+Batch Normalization [2]	88.63
LSPE-CNN	89.89

full connection layer, achieving 88.90% accuracy. Using Batch Normalization in the convolutional layer dose not improve the accuracy. However, after embedding manifold regularization (our method), the classification accuracy has been significantly improved. It can be seen that the manifold regularization takes into account the structural features of the sample, combining the advantages of locality and sparsity, so that the network can better learn the samples in the training process, and the learned network weights are more suitable.

We also show the accuracy of each class in Fig. 2, and found that the accuracy of the 10 categories is relatively stable. It shows that our proposed method has good feature representation ability and generalization ability in different image samples.

4.3   Analysis of Experimental Results of CIFAR-100

In order to verify the image classification ability of the proposed model on multi-category big datasets, the CIFAR-100 dataset was used for comparative experiments. We test the classification performance by setting the number of output channels to 100. We set the manifold patameter as 0.3. The learning rate is initially set as 10^{-3}. We use softmax as a classifier and calculate the loss with cross entropy, and train the data using stochastic gradient descent. The results of experiments on CIFAR-100 by the following

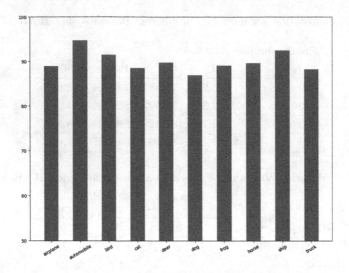

Fig. 2. Accuracy rate for each class of cifar-10 dataset

comparative methods: Learning Pooling [15], Stochastic Pooling [25], Tree based priors [21], NIN [13], AlexNet+Fine-tuning [23], AlexNet+FC.maxout [2], AlexNet+FC.maxout+Batch Normalization [2], are as shown in Table 4.

It can be seen from the experimental results that the proposed LSPE-CNN framework has good generalization ability, the accuracy is higher than the benchmark AlexNet+FC.maxout+Batch Normalization [2], and it is better than the current best image classification method.

Table 4. Classification accuracy (%) of comparison methods on CIFAR-100 dataset

Method	Accuracy (%)
Learning Pooling [15]	56.29
Stochastic Pooling [25]	57.49
Tree based priors [21]	63.15
NIN+Dropout [13]	64.32
AlexNet+Fine-tuning [23]	67.38
AlexNet+FC.maxout [2]	68.25
AlexNet+FC.maxout+Batch Normalization [2]	67.66
LSPE-CNN	69.21

5   Conclusion

In this paper, we present how to improve feature representation for deep architectures by incorporating the manifold into the training process. The proposed novel framework can better characterize similarities between samples and more rationally extract correlation features between samples. Experimental results of CIFAR-10 and CIFAR-100 show the effectiveness of Locality and Sparsity Preserving Embedding CNN. Future work will further adjust the network framework based on the current image classification, and try to apply it to multimedia analysis fields such as video retrieval and target tracking.

Acknowledgments. This work was supported by the National Natural Science Foundation of China (No. 61672268) and the Primary Research & Development Plan of Jiangsu Province (No. BE2015137).

References

1. Ahonen, T., Hadid, A., Pietikainen, M.: Face description with local binary patterns: application to face recognition. IEEE Trans. Pattern Anal. Mach. Intell. **12**, 2037–2041 (2006)
2. Bai, C., Huang, L., Chen, J., Pan, X., Chen, S.: Optimization of deep convolutional neural network for large scale image classification. Ruan Jian Xue Bao/J. Softw. **29**(4), 137–146 (2018)
3. Cai, D., He, X., Zhou, K., Han, J., Bao, H.: Locality sensitive discriminant analysis. IJCAI **2007**, 1713–1726 (2007)
4. Cireşan, D., Meier, U., Schmidhuber, J.: Multi-column deep neural networks for image classification. arXiv preprint arXiv:1202.2745 (2012)
5. Dalal, N., Triggs, B.: Histograms of oriented gradients for human detection. In: International Conference on Computer Vision & Pattern Recognition (CVPR'05), vol. 1, pp. 886–893. IEEE Computer Society (2005)
6. Fang, J., Zhan, Y., Shen, X.: Multi-graph regularized deep auto-encoders for multi-view image representation. In: Hong, R., Cheng, W.-H., Yamasaki, T., Wang, M., Ngo, C.-W. (eds.) PCM 2018. LNCS, vol. 11164, pp. 797–807. Springer, Cham (2018). https://doi.org/10.1007/978-3-030-00776-8_73
7. Glorot, X., Bordes, A., Bengio, Y.: Deep sparse rectifier neural networks. In: Proceedings of the Fourteenth International Conference on Artificial Intelligence And Statistics, pp. 315–323 (2011)
8. He, K., Zhang, X., Ren, S., Sun, J.: Deep residual learning for image recognition. In: Proceedings of the IEEE Conference on Computer Vision and Pattern Recognition, pp. 770–778 (2016)
9. He, X., Niyogi, P.: Locality preserving projections. In: Advances in Neural Information Processing Systems, pp. 153–160 (2004)
10. Hinton, G.E., Salakhutdinov, R.R.: Reducing the dimensionality of data with neural networks. Science **313**(5786), 504–507 (2006)
11. Ioffe, S., Szegedy, C.: Batch normalization: accelerating deep network training by reducing internal covariate shift. arXiv preprint arXiv:1502.03167 (2015)
12. Krizhevsky, A., Sutskever, I., Hinton, G.E.: ImageNet classification with deep convolutional neural networks. In: Advances in Neural Information Processing Systems, pp. 1097–1105 (2012)

13. Lin, M., Chen, Q., Yan, S.: Network in network. arXiv preprint arXiv:1312.4400 (2013)
14. Lowe, D.G.: Distinctive image features from scale-invariant keypoints. Int. J. Comput. Vis. **60**(2), 91–110 (2004)
15. Malinowski, M., Fritz, M.: Learnable pooling regions for image classification. arXiv preprint arXiv:1301.3516 (2013)
16. Qiao, L., Chen, S., Tan, X.: Sparsity preserving projections with applications to face recognition. Pattern Recogn. **43**(1), 331–341 (2010)
17. Redmon, J., Divvala, S., Girshick, R., Farhadi, A.: You only look once: Unified, real-time object detection. In: Proceedings of the IEEE Conference on Computer Vision and Pattern Recognition, pp. 779–788 (2016)
18. Roweis, S.T., Saul, L.K.: Nonlinear dimensionality reduction by locally linear embedding. Science **290**(5500), 2323–2326 (2000)
19. Simonyan, K., Zisserman, A.: Very deep convolutional networks for large-scale image recognition. arXiv preprint arXiv:1409.1556 (2014)
20. Snoek, J., Larochelle, H., Adams, R.P.: Practical Bayesian optimization of machine learning algorithms. In: Advances in Neural Information Processing Systems, pp. 2951–2959 (2012)
21. Srivastava, N., Salakhutdinov, R.R.: Discriminative transfer learning with tree-based priors. In: Advances in Neural Information Processing Systems, pp. 2094–2102 (2013)
22. Szegedy, C., et al.: Going deeper with convolutions. In: Proceedings of the IEEE Conference on Computer Vision and Pattern Recognition, pp. 1–9 (2015)
23. Yang, H.F., Lin, K., Chen, C.S.: Supervised learning of semantics-preserving hash via deep convolutional neural networks. IEEE Trans. Pattern Anal. Mach. Intell. **40**(2), 437–451 (2018)
24. Yuan, Y., Mou, L., Lu, X.: Scene recognition by manifold regularized deep learning architecture. IEEE Trans. Neural Netw. Learn. Syst. **26**(10), 2222–2233 (2015)
25. Zeiler, M.D., Fergus, R.: Stochastic pooling for regularization of deep convolutional neural networks. arXiv preprint arXiv:1301.3557 (2013)

Person Re-identification Using Group Constraint

Ling Mei[1], Jianhuang Lai[2,3]([envelope]), Zhanxiang Feng[2,3], Zeyu Chen[2], and Xiaohua Xie[2,3]

[1] School of Electronics and Information Technology, Sun Yat-sen University, Guangzhou, China
meil3@mail2.sysu.edu.cn
[2] School of Data and Computer Science, Sun Yat-sen University, Guangzhou, China
{stsljh,fengzhx7,xiexiaoh6}@mail.sysu.edu.cn, chenzy5@mail2.sysu.edu.cn
[3] Guangdong Key Laboratory of Information Security Technology, Sun Yat-sen University, Guangzhou, China

Abstract. The group refers to several pedestrians gathering together with a high motion collectiveness for a sustained period of time. The existing person re-identification (re-id) approaches focus on extracting individual appearance cues, but ignores the correlations of different persons in a group. In this paper, we propose a group-guided re-id method named group retrieval correlation (GRC) to address the above problem, which pays more attention to the correlations of surrounding pedestrians in the same group and reliefs the interference caused by the dependence of appearance cues. Compared with traditional re-id methods which compute naive similarity merely by the appearance characteristics, the GRC based re-id method proposes a novel and optimal person similarity by considering the relationships among groups. Therefore, the proposed approach provides sufficient hints of group relationships, which are supplementary to the appearance features and contributes to constructing a more reliable re-id system. Experimental results demonstrate that the group information promotes person re-id by a large margin on the proposed Group-reID dataset in terms of both hand-crafted descriptors and deep features.

Keywords: Person re-identification · Group retrieval correlation · Person identity · Group-reID dataset

1 Introduction

Person re-identification (re-id) [3] has been an important topic in pattern recognition and computer vision. The person re-id techniques aim to recognize pedestrians captured by non-overlapping camera views. Person re-id is a challenging task because of the dramatic variants between the appearances of pedestrians from changing camera fields, including illuminations, viewpoints, and poses. Although

© Springer Nature Switzerland AG 2019
Z. Cui et al. (Eds.): IScIDE 2019, LNCS 11935, pp. 459–471, 2019.
https://doi.org/10.1007/978-3-030-36189-1_38

recent studies have achieved great progresses for person re-id, the current litera-ture is based on the assumption of the ideal situation that the target pedestrians are separately captured by the surveillance systems, without other anonymous pedestrians. The existing works are focused on extracting discriminative indi-vidual appearance features, whose performance may drop significantly in the crowded situation where occlusions and identity confusions occur.

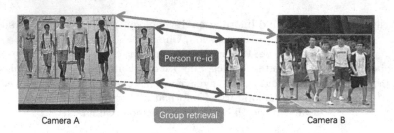

Fig. 1. Illustration of person re-id and group retrieval.

As reported in [17], more than 70% of people tend to move in groups in a crowded public space, either with friends or with strangers. Since humans have such social characteristics, most realistic scenarios are people moving within a group rather than by an individual. E.g., in the subway, a person may be surrounded by other pedestrians and the body parts may be occluded. In these situations, as shown in Fig. 1, if the person was part of a group, the identity of the target pedestrian may be less confusing because the person has additional social constraints in a specific group, which are independent of the camera topology [10, 27]. Thus, learning an efficient representation for the group becomes an essential issue for improving the current literature.

However, there still exist some challenging bottlenecks of associating the group constrains. On one hand, this work needs to collect and annotate large amounts of data, which should include the association between group and person information. On the other hand, the topic of exploring how to exploit the group information to improve the re-id performance is largely under-study, and we should find a rational way to tackle with this issue.

In this paper, we propose a novel person re-id method which associates and benefits from the intrinsic group constrains. As far as we know, this is the first work which exploits group constraints to improve the person re-id methods. The paper makes three major contributions: (1) We propose and annotate a large group-guided re-id dataset named "Group-reID" which is collected from the images of pedestrian groups. The Group-reID dataset contains 524 pedes-trians who constitute 208 unrepeated groups, with group-associated informa-tion between groups and persons. (2) We propose an efficient group-associated method named Group Retrieval Correlation (GRC) to improve the performance of re-id by using the underlying group information. Particularly, We use the annotation of person identity (Person-ID) rather than group identity (Group-ID) to extract naive re-id features, and then associate the constraints of GRC with person re-id. (3) Experimental results show that the additional GRC information with Person-ID promotes the performance of person re-id methods with a large margin, including both hand-crafted features and deep models, which validates the significance and effectiveness of the proposed method.

The paper is organized as follows. In Sect. 1, we first introduce the motivation of group-associated pedestrian re-id problem. In Sect. 2, we review some related works about the proposed method. In Sect. 3, we describe the collected pedestrian re-id dataset and introduce the proposed group-associated algorithm. In Sect. 4, experiments are implemented and the experimental results are presented. In Sect. 5, we give the conclusion.

2 Related Work

2.1 Traditional Re-id Methods

The majority of existing person re-id methods highly rely on individual appearance cues, such appearance-based re-id methods seek to extract a variety of local or global features from the whole body that are distinctive and robust to viewpoints [2] and illumination changes [15].

On one hand, previous handcrafted methods are used widely for feature extraction. WHOS [11] is a handcrafted method of feature representation to mitigate problems caused by background clutter and noise. Similarly, GOG [14] weights patches according to the distance to the central line, but it is not reasonable for the issues like occlusion. In addition, LDFV [9] uses local descriptors that include spatial, gradient, and intensity information to encode the Fisher vector [21] representation. Moreover, gBiCov [21] is a multi-scale biologically inspired feature that uses covariance descriptors to computes Euclidean distance between persons. In texture based models, LOMO [9] uses HSV color histograms and maximizes the occurrence to make a robust representation against viewpoint changes. ELF [5] combines color histograms in the RGB, HSV and YCbCr color spaces to obtain the texture histograms. Furthermore, HistLBP [28] also encodes the RGB, HSV, YCbCr color spaces to color histograms, and combines them with the texture histograms of LBP features.

On the other hand, the deep learning based methods play more useful roles in re-id. Zheng et al. [33] apply a siamese network which combines identification loss and verification loss on the re-id task. Feng et al. [2] propose a deep neural network-based framework which utilizes the view information in the feature extraction stage. Sun et al. [24] learn robust features by a refined part pooling (RPP), which promotes the performance of Part-based Convolutional Baseline (PCB) network. In addition, Zhang et al. [30] propose a crossing Generative Adversarial Network (Cross-GAN) for learning a joint distribution for cross-image representations in an unsupervised manner. Sun et al. [23] introduce a large-scale synthetic data engine to synthesize pedestrians for re-id. Yu et al. [29] propose a deep model for the soft multilabel learning for unsupervised re-id.

However, all above methods highly rely on individual appearance cues and ignore to deal with the correlations of different persons in a group.

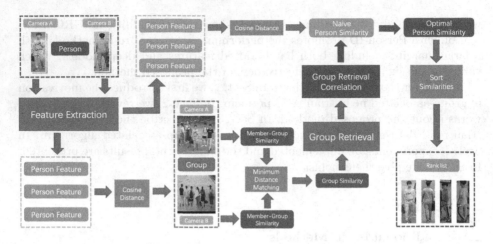

Fig. 2. Framework of person re-id in groups.

2.2 Pedestrian Detection

For the sake of associating groups with individual persons, it is necessary to detect pedestrians from groups. Traditional pedestrian detection methods locate a person by means of appearance cues [12], an early work uses wavelet templates [18] as the feature representation for pedestrians. The covariance matrix [26] are also used for pedestrian detection. Besides, many other strategies, such as filtering on the channel maps [31], optical flow [16], and semantic segmentation [1] can also improve the performance for the pedestrian detection by enhancing discriminative information.

Recent deep learning methods like Faster-RCNN [4] and YOLO [19] improve the performance of pedestrian detection by the advance of object detection. Faster-RCNN detects piles of sliding windows within two stages, while YOLO focus more on the global view in an end-to-end way. YOLO achieves an excellent efficiency among these methods by splitting an image into grids to obtain possible bounding box (bbox) with a fully CNN. Furthermore, YOLOv3 [20] replaces the softmax function with a logistic regression threshold, which retains an outstanding speed and reaches a higher accuracy than previous versions.

3 Framework of Person Re-identification in Groups

In this section, we present the details of our approach which assists re-id by using the group retrieval correlation (GRC). The framework of the proposed method is shown in Fig. 2. First, we extract the naive features of single-person re-id by traditional feature extraction methods, including both hand-crafted methods and deep learning methods. Secondly, we combine these naive features to adapt the group association problem. Notably, we define the member-group similarity to find the association between persons and groups. Thirdly, we describe the

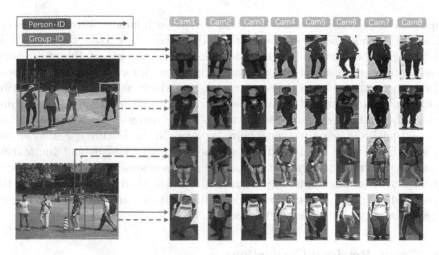

Fig. 3. An associated example of persons and their groups in the Group-reID dataset. The solid lines and the dotted lines with different colors denote the person identity and group identity of each person, respectively. (Color figure online)

relationships between different groups by measuring their similarity with the member-group similarity. Particularly, we apply the minimum distance matching principle to achieve the representations of group retrieval. Finally, the GRC introduces a fusion strategy which combines the group relationship and the single-person relationship to enhance the performance of re-id techniques.

3.1   Feature Extraction

Since the group image contains several pedestrians, as shown in Fig. 3, we use YOLOv3 pedestrian detector [20] to detect all the bboxes of individual persons from group images and annotate them with Group-IDs, which constitutes the

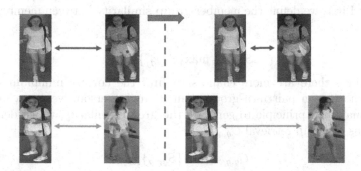

Fig. 4. The verification module of siamese network is designed to enlarge the inter-class distance (green lines) and narrow the intra-class distance (red-lines). (Color figure online)

proposed Group-reID dataset. After that, we use both hand-crafted re-id feature extraction methods and deep learning based methods to extract the naive re-id features. For the hand-crafted methods, they tend to extract the appearance based features such as intensity, texture, gradient information. For the deep learning methods, we use a siamese network [33] which combines the verification module and the identification module to train a more robust feature learning model. The verification module pull the positive image pairs with same class close and push the negative image pairs far away in the feature space, as shown in Fig. 4. Eventually, the identification module is attached behind the feature extraction module to learn distinguishing representations. Notably, all above feature extraction methods only extract the naive re-id features without the corresponding group information, these naive features are the original feature representations of individual pedestrians.

3.2   Group Retrieval Correlation

After feature extraction, we propose to use the naive features to generate group representations. Generally, the similarity between two groups is based on the similarity of different individuals among these groups. Furthermore, the existing re-id methods ignore the correlations between different persons in a group, which is a major consideration of our paper.

Consequently, we propose the group retrieval constraint to describe the correlations among different members in a group. The group retrieval re-identities a specific group of persons as a whole in the regions of other cameras, which is valuable for avoiding the mismatch of a single pedestrian and improving the re-id by exploiting the information of other members in the same group. Therefore the risk of mismatching a group is much lower than that of a person. In order to find the most similar member pairs, we measure the similarity between pedestrian pairs by the cosine distance. Given a query group q with M members $\{i\}_{i=1}^M$ and a gallery group g with N members $\{j\}_{j=1}^N$, we denote S_{q_i,g_j} as the pedestrian similarity between member i in q and member j in g by computing the cosine distance between the pedestrian features. The generation of group similarity is two steps: First, we define the member-group similarity between member q_i and group g as

$$S_{q_i,g} = \max\{S_{q_i,g_j}\}_{j=1}^N \tag{1}$$

Second, since there are more chances to find the correct minimum distance between the person pair in a group than a single person, we use a minimum distance matching principle to generate the group similarity, which defines the correlations of group retrieval $C_{q,g}$ as

$$C_{q,g} = \max\{S_{q_i,g}\}_{i=1}^M \tag{2}$$

where $C_{q,g}$ is the group similarity between q and g. In this way, no matter the size of gallery group is equal to the query group or not, we can measure the group similarity between them, which generalizes the searching range of

---

Algorithm 1. Person Re-identification Using Group Constraint

Input: Query person bboxes $\{q_i\}_{i=1}^{M}$ in groups $\{q\}$ and gallery person bboxes $\{g_j\}_{j=1}^{N}$ in groups $\{g\}$.

Output: Optimal rank results $\mathcal{R}$ of person re-id.

1 **for** *each query person q_i in group q* **do**
2   **for** *each gallery person g_j in group g* **do**
3   Compute the naive member similarity S_{q_i,g_j} between q_i and g_j by cosine distance;
4   Compute the member-group similarity $S_{q_i,g}$ between member q_i and group g by Eq. 1;
5   Using minimum distance matching to get the correlations $C_{q,g}$ of group retrieval by Eq. 2;
6   Obtain the group retrieval constraint $\mathbf{C}_q$ to all gallery groups;
7   Update the naive similarity S_{q_i,g_j} with GRC to the optimal similarity S_{q_i,g_j}^{*} by Eq. 3;
8   Sort the similarity S_{q_i,g_j}^{*} to generate the rank list $\mathcal{R}_i$ for q_i;

9 Return all rank lists $\mathcal{R}$ as the results of group constraint based person re-id .

---

matching groups. After that, we obtain the vector of group retrieval constraint as $\mathbf{C}_q = (C_{q,1}, \ldots, C_{q,g}, \ldots, C_{q,G})$, where G is the number of gallery groups. $C_{q,g}$ is the GRC which measure the similarities of groups.

3.3 Fusion of Naive Features and GRC

After getting the group retrieval constraint, we try to associate this group context information to improve the re-id technique. We update the similarity of individual persons by multiplying the value of group retrieval constraint:

$$S_{q_i,g_j}^{*} = C_{q,g} \times S_{q_i,g_j} \tag{3}$$

where $C_{q,g}$ is the group correlation similarity obtained by Eq. 2, the correlation updates the naive person re-id similarity S_{q_i,g_j} to a group-associated similarity S_{q_i,g_j}^{*}. The naive similarity is obtained via kinds of feature extractors, such as hand-crafted methods like WHOS [11] or deep learning models like ResNet50 [6], *etc.* Algorithm 1 presents the procedures about how GRC optimizes person re-id. Finally, we sort the similarities to get the rank list $\mathcal{R}_i$ of i, then repeat the operation to get the rank lists $\mathcal{R}$ of the whole query set.

Group retrieval correlation (GRC) combines the group constraints with the single person feature representations, GRC make the re-identification constraint to a group rather than all gallery persons, which makes the search range much smaller. In addition, the mismatch possibility of group retrieval is much lower than person re-id, thus by using Eq. 3, the real correct match with a lower rank can be pulled to a higher place, while the wrong match with a higher rank can be pushed to a lower place, which promotes the overall accuracy of person re-id. In this way, GRC can benefit the person re-id by the constraints. As shown

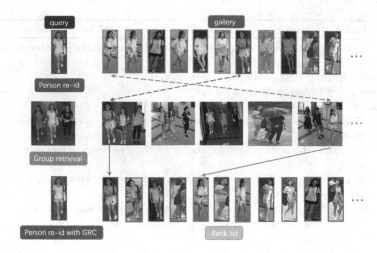

Fig. 5. Illustration of the effectiveness of GRC in terms of improving person re-id with group constraints. Red and green bbox denote the correct and wrong matching results in the list, respectively. (Color figure online)

in Fig. 5, the query changes her clothes which leads the correct rank in gallery lower, but the rank is pulled up to the top combined with the proposed GRC. Figure 5 also shows GRC can relieve the interference of occlusion. Therefore, the proposed additional group information relieves the interference of person re-id caused by the dependence of appearance cues.

Furthermore, our group feature representations are invariant to the relative displacements of individuals within a group, and are independent of the group size. Consequently, GRC provides convenience to person re-id to handle challenges such as changing clothes and occlusion.

4 Experiments

In this section, we first introduce the group-associated person re-id dataset. Then, we give the experimental results. Since this is the first work to use group retrieval information to improve re-id and there is no precedent dataset, we mainly conduct experiments on the proposed Group-reID dataset.

4.1 Group-Associated Person Re-id Dataset

We collect and annotate a person re-id dataset named Group-reID from the data of group images, which contains 17,492 bboxes cropped from the corresponding pedestrian images of 524 persons. The Group-reID dataset constitutes 208 unrepeated groups which are collected under 8 non-overlapping cameras, the dataset is designed to simulate the diverse situations in practical applications. There are 2–6 persons in a group in our dataset.

Table 1. The composition of Group-reID dataset in terms of different cameras.

Camera	Training set			Testing set			Whole dataset		
	Groups	Persons	Samples	Groups	Persons	Samples	Groups	Persons	Samples
1	102	247	1,226	102	269	1,338	204	516	2,564
2	103	250	1,238	103	271	1,346	206	521	2,584
3	102	248	1,222	103	271	1,353	205	519	2,575
4	83	198	958	78	203	999	161	401	1,957
5	71	162	808	68	171	852	139	333	1,660
6	74	168	839	77	189	944	151	357	1,783
7	91	222	1,094	85	221	1,094	176	443	2,188
8	87	212	1,047	87	227	1,134	174	439	2,181
Total	104	251	8,432	104	273	9,060	208	524	17,492

Group-reID dataset is challenging for its diversity of a specific person in terms of pose, illumination, occlusion and changing clothes *etc.*, which is similar to scenarios in real life. For better presentation, in Fig. 3, we choose some re-id samples from two groups in the dataset as exemplar to show the association between re-id and groups.

The bboxes are detected and extracted by using YOLOv3 [20] in each group image. Then we annotate the cropped images with both Group-IDs and Person-IDs. For the sake of avoiding ambiguity, all the groups in our dataset are unique. Therefore, we can focus on exploring the association between individual pedestrians and the corresponding groups.

4.2 Experimental Settings

We divide the persons of Group-reID dataset into training and testing set according to their Group-IDs, the detail is shown in Table 1. According to Table 1, the majority of both persons and groups have appeared in each cameras, which indicates the generalized distribution of the dataset.

For training and testing, as shown in Table 1, there are 8432 and 9060 pedestrian images which come from 251 and 273 persons, respectively. There are about 5 images of each camera in which a person has appeared, then the testing set is split to 1814 probe images and 7246 gallery images according to the ratio of 1:4 approximately.

Following the traditional person re-id evaluation protocol [32], we use the mean average precision (mAP) and the Cumulated Matching Characteristics (CMC) curve [8] to evaluate the performance of all person re-id algorithms. mAP considers both precision and recall of an algorithm, and CMC curve shows the probability that a query identity appears in different-sized candidate lists.

Table 2. Results of person re-id w/o GRC by different annotations on Group-reID dataset (%). "GRC" and "Person Re-id" are results with or without GRC in the person re-id methods, respectively.

Method	Group-ID				**Person-ID**			
	Person Re-id		GRC		Person Re-id		GRC	
	mAP	Rank-1	mAP	Rank-1	mAP	Rank-1	mAP	Rank-1
HistLBP [28]	4.0	9.5	**38.0**	**44.2**	10.5	22.5	43.9	55.3
ELF [5]	6.0	17.4	**43.1**	**62.2**	11.0	26.4	44.5	61.2
LDFV [13]	9.1	21.5	**52.4**	**61.5**	20.5	39.4	64.8	73.4
gBiCov [21]	9.3	27.0	**51.0**	**73.4**	13.7	34.3	51.1	74.3
LOMO [9]	13.9	34.3	**62.6**	**74.3**	24.1	47.3	69.0	80.9
WHOS [11]	26.6	55.8	**79.4**	**89.5**	32.6	62.2	80.0	89.3
GOG [14]	28.0	58.5	**78.1**	**87.8**	39.1	67.1	82.5	90.8
RPP [24]	27.9	55.5	**78.4**	**88.4**	40.1	67.4	85.0	92.7
PCB+RPP [24]	39.0	67.7	**82.9**	**92.3**	43.0	72.9	85.2	93.9
CaffeNet [7]	38.6	68.3	**80.8**	**90.3**	51.5	76.5	91.1	94.8
VGG16 [22]	47.4	74.4	**86.4**	**91.6**	54.3	80.9	91.5	95.3
GoogleNet [25]	55.8	80.7	**88.9**	**93.7**	66.3	86.4	94.2	96.9
ResNet50 [6]	61.9	84.9	**89.6**	**93.0**	66.1	87.2	93.6	95.8

4.3   Results

Since our framework is based on group-associated person re-id, in order to validate whether the group retrieval is helpful to the person re-id task, we evaluate all methods on our proposed Group-reID dataset.

We use kinds of person re-id methods to validate the effectiveness of group retrieval constraint, including traditional hand-crafted methods such as WHOS [11], gBiCov [21], GOG [14], HistLBP [28], LOMO [9], ELF [5], and LDFV [13], the state-of-the-art Convolutional Neural Network (CNN) based methods such as RPP, PCB+RPP [24], and our proposed siamese network based backbone CNN methods such as CaffeNet, VGG16, ResNet50, GoogleNet.

The comparison between single person re-id and using constraint of GRC on all state-of-the-art methods are shown in Table 2, both hand-crafted and deep learning based methods have been improved with GRC significantly, especially for the hand-crafted methods, such as gBiCov raises its mAP and Rank-1 with 41.7% and 46.4% by Group-ID. Moreover, using Person-ID can promote the results better than using Group-ID, since the annotation of Person-ID provides more concrete info of the personal identity to the classifier than Group-ID.

Different from person re-id, group retrieval introduces the relationship among its individuals for further research. Group retrieval identifies persons in a group not only by their appearance but also via their correlations, which makes up for the shortage of person re-id due to its dependence on the appearance cue of each individual. Thus GRC provides useful group constraint to promote person re-id.

Additional experimental results of CMC is shown in Fig. 6, the results indicate that additional group information of GRC promotes the matching rate of all methods effectively with a large margin in Fig. 6, which reveals the superiority of the proposed GRC method.

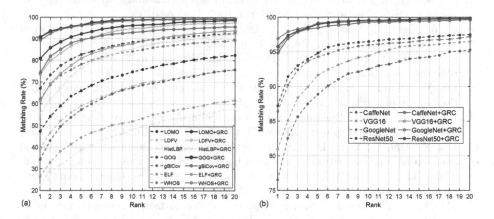

Fig. 6. CMC of all the hand-crafted and deep learning methods w/o GRC by using Person-ID. (a) Handcrafted methods. (b) Four siamese network based CNN backbone methods.

5   Conclusion

In this paper, we recommend a novel GRC method to improve the performance of traditional person re-id methods by using group information. The proposed GRC uses the correlations among groups to solve the shortage caused by the excess dependence of appearance cues, which can relieve the interference of changing clothes and occlusion. In the future, we'll try to find more comprehensive strategy to enhance the robustness of GRC.

Acknowledgements. This work was supported by NSFC (61573387, 61876104).

References

1. Daniel Costea, A., Nedevschi, S.: Semantic channels for fast pedestrian detection. In: CVPR, pp. 2360–2368 (2016)
2. Feng, Z., Lai, J., Xie, X.: Learning view-specific deep networks for person re-identification. TIP **27**(7), 3472–3483 (2018)
3. Gheissari, N., Sebastian, T., Hartley, R.: Person reidentification using spatiotemporal appearance. In: CVPR, vol. 2, pp. 1528–1535. IEEE (2006)

4. Girshick, R.: Fast R-CNN. In: ICCV, pp. 1440–1448 (2015)
5. Gray, D., Tao, H.: Viewpoint invariant pedestrian recognition with an ensemble of localized features. In: Forsyth, D., Torr, P., Zisserman, A. (eds.) ECCV 2008. LNCS, vol. 5302, pp. 262–275. Springer, Heidelberg (2008). https://doi.org/10.1007/978-3-540-88682-2_21
6. He, K., Zhang, X., Ren, S., Sun, J.: Deep residual learning for image recognition. In: CVPR, pp. 770–778 (2016)
7. Krizhevsky, A., Sutskever, I., Hinton, G.E.: ImageNet classification with deep convolutional neural networks. In: NIPS, pp. 1097–1105 (2012)
8. Li, W., Zhao, R., Xiao, T., Wang, X.: DeepReID: deep filter pairing neural network for person re-identification. In: CVPR, pp. 152–159 (2014)
9. Liao, S., Hu, Y., Zhu, X., Li, S.Z.: Person re-identification by local maximal occurrence representation and metric learning. In: CVPR, pp. 2197–2206 (2015)
10. Lin, W., et al.: Group re-identification with multi-grained matching and integration. arXiv preprint arXiv:1905.07108 (2019)
11. Lisanti, G., Masi, I., Bagdanov, A.D., Del Bimbo, A.: Person re-identification by iterative re-weighted sparse ranking. TPAMI **37**(8), 1629–1642 (2015)
12. Liu, X., Mei, L., Yang, D., Lai, J., Xie, X.: Feature visualization based stacked convolutional neural network for human body detection in a depth image. In: Lai, J.H., et al. (eds.) PRCV 2018. LNCS, vol. 11257, pp. 87–98. Springer, Cham (2018). https://doi.org/10.1007/978-3-030-03335-4_8
13. Ma, B., Su, Y., Jurie, F.: Local descriptors encoded by fisher vectors for person re-identification. In: Fusiello, A., Murino, V., Cucchiara, R. (eds.) ECCV 2012. LNCS, vol. 7583, pp. 413–422. Springer, Heidelberg (2012). https://doi.org/10.1007/978-3-642-33863-2_41
14. Matsukawa, T., Okabe, T., Suzuki, E., Sato, Y.: Hierarchical gaussian descriptor for person re-identification. In: CVPR, pp. 1363–1372 (2016)
15. Mei, L., Lai, J., Xie, X., Zhu, J., Chen, J.: Illumination-invariance optical flow estimation using weighted regularization transform. TCSVT (2019)
16. Mei, L., Chen, Z., Lai, J.: Geodesic-based probability propagation for efficient optical flow. Electron. Lett. **54**, 758–760 (2018)
17. Moussaïd, M., et al.: The walking behaviour of pedestrian social groups and its impact on crowd dynamics. PLoS ONE **5**(4), e10047 (2010)
18. Oren, M., Papageorgiou, C., Sinha, P., Osuna, E., Poggio, T.: Pedestrian detection using wavelet templates. In: CVPR, vol. 97, pp. 193–199 (1997)
19. Redmon, J., Divvala, S., Girshick, R., Farhadi, A.: You only look once: Unified, real-time object detection. In: CVPR, pp. 779–788 (2016)
20. Redmon, J., Farhadi, A.: Yolov3: an incremental improvement. arXiv preprint arXiv:1804.02767 (2018)
21. Sánchez, J., Perronnin, F., Mensink, T., Verbeek, J.: Image classification with the fisher vector: theory and practice. IJCV **105**(3), 222–245 (2013)
22. Simonyan, K., Zisserman, A.: Very deep convolutional networks for large-scale image recognition. arXiv preprint arXiv:1409.1556 (2014)
23. Sun, X., Zheng, L.: Dissecting person re-identification from the viewpoint of viewpoint. In: CVPR, pp. 608–617 (2019)
24. Sun, Y., Zheng, L., Yang, Y., Tian, Q., Wang, S.: Beyond part models: person retrieval with refined part pooling (and a strong convolutional baseline). In: Ferrari, V., Hebert, M., Sminchisescu, C., Weiss, Y. (eds.) ECCV 2018. LNCS, vol. 11208, pp. 501–518. Springer, Cham (2018). https://doi.org/10.1007/978-3-030-01225-0_30

25. Szegedy, C., Liu, W., Jia, Y., et al.: Going deeper with convolutions. In: CVPR, pp. 1–9 (2015)
26. Tuzel, O., Porikli, F., Meer, P.: Pedestrian detection via classification on rieman-nian manifolds. TPAMI **30**(10), 1713–1727 (2008)
27. Wei-Shi, Z., Shaogang, G., Tao, X.: Associating groups of people. In: BMVC, pp. 23–1 (2009)
28. Xiong, F., Gou, M., Camps, O., Sznaier, M.: Person re-identification using kernel-based metric learning methods. In: Fleet, D., Pajdla, T., Schiele, B., Tuytelaars, T. (eds.) ECCV 2014. LNCS, vol. 8695, pp. 1–16. Springer, Cham (2014). https://doi.org/10.1007/978-3-319-10584-0_1
29. Yu, H.X., Zheng, W.S., Wu, A., Guo, X., Gong, S., Lai, J.H.: Unsupervised person re-identification by soft multilabel learning. In: CVPR, pp. 2148–2157 (2019)
30. Zhang, C., Wu, L., Wang, Y.: Crossing generative adversarial networks for cross-view person re-identification. Neurocomputing **340**, 259–269 (2019)
31. Zhang, S., Benenson, R., Schiele, B., et al.: Filtered channel features for pedestrian detection. In: CVPR, vol. 1, p. 4 (2015)
32. Zheng, L., Shen, L., Tian, L., Wang, S., Wang, J., Tian, Q.: Scalable person re-identification: a benchmark. In: CVPR, pp. 1116–1124 (2015)
33. Zheng, Z., Zheng, L., Yang, Y.: A discriminatively learned CNN embedding for person reidentification. TOMM **14**(1), 13 (2018)

A Hierarchical Student's t-Distributions Based Unsupervised SAR Image Segmentation Method

Yuhui Zheng[1]($\boxtimes$), Yahui Sun[1], Le Sun[1], Hui Zhang[2],
and Byeungwoo Jeon[3]

[1] School of Computer and Software, Nanjing University of Information Science
and Technology, Nanjing 210044, People's Republic of China
zhengyh@vip.126.com
[2] School of Technology, Nanjing Audit University,
Nanjing 211815, People's Republic of China
[3] College of Information and Communication Engineering,
Sungkyunkwan University, Suwon 440-746, Korea

Abstract. We introduce a finite mixture mode using hierarchical Student's distributions, called hierarchical Student's t-mixture model (HSMM), for SAR images segmentation. The main advantages of the proposed method are as follows: first, in HSMM, the clustering problem is reformulated as a set of subclustering problems each of which can be solved by the traditional SMM algorithm. Second, a novel image content-adaptive mean template is introduced into HSMM to increase its robustness. Third, an expectation maximization algorithm is utilized for HSMM parameters estimation. Finally, experiments show that the HSMM is effective and robust.

Keywords: SAR image segmentation · Hierarchical student's-t distributions · Structure tensor · Nonlocally weighted mean template

1 Introduction

Segmentation is an inevitable technique for understanding SAR imagery [1]. The approach amounts to dividing an image into regions with consistent attributes and distinct statistical features [2–4]. As an important part of image interpretation, SAR image segmentation can inform about the entire structure for subsequent interpretation, highlight the areas of interest, and reveal the essence of the image. However, speckle noise, which is manifested as a granular, black-and-white texture in SAR images, randomizes gray values of the image's pixels, posing a significant challenge to image segmentation task [5, 6].

Recently, the finite mixture model (FMM) has drawn more attentions for clustering and image segmentation. A representative model is the Gaussian mixture model (GMM), in which Gaussian distributions are selected to calculate conditional probability [7]. Since the FMM approach is based on statistical modelling, the GMM approach is flexible and efficient for clustering of multivariate data. However, the existing GMM approaches are often sensitive to contrast levels and noise, probably due to the facts that they fail to exploit the spatial dependences of neighborhood of image pixel.

© Springer Nature Switzerland AG 2019
Z. Cui et al. (Eds.): IScIDE 2019, LNCS 11935, pp. 472–483, 2019.
https://doi.org/10.1007/978-3-030-36189-1_39

In order to further improve the performance of FMM on clustering, Student's t-distribution has been extensively employed in image segmentation [8–10], owing to the heavy tail of its distribution and robustness to noise and outliers. A mixture model with student's t-distribution has been considered as a good alternative to the GMM based image segmentation approach. Compared to the GMM, Student's t-mixture model (SMM) can better model and represent multivariate data. However, since FMMs are commonly based on histogram modelling which does not consider neighborhood information of image pixels, traditional SMMs still fail to overcome the limitation as GMMs. To address this problem, FMMs with Markov random field (MRF) have been proposed in the works [11–18] to utilize locally spatial information. The above-mentioned FMM based image segmentation methods often utilize Gaussian distribution. To overcome the drawback of GMM, Zhang suggested a SAR image segmentation method using the Student's t-mixture model (MSMM) [19] through a mean template for imposing spatial constraints on SMM, in order to explore the relevance in neighborhood of image pixels. Recently, a variant of the work [19] was introduced by Kong [20] with the aid of hierarchical Student's mixture model (HSMM) for natural image segmentation, inspired by the hierarchical mixture of experts (HME) algorithm in [15]. However, the mean templates fail to consider image contents. These templates are image-structure independent and isotropic.

In this paper we attempt to define a new weighted mean template with the aid of structure tensor [21, 22], in order to capture image structure information and enhance the robustness of our method. We employ it to compute the weighted mean template, where the weighting coefficients are adaptive to image content. We incorporate the weighted mean template into the Student's t-mixture model (SMM) and calculated the conditional and prior/posterior probability. The remainder of this paper is organized as follows. We briefly present the background of student's t-distribution and finite mixture model in Sect. 2. A nonlocally weighted mean template is described in Sect. 3. Section 4 introduces our proposed new model, i.e., hierarchical Student's mixture model (HSMM). The parameter learning algorithm is fully given in Sect. 5. Section 6 shows the SAR image segmentation results of various related methods. Section 7 draws our conclusions, finally.

2 Preliminaries

2.1 Student's t-Distribution

Student's t-distribution can be written as a sum of multiple Gaussian distributions with different variances and equal means as

$$S(y|\mu, \lambda, v) = \frac{\Gamma\left(\frac{v}{2} + \frac{1}{2}\right)}{\Gamma\left(\frac{v}{2}\right)} \left(\frac{\lambda}{\pi v}\right) \left[1 + \frac{\lambda(y - v)^2}{v}\right]^{-\left(\frac{v-1}{2}\right)} \tag{1}$$

where μ is mean parameter, λ denotes the precision parameter, and v describes the degrees of freedom. In (1), $\Gamma(\bullet)$ is the gamma function. For a D-dimensional variable y, the multivariate probability density function (PDF) can be calculated by integrating a multivariate Gaussian distribution with respect to a conjugate Wishart prior:

$$S(y|\mu, \Sigma, v) = \frac{\Gamma\left(\frac{v}{2} + \frac{D}{2}\right)}{\Gamma\left(\frac{v}{2}\right)} \frac{|\Sigma|^{\frac{1}{2}}}{(v\pi)^{\frac{D}{2}}} \times \left[1 + \frac{(y - \mu)^T \Sigma(y - \mu)}{v}\right]^{-\left(\frac{v}{2} + \frac{D}{2}\right)} \tag{2}$$

when v tends to infinity, i.e., $v \rightarrow \infty$, Student's t-distribution can be simplified as a Gaussian distribution.

2.2   Finite Mixture Model

A SAR image with N pixels can be divided into K classes in a segmentation task. Let us consider the following sets: $Q = \{1, 2,..., M\}$ represents the class labels of an image; $D = \{1, 2,..., d\}$ denotes the intensity values of an image; $L = \{1, 2,..., N\}$ represents the index set of an image; $X = \{x_i: (x_i \in Q$ and $i \in L)\}$ stands for a random field, describing the labels of an unknown image; $Y = \{y_i: (y_i \in Q$ and $i \in L)\}$ denotes a random field, representing the observation data (intensity values) of an image. y_i represents the observed value (intensity value) at the i-th pixel in an SAR image, and x_i stands for the responding class label to the i-th image pixel. Assume that each $j \in Q, i \in L$ satisfy

$$p(x_i = j) = \pi_j. \tag{3}$$

where π_j denotes the prior probability of the i-th pixel, which belongs to class j and satisfies the following constraints as

$$0 \leq \pi_j \leq 1 \text{ and } \sum_{j=1}^{M} \pi_j = 1. \tag{4}$$

Assume that Ω is the set of the FMM model parameters, i.e., $\Omega = \{\pi_j; \theta_j | j \in Q\}$. Then, considering $y \in Y$, $x \in X$, the joint distribution function of label x and observed SAR image pixel y can be computed by

$$p(x, y|\Omega) == \prod_{j=1}^{M} \left(\pi_{x_i} p(y_i|\theta_j)\right) \tag{5}$$

When $Y_i = y_i$, the marginal distribution function can be written as follows:

$$p(y_i|\Omega) = \sum_{j=1}^{M} \pi_j p(y_i|\theta_j). \tag{6}$$

3 Weighted Mean Template Using Structure Tensor

We apply a weighed geometric mean template to the standard FMM, which potentially increases the amount of spatial information captured by the model, to calculate the conditional probability for the *i-th* pixel owing to its neighborhood pixels. So, the marginal distribution of (6) can be modified as

$$p(y_i|\Omega) = \sum_{j=1}^{M} \pi_j \prod_{m \in \partial_i} p(y_i|\theta_j)^{\frac{w_m}{R_i}}. \tag{7}$$

Where ∂_i are neighborhood pixels of the *i-th* pixel excluding itself, and R_i is a normalization factor

$$R_i = \sum_{m \in \partial_i} w_m. \tag{8}$$

while w_m is a weighting coefficient. Because of the combination of the component and spatial information, the weight becomes smaller as the distance between the i_{th} and m_{th} pixels increases, so it can be defined as the function L_{mi} of Euclidean distance between the *i*-th and *m*-th pixels:

$$w_m = \frac{1}{1 + L_{mi}^2}. \tag{9}$$

4 FMM Using Hierarchical Student's T-Distributions

The standard FMM is based on the histogram statistics model, and existing methods such as GMM and SMM consider the image histogram to be symmetric. We propose a new method to fit the image's histogram using a non-symmetric distribution:

$$p(y_i|\pi, \kappa, \theta) = \sum_{j=1}^{M} \pi_{ij} \sum_{k=1}^{K} \kappa_{ijk} \prod_{m \in \partial_i} S(y_i|\theta_{jk})^{\frac{w_m}{R_i}}. \tag{10}$$

where $p(y_i|\theta_{jk})$ is defined as a Student's t-distribution using (2) with the parameter $\theta_{jk} = \{\mu_{jk}, \Sigma_{jk}, v_{jk}\}$. Let us consider the likelihood function of N data pixels. Using (10), we have:

$$P(Y|\pi, \kappa, \theta) = \prod_{i=1}^{N} \sum_{j=1}^{M} \pi_{ij} \sum_{k=1}^{K} \kappa_{ijk} \prod_{m \in \partial_i} S(y_i|\theta_{jk})^{\frac{w_m}{R_i}}. \tag{11}$$

The logarithmic likelihood function of (11) is:

$$\begin{aligned} L &= log\, P(Y|\pi, \kappa, \theta) \\ &= \sum_{i=1}^{N} log \left[\sum_{j=1}^{M} \pi_{ij} \sum_{k=1}^{K} \kappa_{ijk} \prod_{m \in \partial_i} S(y_i|\theta_{jk})^{\frac{w_m}{R_i}} \right]. \end{aligned} \tag{12}$$

It is easy to optimize the complete data function of (12). Setting the parameters $\Theta = \{\pi_{ij}, \kappa_{ijk}, \eta_{ij}, z_{ijk}, \mu_{jk}, \Sigma_{jk}, v_{jk}\}$, we obtain the objective function $J(\Theta)$ as follows:

$$
\begin{aligned}
J(\Theta) &= \sum_{i=1}^{N} \sum_{j=1}^{M} \eta_{ij} E_{\Theta} \left\{ log\, \pi_{ij} + log \left[\sum_{k=1}^{K} \kappa_{ijk} \prod_{m \in \partial_i} S(y_i | \theta_{jk})^{\frac{w_m}{R_i}} \right] \right\} \\
&= \sum_{i=1}^{N} \sum_{j=1}^{M} \eta_{ij} \left\{ log\, \pi_{ij} + \sum_{k=1}^{K} z_{ijk} E_{\Theta} \left[log\, \kappa_{ijk} + log \prod_{m \in \partial_i} S(y_i | \theta_{jk})^{\frac{w_m}{R_i}} \right] \right\} \qquad (13) \\
&= \sum_{i=1}^{N} \sum_{j=1}^{M} \eta_{ij} \left\{ log\, \pi_{ij} + \sum_{k=1}^{K} z_{ijk} \left[log\, \kappa_{ijk} + E_{\Theta} \left(\sum_{m \in \partial_i} \frac{w_m}{R_i} log\, S(y_i | \theta_{jk}) \right) \right] \right\}
\end{aligned}
$$

An advantage of the proposed method is the establishment of (10) incorporating the pixel intensity values and spatial information of an image.

5   Parametric Learning Algorithm

Using the EM algorithm, we estimate optimal parameter values by considering the objective function (13).

In E-step, to reduce the influence of the neighborhood pixels, we incorporate a weight mean template both into posterior probability and sub-posterior probability. Thus, we have

$$
\eta_{ij} = \frac{\pi_{ij} \prod_{m \in \partial_i} p(y_i | \pi, \kappa, \theta)^{\frac{w_m}{R_i}}}{\sum_{h=1}^{M} \pi_{ih} \prod_{m \in \partial_i} p(y_i | \pi, \kappa, \theta)^{\frac{w_m}{R_i}}}. \qquad (14)
$$

$$
z_{ijk} = \frac{\kappa_{ijk} \prod_{m \in \partial_i} S(y_i | \mu_{jk}, \Sigma_{jk}, v_{jk})^{\frac{w_m}{R_i}}}{\sum_{n}^{K_j} \kappa_{ijn} \prod_{m \in \partial_i} S(y_i | \mu_{jn}, \Sigma_{jn}, v_{jn})^{\frac{w_m}{R_i}}}. \qquad (15)
$$

In fact, the Student's t-distribution can be considered as a Gaussian distribution $g(\mu_{jk}, \Sigma_{jk}, /u_{ijk})$ with the accurate scale factor u. The latent parameter of factor u obeys the gamma distribution:

$$
u_{ijk} = \frac{v_{jk} + D}{v_{jk} + (y_i - \mu_{jk})^T \Sigma_{jk}^{-1} (y_i - \mu_{jk})} \qquad (16)
$$

In M-step, we first estimate the prior probability π_{ij}

$$
\pi_{ij} = \frac{\prod_{m \in \partial_i} (\eta_{ij})^{\frac{w_m}{R_i}}}{\prod_{m \in \partial_i} (\sum_{h=1}^{M} \eta_{ih})^{\frac{w_m}{R_i}}} \qquad (17)
$$

Using the Lagrange multiplier β_{ij}, the sub-prior probability κ_{ijk} is

$$\kappa_{ijk} = \frac{\eta_{ij} z_{ijk}}{\sum_{h=1}^{m} \eta_{ih} \sum_{n=1}^{K} z_{ihn}}. \tag{18}$$

We also incorporate the weight mean template and obtain

$$\kappa_{ijk} = \frac{\prod_{m \in \partial_i} \left(\eta_{ij} z_{ijk}\right)^{\frac{w_m}{R_i}}}{\prod_{m \in \partial_i} \left(\sum_{h=1}^{m} \eta_{ih} \sum_{n=1}^{K} z_{ihn}\right)^{\frac{w_m}{R_i}}} \tag{19}$$

We optimize the parameters $\left(\mu_{jk}, \Sigma_{jk}, \nu_{jk}\right)$ by considering the objective function as

$$
\begin{aligned}
J(\Theta) &= \sum_{i=1}^{N} \sum_{j=1}^{M} \eta_{ij} \left\{ \log \pi_{ij} + \sum_{k=1}^{K} z_{ijk} \times \left[\log \kappa_{ijk} + E_{\Theta} \left(\sum_{m \in \partial_i} \frac{w_m}{R_i} \log S(y_i | \theta_{jk}) \right) \right] \right\} \\
&= \sum_{i=1}^{N} \sum_{j=1}^{M} \eta_{ij} \Big\{ \log \pi_{ij} + \sum_{k=1}^{K} z_{ijk} \Big\{ \log \kappa_{ijk} + \frac{\nu_{jk}}{2} \log \frac{\nu_{jk}}{2} - \frac{D}{2} \log(2\pi) \\
&\quad - \log \Gamma\left(\frac{\nu_{jk}}{2}\right) - \frac{1}{2} \log |\Sigma| + \sum_{m \in \partial_i} \frac{w_m}{R_i} \Big[\frac{D + \nu_{jk}}{2} E_{\Theta}(\log u_{ijk}) \\
&\quad - \frac{\nu_{jk}}{2} u_{ijk} - \frac{1}{2} u_{ijk} (y_i - \mu_{jk})^T \Sigma_{jk}^{-1} (y_i - \mu_{jk}) \Big] \Big\} \Big\}
\end{aligned}
\tag{20}
$$

The objective function with parameters $\left(\mu_{jk}, \Sigma_{jk}, \nu_{jk}\right)$ can be extended by omitting some constant entries as shown in following Equations:

$$
\begin{aligned}
J_{\mu_{jk}} &= \sum_{i=1}^{N} \sum_{j=1}^{M} \eta_{ij} \sum_{k=1}^{K} z_{ijk} \Big[\sum_{m \in \partial_i} \frac{w_m}{R_i} \left(\frac{D + \nu_{jk}}{2} E_{\Theta}(\log u_{ijk}) \right. \\
&\quad \left. - \frac{\nu_{jk}}{2} u_{ijk} - \frac{1}{2} u_{ijk} (y_i - \mu_{jk})^T \Sigma_{jk}^{-1} (y_i - \mu_{jk}) \right) \Big]
\end{aligned}
\tag{21}
$$

$$
\begin{aligned}
J_{\Sigma_{jk}^{-1}} = J_{\mu_{jk}} &= \sum_{i=1}^{N} \sum_{j=1}^{M} \eta_{ij} \sum_{k=1}^{K} \Big[-\frac{1}{2} \log |\Sigma| + \sum_{m \in \partial_i} \frac{w_m}{R_i} \left(\frac{D + \nu_{jk}}{2} E_{\Theta}(\log u_{ijk}) \right. \\
&\quad \left. - \frac{\nu_{jk}}{2} u_{ijk} - \frac{1}{2} u_{ijk} (y_i - \mu_{jk})^T \sum_{jk}^{-1} (y_i - \mu_{jk}) \right) \Big]
\end{aligned}
\tag{22}
$$

$$
\begin{aligned}
J_{\nu_{jk}} &= \sum_{i=1}^{N} \sum_{j=1}^{M} \eta_{ij} \Big\{ \sum_{k=1}^{K} z_{ijk} \Big\{ \frac{\nu_{jk}}{2} \log \frac{\nu_{jk}}{2} - \log \Gamma\left(\frac{\nu_{jk}}{2}\right) \\
&\quad + \sum_{m \in \partial_i} \frac{w_m}{R_i} \Big[\frac{D + \nu_{jk}}{2} E_{\Theta}(\log u_{ijk}) - \frac{\nu_{jk}}{2} u_{ijk} - \frac{1}{2} u_{ijk} (y_i - \mu_{jk})^T \Sigma_{jk}^{-1} (y_i - \mu_{jk}) \Big] \Big\} \Big\}
\end{aligned}
\tag{23}
$$

With the objective functions (21) and (22), we obtain the derivatives of the mean μ_{jk} and the covariance matrices Σ_{jk}, respectively, and then set each value to zero at step $t + 1$.

$$\mu_{jk}^{(t+1)} = \frac{\sum_{i=1}^{N} \eta_{ij}^{(t)} \sum_{m\in\partial_i} \frac{w_m}{R_i} z_{ijk}^{(t)} u_{ijk}^{(t)} y_i}{\sum_{i=1}^{N} \eta_{ij}^{(t)} z_{ijk}^{(t)} u_{ijk}^{(t)}} \tag{24}$$

$$\Sigma_{jk}^{(t+1)} = \frac{\sum_{i=1}^{N} \eta_{ij}^{(t)} \sum_{m\in\partial_i} \frac{w_m}{R_i} z_{ijk}^{(t)} u_{ijk}^{(t)} (y_i - \mu_{jk})^T \Sigma_{jk}^{-1} (y_i - \mu_{jk})}{\sum_{i=1}^{N} \eta_{ij}^{(t)} z_{ijk}^{(t)}} \tag{25}$$

Finally, we set the value of derivative $\partial J_{v_{jk}} / \partial v_{jk}$ to zero, and have

$$\log\left(\frac{v_{jk}^{(t+1)}}{2}\right) - \psi\left(\frac{v_{jk}^{(t+1)}}{2}\right) + 1 + \frac{\sum_{i=1}^{N} \eta_{ij} z_{ijk} \sum_{m\in\partial_i} \frac{w_m}{R_i} \left(\log u_{ijk} - u_{ijk}\right)}{\sum_{i=1}^{N} \eta_{ij} z_{ijk}}$$
$$+ \psi\left(\frac{v_{jk}^{(t)} + D}{2}\right) - \log\left(\frac{v_{jk}^{(t)} + D}{2}\right) = 0 \tag{26}$$

6   Experimental Results

In this section, we tested the proposed HSMM on the set of synthetic and authentic SAR images. Experiments using the SVFMM [18], HMRF-FCM [23], and MSMM [19] approaches were also performed for comparison. Numerical experiments were performed using Matlab R2009b running on the Intel Pentium Dual-Core 2.2 GHZ CPU, with 2 GB RAM.

6.1   Synthetic Images

In this experiment, we used a synthetic SAR image (512×521) with four classes to evaluate the performance of our HSMM compared with those of the SVFMM, HMRF-FCM, and MSMM approaches. Original synthetic SAR image is given in Fig. 1(a). It was initialed by using k-means for classification. Figure 1(b) shows the corresponding simulated SAR image with 5.15 dB peak signal to noise ratio (PSNR). The overall accuracy ratio (OAR) is the ratio of the number of correctly labeled pixels to the overall number of pixels. In our experiment, we used OAR as a measure of the performance accuracy, and compared OAR values across several methods. OAR is defined as

$$OAR = \frac{\text{number of classified pixels}}{\text{total number of pixels}} \times 100\% \tag{27}$$

Methods that yield higher OARs demonstrate better partition performance. OAR values are in the [0–100] range.

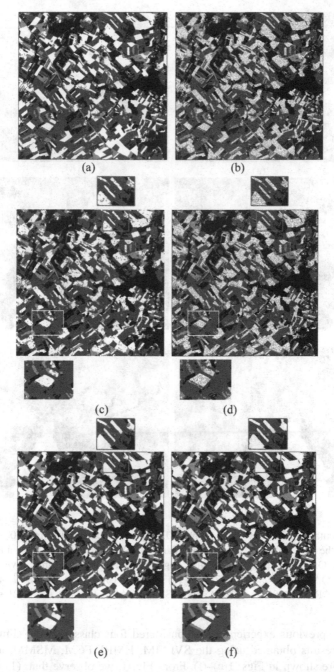

Fig. 1. (a) Original four-class image (512 × 512 pixels). (b) Simulated SAR image, PSNR = 5.15 dB. (c) HMRF-FCM, OAR = 87.52%, t = 153.87 s. (d) SVFMM, OAR = 89.48%, t = 566.1 s. (e) MSMM, OAR = 92.18%, t = 22.38 s. (f) HSMM, OAR = 94.29%, t = 31.17 s. (Color figure online)

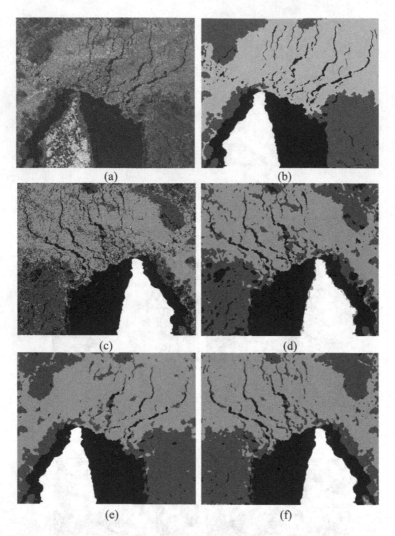

Fig. 2. Segmentation of a SAR image (C-band, HH, pixel spacing of 100 m, 480 × 408 pixels) captured over the Gulf of St. Lawrence on February 20, 1998, and acquired using a RADARSAT ScanSAR. (a) Original image. (b) Manual segmentation (ground truth, provided by CIS). (c) SVFMM, OAR = 80.92%, t = 47.41 s. (d) HMRF-FCM, OAR = 83.88%, t = 597.75 s. (e) MSMM, OAR = 91.29%, t = 16.80 s. (f) HSMM, OAR = 94.11%, t = 19.21 s.

Based on previous experience, we considered four classes in the simulated SAR image. The results obtained using the SVFMM, HMRF-FCM, MSMM, and HSMM approaches are shown in Figs. 1(c)–(f). From Fig. 1, we observe that: (1) all methods yield good segmentation, but the HMRF-FCM and SVFMM have shortcomings for reduction of image noise (from yellow and red boxes); (2) the MSMM performs better than the above two methods as well as the HSMM. However, comparing the color boxes, we see that the HSMM approach yields more accurate classification. From the caption of Fig. 1, although the HSMM approach is somewhat more computationally intensive than the MSMM, it yields the highest OAR value of 94.29%.

6.2 SAR Sea-Ice Images

In this experiment, we used a RADARSAT ScanSAR image (C-band, HH, 100-m pixel spacing) of the Gulf of St. Lawrence, to assess the performance accuracy of the proposed HSMM. This image was acquired on February 20, 1998, shown in Fig. 2(a). This image is well known to be generally divided into four classes: (1) land is the white area at the bottom; (2) water is the dark area around the image; (3) gray ice is in the upper left of the image to the bright area at the center of the image; and (4) the rest is gray-white ice. This image has many long narrow leads clearly visible in the gray ice region, and thus is suitable for testing the proposed model. Figure 2(b) shows the ground truth of the image obtained by manual segmentation; this allows fair performance comparison. Figures 2(c)–(f) show the segmentation results using the SVFMM, HMRF-FCM, MSMM, and HSMM approaches, respectively. The segmentation by the SVFMM is very poor, with much noise, which is unacceptable. The HMEF-FCM and MSMM both demonstrate good performance, but the HMRF-FCM is too computationally intensive. The MSMM demonstrates a slightly better performance, with OAR of 91.29%. It is worth noting that our HSMM yielded the highest OAR (94.11%), and its computation time was somewhat shorter than that of the MSMM, within acceptable limits.

We compared the computation time of the different methods, we can clearly see that our proposed HSMM method is faster than both the SVFMM and HMRF-FCM, and acceptably slower than the MSMM.

7 Conclusions

In this paper, we proposed a simple and efficient algorithm (i.e., the HSMM) to improve the robustness with respect to speckle noise of the traditional SMM, for SAR image segmentation applications. In the HSMM, the clustering problem is solved in two steps. In the first step, we break up the clustering problem into smaller subproblems; in the second step, we propose solutions to these smaller clustering subproblems. Therefore, the traditional probability is regarded as a sub-conditional one and the traditional SMM is deemed as a special case with one sub-class. In addition, we adopted a template for conditional probability and prior/posterior probability. Finally, the EM algorithm was used to maximize the data log-likelihood function and estimate the model's parameters.

References

1. Gemme, L., Dellepiane, S.G.: An automatic data-driven method for SAR image segmentation in sea surface analysis. IEEE Trans. Geosci. Remote Sens. **56**(5), 2633–2646 (2018)
2. Guo, Y., Jiao, L., Wang, S., Liu, F., Hua, W.: Fuzzy-superpixels for polarimetric SAR images classification. IEEE Trans. Fuzzy Syst. **26**(5), 2846–2860 (2018)
3. Zhang, D., Meng, D., Han, J.: Co-saliency detection via a self-paced multiple-instance learning framework. IEEE Trans. Pattern Anal. Mach. Intell. **39**(5), 865–878 (2017)
4. Wang, F., Wu, Y., Zhang, P., Li, M.: Synthetic aperture radar image segmentation using non-linear diffusion-based hierarchical triplet Markov fields model. IET Image Process. **11**(12), 1302–1309 (2017)
5. Akbarizadeh, G., Rahmani, M.: Unsupervised feature learning based on sparse coding and spectral clustering for segmentation of synthetic aperture radar images. IET Comput. Vis. **9**(5), 629–638 (2015)
6. Luo, S., Tong, L., Chen, Y.: A multi-region segmentation method for SAR images based on the Multi-texture model. IEEE Trans. Image Process. **27**(5), 2560–2574 (2018)
7. Thangarajah, A., Wu, Q.M., Yang, J.Y.: Fusion-based foreground enhancement for background subtraction using multivariate multi-model Gaussian distribution. Inf. Sci. **430**, 414–431 (2018)
8. Nguyen, T.M., Wu, Q.M.J.: Robust student's-t mixture model with spatial constraints and its application in medical image segmentation. IEEE Trans. Med. Imaging **31**(1), 103–116 (2012)
9. Peel, D., McLachlan, G.: Robust mixture modeling using the t-distribution. Stat. Comput. **10**(4), 339–348 (2000)
10. Zhang, H., Wu, Q.M.J., Nguyen, T.M.: A robust fuzzy algorithm based on student's t-distribution and mean template for image segmentation application. IEEE Signal Process. Lett. **20**(2), 117–120 (2013)
11. Clifford, P.: Markov random fields in statistics. Disord. Phys. Syst. A **14**(1), 128–135 (1990)
12. Sanjay-Gopal, S., Hebert, T.J.: Bayesian pixel classification using spatially variant finite mixtures and the generalized EM algorithm. IEEE Trans. Image Process. **7**(7), 1014–1028 (1998)
13. Zhang, Y., Brady, M., Smith, S.: Segmentation of brain MR images through a hidden Markov random field model and the expectation-maximization algorithm. IEEE Trans. Med. Imaging **20**(1), 45–57 (2001)
14. Michael, I., Robert, A.: Hierarchical mixtures of experts and the EM algorithm. Neural Comput. **6**(2), 181–214 (1994)
15. Chatzis, S.P., Kosmopoulos, D.I., Varvarigou, T.A.: Robust sequential data modeling using an outlier tolerant hidden Markov model. IEEE Trans. Pattern Anal. Mach. Intell. **31**(9), 1657–1669 (2009)
16. Gerogiannis, D., Nikou, C., Likas, A.: The mixtures of student's t-distributions as a robust framework for rigid registration. Image Vis. Comput. **27**(9), 1285–1294 (2009)
17. Zhang, H., Wu, Q.M.J., Nguyen, T.M.: Incorporating mean template into finite mixture model for image segmentation. IEEE Trans. Neural Netw. Learn. Syst. **24**(2), 328–335 (2013)
18. Blekas, K., Likas, A., Galatsanos, N.P., Lagaris, I.E.: A spatially constrained mixture model for image segmentation. IEEE Trans. Neural Netw. **16**(2), 494–498 (2005)

19. Zhang, H., Wu, Q.M.J., Nguyen, T.M., Sun, X.: Synthetic aperture radar image segmentation by modified student's t-mixture model. IEEE Trans. Geosci. Remote Sens. **52**(7), 4391–4403 (2014)
20. Kong, L., Zhang, H., Zheng, Y., Chen, Y., Zhu, J., Wu, Q.M.J.: Image segmentation using a hierarchical student's-t mixture model. IET Image Process. **11**(11), 1094–1102 (2017)
21. Estellers, V., Soatto, S., Bresson, X.: Adaptive regularization with the structure tensor. IEEE Trans. Image Process. **24**(6), 1777–1790 (2015)
22. Zheng, Y., Beon, J., Zhang, W., Chen, Y.: Adaptively determining regularization parameters in nonlocal total variation regularization for image denoising. Electron. Lett. **51**(2), 144–145 (2015)
23. Chatzis, S.P., Varvarigou, T.A.: A fuzzy clustering approach toward hidden Markov random field models for enhanced spatially constrained image segmentation. IEEE Trans. Fuzzy Syst. **16**(5), 1351–1361 (2008)

Multi-branch Semantic GAN for Infrared Image Generation from Optical Image

Lei Li[1], Pengfei Li[1], Meng Yang[1,2(✉)], and Shibo Gao[3]

[1] School of Data and Computer Science, Sun Yat-sen University, Guangzhou, China
yangm6@mail.sysu.edu.cn
[2] Key Laboratory of Machine Intelligence and Advanced Computing(SYSU),
Ministry of Education, Guangzhou, China
[3] Beijing Aerospace Automatic Control Institute, Beijing 100854, China

Abstract. Infrared remote sensing images capture the information of ground objects by their thermal radiation differences. However, the facility required for infrared imaging is not only priced high but also demands strict testing conditions. Thus it becomes an important topic to seek a way to convert easily-obtained optical remote sensing images into infrared remote sensing images. The conventional approaches cannot generate satisfactory infrared images due to the challenge of this task and many unknown parameters to be determined. In this paper, we proposed a novel multi-branch semantic GAN (MBS-GAN) for infrared image generation from the optical image. In the proposed model, we draw on the idea from Ensemble Learning and propose to use more than one generator to synthesize the infrared images with different semantic information. Specially, we integrate scene classification into image transformation to train models with scene information, which assists learned generation models to capture more semantic characteristics. The generated images are evaluated by PSNR, SSIM and cosine similarity. The experimental results prove that this proposed method is able to generate images retaining the infrared radiation characteristics of ground objects and performs well in converting optical images to infrared images.

Keywords: Infrared image generation · Generative adversarial networks · Residual neural network

1 Introduction

Infrared imaging is a technique of capturing the infrared light from objects and converting them into visible images interpretable by a human eye. Near Infrared light is the portion of the electromagnetic spectrum that just past the Red light. The far infrared radiation that is also called thermal infrared radiation is heat emitted by any object that has a temperature above absolute zero. Different objects have different infrared reflectance which makes them look brighter or darker in infrared images.

© Springer Nature Switzerland AG 2019
Z. Cui et al. (Eds.): IScIDE 2019, LNCS 11935, pp. 484–494, 2019.
https://doi.org/10.1007/978-3-030-36189-1_40

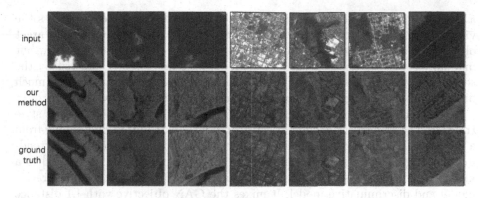

Fig. 1. The transformation results of our method. As shown, our method can generate results that are very close to groundtruth.

Compared with optical remote sensing images, infrared remote sensing images indicate more information about the essence and distribution of ground objects. However, the facility required for infrared imaging is not only priced high but also demands strict testing conditions. Converting easily-obtained optical remote sensing images into infrared remote sensing images helps overcome these restrictions.

Optical and infrared spectra provide different message. While optical images can provide information similar to what the human eye would see, optical images are incapable of providing useful information in situations where the illumination is poor or the weather is bad [1]. Infrared remote sensing images capture the information of ground objects by their thermal radiation differences. So in certain types of situations, infrared remote sensing images are useful than optical images because infrared information is independent of the quality of the environment. Moreover, infrared remote sensing images are widely applied to various fields such as military reconnaissance, climatology, and environmental monitoring. Therefore, we try to generate costly infrared images by more readily available optical images.

The challenge in our task is how to capture the infrared information from optical images. Most current researches utilize physical features, physical modeling and manual setting of environmental parameters to generate infrared images. Luo et al. [4] proposed a method that converts the infrared image into a grayscale image and then divides it into small parts. However, since the target object cannot be segmented completely, it is necessary to manually segment the target object from the background. After that, the temperature and related atmospheric parameters, which are used to calculate the amount of infrared radiation of the target object, of each segmentation area should be set manually. Finally, the infrared image is obtained by physical modeling. Wu et al. [5] use the histogram to convert optical images to infrared images by learning the characteristics of optical/infrared image pairs, but the method is mainly for the conversion of specific target objects (plants, buildings). Li et al. [6] proposed a neural network-based infrared image generation method, which segments the visible light image into different regions, predicts the temperature of the target object of different materials, and then performs the radiation calculation, but manually segmentation is needed. And the results are directly affected by the segmentation of image. It must be mentioned that these methods have much difficulty in processing large quantities of images simultaneously.

Generative adversarial networks [13] is an effective model for image style transfer. The framework is good at capturing data distribution by learning from a big dataset. Recently, many generative adversarial models come out by extending the original GANs in different ways. For example, Pix2pixGAN [14] is a kind of conditional GAN requiring to input the real image into the generative model and discriminative model. It mixes the GAN objective with L1 distance to make the output near the ground truth output in an L1 sense. Zhu et al. [11] creatively proposed the framework that contains two generative adversarial networks. CycleGAN [11] only requires unpaired images for training rather than paired images. The training results make it possible to translate an image from each domain to another. StarGAN [12] is a novel model for multi-domain image-to-image translation. It combines domain information with image information to train just one model for multi-domain translation. With more and more research, the existing GAN methods have been able to generate higher quality images. But this is generally only for certain scenarios under the big data set. For example, learning from the natural scene image in the ImageNet [15] dataset, BigGAN [16] has been able to generate realistic and amazing results. However, for many applications in real-life scenarios, such as infrared remote sensing and visible light image dataset, corresponding adjustments are needed to generate satisfactory images. GAN [13] is a powerful framework for image style transfer. But for infrared images, different ground objects have their own infrared radiation characteristics. By simply using a framework based on GAN, e.g. pix2pixGAN, the transformation model would ignore some characteristics.

In this paper, we seek to find a method converting optical images into infrared images based on image style transfer. In order to retain the infrared radiation characteristics of different objects, we proposed a model that combines residual neural network and generative adversarial networks, which we named MBS-GAN. We draw on the idea from Ensemble Learning and propose to use more than one generator to synthesize the infrared images with different semantic information. Specially, we integrate scene classification into image transformation to train models with scene information, which assists learned generation models to capture more semantic characteristics. Our proposed model overcome this weakness and our result proves that this proposed method is able to generate images retaining the infrared radiation characteristics of ground objects and performs well in converting optical images to infrared images. Some results are demonstrated in Fig. 1.

2 Related Work

The problem to be solved in this paper is how to convert optical remote sensing images into more realistic infrared remote sensing images. As mentioned before, different ground objects have different infrared radiation characteristics. The proposed method in this paper combines scene classification and image style transfer. In this section, we review two methods of tasks above.

ResNet. In image classification, there are many classification models based on deep convolutional neural networks, e.g. AlexNet [7], GooLeNet [8] and VGGNet [9]. In most situations, the more the layers, the better the training results. Excessive number of layers may cause the problem of vanishing/exploding gradients. This problem is solved by normalized initialization and intermediate normalization layers. But there is another problem that as the depth of the network deepens, the training accuracy rate will gradually decline after getting saturated. He et al. [10] proposed a deep residual learning framework to address the degradation of training accuracy. ResNet uses building blocks to replace original convolution layers. Short connections are used in each block to pass through all information of the network input. Experiments on datasets, e.g. ImageNet and CIFAR-10, shows that ResNet are easy to optimize and can solve the degradation problem.

GANs. Goodfellow [13] proposed a novel framework that simultaneously trains two models via an adversarial process. The framework includes a generative model and a discriminative model. The generative model is like a team of counterfeiters that produces fake currency while the discriminant model is like police who detect the counterfeit currency. In the training process, the generative model tries to cheat the discriminant model and the discriminant tries to identify the fake. The competition between the two models drives them to improve their methods until the fake currency generated by the generative model can't be distinguished. Pix2pixGAN [14] is a conditional GAN requiring to input the real image into the generative model and discriminative model. It mixes the GAN objective with L1 distance to make the output near the ground truth output in an L1 sense. The generator of pix2pixGAN uses skip connections to share low-level information between input and output. Relying on an L1 term to force low-frequency correctness, the discriminator of pix2pixGAN uses PatchGAN that only penalizes small patches of an images.

3 Method

In this section, we draw on the idea from the machine learning algorithm Ensemble Learning to propose our GAN promotion method. We propose to use more than one generator to generate images to improve the performance of GAN, similar to the idea of using multiple classifiers to make decisions in the machine learning algorithm Boosting. We will start our research based on the current popular pix2pixGAN.

3.1 Multi-generator Training

Ensemble Learning [17] is a machine learning method that uses a series of classifiers to learn and uses a certain rule to integrate individual learning results to achieve better learning outcomes than a single classifier. Based on this idea, We propose to use multiple generators to improve the image generation quality of GAN. The basis for dividing multiple generators and how to use multiple generators are two issues that need to be addressed. For the first question, the first solution we think of is the semantic category. Obviously, we need to generate images in multiple scenes, and each scene has its own unique characteristics, so it is reasonable and simple to use different generators for different scenes according to semantics. For the second problem, one method that is very easy to solve is to use a classifier to classify the input and then pass the image to the appropriate generator. So as stated above, we propose our model as shown in Fig. 2. As the figure shows, the input image is first passed through a resnet-50 neural network for classification, and then the image is input to the corresponding generator for image generation and finally get the output.

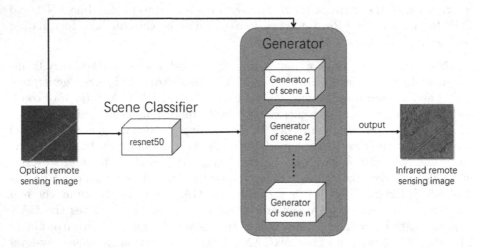

Fig. 2. The architecture of our model. The input image will be first inputted to the scene classifier and then go through the corresponding generator.

However, when a dataset has many different semantic classes and using a generator for each semantic category, the model structure will become very large. So in order to optimize our model, we have to mask some constraint on the semantics. One way to reduce semantic categories is cluster semantics because similar semantics have some common features which can be learned by a common generator. Another simpler approach is to train a generator to find the semantic categories that are difficult to train, ie, the generated categories with lower metrics, and then use separate generator training for difficult categories.

We recommend that the cluster method can be used when there are many semantic categories. Otherwise, the latter can be used when there are few semantic categories, which directly solves the problem of poor semantic generation. With reference to the idea of Ensemble Learning and solving difficulties separately, the quality of image generation of GAN can be improved by our multi-generator training method.

3.2 Objective

We use the loss function of conditional GAN to guide the training process. And the objective of a Conditional GAN in pix2pix [14] can be expressed as

$$L_{cGAN}(G, D) = E_{x,y}[\log D(x, y)] + E_{x,z}[\log(1 - D(x, G(x, z)))] \qquad (1)$$

where z means a random noise, x and y represent conditional image and target image, respectively. Meanwhile, generator G tries to minimize this objective against an adversarial discriminator D that tries to maximize it.

For our method, let X and Y be two image domains (e.g., the optical and infrared image domains). And the training samples from domain X and Y are denote by $\{x^i\}_{i=1}^M \subset X$ and $\{y^i\}_{i=1}^M \subset Y$ respectively, where M is the number of semantics and i means the ith semantic category; this shows that we will have M generators but only one discriminator for all inputs. Here $x^i \subset X$ will be inputted to corresponding generator G_i. G is expected to well learn the scheme of image style transfer and output the images from target domain Y by the feedback from D. So we propose our L_{GAN} objective as:

$$L_{GAN}(G, D) = E_Y[\log D(y^i)] + E_X[\log(1 - D(G_i(x^i)))] \qquad (2)$$

According to [14], previous studies have found it beneficial to mix the GAN objective with a more traditional loss, such as L2 distance [18]. And [14] propose that using L1 distance rather than L2 as L1 encourages less blurring. Therefore, we introduce L1 loss as:

$$L_{l1}(G) = E_{x,y}[\|y^i - G_i(x^i)\|_1] \qquad (3)$$

The full objective of our proposed method is:

$$L = L_{GAN}(G, D) + \lambda L_{l1}(G) \qquad (4)$$

where λ is used to control the weights of L1 loss.

4 Implementation

In this work, we focus on converting optical remote sensing images to near infrared remote sensing images. we consider to classify the optical images into three categories, such as water, habitation and other. We simply divide our work into two parts for scene classification and image transformation.

We train a scene classification model by ResNet50 and image transformation model by pix2pixGAN.

For the scene classification model, we train it based on transfer learning. A ResNet50 model pretrained on ImageNet dataset is preferred. Pre-trained model can reduce training time and greatly improve the accuracy rate of training. There are two points that we have to pay attention to in the training process. The first one is that we have to preprocess images in our dataset. The input image size required for ResNet50 is 224×224 while the size of our images is 256×256. Thus, we have to crop, flip and normalize images before training. A 224×224 crop is centrally sampled from an image or its horizonal flip. The second one is that we have to change fully connected layers in the ResNet50 model to fit our classification model. Cause the number of categories in ImageNet dataset is 1000 and the number of categories in our dataset is 3. We use cross entropy loss function and SGD. The learning rate starts from 0.001 and is divided by 10 after several epoch. Each epoch contains training and validation. The model is set training mode in training and compute loss by the output of forward propagation and the real labels of images. Then the parameters in the model are updated by back propagation. The parameters of the model which gets the highest accuracy rate in validation are returned after the whole training process.

For training of the image transformation model, we have to concatenate the paired images into an 256×512 image. We alternately train the generator and the discriminator in training. The weight of L1 loss in the objective is 100. The models are trained using minibatch SGD and Adam [19]. The learning rate starts from 0.0002 and is linearly decayed after the first 100 epochs. Momentum parameters are set as $\beta_1 = 0.5$ and $\beta_2 = 0.999$.

5   Experiment

5.1   Dataset

A satellite imaging company called Planet recently released a dataset about remote sensing images of Amazon basin. The dataset is used in a Kaggle competition for labeling the ground feature [20]. The chips were derived from Planet's full-frame analytic scene products using 4-band satellites in sun-synchronous orbit and International Space Station orbit. Each image is 256×256 pixels and contains four bands of data: red, green, blue, and near infrared. In our experiment, we use the data which for training in the Kaggle competition to train our model. There are 40479 tiff image files which contains both optical information and near infrared information.

5.2   Experimental Results of Scene Classification

In this section, we simply divide the images into three categories: water, habitation and other based on their own labels. For each category, there are 2000 images for training. Here we use residual neural networks of 18, 34 and 50 layers to train our scene classification models. Table 1 shows the accuracy rates of scene classification using different models and we can observe that ResNet50 produces the highest accuracy rate.

Table 1. The accuracy rates of different ResNet models.

Model	The accuracy rate
ResNet18	0.87
ResNet34	0.92
ResNet50	0.95

5.3 Experimental Results of Image Transformation

Qualitative Evaluation. Figure 1 shows the results of image transformation from optical remote sensing images to near infrared remote sensing images. Compared with the original pix2pixGAN model, the method that training models for each scene produces better results. This method allows models to learn more information of the scene feature. For example, the color of water in infrared images is black. As the Fig. 3 shows, when we train our models with scene information, the color of water is close to ground truth, and results are not well when we simply use pix2pixGAN to train our model.

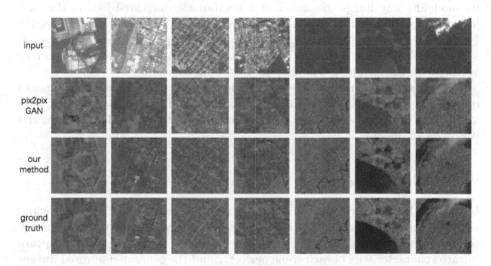

Fig. 3. Comparison of transformation between our method and pix2pixGAN.

Table 2. Evaluation on different image transformation models

Model	Category	PSNR	SSIM	Cosine similarity
pix2pix	water	24.5083	0.7448	0.9863
	habitation	23.5000	0.6996	0.9903
	other	24.8334	0.7258	0.9923
Our method	water	25.4383	**0.7580**	0.9866
	habitation	**26.6665**	0.7502	0.9906
	other	25.7735	0.7547	**0.9947**
pix2pix	overall	24.2805	0.7234	0.9897
Our method	**overall**	**25.9596**	**0.7543**	**0.9906**

Quantitative Evaluation. For quantitative evaluations, we perform to evaluate our experiments by using standard image quality assessment, e.g. PSNR, SSIM [21] and cosine similarity. The peak-signal-to-noise ratio (PSNR) and the structural similarity index measure (SSIM) are widely used objective metrics due to their low complexity and clear physical meaning. The PSNR value approaches infinity as the MSE approaches zero. This shows that a higher PSNR value provides a higher image quality. Also means, a small value of the PSNR implies high numerical differences between images. The SSIM is a well-known quality metric used to measure the similarity between two images. And it is designed by modeling any image distortion as a combination of three factors that are loss of correlation, luminance distortion and contrast distortion [22]. The SSIM is considered to be correlated with the quality perception of the human visual system (HVS). Cosine similarity is a commonly used approach to match similar vector, with the advantages of simplicity and effectiveness.

Table 2 shows the results of our method and pix2pixGAN generating images of each category. Obviously, the results of our method have higher PSNR, SSIM and cosine similarity scores and our method performs better than the original pix2pixGAN model.

6 Conclusion

For the task that converts optical remote sensing images into infrared remote sensing images, our method is able to retain features of different scenes. By combining scene classification with image transformation, the models capture infrared characteristics of each scene perfectly and the generated infrared images are close to ground truth. The experimental results shows our method works well. Although our method can achieve better results, there are cases where the classification is wrong during experiments. In this case, the results are different from the real infrared images and that needs further study.

Acknowledgement. This work is partially supported by the National Natural Science Foundation of China (Grant no.61772568, Grant no.61603364), the Guangzhou Science and Technology Program (Grant no. 201804010288), and the Fundamental Research Funds for the Central Universities (Grant no.18lgzd15).

References

1. Rodhouse, K.N.: A comparison of near-infrared and visible imaging for surveillance applications (2012)
2. Siesler, H.W.: Near-Infrared Spectroscopy: Principles, Instruments, Applications. Wiley, Hoboken (2008)
3. Reich, G.: Near-infrared spectroscopy and imaging: basic principles and pharmaceutical applications. Adv. Drug Delivery Rev. **57**(8), 1109–1143 (2005)
4. Luo, X., Sun, J., Liu, J., Xia, J.: Realization of infrared image acquisition by inversion of visible light image. J. Infrared Laser Eng. (2008). (in Chinese)
5. Wu, G., Bai, T., Bai, F.: Infrared image inversion based on visible light image. J. Infrared Technol. **33**(10), 574 (2011). (in Chinese)
6. Li, M., Xu, Z., Xie, H., Xing, Y.: Infrared image generation method based on visible light image and its detail modulation. J. Infrared Technol. **40**(1), 34–38 (2018). (in Chinese)
7. Krizhevsky, A., Sutskever, I., Hinton, G.E.: Imagenet classification with deep convolutional neural networks. Commun. ACM **60**(6), 84–90 (2017)
8. Szegedy C, Liu W, Jia Y.: Going deeper with convolutions. In: Proceedings of the IEEE conference on computer vision and pattern recognition, pp. 1–9 (2015)
9. Simonyan, K, Zisserman, A.: Very deep convolutional networks for large-scale image recognition. arXiv preprint arXiv:1409.1556 (2014)
10. He, K., Zhang, X., Ren, S.: Deep residual learning for image recognition. In: Proceedings of the IEEE conference on computer vision and pattern recognition, pp. 770-778 (2016)
11. Zhu, J.Y., Park, T., Isola, P.: Unpaired image-to-image translation using cycle-consistent adversarial networks. In: Proceedings of the IEEE International Conference on Computer Vision, pp. 2223-2232 (2017)
12. Choi, Y., Choi, M., Kim, M.: Stargan: unified generative adversarial networks for multi-domain image-to-image translation. In: Proceedings of the IEEE Conference on Computer Vision and Pattern Recognition, pp. 8789-8797 (2018)
13. Goodfellow, I., et al.: Generative adversarial nets. In: Advances in Neural Information Processing Systems, pp. 2672–2680 (2014)
14. Isola, P., Zhu, J.-Y., Zhou, T., Efros, A.A.: Image-to-image translation with conditional adversarial networks. In: Proceedings of the IEEE conference on computer vision and pattern recognition, pp. 1125–1134 (2017)
15. Deng, J., Dong, W., Socher, R., Li, L.-J., Li, K., Fei-Fei, L.: ImageNet: a large-scale hierarchical image database. In: CVPR (2009)
16. Brock, A., Donahue, J., Simonyan, K.: Large scale gan training for high fidelity natural image synthesis. arXiv preprint arXiv:1809.11096 (2018)
17. Dietterich, T.G., et al.: Ensemble learning. In: The handbook of brain theory and neural networks(2), pp. 110–125 (2002)
18. Pathak, D., Krahenbuhl, P., Donahue, J., Darrell, T., Efros, A.A.: Context encoders: Feature learning by inpainting. In: Proceedings of the IEEE conference on computer vision and pattern recognition, pp. 2536–2544 (2016)

19. Kingma, D.P., Ba, J..: Adam: a method for stochastic optimization. In: ICLR (2015)
20. Kaggle competition. planet: Understanding the amazon from space. https://www.kaggle.com/c/planet-understanding-the-amazon-from-space. Accessed 29 Apr 2019
21. Sheikh, H.R., Sabir, M.F., Bovik, A.C.: A statistical evaluation of recent full reference image quality assessment algorithms. IEEE Trans. Image Process. **15**(11), 3440–3451 (2006)
22. Wang, Z., Bovik, A.C., Sheikh, H.R., Simoncelli, E.P.: Image quality assessment: from error visibility to structural similarity. IEEE Trans. Image Process. **13**(4), 600–612 (2004)

Semantic Segmentation for Prohibited Items in Baggage Inspection

Jiuyuan An[1], Haigang Zhang[2(✉)], Yue Zhu[1], and Jinfeng Yang[2]

[1] Tianjin Key Lab for Advanced Signal Processing,
Civil Aviation University of China, Tianjin, China
[2] Institute of Applied Artificial Intelligence of the Guangdong-Hong Kong-Macao
Greater Bay Area, Shenzhen Polytechnic, Shenzhen 518055, China
zhg2018@sina.com

Abstract. The X-ray screening system is crucial to protecting the safety of public spaces. However, automated detection in baggage inspection is still far from practical application. Most detection tasks rely mainly on humans. In this paper, the detection of prohibited items is regarded as a semantic segmentation task. Considering some characters of security imageries, we propose a segmentation net with novel dual attention, which could capture richer features for refining the segmentation results. Our model could not only automatically recognize the class of prohibited items but also locate prohibited items in baggage. It could facilitate the security staffs to carry out inspection. To validate the effectiveness of our proposed model, extensive experiments have been conducted on the real X-ray security imageries datasets. The experimental results show the net achieves super performance (mIoU of 0.683).

Keywords: Prohibited items · Semantic segmentation · Attention

1 Introduction

X-ray screening technique plays an extremely important role in security inspection. In some public spaces, such as airport and train station, X-ray screening technique could help to detect prohibited items, preventing terrorist attacks and other vicious incidents. However, the detection for prohibited items relies mainly on the human. If the security staffs work for a long time, they are possible to become less attentive and easily distracted, allowing for the possibility that prohibited items may be overlooked. In addition, X-ray security imageries are different from natural images. They are monotonous and not have as rich features as natural images. For example, X-ray security imageries have poor colors and textures, lower contrast. So the efficiency is relatively low to recognize prohibited items by the human eyes. Therefore, this task is more suitable for computer processing, freeing human from this heavy work.

With the appearance of various CNN (Convolution Neural Networks), such as AlexNet, GoogleNet, ResNet [8,9,16], etc., the representation ability of networks

© Springer Nature Switzerland AG 2019
Z. Cui et al. (Eds.): IScIDE 2019, LNCS 11935, pp. 495–505, 2019.
https://doi.org/10.1007/978-3-030-36189-1_41

becomes stronger and stronger. Driven by the success of CNN, scholars have launched researches on the intelligence of security inspection. Mery [12] compares ten methods based on computer vision and finds deep learning method is superior to most methods. Afterwards, Akcay [1] explores the use of CNN in the tasks of classification and detection within X-ray baggage imageries. Xu [19] achieves the segmentation of prohibited items.

Semantic segmentation is a branch of computer vision, which is essentially a more rigorous classification task. By segmenting the X-ray security imageries, we could not only recognize the prohibited items but also visually see the location, which facilitates the security staffs to open the baggage for inspection. Comparing with object detection, we think the results of detection are messy, but the segmentation results are more intuitive. In addition, different categories of prohibited items need different disposal methods, which inherently requires the profile information of the prohibited items. Hence, the auto-detection of the prohibited items is regarded as a semantic segmentation task. Considering the poor features of X-ray security imageries, the attention is introduced to enhance the representation of the neural net.

In this paper, we build a semantic segmentation model with novel dual attention for prohibited items, which is illustrated in Fig. 1. To the best of our knowledge, this is the first study introducing semantic segmentation to the prohibited items detection task. By reviewing the convolution neural network, we think the convolution neural network could identify items by extracting both the channel and spatial information. Since the features of X-ray security imageries are poor comparing with natural images, the channel attention and spatial attention are simultaneously adopted to get richer semantic information. To achieve this, we propose a novel dual attention to control the flow of information within the segmentation net. Channel attention module could learn what the item is. And the spatial attention module could learn where the item is. The two modules could enforce some important activations while suppressing feature activations in irrelevant regions.

Our main contributions can be summarized as follows:

1. We build a semantic segmentation model to address the detection of prohibited items in X-ray security imageries.
2. A novel dual attention is introduced to the semantic segmentation to improve the segmentation results.
3. We achieve state-of-the-art performance on the X-ray security imageries datasets.

2 Related Work

Semantic Segmentation. In the past years, fully convolutional networks [10] have pushed the performance of semantic segmentation to soaring heights on a broad array of high-level problems. Following the structure of FCN, there are several high-performance models which could extract more detailed features. Peng [13] proposes large kernels to improve semantic segmentation. DeeplabV1

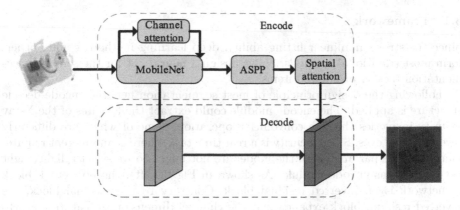

Fig. 1. The semantic segmentation net for prohibited items

[3] could enlarge the receptive field of neural networks without additional parameters by atrous convolution. SegNet [2] with encode-decode structure designs an ingenious connection between different feature maps to obtain richer detailed information. PSPNet [20] proposes a pyramid pooling module to aggregate different scale information. DeeplabV2-V3+ [4–6] designs atrous spatial pyramid pooling module to utilize global context information.

Attention. Inspired by human attention, attention mechanism is proposed to help neural net focus on the key information while ignoring irrelevant information. Attention is originally applied to NLP. And Luong [11] proposes global attention and local attention in machine translation. To get global dependencies, self-attention is proposed by [17]. Since the effect of attention, attention mechanism begins to expand into the field of computer vision. For example, Woo [18] proposes CBAM (convolution block attention module), making it more convenient to embed attention in CNN. The work [7] applies attention to scene segmentation.

Our approach is motivated by the success of segmentation models and attention modules in the above works. In our work, we consider the characters of security imageries and design a suitable model for the semantic segmentation task of prohibited items. Meanwhile, the potential of attention is exploited. We propose a novel dual attention to refine the segmentation results, which is more suitable to the security imageries.

3   Method

In this section, we briefly introduce our net for semantic segmentation of prohibited items. Firstly, we discuss the framework. Secondly, the attention modules embedded in the net is shown.

3.1 Framework

Since the strong nonlinear fitting ability, deep learning methods could outperform most traditional methods. So we design a neural net for the semantic segmentation task of prohibited items.

Following the design principle of most segmentation net, the encode-decode structure is applied. The encode module could extract the features of the X-ray security imageries through convolution operation, some of which are difficultly recognized by eyes. Since security is a real-time task, there is an inherent requirement that the parameters of the model are not huge. So we select a lightweight network [15] as encode module. As shown in Fig. 2(Left), the bottleneck block in network is an inverted residual block. Contrary to the residual block, the inverted residual block expands first the channel dimension of feature maps to increase redundancy of information. Then, the ordinary convolution is replaced by depthwise separable convolution, reducing the compute cost. The last 1×1 convolution adjusts the output channels. In the encode module, the input is firstly fed in 3×3 convolution. Subsequently, the features flow into 4 bottleneck blocks for extracting the hierarchical features. Furthermore, semantic segmentation is a pixel-level classification task. Only when the net "sees" the integrated prohibited items as much as possible, can the items be segmented more accurately. So a larger receptive field is needed. Considering the different methods of enlarging receptive field, we think the atrous convolution is suitable for solving the problem, which not introduces too many parameters. So in next 4 bottleneck block, the atrous convolution is used to capture more information. As a result, the feature map is downsampled 32 times.

Afterwards, in order to deal with the scale problem, the atrous spatial pyramid pooling is applied, which is schematically depicted in Fig. 2(Right). The ASPP module is comprised of 4 convolution layers and an average pooling layer. Especially, the convolution layers are also different. They are respectively 1×1 convolution and 3×3 convolution with different atrous rate. The module design follows a parallel pattern. The output feature map at the last stage is fed into the ASPP module. This operation could mix different scale information to address the scale problem.

The decode module could restore the resolution of segmentation results. In our net, the decode module is very simple. They are just some simple upsampling modules. As the Fig. 1 shows, the low-level feature map and high-level feature map are simultaneously utilized to get richer semantic information.

3.2 Channel Module

After convolution operation, the feature maps are mapped into high dimensional space. Rich information permeates the channels of feature maps. However, the channels are not equally important. So a channel attention module is introduced to decide which channel is important. Our channel attention module is shown in Fig. 3. To explore the inner relationship of channels, the spatial dimension of feature maps is squeezed. So, a way to aggregate spatial information need to be

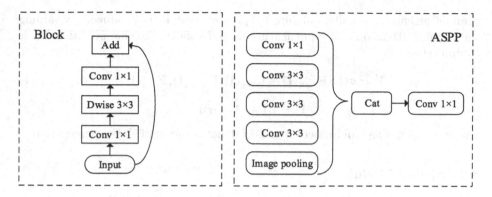

Fig. 2. The segmentation net framework

found. In previous work, the average-pooling has been commonly adopted. Some important feature could be represented by the feature maps averaged. However, we think this representation is not enough. Thus, the max-pooling is introduced into our channel module. In addition, the "sum" and "cat" connection ways are used to increase the information redundancy simultaneously. The redundancy could help to generate more accurate channel attention. The detailed operation is described below.

Given the input feature map $\mathbf{I} \in \mathbb{R}^{C \times H \times W}$. To aggregate the spatial information, the max-pooling F_{max} and average-pooling F_{avg} are respectively applied, generating the two different spatial semantic descriptors: $\mathbf{C}_m$ and $\mathbf{C}_a$, where $\{\mathbf{C}_m, \mathbf{C}_a\} \in \mathbb{R}^{C \times 1 \times 1}$. At the same time, the sum $\mathbf{C}_s$ of two descriptors could be obtained. Afterwards, they are concatenated to form the $3 \times C$ raw channel attention. The raw channel attention is fed into the convolution layer to generate channel attention. The convolution layer is comprised of multi-layer perceptron with a hidden layer. Specially, we set a parameter α to adjust the weight of the

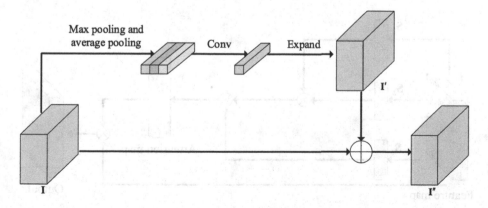

Fig. 3. The channel attention module

channel attention. Finally, enhanced representation $\mathbf{I}''$ is produced by adding the channel attention and input feature map. In short, the channel attention is computed as:

$$\mathbf{I}' = \sigma(\mathrm{M}(\mathrm{F}_{\max}(\mathbf{I}) + \mathrm{F}_{\mathrm{avg}}(\mathbf{I}), \mathrm{F}_{\max}(\mathbf{I}), \mathrm{F}_{\mathrm{avg}}(\mathbf{I}))) \tag{1}$$

$$\mathbf{I}'' = \alpha\mathbf{I} + (1 - \alpha)\mathbf{I}' \tag{2}$$

where σ denotes sigmoid function and M denotes the multi-layer perceptron.

3.3 Spatial Module

Channel attention could learn what the prohibited item is, but the spatial attention could learn where it is. In the previous study, the context relationship has been proved to be essential for semantic segmentation net. Thus, a spatial attention module is designed to extract the context information from the feature maps. In this paper, the long-range context dependency is modeled to enhance the representation of the feature map. It is worth noting that spatial attention is placed after the encode module to reduce the need for computing resource. The spatial module is shown as follows (Fig. 4):

The spatial position relation is fully utilized to get the context dependency, forming the spatial attention. Given the encoded feature map $\mathbf{I}$, it is firstly duplicated to three copies. Then, two of them are fed into a convolution layer to reduce dimension respectively, generating two feature maps $\mathbf{S}_1$ and $\mathbf{S}_2$, where $\{\mathbf{S}_1, \mathbf{S}_2\} \in \mathbb{R}^{C' \times H \times W}$. Next, the shape of $\mathbf{S}_1$ and $\mathbf{S}_2$ is reshaped, the size of $\mathbf{S}_1$ is $HW \times C'$, and the size of $\mathbf{S}_2$ is $C' \times HW$. After that, a matrix multiplication between $\mathbf{S}_1$ and $\mathbf{S}_2$ is performed. We consider the relation of every position and others, obtaining the long-range context dependency information $\mathbf{A}$, where $\mathbf{A} \in \mathbb{R}^{HW \times HW}$. The last copy is fed into a convolution layer. Then multiplication between the output and the context dependency information is performed, producing the new feature map $\mathbf{B}$. The size of $\mathbf{B}$ is $C' \times HW$. Next, the feature

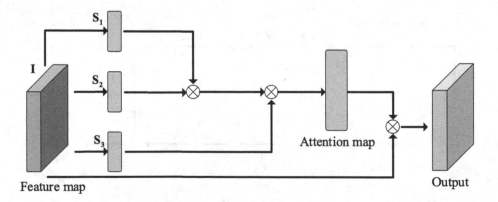

Fig. 4. The spatial attention module

map $\mathbf{B}$ is reshaped to $C' \times H \times W$ and fed into a convolution layer to generate the spatial attention $\mathbf{S} \in \mathbb{R}^{1 \times H \times W}$. Finally, the spatial attention is multiplied into the input feature map $\mathbf{I}$. In short, the spatial attention is computed as:

$$\mathbf{A} = R(\mathbf{S}_1) \otimes R(\mathbf{S}_2) \tag{3}$$

$$\mathbf{B} = Conv(\mathbf{I}) \otimes \mathbf{A} \tag{4}$$

$$\mathbf{I}' = Conv(\mathbf{I} \odot \mathbf{S}) \tag{5}$$

where R denotes the reshape operation, $\otimes$ denotes the matrix multiplication, and $\odot$ denotes the dot product.

4   Experiments

In this section, our model is evaluated in the X-ray security imageries datasets. Experimental results demonstrate our model could achieve state-of-the-art performance. In the next subsections, our datasets are first introduced. Then, we perform a series of contrast experiment. Finally, the ablation experiments are introduced.

4.1   Datasets

We establish the X-ray security imageries datasets based on semantic segmentation task. The style of datasets is similar to PASCAL VOC datasets, which contains 1506 images for training and 377 images for testing. In our datasets, the size of every image is 512×512, and so is the ground truth image. The datasets have high quality pixel-level labels of 7 semantic classes, which include power bank, lighter, fork, knife, gun, pliers and scissors. Due to the scale of the datasets is small, we do some data augmentation operations, such as flip, crop, Gaussian Blur and so on. Finally, augmented datasets have 7532 images.

4.2   Contrast Experiment

In this section, we compare the segmentation performance of our model with other methods. Note that all models are retrained in our datasets for a fair comparison. Our experimental results are listed in Table 1.

As shown in Table 1, our model improves performance remarkably. Unet [14] is often used for medical image segmentation tasks. Here, We design the Unet for the semantic segmentation of prohibited items, and it yields the results of accuracy 0.991 and mIoU 0.558. Next, considering the multiple scale problem and receptive field of the net, the Deeplab is used for semantic segmentation of prohibited items. It yields the result of 0.652 in mIoU, with an improvement of 0.094. Finally, the attention is introduced into our net. Our net significantly improves the segmentation performance, which further brings the improvement of 0.002 in accuracy and 0.031 in mIoU. Our segmentation results are shown in

Fig. 5. Obviously, Unet incorrectly treats the headphone as the fork and other prohibited items segmented are not complete. Deeplab could correctly segment the prohibited items, but the profile is also inaccurate. In contrast, our model could get more accurate segmentation results.

Table 1. Segmentation performance comparison between different models

Model	Accuracy	mIoU
Unet	0.991	0.558
Deeplab	0.991	0.652
Our	0.993	0.683

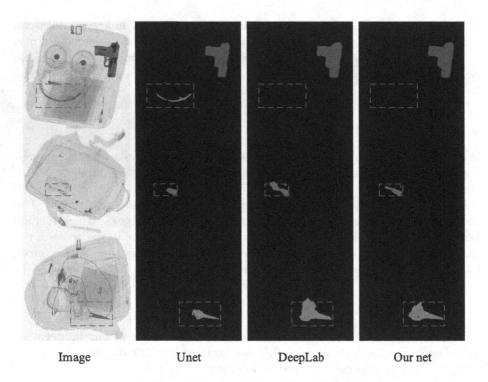

Image	Unet	DeepLab	Our net

Fig. 5. Visualization results for the semantic segmentation of prohibited items.

4.3   Ablation Experiment

To select the best model, different structures of the net are tested. We apply the same training strategies. For training a better model, the poly learning rate policy is achieved. Furthermore, all models are trained using the stochastic gradient descent optimizer. The experimental results are shown in Table 2.

Table 2. Comparison of different structures

Model	Accuracy	mIoU
Deeplab (backbone MobileNet)	0.991	0.655
Deeplab (backbone MobileNet) with channel attention	0.992	0.665
Deeplab (backbone MobileNet) with spatial attention	0.992	0.671
Deeplab (backbone MobileNet) with dual attention	0.993	0.683

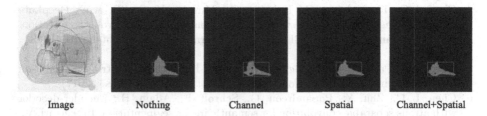

| Image | Nothing | Channel | Spatial | Channel+Spatial |

Fig. 6. Visualization results sample for the ablation experiment.

The MobileNet is adopted as our net backbone. It yields the results of 0.991 in accuracy and 0.655 in mIoU. This structure of the net could consume little computer resource. Afterwards, the performance improvement of the model with attention is verified. The net with channel attention shows the importance of channel attention, which brings the improvement of accuracy and mIoU. Meanwhile, employing spatial attention individually outperforms the net without attention. When we integrate the two attention modules together, the network significantly improves the segmentation performance. The accuracy improves to 0.993, and the mIoU improves to 0.683. The Fig. 6 shows the sample. Obviously, our network could achieve the desired performance.

5 Conclusion

In this paper, we propose a deep learning method, for the semantic segmentation of prohibited items. To address the segmentation of prohibited items, we build a segmentation model with dual attention. The channel attention could learn the importance of feature information, which adaptively weight the input feature map. The spatial attention could learn the long-range semantic dependency. By combining the two modules, the performance of the model is further improved. The experiment results show our model could obtain more precise segmentation results and cost little inference time which is about 0.2 s for an image. Note that this is the first use case of semantic segmentation in the detection of prohibited items. We hope our work could help to advance the auto-detection of prohibited items.

References

1. Akcay, S., Kundegorski, M.E., Willcocks, C.G., Breckon, T.P.: Using deep convolutional neural network architectures for object classification and detection within x-ray baggage security imagery. IEEE Trans. Inf. Forensics Secur. **13**(9), 2203–2215 (2018)
2. Badrinarayanan, V., Kendall, A., Cipolla, R.: SegNet: a deep convolutional encoder-decoder architecture for image segmentation. IEEE Trans. Pattern Anal. Mach. Intell. **39**(12), 2481–2495 (2017)
3. Chen, L.C., Papandreou, G., Kokkinos, I., Murphy, K., Yuille, A.L.: Semantic image segmentation with deep convolutional nets and fully connected CRFs. arXiv preprint arXiv:1412.7062 (2014)
4. Chen, L.C., Papandreou, G., Kokkinos, I., Murphy, K., Yuille, A.L.: DeepLab: semantic image segmentation with deep convolutional nets, atrous convolution, and fully connected CRFs. IEEE Trans. Pattern Anal. Mach. Intell. **40**(4), 834–848 (2018)
5. Chen, L.C., Papandreou, G., Schroff, F., Adam, H.: Rethinking atrous convolution for semantic image segmentation. arXiv preprint arXiv:1706.05587 (2017)
6. Chen, L.-C., Zhu, Y., Papandreou, G., Schroff, F., Adam, H.: Encoder-decoder with atrous separable convolution for semantic image segmentation. In: Ferrari, V., Hebert, M., Sminchisescu, C., Weiss, Y. (eds.) ECCV 2018. LNCS, vol. 11211, pp. 833–851. Springer, Cham (2018). https://doi.org/10.1007/978-3-030-01234-2_49
7. Fu, J., Liu, J., Tian, H., Fang, Z., Lu, H.: Dual attention network for scene segmentation. arXiv preprint arXiv:1809.02983 (2018)
8. He, K., Zhang, X., Ren, S., Sun, J.: Deep residual learning for image recognition. In: Proceedings of the IEEE Conference on Computer Vision and Pattern Recognition, pp. 770–778 (2016)
9. Krizhevsky, A., Sutskever, I., Hinton, G.E.: ImageNet classification with deep convolutional neural networks. In: Advances in Neural Information Processing Systems, pp. 1097–1105 (2012)
10. Long, J., Shelhamer, E., Darrell, T.: Fully convolutional networks for semantic segmentation. In: Proceedings of the IEEE Conference on Computer Vision and Pattern Recognition, pp. 3431–3440 (2015)
11. Luong, M.T., Pham, H., Manning, C.D.: Effective approaches to attention-based neural machine translation. arXiv preprint arXiv:1508.04025 (2015)
12. Mery, D., Svec, E., Arias, M., Riffo, V., Saavedra, J.M., Banerjee, S.: Modern computer vision techniques for x-ray testing in baggage inspection. IEEE Trans. Syst. Man Cybern. Syst. **47**(4), 682–692 (2017)
13. Peng, C., Zhang, X., Yu, G., Luo, G., Sun, J.: Large kernel matters-improve semantic segmentation by global convolutional network. In: Proceedings of the IEEE Conference on Computer Vision and Pattern Recognition, pp. 4353–4361 (2017)
14. Ronneberger, O., Fischer, P., Brox, T.: U-Net: convolutional networks for biomedical image segmentation. In: Navab, N., Hornegger, J., Wells, W.M., Frangi, A.F. (eds.) MICCAI 2015. LNCS, vol. 9351, pp. 234–241. Springer, Cham (2015). https://doi.org/10.1007/978-3-319-24574-4_28
15. Sandler, M., Howard, A., Zhu, M., Zhmoginov, A., Chen, L.C.: MobileNetV2: inverted residuals and linear bottlenecks. In: Proceedings of the IEEE Conference on Computer Vision and Pattern Recognition, pp. 4510–4520 (2018)
16. Szegedy, C., et al.: Going deeper with convolutions. In: Proceedings of the IEEE Conference on Computer Vision and Pattern Recognition, pp. 1–9 (2015)

17. Vaswani, A., et al.: Attention is all you need. In: Advances in Neural Information Processing Systems, pp. 5998–6008 (2017)
18. Woo, S., Park, J., Lee, J.-Y., Kweon, I.S.: CBAM: convolutional block attention module. In: Ferrari, V., Hebert, M., Sminchisescu, C., Weiss, Y. (eds.) ECCV 2018. LNCS, vol. 11211, pp. 3–19. Springer, Cham (2018). https://doi.org/10.1007/978-3-030-01234-2_1
19. Xu, M., Zhang, H., Yang, J.: Prohibited item detection in airport x-ray security images via attention mechanism based CNN. In: Lai, J.-H., et al. (eds.) PRCV 2018. LNCS, vol. 11257, pp. 429–439. Springer, Cham (2018). https://doi.org/10.1007/978-3-030-03335-4_37
20. Zhao, H., Shi, J., Qi, X., Wang, X., Jia, J.: Pyramid scene parsing network. In: Proceedings of the IEEE Conference on Computer Vision and Pattern Recognition, pp. 2881–2890 (2017)

Sparse Unmixing for Hyperspectral Image with Nonlocal Low-Rank Prior

Feiyang Wu[✉], Yuhui Zheng, and Le Sun

School of Computer and Software, Nanjing University of Information Science
and Technology, Nanjing 210044, China
feiyangwu@nuist.edu.cn

Abstract. In this paper, a nonlocal low rank prior associated with spatial smoothness and spectral collaborative sparsity are integrated together for unmixing the hyperspectral data. Based on a fact that the hyperspectral images have self-similarity in nonlocal sense and smoothness in the local sense. To explore the spatial self-similarity, the nonlocal cubic patches are grouped together to form a low-rank matrix. Then, in the framework of linear mixture model, the nuclear norm is constrained to the abundance matrix of these similar patches to enforce low-rank property. In addition, the spectral and local spatial information is also taken into account by introducing collaborative sparsity and TV regularization terms, respectively. Finally, the proposed method is tested on two simulated data sets and a real data set and the results show that the proposed algorithm produces better performance than other state-of-the-art algorithms.

Keywords: Hyperspectral images · Sparse unmixing · Nonlocal self-similarity · Low-rank prior

1 Introduction

Hyperspectral remote sensing generally uses an imaging spectrometer to obtain tens or even hundreds of consecutive narrow-band information (generally less than 10 nm) per pixel in the spectral domain. Therefore, hyperspectral images have the property of high spectral resolution. However, due to the limitations of optical instrument performance and imperfect spectral acquisition techniques, the spatial resolution of hyperspectral images is low and it results in a pixel that may contain more than one type of land cover information, called mixed pixels [1]. Because the existence of many mixed pixels, the accuracy of hyperspectral images processing has been greatly affected. Therefore, hyperspectral unmixing is a key preprocessing technique for hyperspectral image analysis. The goal of hyperspectral unmixing is to extract the endmembers in the scene and estimate the corresponding abundances. Hyperspectral unmixing can be divided into two main models, i.e., linear mixture model (LMM) [2] and nonlinear mixture model (NLMM) [3, 4]. Compared with the NLMM, the LMM has the advantages of simplicity, high efficiency, and clear physical meaning. With LMM, many techniques are introduced to hyperspectral unmixing. Based on the geometrical and statistical methods, a lot of endmember extraction algorithms have been proposed. However, those algorithms perform poorly in highly mixed or noisy hyperspectral image data and

© Springer Nature Switzerland AG 2019
Z. Cui et al. (Eds.): IScIDE 2019, LNCS 11935, pp. 506–516, 2019.
https://doi.org/10.1007/978-3-030-36189-1_42

require pure pixel existence assumption [5], which does not hold in many datasets. To avoid this problem, with the spectral library released by the United States Geological Survey (USGS), the sparse unmixing has attracted more and more researcher's attention. The advantage of the sparse unmixing is that it does not require endmember extraction, but directly uses spectral signals in a given spectral library to construct endmember matrix, and then estimates the abundance coefficient. To increase the accuracy of unmixing, some additional information as priors should be explored. Nevertheless, some algorithms, such as collaborative sparse unmixing by variable splitting and augmented Lagrangian (CLSUnSAL) [6], only considered the sparsity of the spectrum. Some algorithms take account of spatial correlation, such as sparse unmixing via variable splitting and augmented Lagrangian and total variation (SUnSAL-TV) [7] and joint local abundance sparse unmixing (J-LASU) [8], but SUnSAL-TV may cause over smoothing in the edge regions, and J-LASU only considered the similarity of the local region, there is still much room to improve the unmixing performance.

In this study, we propose a nonlocal low rank prior to hyperspectral sparse unmixing (NLLRSU). The hyperspectral image has self-similarity in nonlocal regions. Hence, the non-local cubic patches are grouped together to form a low-rank matrix. Qu et al. [9] pointed out that this high spatial correlation in the original spatial domain of hyperspectral image means that there is also a high correlation in the abundance matrix, which is reflected by the linear correlation between the abundance vectors. Based on this prior, we utilize the low-rank property to estimate abundance map. In addition, we take the spectral and spatial information into account by introducing collaborative sparse and TV regularization terms, respectively.

The contributions are made as follows.

First, we considered the non-local self-similarity property of the hyperspectral image, which can extract spatial information better than state-of-the-art algorithms. Furthermore, we utilized spectral and spatial information simultaneously to deliver better unmixing results. Second, without extracting the endmembers from the data, we used a large spectral library to be the endmember matrix. Moreover, collaborative sparsity was employed to promote the row sparsity, which will result in a more accurate result in the framework of sparse unmixing. Third, we used the alternating direction multipliers method (ADMM) to effectively solve the optimization problem of the proposed model with all convex terms. In addition, the proposed algorithm was finally tested on two simulated data sets and a real hyperspectral image data set, and the results showed that the proposed algorithm outperformed several state-of-the-art algorithms.

In Sect. 2, we discuss the related work of hyperspectral unmixing. In Sect. 3, we describe the proposed NLLRSU algorithm. In Sect. 4, we tested the proposed algorithm and other sparse unmixing algorithms. Finally, we summarize this paper in Sect. 5.

2 Related Work

2.1 Linear Spectral Unmixing

The linear mixed model (LMM) assumes that each pixel spectrum can be linearly combined by all pure spectral signatures (called endmembers) and corresponding abundance. Let $Y \in \mathbb{R}^{l \times s}$ be an observed hyperspectral image data, $M \in \mathbb{R}^{l \times q}$ be an endmember matrix and $X \in \mathbb{R}^{q \times s}$ be the corresponding abundance matrix, where l is the number of bands, q is the number of the endmembers and s is the total number of pixels in the scene. The LMM can be formulated in a matrix form as:

$$Y = MX + N \tag{1}$$

where $N \in \mathbb{R}^{l \times s}$ represents the noise and model error.

Considering the physical meaning of the abundance, the LMM introduces two constraints, named the abundance non-negativity constraint (ANC): $X \geq 0$, and the abundance sum-to-one constraint (ASC) [10]: $\mathbf{1}^T X = \mathbf{1}^T$ (where $\mathbf{1}^T$ is a line vector of 1's).

2.2   Sparse Unmixing

The sparse unmixing utilizes a known spectral library $A \in \mathbb{R}^{l \times t}$ to replace the endmember matrix, where t is the number of spectral signatures of ground objects contained in the spectral library A. Generally, the number of endmembers contained in a hyperspectral image is much smaller than the number of signatures in A, therefore, the abundance matrix X usually contains many zero values, that is, the abundance matrix can be regarded as sparse. Therefore, the sparse unmixing model can be expressed as:

$$\min_{X} \frac{1}{2} \|AX - Y\|_F^2 + \lambda \|X\|_0 \quad \text{s.t.} \quad X \geq 0 \tag{2}$$

where $\|X\|_0$ is the L_0 norm of the abundance matrix X, which cumulates the number of non-zero elements in X.

Because the L_0 norm is non-convex, the problem of minimizing Eq. (2) is an NP-hard problem. With restricted isometry property (RIP) condition [11], the L_0 norm problem can be converted to the convex optimization problem of the L_1 norm, the model can be described as:

$$\min_{X} \frac{1}{2} \|AX - Y\|_F^2 + \lambda \|X\|_1 \quad \text{s.t.} \quad X \geq 0 \tag{3}$$

where $\|X\|_1 = \sum_{i=1}^{k} \|x_i\|_1$ is the sum of the elements in the abundance matrix X.

In practical application, a hyperspectral image often has only a few endmembers out of a large spectral library. CLSUnSAL replaces the L_1 norm by $L_{2,1}$ norm, which exploits the collaborative sparsity of the abundance, and the model can be written as:

$$\min_{X} \frac{1}{2} \|AX - Y\|_F^2 + \lambda \|X\|_{2,1} \quad \text{s.t.} \quad X \geq 0 \tag{4}$$

where $\|X\|_{2,1} = \sum_{i=1}^{k} \|x_i\|_2$, x_i represents the i-th row of X.

In addition to considering spectral information, the spatial correlation of hyperspectral images is also beneficial for improving the performance of unmixing. The

SUnSAL-TV algorithm using TV regularization term to promote piecewise smoothness between adjacent pixels, the model can be formulated as:

$$\min_{X} \frac{1}{2} \|AX - Y\|_F^2 + \lambda\|X\|_1 + \lambda_{TV} TV(X) \quad \text{s.t.} \quad X \geq 0 \tag{5}$$

where $TV(X) = \sum_{\{i,j\} \in \varepsilon} \|x_i - x_j\|_1$ defines a nonisotropic total variation (TV) [7].

However, CLSUnSAL only considers the sparsity of the abundance from a perspective in the spectral domain while SUnSAL-TV only exploits the local smoothness in the spectral domain, it may cause over smoothing in the edge regions.

Based on the above analysis, there is still much room for improving the unmixing performance. Therefore, the NLLRSU algorithm paves a way to obtain more precise spatial information by exploiting the nonlocal self-similarity.

3 Proposed Algorithm

3.1 Nonlocal Self-similarity

As a natural image, the hyperspectral image has smoothness in local regions between adjacent pixels and self-similarity in nonlocal regions. The adjacent pixels are continuously varying to some extent. Therefore, there is a high correlation between the pixels of the local regions in a hyperspectral image, that is to say, the pixels are very likely to contain the same material in a local region. In addition, there are a large number of similar structural information in different regions in a hyperspectral image. These similar structures consist of smooth regions, texture regions and edge regions. For any local region, we can find many similar regions in a hyperspectral image. The information of the hyperspectral image itself is redundant and the pixels in these similar regions are also very likely to contain the same material. Qu et al. [9] indicated that this high spatial correlation of hyperspectral images means that there is also a high correlation in the abundance matrix, which is reflected by the linear correlation between the abundance vectors. Based on this prior, the abundance matrix can be estimated by low-rank property.

In addition, the result of sparse unmixing is affected by the correlation between spectral signatures in the spectral library. In general, a spectral library contains many spectral signatures of the same land-covers in different situations. The similarity between these spectral signatures is very high, which leads to the solution of the sparse unmixed model is not unique. To overcome this disadvantage, it needs to ensure that the linear correlation between the spectral signatures in the spectral library is as small as possible. Therefore, we use a strategy to precondition the spectral library. Given a spectral library A, we calculate the spectral angle $\theta_{i,j}$ between any two different spectral signatures A_i and A_j. The spectral angle $\theta_{i,j}$ is defined as:

$$\theta_{i,j} = \cos^{-1}\left(\frac{A_i^T A_j}{|A_i||A_j|}\right) \tag{6}$$

These spectral signatures are regarded as linear correlation if the spectral angle $\theta_{i,j}$ is less than a given threshold. For these spectral signatures, we replace them by using one spectral signature in them. With this operation, we obtain a pruned spectral library and then we utilize it for sparse unmixing.

3.2   Proposed Model

We restore the abundance matrix to 3D form to find similar structural information. Let $\hat{X} \in \mathbb{R}^{sr \times sc \times t}$ be the 3D form of the abundance data, where sr and sc represent the number of rows and columns, respectively, and $s = sr \times sc$, s is the number of pixels in the abundance matrix and t is the number of signatures in the spectral library. Assume that the number of patches is P in $\hat{X}$ and $\hat{X}_p \in \mathbb{R}^{sp \times sp \times tp}$ is the p-th patch, where $p = 1, 2, \ldots, P$. Take the p-th patch as an example, we use a block matching algorithm [12] to find the r nonlocal patches which are similar to the p-th patch. Then, we stack the $r+1$ patches to a patch group, denoted as $\hat{X}_{r+1} \in \mathbb{R}^{sp \times sp \times u}$, where $u = tp \times (r + 1)$. For each matrix in the spectral dimension of the patch group, it is denoted as $\hat{X}_{i,r+1} \in \mathbb{R}^{sp \times sp}$, where $i = 1, 2, \ldots, u$, we convert it into the vector, denoted as $\hat{X}_{i,r+1} \in \mathbb{R}^w$, where $w = sp \times sp$. Hence, we obtain the abundance matrix of the patch group:

$$D_{\hat{X}_{r+1}} = \left(\hat{x}_{1,r+1}, \hat{x}_{2,r+1}, \ldots, \hat{x}_{u,r+1} \right) \in \mathbb{R}^{w \times u} \tag{7}$$

The nuclear norm is then utilized to enforce the low-rank property of the abundance map $D_{\hat{X}_{r+1}}$ of the patch group $\hat{X}_{r+1}$:

$$\left\| D_{\hat{X}_{r+1}} \right\|_* = \sum_{i=1}^{rank\left(D_{\hat{X}_{r+1}} \right)} \sigma_i \left(D_{\hat{X}_{r+1}} \right) \tag{8}$$

where σ_i represents the i-th singular value. Therefore, we obtain the estimated abundance matrix of this patch group by the low-rank constraint. At last, we restore the estimated abundance matrix to the corresponding position of each patch in the patch group. By this analogy, we do this operation for all P number of patches. For the whole abundance matrix X, we use $\|X\|_{NL*}$ to represent the nonlocal abundance regularization term. Figure 1 illustrates this process.

Based on the sparse unmixing model, the proposed NLLRSU algorithm has three regularizers, named collaborative sparsity, TV and nonlocal abundance low-rankness. The model is formulated as:

$$\min_X \frac{1}{2} \|AX - Y\|_F^2 + \lambda \|X\|_{2,1} + \lambda_{TV} \|HX\|_{1,1} + \lambda_{NL} \|X\|_{NL*} + l_{R+} (X) \tag{9}$$

where $\lambda, \lambda_{TV}, \lambda_{NL}$ are the parameters of the collaborative sparsity, TV and nonlocal low-rank regularization terms, respectively. $l_{R+}(x)$ is an indicator function, when $x \geq 0$, $l_{R+}(x) = 0$; otherwise $l_{R+}(X) = +\infty$, it is utilized to ensure ANC constraint.

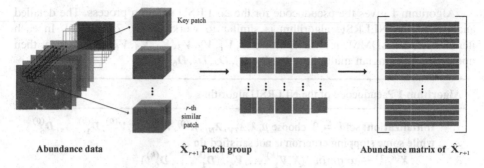

Fig. 1. The process of application nonlocal low-rank property to abundance data.

It is difficult to optimize the model (9) directly, so we employ the ADMM [13] to solve it. The core idea of ADMM is that by introducing some new variables cautiously, a difficult problem can be transformed into several simple subproblems. The subproblems are solved one by one and alternately updated. Here, we introduce the variables $V_1, V_2, V_3, V_4, V_5, V_6$. The model (9) can be formulated as:

$$\min_{X,V_1,V_2,V_3,V_4,V_5,V_6} \frac{1}{2}\|V_1 - Y\|_F^2 + \lambda\|V_2\|_{2,1} + \lambda_{TV}\|V_4\|_{1,1} + \lambda_{NL}\|V_5\|_{NL*} + l_{R+}(V_6)$$

$$\text{s.t.} V_1 = AX$$
$$V_2 = X$$
$$V_3 = X$$
$$V_4 = HV_3$$
$$V_5 = X$$
$$V_6 = X$$

$$(10)$$

The Lagrangian function of Eq. (10) is:

$$\mathcal{L}(X, V_1, V_2, V_3, V_4, V_5, V_6, D_1, D_2, D_3, D_4, D_5, D_6)$$
$$= \frac{1}{2}\|V_1 - Y\|_F^2 + \lambda\|V_2\|_{2,1} + \lambda_{TV}\|V_4\|_{1,1} + \lambda_{NL}\|V_5\|_{NL*}$$
$$+ l_{R+}(V_6) + \frac{\mu}{2}\|V_1 - AX + D_1\|_F^2 + \frac{\mu}{2}\|V_2 - X + D_2\|_F^2 \qquad (11)$$
$$+ \frac{\mu}{2}\|V_3 - X + D_3\|_F^2 + \frac{\mu}{2}\|V_4 - HX + D_4\|_F^2$$
$$+ \frac{\mu}{2}\|V_5 - X + D_5\|_F^2 + \frac{\mu}{2}\|V_6 - X + D_6\|_F^2$$

where $D_1, D_2, D_3, D_4, D_5, D_6$ are Lagrangian multipliers and μ is the Lagrangian parameter.

Algorithm 1 gives the pseudocode for the NLLRSU solution process. The detailed analysis of the NLLRSU algorithm is similar to works in [6, 7] and [8]. In each iteration of the ADMM, it is to optimize $X, V_1, V_2, V_3, V_4, V_5, V_6$ in sequence, then update the Lagrangian multipliers $D_1, D_2, D_3, D_4, D_5, D_6$.

Algorithm 1 Pseudocode of the NLLRSU algorithm

---

1. **Initialization**: set $k = 0$, choose $\mu, \lambda, \lambda_{TV}, \lambda_{NL}, X^{(0)}, V_1^{(0)}, \ldots, V_6^{(0)}, D_1^{(0)}, \ldots, D_6^{(0)}$
2. **while** some stopping criterion is not satisfied **do**
3. $\qquad X^{(k+1)} \leftarrow \underset{X}{argmin}\square(X, V_1^{(k)}, \ldots, V_6^{(k)}, D_1^{(k)}, \ldots, D_6^{(k)})$
4. $\qquad$ **for** $i = 1, \ldots, 6$ **do**
5. $\qquad\qquad V_i^{(k+1)} \leftarrow \underset{V_i}{argmin}\square(X^{(k)}, V_1^{(k)}, \ldots, V_i, \ldots, V_6^{(k)})$
6. $\qquad$ **end for**
7. $\qquad$ **Update Lagrange multipliers**
8. $\qquad D_1^{(k+1)} \leftarrow D_1^{(k)} - AX^{(k+1)} + V_1^{(k+1)}$
9. $\qquad D_4^{(k+1)} \leftarrow D_4^{(k)} - HV_3^{(k+1)} + V_4^{(k+1)}$
10. $\qquad D_i^{(k+1)} \leftarrow D_i^{(k)} - X^{(k+1)} + V_i^{(k+1)}, i = 2,3,5,6$
11. $\qquad$ **Update iteration** $k = k + 1$
12. **end while**

---

4 Experimental Results

In this section, we used two simulated data sets and a real hyperspectral image data set to validate the unmixing performance of the proposed algorithm. For each simulated data set, our experiments were performed in three different signal-to-noise ratio (SNR) levels, i.e., 10 dB, 20 dB, and 30 dB. We also compared the proposed algorithm with CLSUnSAL [6], SUnSAL-TV [7] and J-LASU [8].

Table 1. SRE (dB) result.

Data	SNR	CLSUnSAL	SUnSAL-TV	J-LASU	NLLRSU
DS1	10	1.6544	3.9803	10.6039	**11.4346**
	20	6.0518	7.1069	15.3860	**16.1892**
	30	8.1140	15.0419	19.6962	**21.2784**
DS2	10	1.3889	2.6690	4.6611	**5.0926**
	20	4.4781	5.2910	6.4600	**7.0514**
	30	9.6948	10.8141	10.7305	**11.8417**

Table 2. RMSE result.

Data	SNR	CLSUnSAL	SUnSAL-TV	J-LASU	NLLRSU
DS1	10	0.0286	0.0218	0.0102	**0.0093**
	20	0.0172	0.0152	0.0059	**0.0054**
	30	0.0136	0.0061	0.0036	**0.0030**
DS2	10	0.0347	0.0299	0.0238	**0.0227**
	20	0.0243	0.0221	0.0194	**0.0181**
	30	0.0133	0.0117	0.0118	**0.0104**

We employed splib06 [14], a USGS spectral library, to our simulated data experiments, and we selected 240 endmember signatures randomly from the splib06 as the spectral library used in the experiment, denoted as $A \in \mathbb{R}^{224 \times 240}$. The simulated data set 1 (DS1) and data set 2 (DS2) are generated by randomly selected five spectral signatures from the spectral library **A**. DS1 have 75×75 pixels. Each pixel has 224 bands and the corresponding abundance is constrained by ASC. Some square regions are pure and some regions are mixed by two to five endmembers. The background pixels of this data set are composed of the same five endmembers with randomly fixed abundance values. DS2 has 58×58 pixels and each pixel has 224 bands without pure pixels. The abundance of DS2 shows the piecewise smoothing transition. To evaluate the performance of the four algorithms, we used two evaluation indicators: signal reconstruction error (SRE) [15], which is defined as $SRE(dB) = 10\, log_{10}\left(E\left[\|X\|_2^2\right]/E\left[\|X - \hat{X}\|\right]_2^2\right)$, and root mean square error (RMSE) [16], defined as $RMSE = \sqrt{\sum_{i=1}^{t}\sum_{j=1}^{s}\left(x_{ij} - \hat{x}_{ij}\right)^2/(t \times s)}$, where **X** is the true abundance matrix and $\hat{X}$ is the estimated abundance matrix, x_{ij} and $\hat{x}_{ij}$ represent each element in the true abundance matrix and the estimated abundance matrix, respectively.

Tables 1 and 2 show the values of SRE (dB) and RMSE of the four algorithms with two simulated data sets, respectively. From these tables, we can observe that the proposed NLLRSU algorithm performs better than other state-of-the-art algorithms in the same SNR level. Figure 2 shows the abundance image reconstructed by the four algorithms in 30 db, respectively. From Fig. 2, we can find that SUnSAL-TV algorithm shows the smoothest results in the edge regions. Nevertheless, the edge transition regions reconstructed by our algorithm is closer to the true abundance distribution. The result means that our proposed algorithm keeps the structure information of the image better than other algorithms.

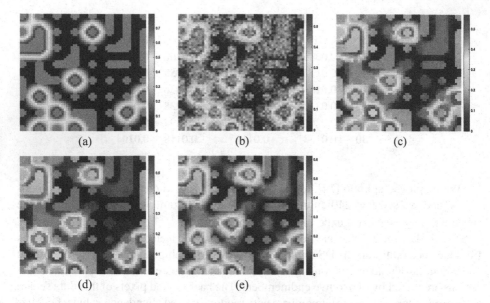

Fig. 2. Estimated abundance of endmember 2 for DS2 with SNR=30 dB. (a) Ground truth; (b) CLSUnSAL; (c) SUnSAL-TV; (d) J-LASU; (e) NLLRSU.

For real hyperspectral data set experiment, we used the famous AVIRIS Cuprite mine map to test four algorithms performance. We utilized a subimage of this data set, with 250 × 191 pixels. This subimage contains 224 spectral bands from 0.4 to 2.5 with a spectral resolution of 10 nm. Due to the low SNR and water absorption, we removed the bands of 1–2, 105–115, 150–170, 223–224, and left 188 bands. The USGS spectral library used in this data set, with 498 endmember signatures. Figure 3 shows the results.

From Fig. 3, we can find that CLSUnSAL and SUnSAL-TV perform poorly in the aspect of noise suppression, and NLLRSU keeps the structure information better than other algorithms. Through these experiments, we can see that the NLLRSU algorithm can help to improve the unmixing results.

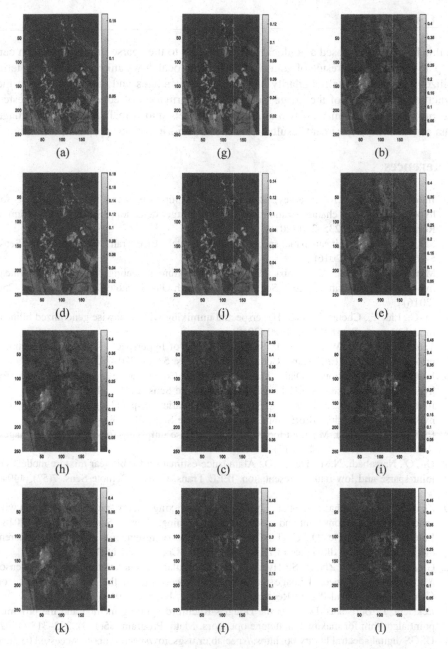

Fig. 3. Estimated abundance of Cuprite data for Alunite, Buddingtonite and Chalcedony (column 1–3). (a-c) CLSUnSAL; (d-f) SUnSAL-TV; (g-i) J-LASU; (j-l) NLLRSU.

5 Conclusions

In this letter, we proposed a nonlocal low rank prior to the sparse unmixing, which can help to improve the result of unmixing. The nonlocal low-rank regularization term utilizes the nonlocal self-similarity of hyperspectral images and can help to keep the structure information of the image, such as the information of edge region. We tested the proposed algorithm with two simulated data sets and a real hyperspectral image data set, and the experiment results show the superiority of our proposed algorithm.

References

1. Jiao, C., Chen, C., McGarvey, R.G., et al.: Multiple instance hybrid estimator for hyperspectral target characterization and sub-pixel target detection. ISPRS J. Photogramm. Remote Sens. **146**, 235–250 (2018)
2. Shi, C., Wang, L.: Linear spatial spectral mixture model. IEEE Trans. Geosci. Remote Sens. **54**(6), 3599–3611 (2016)
3. Marinoni, A., Plaza, A., Gamba, P.: Harmonic mixture modeling for efficient nonlinear hyperspectral unmixing. IEEE J. Sel. Top. Appl. Earth Obs. Remote Sens. **9**(9), 4247–4256 (2016)
4. Li, C., Liu, Y., Cheng, J., et al.: Hyperspectral unmixing with bandwise generalized bilinear model. Remote Sens. **10**(10), 1600 (2018)
5. Tang, W., Shi, Z., Wu, Y., et al.: Sparse unmixing of hyperspectral data using spectral a priori information. IEEE Trans. Geosci. Remote Sens. **53**(2), 770–783 (2016)
6. Iordache, M.D., Bioucas-Dias, J.M., Plaza, A.: Collaborative sparse regression for hyperspectral unmixing. IEEE Trans. Geosci. Remote Sens. **52**(1), 341–354 (2013)
7. Iordache, M.D., Bioucas-Dias, J.M., Plaza, A.: Total variation spatial regularization for sparse hyperspectral unmixing. IEEE Trans. Geosci. Remote Sens. **50**(11), 4484–4502 (2012)
8. Rizkinia, M., Okuda, M.: Joint local abundance sparse unmixing for hyperspectral images. Remote Sens. **9**(12), 1224 (2017)
9. Qu, Q., Nasrabadi, N.M., Tran, T.D.: Abundance estimation for bilinear mixture models via joint sparse and low-rank representation. IEEE Trans. Geosci. Remote Sens. **7**(52), 4404–4423 (2014)
10. Zhang, X., Li, C., Zhang, J., et al.: Hyperspectral unmixing via low-rank representation with space consistency constraint and spectral library pruning. Remote Sens. **10**(2), 339 (2018)
11. Lou, Y., Yin, P., He, Q., et al.: Computing sparse representation in a highly coherent dictionary based on difference of L1 and L2. J. Sci. Comput. **64**(1), 178–196 (2015)
12. Chang, Y., Yan, L., Zhong, S.: Hyper-Laplacian regularized unidirectional low-rank tensor recovery for multispectral image denoising. In: Proceedings of the IEEE Conference on Computer Vision and Pattern Recognition, Honolulu, HI, pp. 5901–5909 (2017)
13. Eckstein, J., Bertsekas, D.P.: On the Douglas-Rachford splitting method and the proximal point algorithm for maximal monotone operators. Math. Program. **55**(1–3), 293–318 (1992)
14. USGS digital spectral library 06. https://speclab.cr.usgs.gov/spectral.lib06. Accessed 08 June 2016
15. Altmann, Y., Pereyra, M., Bioucas-Dias, J.: Collaborative sparse regression using spatially correlated supports-application to hyperspectral unmixing. IEEE Trans. Image Process. **24**(12), 5800–5811 (2015)
16. Guerra, R., Santos, L., López, S., et al.: A new fast algorithm for linearly unmixing hyperspectral images. IEEE Trans. Geosci. Remote Sens. **53**(12), 6752–6765 (2015)

Saliency Optimization Integrated Robust Background Detection with Global Ranking

Zipeng Zhang[1], Yixiao Liang[1], Jian Zheng[1], Kai Li[1], Zhuanlian Ding[1,2], and Dengdi Sun[2(✉)]

[1] School of Internet, Anhui University, Hefei 230039, China
bigzipeng@gmail.com, liangyixiao@stu.ahu.edu.cn, ahu_zhengjian@163.com,
likaiuser@qq.com, dingzhuanlian@163.com
[2] Key Lab of Intelligent Computing and Signal Processing of Ministry of Education,
School of Computer Science and Technology, Anhui University, Hefei 230601, China
sundengdi@163.com

Abstract. Image saliency detection plays an important role in the field of computer vision. In order to make the saliency detection achieve better results, this paper proposes an algorithm that combines the global cue with the robust background prior measurement. Specifically, we first get a global ranking of superpixels by absorbing time in Markov chain, which is the absorbing nodes represent the superpixels of the fictitious boundary, and the transient nodes denote the others. Then, the global similarity of transient node can be measured by its absorbing time, so a global ranking for each transient node can be calculated, which is called as global cue in this paper. Finally, considering the remarkable energy optimization model, we integrate the robust background prior measurement with calculated global cue to form a new optimization model for saliency detection. In conclusion, our method is better than some typical significant detection algorithms on several datasets through the experimental verification.

Keywords: Saliency detection · Markov chain · Robust background

1 Introduction

Saliency detection in computer vision aims to find the most informative part in an image. It has been rapidly increased for many years and widely applied in numerous aspects of vision problems, such as image segmentation [1], image cropping [2], object recognition [3], video summarization [4], etc. Usually, these methods can be generally divided into bottom-up methods, top-down approaches [2,7–12] and mixed models [3,5,6].

In recent years, the bottom-up method based on graph model is widely researched and applied. These methods modeling with a neighbor graph model in which the superpixels of an image are represented by the nodes, the edges stand

© Springer Nature Switzerland AG 2019
Z. Cui et al. (Eds.): IScIDE 2019, LNCS 11935, pp. 517–528, 2019.
https://doi.org/10.1007/978-3-030-36189-1_43

for the neighbor relationships and visual appearance similarities between the superpixels. Then, these models utilize some graph-based learning algorithms, such as manifold ranking [13], Markov random walks [14], etc., to obtain the metric of each superpixel. For example, Yang et al. [13] select some foreground and background clues via graph-based manifold ranking and sort the remaining superpixels with multiple sorting models to get the saliency of image. Jiang et al. [14] formulate saliency detection via absorbing Markov chain on an image graph model and further improve the methods in some smooth regions in the long-range. Zhu et al. [15] proposed an energy optimization framework to comprehensively consider the background, foreground and neighboring smoothing items of the image, and adopt a unified optimization framework and algorithm to calculate the saliency of the image. Xiao et al. [16] introduce a consistent measurement ranking model. Similar to Yang's study, they adopt a two-stage method to calculate the saliency value.

As we know, there is no doubt that graph-based ranking algorithm play an efficient role in saliency detection for images. But there exists a obviously defect that usually these methods ignore the global cue for images. To solve this problem, we put up an idea to combine the advantage of absorbing Markov chain and robust background detection. Compared to traditional saliency detection methods, we use robust background measurement model to get the main model theme. In addition, we set a parameter to control the balance between global cue and background. The experimental results show that compared with the traditional method, the proposed algorithm can obtain higher significance of detection accuracy.

2   Global Ranking via Absorbed Time in Markov Chain

Recently, the absorption of Markov chain is a popular method in saliency detection and has been widely used in data labeling problem. Given a set of data $X = \{x_1, \cdots, x_t, x_{t+1}, \cdots, x_n\} \in R^{m \times n}$ containing n data points. The former t data points are labeled query data points and the rest are unlabeled. Our goal is to determine a sorting function based on the relationship between query nodes and unlabeled node. In Markov chain, we use the absorbed time to complete our goal. The query data points are transient nodes and unlabeled data points are absorbing nodes when we make use of Markov random walks on a graph in the experiment.

At first, we use the method we define a graph $G = (V, E)$ in the dataset, where nodes V are the set of samples and E present the edges weighted by an affinity matrix $W = [w_{ij}]_{n \times n}$. Then we can construct another matrix $A = [a_{ij}]_{n \times n}$ with the following operations. If node i is a transient node and it's not connected to its neighbor node or neighbor's neighbor node j, then $a_{ij} = w_{ij}$, $a_{ii} = 1$, otherwise $a_{ij} = 0$. In addition, the degree matrix is denoted as $D = diag\{d_{11}, \cdots, d_{nn}\}$, where $d_{ii} = \sum_j w_{ij}$.

Then we can compute the expected absorbing time for each transient node. This time can represent the relevance between transient nodes and absorbing nodes. The transient nodes have less expected absorbing time when they are more similar with the absorbing nodes. Therefore, we use the expected absorbing time to rank the transient nodes (just the unlabeled points). Here the steps to calculate the expected time for each transient node as follows:

- Step 1. Compute the transition matrix $P = D^{-1}A$.
- Step 2. Adjust the matrix to make sure that the first t nodes are transient nodes and the last $n-t$ nodes are absorbing nodes. Then the transition matrix P has the following canonical form

$$P \rightarrow \begin{pmatrix} Q & R \\ 0 & I \end{pmatrix}$$

- Step 3. Compute the expected absorbing time:

$$T = (I - Q)^{-1} \times c, \tag{1}$$

where I is the $t \times t$ identity matrix, $Q \in [0,1]^{t \times t}$ contains the transition probabilities between any pair of transient states and c is a $n-t$ dimensional column vector all of whose elements are 1.

3   Robust Background Measure

Robust background prior measure plays an important role in the current image saliency detection. Zhu et al. [15] propose a concept to quantify the degree of connection between an image region and a boundary, called boundary connectivity, which can be used to calculate specific background prior probability. Border connectivity is typically larger for background areas, while object areas are typically smaller.

First, to get background prior probability, some concepts are related:

Between two superpixels (p, q), the weight $d_{app}(p, q)$ is assigned as the Euclidean distance between their average colors in the CIE-Lab color space. $d_{geo}(p, q)$ is the geodesic distance defined as the accumulated edge weights along their shortest path on the graph:

$$d_{geo}(p, q) = \min_{p_1=p, \cdots, p_n=q} \sum_{i=1}^{N-1} d_{app}(p_i, p_{i+1}), \tag{2}$$

the "spanning areas" of each superpixel p is defined as:

$$Area(p) = \sum_{i=1}^{N} \exp(-\frac{d_{geo}^2(p, p_i)}{2\sigma_{clr}^2}) = \sum_{i=1}^{N} S(p, p_i), \tag{3}$$

the length along the boundary is defined as:

$$Len_{bnd}(p) = \sum_{i=1}^{N} S(p, p_i) \cdot \delta(p_i \in Bnd), \tag{4}$$

where σ_{clr} is a parameter which we set as 10, Bnd is the set of image boundary superpixels, δ is 1 for superpixels on the image boundary and 0 otherwise.

Therefore, for any given image superpixel p, the probability that p belongs to the background is calculated:

$$BndCon\,(p) = \frac{Len_{bnd}(p)}{\sqrt{Area(p)}}. \tag{5}$$

Then, based on $BndCon\,(p)$, the background prior can be calculated:

$$bg_i = 1 - \exp(-\frac{BndCon^2\,(p_i)}{2\sigma_{bndCon}^2}), \tag{6}$$

the bg_i is close to 1 when it's boundary connectivity is relatively large, and close to 0 on the contrast, we empirically set $\sigma_{bndCon}^2 = 1$. For the background weighted contrast, many related works enhance the traditional contrast computation via using the region contrast against its surroundings as a saliency cue.

Next, to get foreground prior probability, a superpixel's contrast can be defined as:

$$Ctr_p = \sum_{i=1}^{N} d_{app}(p, p_i) W_{spa}(p, p_i), \tag{7}$$

where $W_{spa}\,(p, p_i) = \exp(-\frac{d_{spa}^2(p,p_i)}{2\sigma_{spa}^2})$, $d_{spa}(p, p_i)$ is the distance between the centers of superpixel p and p_i.

On the basis of background prior, the foreground prior can be calculated as:

$$fg_i = Ctr_p \cdot bg_i. \tag{8}$$

Finally, summarizing the two factors above, the objective cost function for salient object detection with robust low level cues is constructed as Eq. (9), which assign the object region value 1 and the background region value 0 automatically.

$$\sum_{i=1}^{N} bg_i s_i^2 + \sum_{i=1}^{N} fg_i s_i (s_i - 1)^2 + \sum_{i=1}^{N} \sum_{j=1}^{N} W_{ij}(s_i - s_j)^2, \tag{9}$$

where $W_{ij} = \exp(-\frac{d_{app}^2(p,p_i)}{2\sigma^2})$, bg_i is the background of the superpixel pi prior, fg_i is the foreground a priori of superpixel pi as mentioned above, s_i is the saliency value of N superpixel pi, W_{ij} is a measure of similarity between adjacent superpixels i and j.

4   Model Construction and Saliency Solution

We find that the saliency optimization model only considers the superpixels of the image and ignore a global ranking model for image saliency computation. In order to refine a remarkable optimization model, we make a combination of the global and background cue with the parameter μ as the equilibrium condition. The specific process is as follows:

Firstly, we use simple linear iterative clustering (SLIC) algorithm [17] to divide the input image into multiple superpixels. Then, we calculate absorbing time from each transient node (superpixel) to boundary absorbing nodes. It provides a global similarity to all absorbing nodes via absorption time measured by this transient node, so a global ranking of each transient node can be calculated. We also get the possibility of background and foreground term and integrate them together to capture a final saliency value of the image. The purposed objective function is defined as:

$$S_i^* = \arg\min_S \sum_{i=1}^{N} bg_i S_i^2 + \sum_{i=1}^{N} fg_i (S_i - 1)^2 + \sum_{i=1}^{N}\sum_{j=1}^{N} W_{ij}(S_i - S_j)^2 + \mu \sum_{i=1}^{N}(S_i - T_i)^2.$$
(10)

Among them, bg_i, fg_i are the background and foreground of the superpixel. S_i is the saliency values of each node and T_i is correlation with the normalized value of the absorbed time, μ is a parameter which controls the balance of two models.

After a series mathematic processes, we obtain the result:

$$S_i^* = (D - W + B + F + \mu I)^{-1}(f_g + \mu T).$$
(11)

In the above formula, D is a degree matrix, W is the affinity matrix between adjacent superpixels. B and F are diagonal matrices of $B_{ii} = b_i^*$ and $F_{ii} = f_i^*$, respectively. f_g represent the foreground of all superpixel and T is the normalized absorption time of unlabeled nodes.

An important benefit of our model is that it integrates global information with the nature of robust background detection in the data ranking process. The parameter μ is a small constant (empirically set to 0.3). As the parameter μ become larger, the influence of the globality increase greater gradually. On the contrary, when the parameter μ is smaller, the contribution of the information with robustness get greater.

5 Experiments

In this part, we select some standard benchmark datasets to evaluate our method: DUT-OMRON [13], MSRA [18] and SOD [19], and test parameters with PASCAL-S and BSD300. MSRA including more salient objectives is commonly used in most saliency detection methods and the other four datasets are also popular in saliency detection area. To show the advantages of our method, we compare it with some other methods we can find including MR [13], GLR [16], GS [20], CU [21], FT [22], SF [23], simpsal [24], SR [25] and HS [26].

5.1 Parameter Setting

There are three parameters in our method: the pixels in each superpixel n, the compactness parameter m, the balance parameter μ. Testing a lot of datasets, we find the values $n = 600$, $m = 20$, $\mu = 0.4$, are with better visual effect in most datasets, but it still exits some special values for several images.

5.2   Evaluation Standard

Precision-Recall Curves (PR-Curves): The standard precision-recall curves are used to show performance evaluation. Normalizing the curve saliency map to [0, 255], the binary mask threshold is generated from 0 to 255, and compared against the ground truth.

F-Measure: Since both precision and recall are important to evaluate, F-measure comes to weight between precision and recall to get an average value. It shows as:

$$F = \frac{(1+\alpha)precision * recall}{\alpha precision + recall}, \tag{12}$$

In our experiments, α is valued 0.3.

5.3   Global and Robust Cue Validation

In order to verify the special effectiveness of the proposed saliency detection method, we use three standard benchmark datasets: MSRA, SOD and DUT-OMRON. For performance evaluation, we use precision recall curves to compare the proposed method with two other methods which only use a global or robust cue. Several recent approaches exploit boundary prior, like image boundary regions are mostly backgrounds, to enhance saliency computation. Such methods achieve state-of-the-art results, suggesting that boundary prior is effective, but they simply treat all image boundaries as background. This is fragile and may fail even when the object only slightly touches the boundary. Just like the method named GLR, in order to improve the robustness of the background prior, the method of boundary connectivity value is used to calculate the probability that the edge super-pixel block belongs to the background, which obviously improves accuracy of the background prior seed images. However, a limitation for this method is that apparently ignores the global structure of images.

It is well known that the integration of multiple different level cues usually can produce better results. Therefore, to verify the effectiveness of our idea, we intend to integrate the background prior methods based on boundary connectivity with global cues. In our experimental calculations, the larger the precision-recall value, the better the effect of this method. Figure 1 shows that our precision-recall curve is always above the other two methods, therefore, our method is better than these methods only using signal global or robust cue ranking. Figure 2 are some visual example images from the above three datasets, which also compare the proposed method with the others, where we can find that the detected salient objects are more specific and clearer. The value of parameter μ is 0.4 in the experiment, which means that the robustness clues are large, so in theory, the detection effect from the proposed method should be more obvious for the image with the

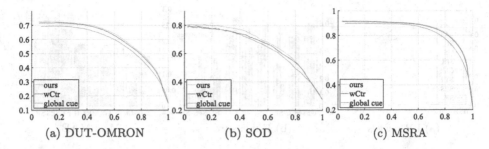

Fig. 1. Precision-recall curves comparing on global and robust cue (x: recall y: precision)

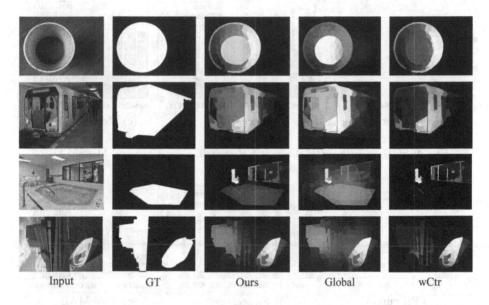

Fig. 2. Some examples from DUT-OMRON, SOD and MSRA comparing ours, global and robust

target leaning against the edges comparing to the previous two methods, and our experimental results show this feature.

Mean Absolute Error (MAE): Considering PR curves only show the object saliency but not background, it's better to introduce the MAE metrics. MAE is defined as the mean absolute difference between the binary GT and the saliency map which is normalized in the range $[0, 1]$. It measures the distance from the saliency map to the actuality:

$$MAE = \frac{1}{H * W} \sum_{i=1}^{H} \sum_{j=1}^{W} |S(i,j) - GT(i,j)|, \tag{13}$$

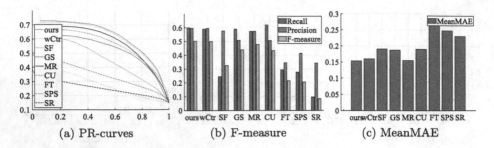

Fig. 3. DUT-OMRON on each method about PR curves, F-measure and Mean-MAE

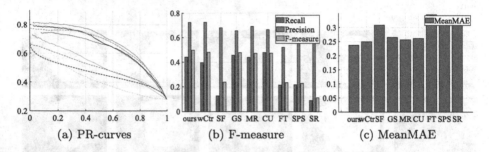

Fig. 4. SOD on each method about PR curves, F-measure and MeanMAE

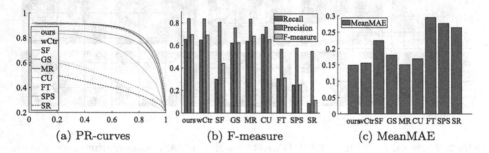

Fig. 5. MSRA on each method about PR curves, F-measure and MeanMAE

In the formula, H and W stand for the height and width of image. $S(i,j)$ is the saliency map of each super pixel, of which $GT(i,j)$ is the ground truth.

5.4 Comparison with Different Methods

Figures 3, 4 and 5 show three comparison metrics with different methods on MSRA, SOD and DUT-OMRON datasets. In each of the figures, the left side shows the comparison of the PR values, the middle is the histogram of the F-measure values, and the right side shows the comparison of the MeanMAE values. In our definition, PR values and F-measure values can only measure the quality of the algorithm from a similar perspective, while MeanMAE's metric is more comprehensive, it's an overall consideration of the two other metrics. Compared with the GLR method from robust cue, the bars of F-measure and MeanMAE are lower, the PR curves is also. We can conclude that the three indicators have achieved more significant effect, even for the MR method with significant effect, our method still has certain improvements. Compared with the other eight methods, these three figures show that our method works well on saliency consistency detection.

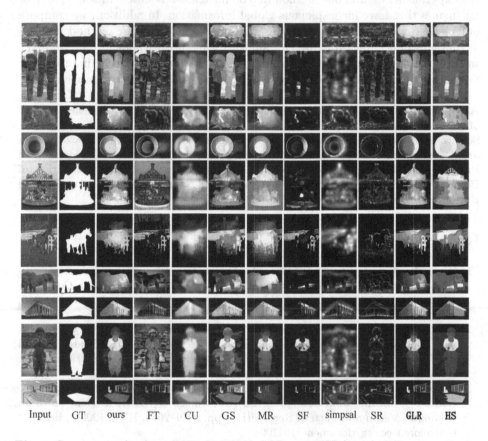

Input   GT   ours   FT   CU   GS   MR   SF   simpsal   SR   **GLR**   **HS**

Fig. 6. Some examples from DUT-OMRON, SOD and MSRA with different methods

In Fig. 6, there are some typical visual comparison examples from DUT-OMRON, SOD and MSRA with different methods. For images with multiple saliency objects, our method can detect more accurately, like the second, third, sixth, and seventh rows, our results are much closer to the real GT map. At the same time, for images with a complex saliency object, our approach also does better than the other nine methods, which is able to highlight prominent targets as soon as possible, and remove backgrounds.

6   Conclusion

In this paper, we propose an image saliency detection algorithm that combines the robust background prior measurement with global cue. Better results are achieved by adjusting the effect of global cue on the overall algorithm. The experimental results show that the accuracy, recall and F-Measure indicators of the proposed algorithm are better than the other eight algorithms. Moreover, in the experiment, we find our method maybe have less advantage in some particular images that have inconspicuous global information. In addition, we compare our method with some deep-learning methods. Obviously, our method has less advantages in accuracy, but we can analysis a picture without any preliminary work thus we can save the pre-training time. In future, we will improve our method and extend it to multi-scale to get better saliency detection results.

Acknowledgments. This work was supported by the Key Natural Science Project of Anhui Provincial Education Department (KJ2018A0023), the Guangdong Province Science and Technology Plan Projects (2017B010110011), the Anhui Key Research and Development Plan (1804a09020101), the National Basic Research Program (973 Program) of China (2015CB351705), the National Natural Science Foundation of China (61906002, 61402002 and 61876002) and 2018 College Students Innovation and Entrepreneurship Training Program (201810357353).

References

1. Wang, L., Xue, J., Zheng, N., Hua, G.: Automatic salient object extraction with contextual cue. In: Proceedings of the IEEE International Conference on Computer Vision, pp. 105–112. IEEE (2011). https://dl.acm.org/citation.cfm?doid=2009916. 2009934
2. Marchesotti, L., Cifarelli, C., Csurka, G.: A framework for visual saliency detection with applications to image thumbnailing. In: Proceedings of the IEEE International Conference on Computer Vision, pp. 2232–2239. IEEE (2009). https://ieeexplore. ieee.org/document/5459467
3. Navalpakkam, V., Itti, L.: An integrated model of top-down and bottom-up attention for optimizing detection speed. In: Proceedings of the IEEE Conference on Computer Vision and Pattern Recognition, pp. 2049–2056. IEEE (2006). https:// ieeexplore.ieee.org/document/1641004
4. Ma, Y.F., Lu, L., Zhang, H.J., Li, M.: A user attention model for video summarization. In: Proceedings of the ACM International Conference on Multimedia, pp. 533–542. ACM (2002). https://dl.acm.org/citation.cfm?doid=641007.641116
5. Borji, A., Sihite, D.N., Itti, L.: Probabilistic learning of task-specific visual attention. In: Proceedings of the IEEE Conference on Computer Vision and Pattern Recognition, pp. 470–477. IEEE (2012). https://ieeexplore.ieee.org/document/ 6247710
6. Itti, L., Koch, C.: Feature combination strategies for saliency-based visual attention systems. J. Electron. Imaging **10**(1), 161–170 (2001)
7. Gao, D., Vasconcelos, N.: Discriminant saliency for visual recognition from cluttered scenes. In: Advances in Neural Information Processing Systems, pp. 481–488 (2005)

8. Frintrop, S., Backer, G., Rome, E.: Goal-directed search with a top-down modulated computational attention system. In: Kropatsch, W.G., Sablatnig, R., Hanbury, A. (eds.) DAGM 2005. LNCS, vol. 3663, pp. 117–124. Springer, Heidelberg (2005). https://doi.org/10.1007/11550518_15

9. Kanan, C., Tong, M.H., Zhang, L., Cottrell, G.W.: SUN: top-down saliency using natural statistics. Vis. Cogn. **17**(6–7), 979–1003 (2009). ttps://www.tandfonline.com/doi/abs/10.1080/13506280902771138

10. Alexe, B., Deselaers, T., Ferrari, V.: What is an object?. In: Proceedings of the IEEE Conference on Computer Vision and Pattern Recognition, pp. 73–80. IEEE (2010). https://ieeexplore.ieee.org/document/5540226

11. Yang, J., Yang, M.-H.: Top-down visual saliency via joint CRF and dictionary learning. In: Proceedings of the IEEE Conference on Computer Vision and Pattern Recognition, pp. 2296–2303. IEEE (2012)

12. Ng, A.Y., Jordan, M.I., Weiss, Y.: On spectral clustering: analysis and an algorithm. In: Advances in Neural Information Processing Systems, pp. 849–856 (2002)

13. Yang, C., Zhang, L., Lu, H., Ruan, X., Yang, M.H.: Saliency detection via graphbased manifold ranking. In Proceedings of the IEEE Conference on Computer Vision and Pattern Recognition, pp. 3166–3173. IEEE (2013)

14. Jiang, B., Zhang, L., Lu, H., Yang, C., Yang, M.H.: Saliency detection via absorbing Markov chain. In Proceedings of the IEEE International Conference on Computer Vision, pp. 1665–1672. IEEE (2013)

15. Zhu, W., Liang, S., Wei, Y., Sun, J.: Saliency optimization from robust background detection. In Proceedings of the IEEE Conference on Computer Vision and Pattern Recognition, pp. 2814–2821. IEEE (2014)

16. Xiao, Y., Wang, L., Jiang, B., Tu, Z., Tang, J.: A global and local consistent ranking model for image saliency computation. J. Vis. Commun. Image Represent. **46**, 199–207 (2017)

17. Achanta, R., Shaji, A., Smith, K., Lucchi, A., Fua, P., Süsstrunk, S.: SLIC superpixels compared to state-of-the-art superpixel methods. IEEE Trans. Pattern Anal. Mach. Intell. **34**(11), 2274–2282 (2012)

18. Liu, T., et al.: Learning to detect a salient object. IEEE Trans. Pattern Anal. Mach. Intell. **33**(2), 353–367 (2011)

19. Movahedi, V., Elder, J.H.: Design and perceptual validation of performance measures for salient object segmentation. In: Proceedings of the IEEE Conference on Computer Vision and Pattern Recognition-Workshops, pp. 49–56. IEEE (2010). https://ieeexplore.ieee.org/document/5543739

20. Wei, Y., Wen, F., Zhu, W., Sun, J.: Geodesic saliency using background priors. In: Fitzgibbon, A., Lazebnik, S., Perona, P., Sato, Y., Schmid, C. (eds.) ECCV 2012. LNCS, vol. 7574, pp. 29–42. Springer, Heidelberg (2012). https://doi.org/10.1007/978-3-642-33712-3_3

21. Tong, N., Lu, H., Zhang, L., Ruan, X.: Saliency detection with multi-scale superpixels. IEEE Signal Process. Lett. **21**(9), 1035–1039 (2014)

22. Achanta, R., Hemami, S., Estrada, F., Süsstrunk, S.: Frequency-tuned salient region detection. In: Proceedings of the IEEE Conference on Computer Vision and Pattern Recognition, pp. 1597–1604. IEEE (2009). https://ieeexplore.ieee.org/document/5206596

23. Perazzi, F., Krähenbühl, P., Pritch, Y., Hornung, A.: Saliency filters: contrast based filtering for salient region detection. In: Proceedings of the IEEE Conference on Computer Vision and Pattern Recognition, pp. 733–740. IEEE (2012)

24. Harel, J., Koch, C., Perona, P.: Graph-based visual saliency. In: Advances in Neural Information Processing Systems, pp. 545–552 (2007)

25. Hou, X., Zhang, L.: Saliency detection: a spectral residual approach. In: Proceedings of the IEEE Conference on Computer Vision and Pattern Recognition, pp. 1–8. IEEE (2007)
26. Yan, Q., Xu, L., Shi, J., Jia, J.: Hierarchical saliency detection. In: Proceedings of the IEEE Conference on Computer Vision and Pattern Recognition, pp. 1155–1162. IEEE (2013)

Improvement of Residual Attention Network for Image Classification

Lu Liang[1], Jiangdong Cao[1], Xiaoyan Li[1(✉)], and Jane You[2]

[1] School of Computer Science, China University of Geosciences,
Wuhan 430074, China
lixy@cug.edu.cn
[2] Department of Computing, The Hong Kong Polytechnic University,
Kowloon, Hong Kong 999077, China

Abstract. The existing residual attention network (RAN) method mainly utilizes the deeper network layer for the image objects which are to be classified. However, when the network depth is simply increased, it will lead to gradient dispersion (or explosion) effect. To address the problem, we propose a new improvement method of residual attention network for image classification, which applies several upsampling schemes to the RAN process, i.e., the stacked network structure extraction, and the bottom-up and top-down feedforward attention for residual learning. In the experiments, we have given comparisons of different network structures and different upsampling methods that are applied to the RAN. The proposed improvement method achieves state-of-the-art image classification performance on two benchmark datasets including CIFAR-10 (4.23% error) and CIFAR-100 (21.15% error). Compared with the traditional RAN method, the proposed improvement method can improve the accuracy of image classification to some extent.

Keywords: Residual attention network · Upsampling methods · Network structure

1 Introduction

Analysis of human perception of an image has shown that people mainly pay more attention to the salient objects rather than the entire scene in the image. After the gaze over time, the people combine the valuable information to get the initial understanding of the image, and then make some decisions, e.g., object classification and the scoring of image quality on the subjective level [1]. Actually, attention only generates different representations of the salient objects at a focused location in the input image. Therefore, the attention mechanism has a wide range of applications, such as natural language processing [2], speech recognition [3], entailment reasoning [4], and image caption [5], and so on.

For the application of image caption, the attention mechanism can be roughly divided into two parts: hard attention and soft attention [5]. Hard attention does

© Springer Nature Switzerland AG 2019
Z. Cui et al. (Eds.): IScIDE 2019, LNCS 11935, pp. 529–539, 2019.
https://doi.org/10.1007/978-3-030-36189-1_44

not consider the inapparent image areas that the generated attention weights are smaller than a certain threshold, which is often set to zero. On the other hand, soft attention serves to extract significant information and make appropriate judgments by introducing different attention weights to different image areas. The larger attention weight is, the more significant information is. Conversely, the smaller attention weight illustrates that the information at the location is insignificant. The purpose of soft attention mechanism is to enhance the salient information and suppress the unvaluable information for the input image.

As is known to us all, the depth of neural network is very crucial to the precision of image feature extraction. However, if the depth of network is simply increased, the performance may rise at the beginning and then fall in the subsequent process. In other words, it can improve the classification accuracy when the number of network layers increases at the beginning. But when the structure of the model gets deeper and deeper, the performance of classification accuracy becomes poor.

To address the above problem, He et al. [6] proposed a residual network for image detection, recognition and classification. Compared with state-of-the-art networks, the residual networks achieve better performance with higher classification accuracy by adding the insertion of identity mapping between modules. That is because it can prevent the model degradation while deepening the network.

In general, as the model deepens, attention will be more concentrated in key areas, which will help eliminate irrelevant information. By increasing the attention module and capturing different types of attention, which consults the characteristics of residual network and attention mechanism, and model's performance has been improved [7]. At the same time, due to its advanced network structure, the depth of the network can be easily extended to hundreds of layers.

In our network structure, by improving some details of residual attention network, the model can capture different types of attention, and at the same time, by the means of improving the method of upsampling, the model can get more key information.

2   Related Work

After [8], attention mechanism has been widely used in neural networks to deal with various problems.

In 2014, the traditional residual attention network (RNN) add attention mechanism, which helps model to learn the position that we should focus on according to the previous state, and process a part of pixels instead of all pixels, which can help reduce the pixels to be processed and reduce the complexity of the model [1].

In CNN, the attention mechanism is mainly divided into two types, one is spatial attention, and the other is channel attention. Among them, spatial attention mainly focuses on different areas of picture, and channel attention mainly focuses on different information of the picture.

In 2016, spatial attention and channel attention are added to CNN. The CNN network can extract features preferably [9]. Finally, in the field of image caption, the results are significantly better than other network models.

In addition, the attention mechanism is also used in the field of machine translation. In 2014, neural machine translation generally uses the encoder-decoder framework, and the encoder part compresses the input information into a fixed-length vector [2]. Therefore, in the face of long sentences, a lot of information will be lost, although there are some small tricks to solve this problem, such as using the LSTM model, but the effect is always very little. Because at the time of decoding, the context information, position information, and the like of the input words corresponding to the current predicted words are basically lost. However, after attention mechanism is introduced, the decoder automatically extracts useful information from the fixed-length vector to decode the current vocabulary, and the performance of machine translation is greatly improved.

Parmar et al. established a model based on self-attention [10]. By restricting the self-attention mechanism to attend to local neighborhoods, the size of image that the model can handle in practice is significantly increased. Not only that, Parmar N et al. also used the attention mechanism for image synthesis, and high-resolution images with low-resolution images.

After He et al. [6] proposed the residual network in 2015, Xie et al. [15] proposed an upgraded version of the residual network, ResNeXt. The innovation of this network is to propose aggregated transformation, which replaces the original ResNet's three-layer convolution block with a parallel stack of blocks of the same topology, which makes network more scalable, improves the accuracy of image classification and reduces the computational complexity of the model.

In the field of image classification, attention mechanism has been applied. Through the stacked network structure, attention residual learning, and bottom-up top-down feedforward attention, the network refers to the idea of attention mechanism and residual network which gets more discriminative features, different types of attentions, deeper network structure [7]. Thus, a network model with higher accuracy in image classification is obtained.

However, we think that residual attention network needs a little improvement, because we want a network with higher image classification accuracy. As a result, we compared several different interpolation methods, as well as different network structures, and then picked the most appropriate method to improve our network.

3 Residual Attention Network

We can implement our residual attention network through Stack network structure, attention residual learning mechanism, and bottom-up top-down feedforward attention [7]. The residual attention network consists of a stack of attention modules, each of which consists of mask branch and trunk branch, as Fig. 1 shows. The trunk branch is responsible for processing features. And the mask branch acts as a feature selector to help screen important information while also being used as a gradient updated filter during back propagation and the mask gradient of the input features is

$$\frac{dM(x,\theta)T(x,\phi)}{d\phi} = M(x,\theta)\frac{dT(x,\phi)}{d\phi} \tag{1}$$

Where the θ are the mask branch parameters, while the ϕ are the trunk branch parameters, $M(x,\theta)$ are the mask branch output with the input x, and $T(x,\phi)$ are the trunk branch output. The function above shows that attention modules are robust to noisy labels because mask branch prevents incorrect gradients (from noisy tags) from updating trunk branch parameters.

The bottom-up top-down structure imitates the fast feedforward and feedback attention process. The output mask is used as control gates for neurons of trunk branch, where i ranges over all spatial positions and c is the index of the channel, so the output of attention module is

$$H_{i,c}(x) = M_{i,c}(x) \times T_{i,c}(x) \tag{2}$$

However, the above structure has limitations. Firstly, the output of mask branch ranges from 0 to 1, and repeated multiplication will weaken the feature gradually. Secondly, the feature map output from the mask branch may damage the output of the trunk branch. Thus, they modify output H of attention module as

$$H_{i,c}(x) = (1 + M_{i,c}(x)) \times T_{i,c}(x) \tag{3}$$

It can be seen that when the approximate value of $M(x)$ is 0, $H(x)$ will approximate the original features $F(x)$, $M(x)$ as feature selector, enhancing useful features and suppressing noise of the trunk features.

According to attention mechanism in [12], the bottom-up attention module is often used to extract the region of interest in the image, to obtain object features, while top-down attention module is used to learn the weights corresponding to the features, so as to achieve a deep understanding of the visual image. This combination of bottom-down and top-down attention mechanism can be used in image classification to help extract features, which allows us to develop end-to-end trainable networks from the top down. Unlike the stacked hourglass network, the bottom-up top-down structure guides feature learning.

In our work, we need to capture different scales of attention. Therefore, after inputting the image, we perform max pooling multiple times to quickly increase the receptive field. When capturing different scales of information, we need to use linear interpolation to enlarge receptive field so that combine the captured attention with the feature map output by the trunk branch. As shown in Fig. 1, the number of linear interpolations and the number of max pooling are equal to ensure the merging of two feature maps. The operation of down sample ensures that the model can capture information on a smaller scale. The number of times of down sample and up sample must be the same to ensure that the two feature maps are merged. Finally, the sigmoid layer normalizes the output range of the mask branch to [0, 1].

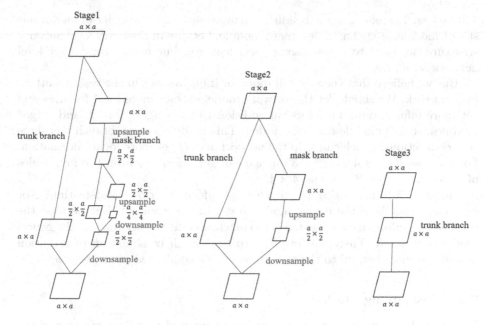

Fig. 1. The receptive field changes of three different attention modules. As we can see, when the input image size is a × a, different attention modules can get information of different scales. In addition, the times of up sample and down sample in the attention module must be equal to ensure that the information is fused together.

3.1   Improved Network Structure

In the residual attention network, there are three types of attention modules, each with two skip connections, one skip connection and no skip connection, called stage1, stage2 and stage3, respectively. Since skip connections can help the network capture different scale information, skip connections include operations such as up sample and down sample, so it is very important for the network. For example, for a 32 × 32 image, the stage1 module can get features on the 32 × 32, 16 × 16, and 8 × 8 scales by down sample, the stage2 module can get features on the 32 × 32 and 16 × 16 scales, and the stage3 module can only get features on the 32 × 32 scale.

Extracting features of multiple scales enables neural networks to obtain information of different scales, thereby improving the classification accuracy of different neural networks. For example, one of the functions of receptive field block is to help neural networks capture features of different scales [13,14]. Szegedy et al. pointed out that one of the functions of inception module is to help neural networks get different scales of information, these two modules are obvious for the improvement of the accuracy of the neural network [6].

We all know that in the shallower layers of the convolutional neural network, some local features, such as edges, corners, etc., are extracted; while the deeper layers in the network generally get higher features, and the local features are will be less. Therefore, in the residual attention network, the order of modules is stage1 module, stage2 module, stage3 module, because in the lower level network structure, we need to extract some local features, but not in the higher level network structure.

But we believe that there is still room for improvement in the residual attention network. We should let the network model extract more local features and get more information. Therefore, in addition to the stage1, stage2, and stage3 modules, stage0 module has been added. This module's mask branch has three skip connections, which can help the network model get more local information. For the stage0 module, a 32×32 image can get information on the four scales of 32×32, 16×16, 8×8, and 4×4.

However, will extracting too much local information be self-defeating? Not only can it not help the neural model to get higher accuracy, but because the useless local information captured is too much, so that the accuracy of the neural model is reversed. Therefore, in order to test which combination of attention modules is more helpful to the neural model, we conducted an experiment.

3.2   Upsampling Method

In our work, we need to use the upsampling method to adjust the size of the feature map. By changing the up sample method, we can get different classification results. We use three types of upsampling methods, including pixel shuffle, nearest neighbor interpolation, bilinear interpolation and pixel shuffle.

The nearest neighbor interpolation algorithm is also called the zero-order interpolation algorithm. The main principle is to make the pixel value of the output pixel equal to the pixel value closest to it in the neighborhood. The advantage of this approach is that the calculation is simple, and the downside is that the image is prone to jaggedness in the changing place.

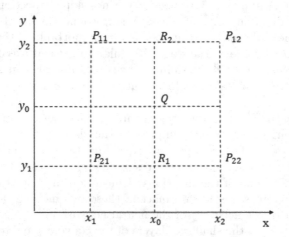

Fig. 2. Discription of pixels' relationship

The bilinear interpolation algorithm is a better image scaling algorithm. It fully utilizes the four real pixel values around the virtual point in the original image to jointly determine a pixel value in the target image. Therefore, the calculation process is more complicated than the nearest neighbor interpolation algorithm. As shown in Fig. 2, the calculation process is as follows:

Firstly, two interpolation calculations are performed in the x direction to calculate pixel values of R_1 and R_2.

$$f(x_0, y_1) \approx \frac{x_2 - x_0}{x_2 - x_1} f(P_{21}) + \frac{x_0 - x_1}{x_2 - x_1} f(P_{22}) \tag{4}$$

$$f(x_0, y_2) \approx \frac{x_2 - x_0}{x_2 - x_1} f(P_{11}) + \frac{x_0 - x_1}{x_2 - x_1} f(P_{12}) \tag{5}$$

Secondly, linear interpolation in the y direction is performed, and pixel values of the Q point are calculated by R_1 and R_2.

$$f(x_0, y_0) \approx \frac{y_2 - y_0}{y_2 - y_1} f(x_0, y_1) + \frac{y_0 - y_1}{y_2 - y_1} f(x_0, y_2) \tag{6}$$

Now we talk about pixel shuffle, generally speaking, pooling will make the resolution of the feature map smaller, but according to [11], when $stride < 1$, the operation of the sub-pixel convolution can make the width and height of the feature map larger, that is, the resolution increases.

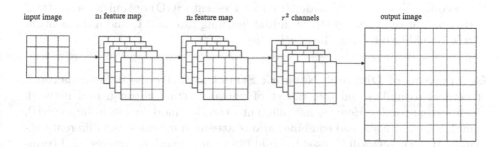

input image n_1 feature map n_2 feature map r^2 channels   output image

Fig. 3. The operation of the sub-pixel convolution

The operation of the sub-pixel convolution is as shown in Fig. 3, the input of the network is a low-resolution original picture with a size of $1 \times H \times W$. After two layers of convolution, the image size is $r^2 \times H \times W$. The r^2 channels of each pixel are rearranged to obtain a high-resolution image of size $1 \times rH \times rW$. The essence of this operation is to periodically insert low-resolution features into high-resolution images at specific locations. The image processing speed of this method is quite fast.

4 Experiment

In this section, we evaluate the performance of proposed Residual Attention Network on a series of benchmark datasets including CIFAR-10 and CIFAR-100. Our experiment has two parts, in the first part, we compare the effects of different network structures and upsampling methods on the accuracy of neural network classification using the CIFAR-10 dataset, and then we will analysis the result. In the second part, we will use the best network structure and upsampling methods for the neural network on the CIFAR-100 dataset to further validate our conclusions.

4.1 CIFAR-10 and Analysis

The CIFAR-10 dataset consists of 60000 32×32 color images in 10 classes, with 6000 images per class, which has 50000 training images and 10000 test images. The dataset has five training batches and one test batch with 10000 images per batch. The test batch includes 1000 randomly-selected images from each class, while the training batches contain the rest of images randomized, which may contain more images from one class than another. The classes don't overlap at all.

We firstly perform image preprocessing, each side of the image will be filled with 4 pixels with a value of 32×32 crop is randomly sampled from an image or its horizontal flip, and the per-pixel RGB mean value will be substracted. We used the Nesterov Stochastic Gradient Descent (SGD) optimization method with a weight decay of 0.0001, an initial learning rate of 0.1, and a momentum of 0.9. In addition to this, the epoch is set to 300.

Comparison of Different Network Structures. We use different network structures to understand the impact of network structure on neural network classification results. Here we use different attention modules, including stage0, stage1, stage2, stage3, and combine various attention modules into different networks. Attention-92 will be used to build different network structures, and training will be done on the CIFAR-10 dataset.

In Table 1, we can know that among network structures, Stage2 reduces 0.34% error rate compare to Stage1 + Stage2 + Stage3 with Attention-92, which has a smaller number of parameters than Stage1 + Stage2 + Stage3, and at the same time, Stage0 + Stage1 + Stage2 can reduce 0.37% error rate compare to Stage1 + Stage2 + Stage3, however, which has more parameters. Thus, we believe that the performance of Stage2 is good, because it has a smaller number of parameters and can reduce error rate effectively. Besides, comparing Stage3 with Stage1 + Stage2 + Stage3, you will find that without mask branch, test error is increased by 0.05%. When we changed all attention module to Stage1, the test error increased by 0.16% compared to Stage2. Stage1 can get smaller scale information, so we suspect that when the scale is too small, the neural network will get some useless information.

Table 1. Test error (%) on CIFAR-10 using different networks with bilinear interpolation

Network structures	Test error (%)	Params×10^6
Stage1	4.42	15.11
Stage2	4.26	12.11
Stage3	4.65	12.11
Stage0+Stage1+Stage2	4.23	12.64
Stage1+Stage2+Stage3	4.60	12.17

Comparison of Different Upsampling Methods. In order to know which interpolation method is more suitable for the residual attention network, we use three different interpolation methods to train on the Attention-92 network and the CIFAR-10 dataset. The results are as follows:

Table 2. Test error (%) on CIFAR-10 using different upsampling methods.

Upsampling methods	Test error (%)	Params × 10^6
Nearest neighbor interpolation	4.51	12.17
Bilinear interpolation	4.60	12.17
Pixel shuffle	4.43	15.12

As shown in Table 2, different upsampling methods result in different training results, which indicates that the replacement of the upsampling method can improve the accuracy of image classification. Among the three upsampling methods, the lowest error rate is pixel shuffle, which can reduce the error rate by 0.17%, followed by nearest neighbor interpolation, but which has the largest number of parameters, indicating that pixel shuffle can improve the accuracy of model classification, but this is an increase in computational complexity as a price. At the same time, nearest neighbor interpolation has a slightly better error rate than bilinear interpolation, but the number of parameters is the same as the latter.

4.2 Image Classification on CIFAR-100

In this section, we conduct experiment on CIFAR-100 dataset, and the evaluation is conducted on Attention-92. We perform image preprocessing, each side of the image will be filled with 4 pixels with a value of 0 and a 32 crop is randomly sampled from an image or its horizontal flip, and the per-pixel RGB mean value will be subtracted. We used the Nesterov Stochastic Gradient Descent (SGD) optimization method with a weight decay of 0.0001, an initial learning rate of 0.1, and a momentum of 0.9. In addition to this, the epoch is set to 300.

Table 3. Test error (%) on CIFAR-100 using different networks

Network	Test error (%)	Params × 10^6
Stage1+Stage2+Stage3+Blinear Intorpolation	21.69	12.26
Stage1+Stage2+Stage3+Pixel Shuffle	21.29	15.21
Stage2+Blinear Intorpolation	21.15	12.20

In Table 3, we can see that upsampling method using pixel shuffle and changing attention modules to stage2 can improve the classification accuracy of the model. Compared with the first item in the table, the third method of replacing the attention modules reduces the error rate by 0.54% and which has less parameters. And pixel shuffle can help decrease the classification error rate of 0.4% but it will increase the number of parameters.

5   Conclusion

In residual attention network, by merging the essence of residual network and attention mechanism, the accuracy of image classification has been significantly improved. In our work, we explored the improvement of upsampling method and network structure of residual attention network. Among three upsampling methods, we believe that although the classification error rate of pixel shuffle is lower than bilinear interpolation and nearest neighbor interpolation, but we do not advocate it because it increases the number of parameters. At the same time, experiments with different network structures tell us that although collecting more different scale information is beneficial to improve classification efficiency, however, too low-scale information may be not good for image classification. In future, we plan to get more improvement of network structure on residual attention network.

References

1. Mnih, V., Heess, N., Graves, A .: Recurrent models of visual attention. In: Advances in Neural Information Processing Systems, pp. 2204–2212 (2014)
2. Bahdanau, D., Cho, K., Bengio, Y.: Neural machine translation by jointly learning to align and translate. arXiv preprint arXiv:1409.0473 (2014)
3. Fan, R., Zhou, P., Chen, W.: An online attention-based model for speech recognition. arXiv preprint arXiv:1811.05247 (2018)
4. Rocktschel, T., Grefenstette, E., Hermann, K.M., et al.: Reasoning about entailment with neural attention. arXiv preprint arXiv:1509.06664 (2015)
5. Xu, K., Ba, J., Kiros, R., et al.: Show, attend and tell: neural image caption generation with visual attention. In: International Conference on Machine Learning, pp. 2048–2057 (2015)
6. He, K., Zhang, X., Ren, S., et al.: Deep residual learning for image recognition. In: Proceedings of the IEEE Conference on Computer Vision and Pattern Recognition, pp. 770–778 (2016)

7. Wang, F., Jiang, M., Qian, C., et al.: Residual attention network for image classification. In: Proceedings of the IEEE Conference on Computer Vision and Pattern Recognition, pp. 3156–3164 (2017)
8. Vaswani, A., Shazeer, N., Parmar, N., et al.: Attention is all you need. In: Advances in Neural Information Processing Systems, pp. 5998–6008 (2017)
9. Chen, L., Zhang, H., Xiao, J., et al.: SCA-CNN: spatial and channel-wise attention in convolutional networks for image captioning. In: Proceedings of the IEEE Conference on Computer Vision and Pattern Recognition, pp. 5659–5667 (2017)
10. Parmar, N., Vaswani, A., Uszkoreit, J., et al.: Image transformer. arXiv preprint arXiv:1802.05751 (2018)
11. Shi, W., Caballero, J., Huszár, F., et al.: Real-time single image and video super-resolution using an efficient sub-pixel convolutional neural network. In: Proceedings of the IEEE Conference on Computer Vision and Pattern Recognition, pp. 1874–1883 (2016)
12. Anderson, P., He, X., Buehler, C., et al.: Bottom-up and top-down attention for image captioning and visual question answering. In Proceedings of the IEEE Conference on Computer Vision and Pattern Recognition, pp. 6077–6086 (2018)
13. Liu, S., Huang, D., Wang, Y., et al.: Receptive field block net for accurate and fast object detection. arXiv preprint arXiv:1711.07767 (2017)
14. Szegedy, C., Liu, W., Jia, Y., et al.: Going deeper with convolutions. In: Proceedings of the IEEE Conference on Computer Vision and Pattern Recognition, pp. 1–9 (2015)
15. Xie, S., Girshick, R., Dollár, P., et al.: Aggregated residual transformations for deep neural networks. In: Proceedings of the IEEE Conference on Computer Vision and Pattern Recognition, pp. 1492–1500 (2017)

Nuclei Perception Network for Pathology Image Analysis

Haojun Xu[1], Yan Gao[1], Liucheng Hu[1], Jie Li[1], and Xinbo Gao[1,2(✉)]

[1] School of Electronic Engineering, Xidian University, Xi'an 710071, China
{damnull,gyy1101}@outlook.com, sword695230@gmail.com,
{leejie,xbgao}@mail.xidian.edu.cn
[2] State Key Laboratory of integrated Services Networks, Xi'an, China

Abstract. Nuclei segmentation is a challenge task in medical image analysis. A digital microscopic tissue image may contain hundreds or even thousands nuclear. Its morphological information provides the biological basis for the diagnosis and classification of diseases. The task requires to detect every nuclear of cells in a densely packed scene and get the segmentation of them for further pathological analysis. Nuclei segmentation can also be described as an instance segmentation task in densely packed scene. In this article, we propose a novel anchor-free dense instance segmentation framework to alleviate the issues. The network detects nuclears and segment them simultaneously. Then the nuclear segmentation mask is aggregated as nuclear instance guided by the offset map generated from the network. The network works by combining target location with pixel-by-pixel classification to distinguish crowded objects. The proposed method performs well on nuclear segmentation dataset.

Keywords: Deep learning · CNN · Nuclei segmentation

1 Introduction

A large number of scanned digital images of stained pathological specimens (H&E images) [1] are the key to histopathology. Analysis of nuclei morphology and characteristics based on histology slide images is helpful in predicting treatment outcomes and assessing cancer grade. For pathologists, manually annotating the nuclei on the entire slide is time consuming and laborious. Accurate automated nuclei segmentation technology can significantly reduce human workload and be more objective and rational. As a basic unit of histopathological images, cells present a disordered and repetitive pattern. Similar to buildings in urban landscapes, or items with similarly packaged on retail shelves. We abstract the nuclei segmentation task into the instance segmentation in dense scenes, and need to classify each object pixel by pixel. In these organizational scenes, the objects are in tight contact. However, due to the large image size, improper processing will lead to the loss of cell-level information. In addition, there was color

© Springer Nature Switzerland AG 2019
Z. Cui et al. (Eds.): IScIDE 2019, LNCS 11935, pp. 540–551, 2019.
https://doi.org/10.1007/978-3-030-36189-1_45

variability in tissues under pathological conditions, such as different organs and the same organ with different cancer grade. Therefore, accurate morphological analysis of pathological images remains challenging.

There are a number of methods for the study of nuclei segmentation, including traditional methods and supervised machine learning methods. Since the emergence of deep learning in ImageNet competition [2] in 2012, most of the image recognition technology has been replaced by deep learning. The same is true for pathological analysis [3,4], and there are some works specifically for cell segmentation [5,6]. Some people consider color decomposition and propose a bilinear network for fine-grained classification [7]. Others have attempted to analyze unlabelled images by adversarial networks [8]. There are also attempts to perform distance regression [9], contour-aware loss [10], and enhanced sparse convolution [11].

We consider combining information from detection with instance mask prediction to analyze the nuclei on the entire slide image. Natural image target detection based on deep learning has been developed to produce not only rectangular bounding boxes [12,13], and also perform instance segmentation [14]. Recently, the new workflow anchor free methods have appeared to solve the rough positioning of the object [15]. The bounding box is convenient for network representation, but they are not suitable for histopathological images. Due to the tight location between cells, the box will inevitably contain partial semantic information of other cells, affecting object localization and prediction.

In this paper, we present an instance segmentation method called nuclei perception network (NPNet), which perceives cells in a finer-grained manner. In the nuclei detection phase, NPNet limits the feature space by learning a set of points of the adaptive fitting object. The model are more focused on the semantic representation of the target region. During the training process, four special points are guided by the ground truth to reconstruct the region of interest. The targets are classified according to the center point of each region. In addition, an optional center correction branch was added to allow for possible deviations from the actual geometric center of the nucleus. In the instance segmentation phase, we added a branch to perform segmentation in parallel with the existing detection branch. NPNet allows end-to-end training to generate fine-grained nuclei segmentation without additional supervision. Our main contributions are reflected in three aspects: First, we propose a fine-grained nuclei detection branch that focuses on the semantic representation of each object region. It can improve the performance of object detection in nuclear crowded area. Second, we use the center correction branch to correct the target center position, and improve the accuracy of nuclei classification and localization. Third, by introducing a new segmentation branch, a unified solution combining anchor free detection with instance segmentation is proposed.

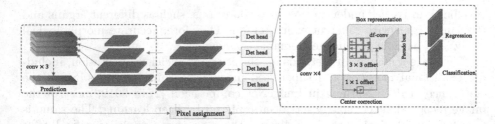

Fig. 1. NPNet overview. FPN [16] is the backbone, and Det head is the detector represented by sampling points. The pixel assignment branch generates the instances segmentation result.

2   Related Work

2.1   Nuclei Segmentation in Digital Pathology

For many years, histopathological analysis has been the focus of computer vision competition. There are many hot topics in the field [17], such as classification of tissue regions, detection of specific cells and so on. The segmentation of nuclei [18,19] can obtain morphology information, providing a biological basis for the diagnosis and classification of diseases [17]. Due to its importance, extensive work has been proposed [6,20]. Traditional methods include morphology [21–24], pixel classification [25], deformable model [26] and graph-based segmentation methods [27,28]. Machine learning-based models rely on manual features to segment nuclear and non-nuclear regions [22,29], providing greater accuracy. However, the above methods are only effective for one or more specific types of nuclei or images, and are usually very sensitive to manual setting parameters.

Neural networks are now considered to be one of the most popular methods in the field of computer vision. It has been successfully applied to the detection and segmentation of histopathological nuclei. Compared with traditional machine learning techniques, this method of learning data representation has greater performance improvement. Although these semantic-based segmentation networks perform well, the object boundaries of cell segmentation are more difficult to completely separate. As discussed earlier, there is direct contact between cells and cells in biomedical data. Because convolution in the network will reduce the resolution of images, the probability map needs to be upsampled to the original size. The boundary of small and dense nuclei is inevitably blurred or overlapped. To solve this problem, many authors adopt post-processing methods, such as bottleneck detection [30], and ellipse fitting [31,32]. Alternatively, strategies such as labeling control watershed [33,34] or regional growth [20] can be used to obtain the nuclear profile directly. Other authors perform instances segmentation, such as giving the contour pixels a large weight [10], and predicting the object and its contour [35].

2.2 Object Detectors

At present, object detectors based on deep convolutional neural networks can be roughly divided into two-stage and one-stage modes. The multi-stage method performs best on the benchmark. R-CNN [36] is one of the basic researches in the field of detection. It uses the selective search method [37] to locate in the input image. SPPNet [12] and Fast-RCNN [38] extract the region of interest (RoIs) from the feature map for object representation. After Faster-RCNN [13] proposed the region proposal network (RPN), the network can perform end-to-end training. The single-shot detectors YOLO [39] and SSD [40] achieve real-time detection at the expense of some performance. There is also some work dedicated to improving detection accuracy. R-FCN [41] uses a position-sensitive score map. Cascaded R-CNN [42] addresses quality mismatch at inference by a series of IoU thresholds. FPN [16] and DSSD [43] connect deep semantic information to shallow layers.

Rectangular bounding boxes appear at every stage of the development of these algorithms. However, using different scales bounding boxes as the anchor to get the final target position is a very complicated process. In order to solve the shortcomings of boxes, many work attempts to develop more flexible object representations. Like the ellipse proposal in pedestrian detection [44], and the rotating bounding box in work [45, 46]. There are also keypoint-based methods Extreme Net, CornerNet and its improved CenterNet. In addition, RepPoints easily simulates geometric changes in objects by adaptively learning semantic points. These anchor free detectors only need to learn smaller representation spaces. For example, CornerNet [47] locates an object through a set of 2-d points instead of a 4-d bounding box.

2.3 Instance Segmentation

Different from the early work about instance segmentation [48,49], Mask R-CNN [14] adds the segmentation branch based on Faster R-CNN. The target prediction is matched with the original image by bilinear interpolation. Path aggregation network [50] obtains results through multiple layers rather than one layer, further improving performance. Inspired by these methods, we performed an Instance segmentation of the nucleus on the entire slide. However, the two-stage approach is not entirely suitable for crowded biological scenarios. By fitting the object with sampling points, the proposed NPNet can express the nucleus in a finer feature space.

3 Methods

The NPNet structure is shown in Fig. 1. The FPN [16] is used as the backbone network, which consists of a points representation detector and an instance segmentation branch. Among them, the instance segmentation is performed by the semantic segmentation branch and the pixel assignment branch. We will describe each module in detail below.

3.1   Point Representation of the Nucleus

The main idea of the framework is to unify the anchor-free detector and instance segmentation. Consider a non-uniform distribution of the shape and location of the nucleus. We believe that the rectangular space learned by the anchor-based detector introduces unnecessary interference and requires a more flexible representation. Following RepPoints [15], the proposed NPNet uses a set of free sampling points to learn the nucleus. Sampling points are not limited by spatial shapes and only care about the texture and semantic background of a single object. Reduce the semantic interference of neighboring objects. The scheme works as follows. Given a digital image I, suppose we get a feature map F_i by backbone network. First, model a set of adaptive sample points $N = \{(x_k, y_k)\}_{k=1}^n$. Where n is the total number of sample points used in the representation. In our work, n is 9 by default. Then gradually refine these points to get the location of the nucleus, which can be simply expressed as

$$N_r = \{(x_k + \triangle x_k, y_k + \triangle y_k)\}_{k=1}^n \tag{1}$$

where $\triangle x_k, \triangle y_k$ are the position offset of the new sample point relative to the old one. The anchor-based detector takes the position on the input image as the center of the box. In contrast, we generate 18 channels for each pixel to learn the coordinates of 9 points. In other words, we directly consider pixels as training samples. If a pixel is classified as a positive sample, we will use these points to generate a pseudo bounding box by *Min-max function* in [15] for subsequent evaluation.

After generating the pseudo-box, we set the top-left and lower-right points of the ground truth boundary box as its regression target. Smooth L1 function [38] is used as the regression loss. The remaining points automatically learn the importantsemantic information of the target.

3.2   Center Point Based Classification Branch

At the initial stage of object detection, we have initially described the nucleus through sampling points. A refined bounding box is generated by pseudo-boundary regression. We call it the nuclear proposal P. Because the construction space of the center point is small and the model can converge faster. We follow YOLO [39] and DenseBox [51] to use the image features of the object center point as the classification feature of the object. In other words, we use the confidence score of the feature to determine whether it is an object.

However, there are some limitations in the center point representation in tissue pathological images. Because of the nature of the image, there are many nuclear dense areas in the feature map. If a position falls within multiple bounding boxes, it is considered a fuzzy sample. Such attribution ambiguity will affect the accuracy of subsequent instance segmentation. The proposed FPN structure [16] has been proven to greatly alleviate this problem. Objects of different scales are assigned to different feature layers for processing separately, shown in Fig. 2.

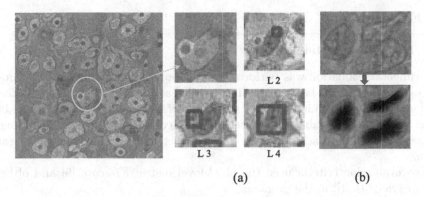

(a) (b)

Fig. 2. (a) Fuzzy samples: adjacent nuclei of different sizes and their FPN grades. L2 represents the second level feature of FPN. (b) Assign offset map of nuclear pixels. Each nuclear pixel has an offset, and the target of offset is to make the point nearby the center of nuclear. We use smooth-L1 loss to regress them.

Small objects have high-resolution feature maps, which avoids smaller objects being swallowed by larger objects. Another problem is that the pseudo box center of small size or large morphological variation nuclear is deviated from the actual geometric center of the object. May affect the confidence score. So we generate an offset field for the center point of each frame through 1×1 convolutions, and use the offset field to correct the center feature by a 1×1 deformable convolution. Making it as close as possible to the geometric center.

The overall workflow of the target recognition branch is shown in Fig. 1 Det head. Starting from the center point, a set of represented points is obtained through 3×3 regular convolution. After the refined P is obtained, the final classification features are obtained through deformable convolution. In the whole process, point learning is driven by both regression loss and target classification loss. Therefore, the reconstruction points can be focused on and target location can be done more accurately.

3.3 Instance Segmenation for Nucleus

We first fuse the hierarchical features of FPN and call it $fuseF$. Then use the $fuseF$ to generate binary segmentation map and the offset field for each pixel. Through the results generated by the detector, the center point is extracted as the id of the instance segmentation. Each pixel that is divided into positive classes is added to its generated offset field to obtain the corrected position of each pixel. These points are then assigned to the nearest id to form the final instance mask.

4   Experiment

4.1   Dataset and Metric

Experimental evaluation was performed on elaborately annotated H&E stained image dataset [1]. The dataset contains detailed segmentation of 30 images from 7 different organs. Each image has a resolution of 1000×1000. More than 21,000 nuclei were manually labeled. The training set contains 16 images, including 4 organs, 4 images for each category. The test set had 14 images, two for each organ.

The evaluation criteria used the pixel-level metric *F1-score* [9] and object-level metric *AJI* [1]. in the literature.

The proposed method uses random cropping and flipping techniques for data enhancement. This method is implemented by mxnet. The optimizer uses SGD. The learning rate strategy uses "poly" [52], initial learning rate is 0.01, momentum is 0.9, and weight decay is 0.0001. The model is training total 200 epochs. The non-maximum suppression threshold was set to 0.7 to ensure higher detection accuracy of edge contact cells. The minimum detection confidence is set to 0.5.

FPN [16] hierarchical allocation problem, considering that most targets are small, and there is no extreme imbalance of length and width. We use the minimum size of the target as the basis for distribution, which is

$$lvl(B) = \lceil \frac{min(w_B, h_B)}{stride} \rceil \tag{2}$$

where B is the refinement bounding box of the target, w_B and h_B are the width and length of box, and the stride of each level in FPN are 4, 8, 16, and 32 respectively. For targets beyond the highest level of the allocation hierarchy, we put it at the last level.

4.2   Results

The result of the instance segmentation is shown in the Fig. 3. The detection branch and the semantic segmentation prediction work together on the model results. Semantic segmentation maps can learn object-intensive regions. The

Table 1. Quantitative results of nuclear segmentation.

Network	AJI	F1-score
[1]	0.5083	0.7623
UNet	0.3833	0.7793
DIST [9]	0.5598	0.7863
Mask R-CNN	0.5002	0.7470
Ours	0.5044	0.7842

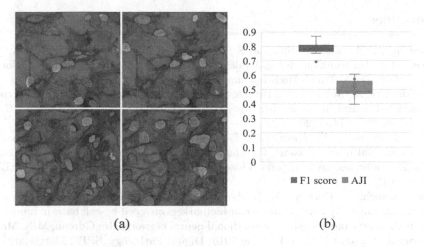

(a)   (b)

Fig. 3. (a) The left column shows the ground truth and the right show our prediction on validating set. From the results, our model shows the ability to handle the instance assignment of connected segmentation area. (b) The box plot of our segmentation result on validating set.

center point of the detection branch separates them into separate instances. As can be seen from Table 1, the F1 of the method is 0.7842 ± 0.0478, and the AJI is 0.5044 ± 0.0625. Based on the mean and standard deviation of the evaluation metrics, the results of the framework are desirable.

5   Conclusions

In this paper, we present a novel nuclei perception network, NPNet. Integrate the anchor free detector into the instance segmentation task. The model performs object representation through sampling points, and adds a semantic segmentation branch to coordinate the determination of the object area. The segmentation branch of the method is based on semantic segmentation, taking into account both the semantic information and background information. In this way, the learning of overlapping regions is supervised by both target detection and semantic segmentation. The target overlap problem can be alleviated to a large extent. Evaluation was performed on the digital pathology challenge dataset, which is comparable to the comparison method.

Acknowledgments. This work was supported in part by the National Natural Science Foundation of China under Grant 61432014, 61772402, U1605252 and 61671339, and in part by National High-Level Talents Special Support Program of China under Grant CS31117200001.

References

1. Kumar, N., Verma, R., Sharma, S., Bhargava, S., Vahadane, A., Sethi, A.: A dataset and a technique for generalized nuclear segmentation for computational pathology. IEEE Trans. Med. Imaging **36**(7), 1550–1560 (2017)
2. Krizhevsky, A., Sutskever, I., Hinton, G.E.: ImageNet classification with deep convolutional neural networks. In: Advances in Neural Information Processing Systems, pp. 1097–1105 (2012)
3. Hou, L., Samaras, D., Kurc, T.M., Gao, Y., Davis, J.E., Saltz, J.H.: Patch-based convolutional neural network for whole slide tissue image classification. In: 2016 IEEE Conference on Computer Vision and Pattern Recognition (CVPR). IEEE, June 2016
4. Sheikhzadeh, F., Carraro, A., Korbelik, J., MacAulay, C., Guillaud, M., Ward, R.K.: Automatic labeling of molecular biomarkers on a cell-by-cell basis in immunohistochemistry images using convolutional neural networks. In: Gurcan, M.N. Madabhushi, A. (eds.) Medical Imaging 2016: Digital Pathology. SPIE, March 2016
5. Greenspan, H., van Ginneken, B., Summers, R.M.: Guest editorial deep learning in medical imaging: overview and future promise of an exciting new technique. IEEE Trans. Med. Imaging **35**(5), 1153–1159 (2016)
6. Irshad, H., Veillard, A., Roux, L., Racoceanu, D.: Methods for nuclei detection, segmentation, and classification in digital histopathology: a review—current status and future potential. IEEE Rev. Biomed. Eng. **7**, 97–114 (2014)
7. Wang, C., Shi, J., Zhang, Q., Ying, S.: Histopathological image classification with bilinear convolutional neural networks. In: 2017 39th Annual International Conference of the IEEE Engineering in Medicine and Biology Society (EMBC). IEEE, July 2017
8. Zhang, Y., Yang, L., Chen, J., Fredericksen, M., Hughes, D.P., Chen, D.Z.: Deep adversarial networks for biomedical image segmentation utilizing unannotated images. In: Descoteaux, M., Maier-Hein, L., Franz, A., Jannin, P., Collins, D.L., Duchesne, S. (eds.) MICCAI 2017. LNCS, vol. 10435, pp. 408–416. Springer, Cham (2017). https://doi.org/10.1007/978-3-319-66179-7_47
9. Naylor, P., Lae, M., Reyal, F., Walter, T.: Segmentation of nuclei in histopathology images by deep regression of the distance map. IEEE Trans. Med. Imaging **38**(2), 448–459 (2019)
10. Chen, H., Qi, X., Yu, L., Heng, P.-A.: DCAN: deep contour-aware networks for accurate gland segmentation. In: 2016 IEEE Conference on Computer Vision and Pattern Recognition (CVPR). IEEE, June 2016
11. Song, J., Xiao, L., Molaei, M., Lian, Z.: Multi-layer boosting sparse convolutional model for generalized nuclear segmentation from histopathology images. Knowl.-Based Syst. **176**, 40–53 (2019)
12. He, K., Zhang, X., Ren, S., Sun, J.: Spatial pyramid pooling in deep convolutional networks for visual recognition. In: Fleet, D., Pajdla, T., Schiele, B., Tuytelaars, T. (eds.) ECCV 2014. LNCS, vol. 8691, pp. 346–361. Springer, Cham (2014). https://doi.org/10.1007/978-3-319-10578-9_23
13. Ren, S., He, K., Girshick, R., Sun, J.: Faster R-CNN: towards real-time object detection with region proposal networks. IEEE Trans. Pattern Anal. Mach. Intell. **39**(6), 1137–1149 (2017)
14. He, K., Gkioxari, G., Dollar, P., Girshick, R.: Mask R-CNN. In: 2017 IEEE International Conference on Computer Vision (ICCV). IEEE, October 2017

15. Yang, Z., Liu, S., Hu, H., Wang, L., Lin, S.: RepPoints: point set representation for object detection. arXiv preprint arXiv:1904.11490 (2019)
16. Lin, T.-Y., Dollar, P., Girshick, R., He, K., Hariharan, B., Belongie, S.: Feature pyramid networks for object detection. In: 2017 IEEE Conference on Computer Vision and Pattern Recognition (CVPR). IEEE, July 2017
17. Xu, Y., et al.: Large scale tissue histopathology image classification, segmentation, and visualization via deep convolutional activation features. BMC Bioinform. **18**(1), 281 (2017)
18. Chen, J.-M., Li, Y., Jun, X., Gong, L., Wang, L.-W., Liu, W.-L., Liu, J.: Computer-aided prognosis on breast cancer with hematoxylin and eosin histopathology images: a review. Tumor Biol. **39**(3), 101042831769455 (2017)
19. Ali, H.R., et al.: Lymphocyte density determined by computational pathology validated as a predictor of response to neoadjuvant chemotherapy in breast cancer: secondary analysis of the ARTemis trial. Ann. Oncol. **28**(8), 1832–1835 (2017)
20. Xing, F., Yang, L.: Robust nucleus/cell detection and segmentation in digital pathology and microscopy images: a comprehensive review. IEEE Rev. Biomed. Eng. **9**, 234–263 (2016)
21. Cheng, J., Rajapakse, J.C.: Segmentation of clustered nuclei with shape markers and marking function. IEEE Trans. Biomed. Eng. **56**(3), 741–748 (2009)
22. Faridi, P., Danyali, H., Helfroush, M.S., Jahromi, M.A.: An automatic system for cell nuclei pleomorphism segmentation in histopathological images of breast cancer. In: 2016 IEEE Signal Processing in Medicine and Biology Symposium (SPMB). IEEE, December 2016
23. Wang, P., Hu, X., Li, Y., Liu, Q., Zhu, X.: Automatic cell nuclei segmentation and classification of breast cancer histopathology images. Sig. Process. **122**, 1–13 (2016)
24. Filipczuk, P., Kowal, M., Obuchowicz, A.: Automatic breast cancer diagnosis based on k-means clustering and adaptive thresholding hybrid segmentation. In: Choraś, R.S. (ed.) Advances in Intelligent and Soft Computing, vol. 120, pp. 295–302. Springer, Heidelberg (2011). https://doi.org/10.1007/978-3-642-23154-4_33
25. Gao, Y., et al.: Hierarchical nucleus segmentation in digital pathology images. In: Gurcan, M.N., Madabhushi, A. (eds.) Medical Imaging 2016: Digital Pathology. SPIE, March 2016
26. Guo, P., Evans, A., Bhattacharya, P.: Segmentation of nuclei in digital pathology images. In: 2016 IEEE 15th International Conference on Cognitive Informatics and Cognitive Computing (ICCI* CC), pp. 547–550. IEEE (2016)
27. Al-Kofahi, Y., Lassoued, W., Lee, W., Roysam, B.: Improved automatic detection and segmentation of cell nuclei in histopathology images. IEEE Trans. Biomed. Eng. **57**(4), 841–852 (2010)
28. Rother, C., Kolmogorov, V., Blake, A.: GrabCut. ACM Trans. Graph. **23**(3), 309 (2004)
29. Ali, H.R., et al.: Computational pathology of pre-treatment biopsies identifies lymphocyte density as a predictor of response to neoadjuvant chemotherapy in breast cancer. Breast Cancer Res. **18**(1), 21 (2016)
30. Liao, M., et al.: Automatic segmentation for cell images based on bottleneck detection and ellipse fitting. Neurocomputing **173**, 615–622 (2016)
31. Kharma, N., et al.: Automatic segmentation of cells from microscopic imagery using ellipse detection. IET Image Proc. **1**(1), 39 (2007)
32. Hai, S., Xing, F., Lee, J.D., Peterson, C.A., Yang, L.: Automatic myonuclear detection in isolated single muscle fibers using robust ellipse fitting and sparse representation. IEEE/ACM Trans. Comput. Biol. Bioinf. **11**(4), 714–726 (2014)

33. Veta, M., van Diest, P.J., Kornegoor, R., Huisman, A., Viergever, M.A., Pluim, J.P.W.: Automatic nuclei segmentation in H&E stained breast cancer histopathology images. PLoS One **8**(7), e70221 (2013)
34. Qu, A., et al.: Two-step segmentation of hematoxylin-eosin stained histopathological images for prognosis of breast cancer. In: 2014 IEEE International Conference on Bioinformatics and Biomedicine (BIBM). IEEE, November 2014
35. Zhang, D., et al.: Panoptic segmentation with an end-to-end cell R-CNN for pathology image analysis. In: Frangi, A.F., Schnabel, J.A., Davatzikos, C., Alberola-López, C., Fichtinger, G. (eds.) MICCAI 2018. LNCS, vol. 11071, pp. 237–244. Springer, Cham (2018). https://doi.org/10.1007/978-3-030-00934-2_27
36. Girshick, R., Donahue, J., Darrell, T., Malik, J.: Rich feature hierarchies for accurate object detection and semantic segmentation. In: Computer Vision and Pattern Recognition (2014)
37. Felzenszwalb, P.F., Huttenlocher, D.P.: Efficient graph-based image segmentation. Int. J. Comput. Vis. **59**(2), 167–181 (2004)
38. Girshick, R.: Fast r-CNN. In: 2015 IEEE International Conference on Computer Vision (ICCV). IEEE, December 2015
39. Redmon, J., Divvala, S., Girshick, R., Farhadi, A.: You only look once: unified, real-time object detection. In: 2016 IEEE Conference on Computer Vision and Pattern Recognition (CVPR). IEEE, June 2016
40. Liu, W., et al.: SSD: single shot MultiBox detector. In: Leibe, B., Matas, J., Sebe, N., Welling, M. (eds.) ECCV 2016. LNCS, vol. 9905, pp. 21–37. Springer, Cham (2016). https://doi.org/10.1007/978-3-319-46448-0_2
41. Dai, J., Li, Y., He, K., Sun, J.: R-FCN: object detection via region-based fully convolutional networks. In: Advances in Neural Information Processing Systems, pp. 379–387 (2016)
42. Cai, Z., Vasconcelos, N.: Cascade R-CNN: delving into high quality object detection. In: 2018 IEEE/CVF Conference on Computer Vision and Pattern Recognition. IEEE, June 2018
43. Fu, C.-Y., Liu, W., Ranga, A., Tyagi, A., Berg, A.C.: DSSD: deconvolutional single shot detector. arXiv preprint arXiv:1701.06659 (2017)
44. Leibe, B., Seemann, E., Schiele, B.: Pedestrian detection in crowded scenes. In: 2005 IEEE Computer Society Conference on Computer Vision and Pattern Recognition (CVPR 2005). IEEE (2005)
45. Huang, C., Ai, H., Li, Y., Lao, S.: High-performance rotation invariant multiview face detection. IEEE Trans. Pattern Anal. Mach. Intell. **29**(4), 671–686 (2007)
46. Zhou, Y., Ye, Q., Qiu, Q., Jiao, J.: Oriented response networks. In: 2017 IEEE Conference on Computer Vision and Pattern Recognition (CVPR). IEEE, July 2017
47. Law, H., Deng, J.: CornerNet: detecting objects as paired keypoints. In: Ferrari, V., Hebert, M., Sminchisescu, C., Weiss, Y. (eds.) Computer Vision – ECCV 2018. LNCS, vol. 11218, pp. 765–781. Springer, Cham (2018). https://doi.org/10.1007/978-3-030-01264-9_45
48. Dai, J., He, K., Sun, J.: Instance-aware semantic segmentation via multi-task network cascades. In: 2016 IEEE Conference on Computer Vision and Pattern Recognition (CVPR). IEEE, June 2016
49. Li, Y., Qi, H., Dai, J., Ji, X., Wei, Y.: Fully convolutional instance-aware semantic segmentation. In: 2017 IEEE Conference on Computer Vision and Pattern Recognition (CVPR). IEEE, July 2017

50. Liu, S., Qi, L., Qin, H., Shi, J., Jia, J.: Path aggregation network for instance segmentation. In: 2018 IEEE/CVF Conference on Computer Vision and Pattern Recognition. IEEE, June 2018

51. Huang, L., Yang, Y., Deng, Y., Yu, Y.: DenseBox: unifying landmark localization with end to end object detection. arXiv preprint arXiv:1509.04874 (2015)

52. Zhao, H., Shi, J., Qi, X., Wang, X., Jia, J.: Pyramid scene parsing network. In: 2017 IEEE Conference on Computer Vision and Pattern Recognition (CVPR). IEEE, July 2017

A k-Dense-UNet for Biomedical Image Segmentation

Zhiwen Qiang, Shikui Tu$^{(\boxtimes)}$, and Lei Xu$^{(\boxtimes)}$

Department of Computer Science and Engineering,
Centre for Cognitive Machines and Computational Health (CMaCH), SEIEE School,
Shanghai Jiao Tong University, Shanghai, China
{q7853619,tushikui,leixu}@sjtu.edu.cn

Abstract. Medical image segmentation is the premise of many medical image applications including disease diagnosis, anatomy, and radiation therapy. This paper presents a k-Dense-UNet for segmentation of Electron Microscopy (EM) images. Firstly, based on the characteristics of the long skip connection of U-Net and the mechanism of short skip connection of DenseNet, we propose a Dense-UNet by embedding the dense blocks into U-Net, leading to deeper layers for better feature extraction. We experimentally show that Dense-UNet outperforms the popular U-Net. Secondly, we combine Dense-UNet with one of the newest U-Net variants called kU-Net into a network called k-Dense-UNet, which consists of multiple Dense-UNet submodules. Skip connections are added between the adjacent submodules, to pass information efficiently, helping the model to identify fine features. Experimental results on the ISBI 2012 EM dataset show that k-Dense-UNet achieves better performance than U-Net and some of its variants.

Keywords: Image segmentation · Electron Microscopy · Skip connection

1 Introduction

High-resolution Electron Microscopy (EM) image segmentation has great value on many medical image applications, and it has shown important value in anatomy, radiation therapy, and biomedical research. Manual labeling the element in the EM images is normally done by a human neuroanatomist. However, since the medical image data is very complicated, it is time-consuming to manually label such data. Thus, artificial intelligence technology has gradually become a popular direction of medical image segmentation.

In recent years, deep learning approaches based on Convolutional Neural Networks [1–7] have been used on the EM image segmentation task. One of the most well-known attempts is U-Net [2]. It consists of a contraction path and a symmetric expansion path. To enable precise localization, high-resolution features from the contracting path are combined with the upsampled output. Such

© Springer Nature Switzerland AG 2019
Z. Cui et al. (Eds.): IScIDE 2019, LNCS 11935, pp. 552–562, 2019.
https://doi.org/10.1007/978-3-030-36189-1_46

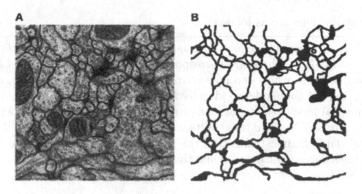

Fig. 1. Examples in the ISBI 2012 EM dataset: an EM image (A) and its ground truth segmentation (B).

skip connections enabled U-Net to work well on biomedical image segmentation tasks.

However, deep learning networks normally suffer the vanishing gradient problem, which limits their depth. He *et al.* [8] proposed ResNet, with skip connection between layers, the network's depth can be improved without damaging its performance. Huang *et al.* [9] proposed DenseNet, by using dense connections. It achieved better performance than ResNet in certain datasets.

Many following works were built on U-Net, and they tend to increase the depth of U-Net by certain metrics. FusionNet [3] applied residual blocks in U-Net to enable the model to have a larger depth to achieve better performance. This model also combined long and short skip connection together. kU-Net [7] consists of multiple submodule U-Net to sequentially extract information at different scales, from the coarsest scale to the finest scale. Each submodule will propagate information to the subsequent submodule to help feature extraction.

Actually, key features of recent models such as U-net, ResNet, and DenseNet have been found in Least Mean Square Error Reconstruction (Lmser) self-organizing network, which was first proposed in 1991 [17,18]. Lmser is a further development of autoencoder (AE) with favorable features, including Duality in Paired Neurons (DPN) and Duality in Connection Weight (DCW), which come from folding AE along the central coding layer. DPN can be regarded as adding shortcut connections between the paired neurons. The feedback links from decoder to encoder can be regarded as the skip connections between two U-Nets in kU-Net. More advances about Lmser are referred to a recent review in [19].

In this paper, Dense-UNet is proposed on segmentation of EM images. Not only we experimentally show that Dense-UNet outperforms U-Net, but also we proceed to a version called k-Dense-UNet that integrates Dense-UNet and kU-Net. Experimental results on the ISBI 2012 EM dataset show that k-Dense-UNet achieves better performance than U-Net and some of its variants.

Our contributions are as follows:

- We present k-Dense-UNet for EM image segmentation. It integrates Dense-UNet and kU-Net. There are corresponding long skip connections between adjacent modules, which can pass coarser-scale information to the next submodule, helping the model to identify finer features.
- Experimental results show that the proposed method can achieve better performance than U-Net and some of its variants in the ISBI 2012 EM challenge. Ablation study demonstrates that the skip connection between the adjacent submodules can enhance and refine the segmentation outputs.

2   Related Work

2.1   Deep Learning Methods for EM Image Segmentation

One of the earliest work was done by Ciresan *et al.* [1]. He implemented a succession of convolutional and max-pooling layers to predict the segmentation. Their work won the ISBI 2012 challenge. Long *et al.* [10] proposed the FCN structure to replace fully connected layers with convolutional layer which can preserve the spatial information. Since then, many variants of FCN have been proposed for EM image segmentation task. Shen *et al.* [11] created a multi-stage and multi-recursive-input FCN. The model can predict outputs at a different level in each stage, and combining all the predictions with the original images to generate the next stage's input. Ronneberger *et al.* [2] proposed U-Net architecture, which consists of four downsampling steps and four corresponding upsampling steps. Long skip connection layers exist between the downsampled feature map and the commensurate upsampled feature map, which can preserve low-level information. This model won the ISBI 2015 challenge. However, it still suffers the vanishing gradient problem, which limits the depth of U-Net. He *et al.* [8] proposed the residual blocks and demonstrated that short skip connections between layers can reduce the influence of vanishing gradients. Quan *et al.* [3] presented FusionNet, which embedded U-Net with residual blocks to combine short and long skip connections.

2.2   DenseNet Architecture

DenseNet [9] was proposed in 2017, by using dense connections, it reaches better results compares to ResNet [8] and pre-activated ResNet [12] on multiple datasets (CIFAR-10, CIFAR-100 [13], SVHN Small-Scale Dataset [14]). In DenseNet, each layer obtains additional inputs from all preceding layers and passes on its own feature-maps to all subsequent layers. This so-called dense block structure enables the network to be thinner and compact, which lead to higher computational efficiency and memory efficiency. We refer the readers to [9] for the detailed architecture of DenseNet.

2.3   kU-Net Structure

kU-Net [7] is the combination of U-Net submodules, it was observed that human experts tend to first zoom out the image to determine the target object and then zoom in to obtain the accurate boundaries of the targets. The kU-Net structure contains two mechanisms which can simulate such human behaviors.

- kU-Net contains a sequence of submodule U-Nets to enable the information extraction carried at different scales sequentially.
- The information extracted by the submodule U-Net in a coarser scale will be propagated to the subsequent submodule U-Net to enable the feature extraction at a finer scale.

We refer the readers to [7] for the detailed structure of kU-Net.

3   Methods

3.1   Overview of the Proposed Network

k-Dense-UNet is the combination of Dense-UNet and kU-Net, an example of its architecture is shown in Fig. 2. It takes advantage of Dense-UNet's feature extraction and combines the idea of kU-Net to gradually extract the features to a finer scale. Similar to U-Net, the upsampling part of the submodule Dense-UNet is skip-connected to the subsequent Dense-UNet's max pooling part, which is equivalent to transferring the coarser information to the next sub-module to achieve more precise image segmentation result.

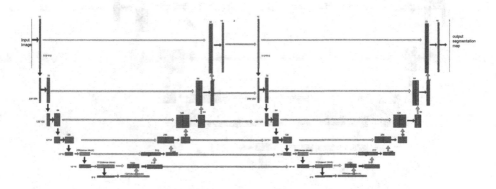

Fig. 2. Architecture of k-Dense-UNet(k = 2)

In practice, all the convolutional layers adopt 3 × 3 kernels with stride size as 1. For all the upsampling layers, 3 × 3 kernels are applied with stride size as 2. Activation functions are set as ReLUs. Batch normalization [16] is implemented to reduce over-fitting and increase the model's learning rate.

3.2   Dense-UNet Architecture

DenseNet has more diverse features and tend to have richer modes since each layer receives all of the previous layers as input: the so-called dense block structure. At the same time, because features of all complexity levels are used. DenseNet performs well when training data is insufficient.

Based on the above advantages of DenseNet, we embedded the dense block into U-Net to obtain more sufficient feature extraction and get a more precise segmentation map. This resulted in the Dense-UNet shown in Fig. 3. The gray arrow indicates the long skip connection between the max pooling layer and the corresponding upsampling layer, the red arrow indicates the 2 × 2 max pooling operation, the green arrow indicates the upsampling operation. It is worth noting that after the max pooling of the network, three dense blocks are embedded instead of the original two-convolution with batch normalization and ReLu activation function. These dense blocks are represented by purple rectangular blocks in the figure, and the corresponding operations are indicated by yellow arrows.

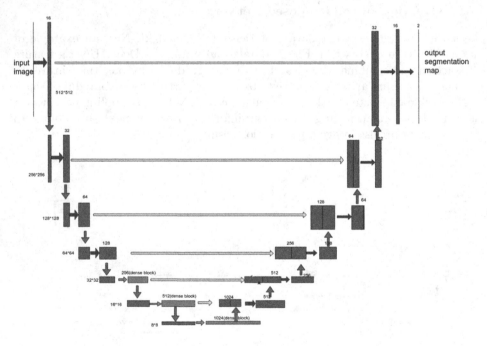

Fig. 3. Dense-UNet architecture (Color figure online)

Let's take the second dense block as an example to describe its operation shown in Fig. 4: Set the input x to obtain the output y through the dense operation. The dense operation is defined as: input through 128 1 × 1 convolution kernels, 128 3 × 3 Convolution kernel, 512 1 × 1 convolution kernels. The network is subject to batch regularization and ReLu activation functions to reduce

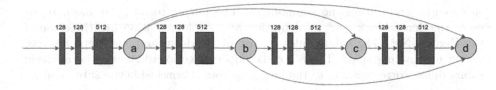

Fig. 4. Dense Block structure

gradient disappearance and over-fitting after each convolution operation. For the second dense block ($16 \times 16 \times 512$), the input is subjected to the dense operation to obtain the output a. a after the dense operation to get the output b, b after the dense operation and add a to get c, c after the dense operation and add a and b to get d, which is the final output.

The dense operation for the three dense blocks are shown in Table 1.

Table 1. Dense operation

Type	Dense operation	Times
Dense Block 1	$(1 \times 1, 64)(3 \times 3, 64)(1 \times 1, 64)$	3
Dense Block 2	$(1 \times 1, 128)(3 \times 3, 128)(1 \times 1, 512)$	4
Dense Block 3	$(1 \times 1, 256)(3 \times 3, 256)(1 \times 1, 1024)$	6

Similar to DenseNet, the kth layer receives the feature-maps of all preceding layers, $x_0, \ldots, x_{k-1}$ as input:

$$x_k = H_k([x_0, x_1, \ldots, x_{k-1}])$$

where $[x_0, x_1, \ldots, x_{k-1}]$ refers to the concatenation of the feature-maps produced in layers $0, \ldots, k - 1$.

It is worth noting that in order to reduce the complexity and size of dense blocks, a 1×1 convolution is added, and then a 3×3 convolution input is performed, which can greatly reduce the amount of calculation without damaging the accuracy of the model. This is also the design of the bottleneck layer of ResNet.

3.3   k-Dense-UNet Formation

In the k-Dense-UNet model, the internal operation of each sub-module is similar to Dense-UNet, and the dark blue arrow indicates the two-convolution with batch normalization and ReLu activation function, the red arrow represents the max pooling of 2×2 scale, and the green arrow represents the upsampling operation, the operation yellow arrow stands is consistent with that described in Dense-UNet, the grey arrow represents skip connections between the adjacent submodules and within the submodules. The purple rectangular block represents the dense block, and its definition and implementation are the same as the Dense-UNet. The submodules consist of six downsampling steps followed by six upsampling steps. There are six skip connections between the adjacent submodules, corresponding to the long skip connection inside the submodules.

4   Experiments and Results

4.1   Dataset and Evaluation Metrics

We use the ISBI 2012 EM Segmentation Dataset to test the effectiveness of our model. Figure 1 shows an example of the dataset. The training part of the dataset contains 30 pairs of EM images and ground truth labels. The testing part contains 30 EM images without the ground truths.

The Evaluation metrics are the Foreground-restricted Rand Scoring after border thinning: V^{Rand}. The details of this metric can be found in [15].

4.2   Experiments on Loss Function

In order to compensate for the different frequency of pixels in a certain class form the training set, We use weighted loss function in all the experiments to force the network to learn the small borders between cells.

In this experiment, we compare three loss functions: weighted-bce, weighted-dice, and weighted dice & weighted-bce loss function. The model is U-Net. The training dataset is 30 pairs of EM images. As this dataset is small, we applied several data augmentation techniques to enlarge the dataset, which includes rotation, horizontal and vertical flip. We training the model using Adam optimizer with a learning rate of 2×10^{-4}. The training metric is IOU score. We also applied the EarlyStopping, ReduceLROnPlateau methods to improve the performance of the model.

The predicted segmentation images obtained form these loss functions are shown in Fig. 5. Among them, A, B, C are the result of the weighted dice loss function, weighted dice& weighted-bce loss function, weighted bce loss function respectively. D, E, F are the same enlarged part of A, B, C respectively.

We observed that the boundary of the predicted image obtained by weighted dice& weighted-bce loss function is more complete than the rest two. The result of the above models is shown in Table 2, which are sorted by V^{Rand}. Since

Table 2. Results of three types of loss functions sorted by V^{Rand}

Model	Loss function	V^{Rand}
U-Net	Weighted-dice	0.886397133
U-Net	Weighted-bce	0.941848881
U-Net	Weighted-bce& weighted-dice	**0.950320002**

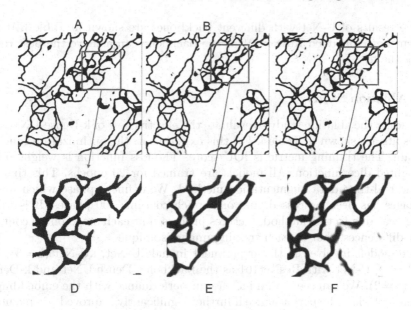

Fig. 5. Comparison of different loss function's result

weighted dice& weighted-bce loss function also has the highest V^{Rand} score, we use this loss function in the rest of the experiments.

4.3   Experiments on Differenet Backbones

This experiment first tested the V^{rand} of U-Net as the benchmark. Then, similar to Dense-UNet, we embedded ResNet50, ResNet101, ResNeXt, and SEResNet into U-Net as the backbone to test the V^{rand} of these networks.

The optimization function of the model is Adam. The training metric is IOU score. The loss function is weighted bce & weighted dice function. All models are trained for 20 epochs. Besides the data augmentation methods used in the previous experiment, we also adopted the elastic transformation method.

Table 3. Results of U-Net with different backbones sorted by V^{Rand}

Model	Backbone	V^{Rand}
U-Net	ResNet50	0.931212484
U-Net	SeresNet50	0.9429939
U-Net	ResNext50	0.948937035
U-Net	None	0.950320002
U-Net	ResNet101	**0.957619972**

The results of U-Net with different backbones are shown in Table 3. It can be seen that embedded ResNet101 as the backbone can achieve the best result among the five.

4.4 Ablation Study

We conduct the ablation study to evaluate the effectiveness of k-Dense-UNet, and the results are shown in Table 4. For this experiment, the optimization function is Adam. The training metric is IOU score. The loss function is weighted bce & weighted dice function. All models are trained for 20 epochs. This time we use the real-time data augmentation method. We define a phase which means 30 images has been processed, and one epoch contains 300 phases. It is worth noting that due to this method, pictures processed at each stage phase contains subtle differences, that is, each training image is unique.

The models involved in this experiment include U-Net, kU-Net (k = 3), kU-Net (k = 2), U-Net with ResNet101 as the backbone, Dense-UNet and k-Dense-UNet (k = 2). We can see a mild increase in performance with the embedding of the dense blocks. The performance if further significantly improved when embedding the dense block structure into kU-Net. We can summarize this experiment as follows:

- kU-Net structure can propagate coarser scales to subsequent modules to assist in finer feature extraction.
- Dense-UNet takes advantage of DenseNet's feature extraction capabilities which can achieve better results than U-Net and U-Net (ResNet101 as backend)
- The parameter k in kU-Net increases the input window size of the network exponentially. The smaller k value is sufficient to process many biomedical images (k = 2): the model of kU-Net (k = 3) is not as good as k = 2. The result might be that the model is too complicated, which lead to the network to a certain degree of over-fitting.

Table 4. Results of U-Net with different backbones sorted by V^{Rand}

Model	Backbone	V^{Rand}
U-Net	None	0.956101213
kU-Net (k = 3)	None	0.959437719
kU-Net (k = 2)	None	0.963030825
U-Net	ResNet101	0.963493141
Dense-UNet	DenseNet	**0.964117979**
k-Dense-UNet (k = 2)	DenseNet	**0.972352852**

5   Conclusion

In this paper, by embedding dense blocks into U-Net, we present Dense-UNet for biomedical image segmentation. It can obtain more sufficient feature extraction and get a more precise segmentation map compared to U-Net. Moreover, by integrating kU-Net and Dense-UNet, we proposed k-Dense-UNet, which takes advantage of Dense-UNet's feature extraction capabilities and combines the idea of kU-Net to gradually extract the features to a finer scale. By harnessing the short skip connection in the dense block, the long skip connection in the Dense-UNet submodules and the skip connection between the adjacent submodules, we can achieve more precise image segmentation maps. Experimental results on the ISBI 2012 EM dataset show that the proposed method can achieve better results compared to U-Net and some of its variants.

Acknowledgement. The first author would like to thank Yuze Guo for helpful discussions. This work was supported by the Zhi-Yuan Chair Professorship Start-up Grant (WF220103010), and Startup Fund (WF220403029) for Youngman Research, from Shanghai Jiao Tong University.

References

1. Ciresan, D., Giusti, A., Gambardella, L.M., Schmidhuber, J.: Deep neural networks segment neuronal membranes in electron microscopy images. In: NIPS (2012)
2. Ronneberger, O., Fischer, P., Brox, T.: U-Net: convolutional networks for biomedical image segmentation. In: Navab, N., Hornegger, J., Wells, W.M., Frangi, A.F. (eds.) MICCAI 2015. LNCS, vol. 9351, pp. 234–241. Springer, Cham (2015). https://doi.org/10.1007/978-3-319-24574-4_28
3. Quan, T.M., Hildebrand, D.G.C., Jeong, W.-K.: FusioNnet: a deep fully residual convolutional neural network for image segmentation in connectomics, CoRR, vol. abs/1612.05360 (2016)
4. Chen, H., Qi, X., Cheng, J.-Z., Heng, P.-A.: Deep contextual networks for neuronal structure segmentation. In: AAAI (2016)
5. Chen, K., Zhu, D., Lu, J., Luo, Y.: An adversarial and densely dilated network for connectomes segmentation. Symmetry **10**, 467 (2018)
6. Drozdzal, M., et al.: Learning normalized inputs for iterative estimation in medical image segmentation. Med. Image Anal. **44**, 1–13 (2018)
7. Chen, J., Yang, L., Zhang, Y., Alber, M., Chen, D.Z.: Combining fully convolutional and recurrent neural networks for 3D biomedical image segmentation. In: NIPS (2016)
8. He, K., Zhang, X., Ren, S., Sun, J.: Deep residual learning for image recognition. In: CVPR (2016)
9. Huang, G., Liu, Z., Van Der Maaten, L., et al.: Densely connected convolutional networks. In: Proceedings of the IEEE Conference on Computer Vision and Pattern Recognition, pp. 4700–4708 (2017)
10. Shelhamer, E., Long, J., Darrell, T.: Fully convolutional networks for semantic segmentation. TPAMI (2017)
11. Shen, W., Wang, B., Jiang, Y., Wang, Y., Yuille, A.L.: Multi-stage multi-recursive-input fully convolutional networks for neuronal boundary detection. In: ICCV (2017)

12. He, K., Zhang, X., Ren, S., Sun, J.: Identity mappings in deep residual networks. In: Leibe, B., Matas, J., Sebe, N., Welling, M. (eds.) ECCV 2016. LNCS, vol. 9908, pp. 630–645. Springer, Cham (2016). https://doi.org/10.1007/978-3-319-46493-0_38

13. Krizhevsky, A., Hinton, G.: Learning multiple layers of features from tiny images. Technical report, University of Toronto (2009)

14. Netzer, Y., Wang, T., Coates, A., Bissacco, A., Wu, B., Ng, A.Y.: Reading digits in natural images with unsupervised feature learning. In: NIPS Workshop on Deep Learning and Unsupervised Feature Learning (2011)

15. Arganda-Carreras, I., et al.: Crowdsourcing the creation of image segmentation algorithms for connectomics. Front. Neuroanat. 9, 142 (2015)

16. Ioffe, S., Szegedy, C.: Batch normalization: accelerating deep network training by reducing internal covariate shift. arXiv preprint arXiv:1502.03167 (2015)

17. Xu, L.: Least MSE reconstruction for self-organization: (i)&(ii). In: Proceedings of 1991 International Joint Conference on Neural Networks, pp. 2363–2373 (1991)

18. Xu, L.: Least mean square error reconstruction principle for self-organizing neural-nets. Neural Netw. 6(5), 627–648 (1993)

19. Xu, L.: An overview and perspectives on bidirectional intelligence: Lmser duality, double IA harmony, and causal computation. IEEE/CAA Journal of Automatica Sinica 6(4), 865–893 (2019)

Gated Fusion of Discriminant Features
for Caricature Recognition

Lingna Dai[1], Fei Gao[1,2]([✉]), Rongsheng Li[3], Jiachen Yu[1], Xiaoyuan Shen[1],
Huilin Xiong[1], and Weilun Wu[1]

[1] Key Laboratory of Complex Systems Modeling and Simulation,
School of Computer Science and Technology, Hangzhou Dianzi University,
Hangzhou 310018, China
dln2014@126.com, gaofei@hdu.edu.cn
[2] State Key Laboratory of Integrated Services Networks,
School of Electronic Engineering, Xidian University, Xi'an 710071, China
[3] State Grid Yantai Power Supply Company,
Yantai 264001, People's Republic of China
lirongshengbupt@126.com

Abstract. Caricature recognition is a challenging problem, because
there are typically geometric deformations between photographs and caricatures. It is nontrivial to learn discriminant large-margin features. To
combat this challenge, we propose a novel framework by using a gated
fusion of global and local discriminant features. First, we employ A-
Softmax loss to jointly learn angularly discriminant features of the whole
face and local facial parts. Besides, we use the convolutional block attention module (CBAM) to further boost the discriminant ability of the
learnt features. Next, we use global features as dominant representation
and local features as supplemental ones; and propose a gated fusion unit
to automatically learn the weighting factors for these local parts and
moderate local features correspondingly. Finally, an integration of all
these features is used for caricature recognition. Extensive experiments
are conducted on the cross-modal face recognition task. Results show
that, our method significantly boosts previous state-of-the-art Rank-
1 and Rank-10 from 36.27% to 55.29% and from 64.37% to 85.78%,
respectively, for caricature-to-photograph (C2P) recognition. Besides,
our method achieves a Rank-1 of 60.81% and Rank-10 of 89.26% for
photograph-to-caricature (P2C) recognition.

Keywords: Caricature · Face recognition · Deep learning ·
Convolutional block attention module · SphereFace

1 Introduction

One caricature is a facial sketch attempting to portray a facial essence by exag-
gerating some prominent characteristics. Caricatures have been popularly used
in news and social media, leading to a wide demand of caricature recognition.

© Springer Nature Switzerland AG 2019
Z. Cui et al. (Eds.): IScIDE 2019, LNCS 11935, pp. 563–573, 2019.
https://doi.org/10.1007/978-3-030-36189-1_47

Although caricatures can be easily recognized by human, caricature recognition is a great challenge for computers, due to serious geometric deformations between photographs and caricatures, as well as the varying styles of caricatures. It is nontrivial to match photographs and caricatures.

For now, there are only few studies on caricature recognition. In the beginning, researchers focus on designing and learning facial attribute features. To name a few, Klare et al. [7] propose to use manually labelled attribute features and machine learning techniques to calculate the similarity between a caricature and a photo. Laterly, Abaci and Akgul [1] extract facial attribute features for photos but manually label attribute features for caricatures, and then use a genetic algorithm (GA) or LR for recognition. In these works, attributes are usually heuristically designed and laboriously labelled, limiting the performance and generalization ability.

Recently, due to the prominent ability of convolutional neural networks (CNNs), researchers use CNNs for caricature recognition [3,4,8]. For example, Garg et al. [3] propose to use two CNNs to extract features from photograph and caricature, respectively, and use a joint of the verification loss and the identification loss to supervise the representation learning. Li et al. [5,8] integrate the global similarity of the whole face and local similarities of four local parts (i.e., eye, nose, mouth and chin parts), and then integrate these similarities together to obtain better similarity measure for photograph and caricature. This work shows that local features are essential for caricature recognition.

Despite such progress in caricature recognition, the performance is still far from satisfactory. The limited performance might due to the open-set protocol of caricature recognition, i.e. the testing identities are usually disjoint from the training set. In this scenario, it is critical to learn discriminative large-margin features. However, existing works [3,8] use softmax loss or pairwise loss, which can only learn separable features that are not necessarily discriminative [10]. It is still challenging to learn discriminant large-margin features, which can as well bridge the gap between photographs and caricatures, for caricature recognition.

In this paper, we propose to jointly learn discriminant features from the whole face and local facial parts, and then use a gated fusion unit to integrate them together for caricature recognition. Inspired by the great success of SphereFace [10] in face recognition (FR) of photographs, we use SphereFace as our network prototype and use the angular softmax (A-Softmax) loss for learning angularly discriminative features. Besides, we employ the convolutional block attention module (CBAM) to further boost the discriminant ability of the learnt features. Afterwards, we use global features as dominant representation and local features as supplemental ones; and propose a gated fusion unit to automatically learn the weighting factors for local facial parts and moderate local features correspondingly. Finally, the moderated features are concatenated together for caricature recognition. Extensive experiments are conducted on the cross-modal face recognition task. Results show that our method significantly outperform previous state-of-the-art. Besides, we conduct thorough ablation study to analyse the reasons why such inspiring performance is achieved.

Our contributions in this work are mainly three-fold:

- We employ A-Softmax loss and CBAM for learning discriminant features for caricature recognition;
- We propose a gated fusion unit to automatically moderate features learnt from local facial parts; and
- Our method achieves significantly performance improvement over previous state-of-the-art in cross-modal face recognition tasks.

2   Proposed

The pipeline of our model is as shown in Fig. 1. Specially, we first pre-process an input image (photograph or caricature) and obtain the corresponding whole face and a number of local parts. Next, we use CNNs to extract global and local features from them, respectively. Afterwards, we use a gated fusion unit to moderate local features and combine them with global features for caricature recognition. Details will be presented below.

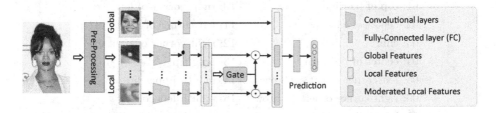

Fig. 1. Pipeline of the proposed caricature recognition framework.

2.1   Preprocessing

In the database we used, i.e. WebCaricature [6], for each image, 17 landmarks have been provided (Fig. 2(a)), including 4 basic face contours, 4 eye brows, 4 eye corners, 1 nose tip, and 4 mouth contours. The indices and name of these landmarks are listed in Table 1.

According to the landmarks, each image is first aligned by affine transformation to guarantee the outer corners of eyes, nose tip, and mouth corners at fixed locations. Afterwards, the image is cropped based on the bounding box of the face landmarks and resized to 112×96 to represent the whole face. In addition, for each landmark, we crop a patch of size $h/112 * 20$ around it. h denotes the height of the input image. Then each patch is resized to 20×20 and adopted as the input of the corresponding local face recognition (FR) network. All the processes are illustrated in Fig. 2.

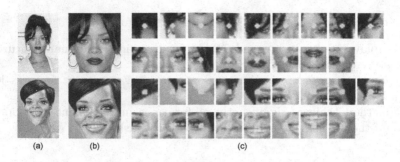

(a)   (b)   (c)

Fig. 2. Illustration of preprocessed images. (a) Original image with 17 landmarks, (b) aligned images, i.e. the global face, and (c) local facial parts corresponding to 17 landmarks.

Table 1. Networks and facial landmarks.

Index: location	Index: location
00: whole image (global)	09: left corner of left eye
01: top of hairline	10: right corner of left eye
02: center of left ear	11: left corner of right eye
03: chin	12: right corner of right eye
04: center of right ear	13: nose tip
05: left corner of left eyebrow	14: mouth upper lip top
06: right corner of left eyebrow	15: mouth left corner
07: left corner of right eyebrow	16: mouth lower lip bottom
08: right corner of right eyeborw	17: mouth right corner

2.2   Network Architecture

In this paper, we adopt SphereFace [10] as the prototype due to its outstanding performance in solving the open-set face recognition problems. Specially, our global FR network is the same as the 20-layer version of SphereFace, but with a convolutional block attention module (CBAM) behind the Conv1.x layer. We use a shallow version of SphereFace with a CBAM as the local FR network. Detailed architectures are as shown in Table 2.

Convolutional Block Attention Module (CBAM). The structure of CBAM [13] is illustrated in Fig. 3. CBAM includes a channel attention module and a spatial attention module. Given a feature map $\mathbf{F} \in \mathbb{R}^{C \times H \times W}$, the channel

Table 2. Architectures of the global and local face recognition networks.

Layer	Global FR network	Local FR network
Conv1.x	$\begin{bmatrix} 3 \times 3, 64 \end{bmatrix}, S2$ $\begin{bmatrix} 3 \times 3, 64 \\ 3 \times 3, 64 \end{bmatrix} \times 1$	$\begin{bmatrix} 3 \times 3, 64 \end{bmatrix}, S2$ $\begin{bmatrix} 3 \times 3, 64 \\ 3 \times 3, 64 \end{bmatrix} \times 1$
CBAM	CBAM	CBAM
Conv2.x	$\begin{bmatrix} 3 \times 3, 128 \end{bmatrix}, S2$ $\begin{bmatrix} 3 \times 3, 128 \\ 3 \times 3, 128 \end{bmatrix} \times 2$	$\begin{bmatrix} 3 \times 3, 128 \end{bmatrix}, S2$ $\begin{bmatrix} 3 \times 3, 128 \\ 3 \times 3, 128 \end{bmatrix} \times 1$
Conv3.x	$\begin{bmatrix} 3 \times 3, 256 \end{bmatrix}, S2$ $\begin{bmatrix} 3 \times 3, 256 \\ 3 \times 3, 256 \end{bmatrix} \times 4$	$\begin{bmatrix} 3 \times 3, 256 \end{bmatrix}, S2$ $\begin{bmatrix} 3 \times 3, 256 \\ 3 \times 3, 256 \end{bmatrix} \times 1$
Conv4.x	$\begin{bmatrix} 3 \times 3, 512 \end{bmatrix}, S2$ $\begin{bmatrix} 3 \times 3, 512 \\ 3 \times 3, 512 \end{bmatrix} \times 1$	$\begin{bmatrix} 3 \times 3, 512 \end{bmatrix}, S2$ $\begin{bmatrix} 3 \times 3, 512 \\ 3 \times 3, 512 \end{bmatrix} \times 1$
FC1	512	512

attention module infers an 1D channel attention vector $\mathbf{M_c} \in \mathbb{R}^{C \times 1 \times 1}$ and the spatial attention model sequentially infers a 2D spatial attention map $\mathbf{M_s} \in \mathbb{R}^{1 \times H \times W}$. The overall process is operated as:

$$\mathbf{F'} = \mathbf{M_c}(\mathbf{F}) \odot \mathbf{F},$$
$$\mathbf{F''} = \mathbf{M_s}(\mathbf{F'}) \odot \mathbf{F'}, \tag{1}$$

where $\odot$ denotes element-wise multiplication. The cth element in $\mathbf{M_c}$ moderates all the values in the cth channel of $\mathbf{F}$; and the element at location (i, j) in $\mathbf{M_s}$ moderates all the activations across all the channels at the same location. Supervised by the final face recognition loss function, CBAM is expected to make the network pay more attention on significant channels and locations.

In the implementation, the channel attention module composes of a max-pooling layer (MaxPool), an average-pooling layer (AvgPool), and two fully-connected layers followed by a ReLU and Sigmoid activation layer respectively (MLP); the spatial attention module composes a max-pooling layer, an average-pooling layer, and a convolutional layer followed by a Sigmoid layer.

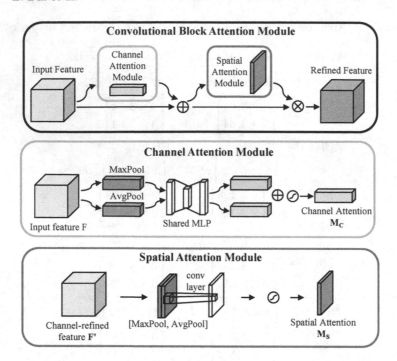

Fig. 3. Pipeline of the convolutional block attention module (CBAM) [13].

2.3 Gated Fusion Unit

Initially, different facial parts contribute diversely to caricature recognition. It is promising to weight the significances of local features, in the computation of the similarity between two images. Inspired by this motivation, we propose a gated fusion unit to automatically learn the weighting factors for moderating local features (Fig. 4).

Our gated fusion unit takes a concatenation of all the local features as input, and composes a average-pooling layer, a fully-connected layer, and a Sigmoid layer. Here, the average-pooling layer pool each local feature vector into a scalar. The outputs of the Sigmoid layer are the weighting factors for local features. Let $\mathbf{w} = [w_1, w_2, ..., w_{17}]$ denote the weight vector. w_k is multiplied to the features vector extracted from the kth local part. Features extracted from all the FR networks are concatenated as:

$$f(x) = \phi_0(x) \oplus w_1\phi_1(x) \oplus \cdots \oplus w_{17}\phi_{17}(x), \tag{2}$$

where $\phi(\cdot)$ denotes the kth FR networks.

In the training stage, $f(x)$ is input into the last fully-connected layer for predicting the identity of the input image x. In the testing stage, the similarity

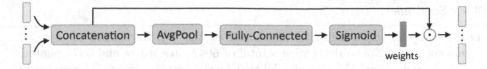

Fig. 4. Pipeline of the gated fusion unit.

between two images (x_1, x_2) are computed as the cosine distance between their corresponding integrated features, i.e.

$$s(x_1, x_2) = \cos(f(x_1), f(x_2)). \tag{3}$$

This similarity can be used for both face identification or verification.

2.4 A-Softmax Loss Function

Inspired by the great success of SphereFace [10], we introduce the angular softmax (A-Softmax) loss to caricature recognition for learning angularly discriminative features. A-Softmax Loss is expressed as:

$$
\begin{aligned}
||\mathbf{x}||(\cos m\theta_1 - \cos \theta_2) = 0 \quad \text{for class 1,} \\
||\mathbf{x}||(\cos \theta_1 - \cos m\theta_2) = 0 \quad \text{for class 2,}
\end{aligned}
\tag{4}
$$

where θ_i is the angle between $\mathbf{W}_i$ and $\mathbf{x}$; $\mathbf{W}_i$ denotes parameters in the last fully-connected layer; $m(m \geq 1)$ is an integer to quantitatively controls the size of angular margin. In the implementation, we set $m = 4$.

2.5 Implementation

During training, we minimize the A-Softmax loss by performing mini-batch stochastic gradient descent (SGD) over a training set of WebCaricature (126 persons) with a batch size of 128. Specifically, we set the values of momentum and weight decay to 0.9 and 5×10^{-4}, respectively; our learning rate is 0.01. Our experiments are implemented via Pytorch and executed on a server with one NVIDIA TITAN 1080Ti GPU.

3 Experiments

To verify the effectiveness of the proposed framework, we conduct extensive experiments on the cross-modal face recognition task. Besides, we conduct thorough ablation study to analyse the impacts of the proposed modules. Details will be introduced below.

3.1 Settings

Datasets. Our experiments are developed on the WebCaricature dataset [6], which concludes 252 subjects with a total of 6042 caricatures and 5974 photos. For each caricature, 17 manually labeled landmarks are given, as previously presented in Table 1 and Fig. 2. Details can be found from the dataset website. In our experiments, the dataset is divided into a training set and a testing set. Each set includes the caricatures and photos related to 126 individuals. There is no overlap in individual between the training and testing sets.

Evaluation Protocols. All the methods presented in this section are evaluated on the caricature recognition task. Given a caricature (photograph), the goal is to search the corresponding photographs (caricatures) from a photograph (caricature) gallery. For the "Caricature-to-Photograph (C2P)" setting, all the caricatures in the testing set (126 individuals) will be used as probes (i.e., 2961 images); and one photograph of each individual will be selected to the gallery (i.e., 126 images). The setting of "Photograph-to-Caricature (P2C)" is similar to C2P. Rank-1 and Rank-10 are chosen as the evaluation criteria.

3.2 Ablation Study

To analyse the impacts of the proposed techniques, we build several variants of our model by changing the modules used in the networks. In Table 3, we list all the corresponding model variants and results. Here, *global* denotes merely using the global FR network *without CBAM*, *local* denotes using a concatenation of all local features *without CBAM*, *Gate* denotes using the gated fusion unit presented in Sect. 2.3, and *CBAM* denotes using the CBAM in the corresponding FR network(s).

Obviously, both the gated fusion unit and CBAM consistently improve the performance for both P2C and C2P tasks. Besides, *global+local* shows much worse performance over *global*. This might due the weak ability of local networks (*local*). In contrast, by using the gated fusion unit, *global+local+Gate* outperform *global* by about 2 points. Such improvement implies that the gated fusion unit can not only prevent local FR networks from damaging the performance, but also make local features benefit the final result. Inspiringly, our full model, *global+local+Gate+CBAM*, achieves the best performance.

3.3 Comparison with State-of-the-art

We further compare our method with previous state-of-the-art, LGDML [8]. As shown in Table 4, our method significantly boost previous state-of-the-art Rank-1 and Rank-10 from 36.27% to 55.29% and from 64.37% to 85.78%, respectively, for C2P recognition. Besides, our method achieve a Rank-1 of 60.81% and Rank-10 of 89.26% for P2C recognition. Such dramatic performance improvements demonstrate the superiority of our method. Figures 5 and 6 illustrate some examples.

Table 3. Ablation study on the WebCaricature database.

	P2C		C2P		Average	
	rank1	rank10	rank1	rank10	rank1	rank10
global	56.59	85.44	52.47	83.49	54.53	84.47
global+CBAM	57.34	87.42	52.19	83.63	54.77	85.53
local	5.34	23.33	5.05	23.51	5.20	23.42
local+Gate	7.94	30.69	7.59	32.39	7.77	31.54
local+Gate+CBAM	8.19	29.73	8.12	34.58	8.16	32.16
global+local	25.90	62.15	28.58	67.40	27.24	64.78
global+local+Gate	59.43	88.24	53.11	84.86	56.27	86.55
global+local+Gate+CBAM (Ours)	**60.81**	**89.26**	**55.29**	**85.78**	**58.05**	**87.52**

Table 4. Comparison with state-of-the-art on the WebCaricature database. The best performance indices among all these models are highlighted in **boldface** fonts.

	P2C		C2P		Average	
	rank1	rank10	rank1	rank10	rank1	rank10
LGDML Binomial [8]	-	-	28.40	67.65	-	-
LGDML Logistic [8]	-	-	29.42	67.00	-	-
LGDML Binomial w/tSNE [8]	-	-	36.14	65.96	-	-
LGDML Logistic w/tSNE [8]	-	-	36.27	64.37	-	-
Ours	**60.81**	**89.26**	**55.29**	**85.78**	**58.05**	**87.52**

Fig. 5. Example cases of photograph-to-caricature (P2C) recognition results by our method. For each probe photograph, top 10 relevant caricatures are exhibited, where the caricatures annotated with red rectangular boxes are the ground-truth. (Color figure online)

Probe
Caricature   Top-10 relevant photos

Fig. 6. Example cases of caricature-to-photograph (C2P) recognition results by our method. For each probe caricature, top 10 relevant photographs are exhibited, where the photographs annotated with red rectangular boxes are the ground-truth.

4   Conclusion

In this paper, we propose a discriminant feature learning and gated feature fusion module for caricature recognition. Experimental results demonstrate our significant superiority over previous state-of-the art. In the future, we will further generalize the proposed frame work to other cross-modal face recognition tasks, such as face photo-sketch recognition [11,14]. Besides, it is necessary to boost the ability of the local FR networks. Finally, it is promising to employ the caricature recognition models in caricature generation [2,9,12] for depicting identity-aware caricatures.

Acknowledgements. This work was supported in part by the National Natural Science Foundation of China under Grants 61601158, 61971172, 61971339, and 61702145, in part by the Education of Zhejiang Province under Grant Y201840785 and Y201942162, and in part by the China Post-Doctoral Science Foundation under Grant 2019M653563.

References

1. Abaci, B., Akgul, T.: Matching caricatures to photographs. SIViP **9**(1), 295–303 (2015)
2. Cao, K., Liao, J., Yuan, L.: CariGANs: unpaired photo-to-caricature translation. arXiv preprint arXiv:1811.00222 (2018)
3. Garg, J., Peri, S.V., Tolani, H., Krishnan, N.C.: Deep cross modal learning for caricature verification and identification (cavinet). arXiv preprint arXiv:1807.11688 (2018)
4. Hill, M.Q., et al.: Deep convolutional neural networks in the face of caricature: identity and image revealed. arXiv preprint arXiv:1812.10902 (2018)

5. Huo, J., Gao, Y., Shi, Y., Yin, H.: Variation robust cross-modal metric learning for caricature recognition. In: Proceedings of the on Thematic Workshops of ACM Multimedia 2017, pp. 340–348. ACM (2017)
6. Huo, J., Li, W., Shi, Y., Gao, Y., Yin, H.: WebCaricature: a benchmark for caricature recognition. In: British Machine Vision Conference (2018)
7. Klare, B.F., Bucak, S.S., Jain, A.K., Akgul, T.: Towards automated caricature recognition. In: IAPR International Conference on Biometrics, pp. 139–146 (2012)
8. Li, W., Huo, J., Shi, Y., Gao, Y., Wang, L., Luo, J.: A joint local and global deep metric learning method for caricature recognition. In: Jawahar, C.V., Li, H., Mori, G., Schindler, K. (eds.) ACCV 2018. LNCS, vol. 11364, pp. 240–256. Springer, Cham (2019). https://doi.org/10.1007/978-3-030-20870-7_15
9. Li, W., Xiong, W., Liao, H., Huo, J., Gao, Y., Luo, J.: Carigan: caricature generation through weakly paired adversarial learning. arXiv preprint arXiv:1811.00445 (2018)
10. Liu, W., Wen, Y., Yu, Z., Li, M., Raj, B., Song, L.: SphereFace: deep hypersphere embedding for face recognition. In: The IEEE Conference on Computer Vision and Pattern Recognition (CVPR), July 2017
11. Peng, C., Gao, X., Wang, N., Li, J.: Graphical representation for heterogeneous face recognition. IEEE Trans. Pattern Anal. Mach. Intell. 39(2), 301–312 (2017)
12. Shi, Y., Deb, D., Jain, A.K.: WarpGAN: automatic caricature generation. In: Proceedings of the IEEE Conference on Computer Vision and Pattern Recognition, pp. 10762–10771 (2019)
13. Woo, S., Park, J., Lee, J.Y., Kweon, I.S.: CBAM: convolutional block attention module. In: ECCV (2018)
14. Zhu, M., Li, J., Wang, N., Gao, X.: A deep collaborative framework for face photo-sketch synthesis. IEEE Trans. Neural Netw. Learn. Syst. 99, 1–13 (2019)

Author Index

An, Jiuyuan   I-495

Bao, Xiao-Rong   II-126
Bao, Xudong   I-103

Cai, Jinghui   II-394
Cai, Ruichu   II-381
Cao, Jiangdong   I-529
Chang, Dongliang   I-273
Chen, Baoxing   II-1
Chen, Bo   II-278
Chen, Haitao   II-278
Chen, JiaMing   II-163
Chen, Kaixuan   I-155
Chen, Si   II-369
Chen, Wen-Sheng   II-278
Chen, Yunjie   I-349
Chen, Zeyu   I-459
Chen, Zhenyu   II-232
Cheng, Lu   I-273
Cheng, Peng   I-299
Cheng, Qihua   I-178
Chu, Honglin   I-142
Cui, Hengfei   I-238
Cui, Hetao   I-424
Cui, Ziguan   I-386

Dai, Lingna   I-563
Dai, Wei   II-219
Ding, Junjie   I-362
Ding, Yupeng   I-80
Ding, Zhuanlian   I-517
Dong, Huailong   II-312
Du, Jing   I-201
Du, Xiaochen   II-335
Duan, Lijuan   II-139

Fan, Haoyi   I-191
Feng, Changju   I-374
Feng, Hailin   II-335
Feng, Zhanxiang   I-459
Fu, Jingru   II-101

Fu, Keren   I-299
Fu, Shou-fu   II-65

Gao, Fei   I-563
Gao, Kangkang   I-299
Gao, Shibo   I-484
Gao, Xinbo   I-285, I-540
Gao, Yan   I-540
Gong, Chen   II-40
Gu, Rong-rong   II-65
Guo, Tangyi   I-362

Hao, Zhifeng   II-381
He, Nannan   II-266
He, Tieke   II-151
He, Yi   I-213
He, Yonggui   I-435
Hei, Tieke   II-219
Hou, Jinze   II-139
Hou, Li-he   II-114
Hou, Lv-lin   II-347
Hou, Zhixin   I-68
Hu, Bin   II-418
Hu, Heng   II-335
Hu, Liucheng   I-540
Huang, Chanying   II-325
Huang, Dong   II-266
Huang, Ling   II-406
Huang, Wenjin   II-197
Huang, Yi   II-53

Jeon, Byeungwoo   I-472
Jiang, Xiang   I-399
Jiao, Zhicheng   I-285
Jin, Bo   II-394
Jin, Mengying   I-349

Katayama, Takafumi   I-35
Ke, Jingchen   II-40

Lai, Jianhuang   I-411, I-459
Lai, Zhihui   I-142
Li, Chaobo   I-80

Li, Hongjun   I-80
Li, Huijie   II-356
Li, Jiajie   II-335
Li, Jie   I-285, I-540
Li, Junchen   II-430
Li, Kai   I-517
Li, Lei   I-484
Li, Minchao   II-15
Li, Pengfei   I-484
Li, Rongsheng   I-563
Li, Weiqing   II-253
Li, Xiaoyan   I-529
Li, Yu   II-151, II-232, II-243
Li, Ziyu   I-249
Li, Zuoyong   I-191
Lian, Zhichao   I-362, I-374, II-325
Liang, Lu   I-529
Liang, Songxin   I-114
Liang, Yixiao   I-517
Lin, Chunhui   I-103
Lin, Minmin   II-1
Lin, Rui   I-336
Lis-Gutiérrez, Jenny Paola   II-289
Liu, Eryun   I-262
Liu, Feng   I-386
Liu, Hua-jun   II-114
Liu, Shaohan   II-175
Liu, ShaoHan   II-187
Liu, Ting   I-399
Liu, Yang   I-128
Liu, Yazhou   I-312
Liu, Yue   I-362
Liu, Zhipu   I-91
Liu, Zhonggeng   II-243
Lu, Guangming   I-336
Lu, JinJun   II-187
Lu, Liang   II-347
Lu, Yan   I-191
Lu, Yao   I-336
Luo, Ming   I-399
Luo, WenFeng   I-324

Ma, Benteng   I-58
Ma, Juncai   I-399
Ma, Rongliang   I-273
Ma, Zhanyu   I-273
Mei, Ling   I-459
Mei, Xue   I-213
Miao, Jun   II-139
Mu, Xuelan   I-68

Pan, Binbin   II-278
Pan, Huadong   II-243
Pan, Jinshan   I-128
Peng, Huan   II-369

Qian, Jiaying   I-155
Qiang, Zhiwen   I-552
Qiao, Yuanhua   II-139
Qiu, Mingyue   I-226
Qu, Xichao   II-253

Ren, Chuan-Xian   I-155

Shen, Jifeng   I-435
Shen, Xiaoyuan   I-563
Sheni, Siyuan   II-219
Shi, He   II-175
Shimamoto, Takashi   I-35
Shu, Xin   II-302
Shuai, Jia   II-163
Song, Tian   I-35
Song, Yaoye   I-435
Su, Yang   I-58
Su, Zhixun   I-114, I-128
Sun, Dengdi   I-517
Sun, Le   I-472, I-506
Sun, Quansen   I-424
Sun, XiuSong   II-187
Sun, Yahui   I-472

Tan, Li   I-324
Tan, Yongwei   II-406
Tang, Jingsheng   II-418
Tang, Longzhe   II-418
Tang, Songze   I-226
Tong, Xinjie   I-249
Tu, Shikui   I-552, II-15, II-197

Viloria, Amelec   II-289

Wang, Bo   II-175
Wang, Boyang   II-394
Wang, Chang-Dong   II-406
Wang, Dong-sheng   II-65
Wang, Hongyan   I-114
Wang, Huan   I-165
Wang, Lu   I-35
Wang, Wei-min   II-65
Wang, Xiangfeng   II-394
Wang, Zhiqing   I-262

Wei, Hanyu   II-209
Wei, Jie   I-58
Wei, Li   I-22
Wei, Shikui   I-399
Wei, Zhihui   I-201
Wei, Zhisen   II-1
Wen, Jiajun   I-142
Wu, Chunsheng   I-273
Wu, Feiyang   I-506
Wu, Qing   II-151
Wu, Weilun   I-563

Xia, Yong   I-58, I-238
Xia, Yu   I-447
Xiang, Zhiyu   I-262
Xiao, Yan-dong   II-347
Xie, Feng   II-381
Xie, Jiyang   I-273
Xie, Xiaohua   I-411, I-459
Xiong, Huilin   I-563
Xu, Chunyan   I-22
Xu, Gengxin   I-155
Xu, Haojun   I-540
Xu, Lei   I-1, I-552, II-15, II-197
Xu, Yang   I-201
Xu, Zhengfeng   II-89
Xu, Zhenxing   II-209

Yan, Ge   II-232
Yan, Jinyao   II-369
Yang, Jinfeng   I-495, II-430
Yang, Jingmin   II-1
Yang, Meng   I-324, I-484, II-163, II-442
Yang, Minqiang   II-418
Yang, Qun   II-175, II-187
Yang, Shanming   I-299
Yang, Wankou   I-249, I-435
Yang, Zhao   II-89
Yao, Yutao   I-386
Ye, Jun   I-46
Yin, Jun   II-243
You, Jane   I-529
Yu, Dong-Jun   II-126
Yu, Hongbin   II-302
Yu, Jiachen   I-563
Yuan, Shenqiang   I-213

Zeng, Weili   II-89
Zeng, Yan   II-381
Zerda Sarmiento, Álvaro   II-289
Zhan, Yongzhao   I-447
Zhang, Daoqiang   II-209, II-356
Zhang, Haigang   I-495, II-430
Zhang, Hui   I-472
Zhang, Jin   I-213
Zhang, Jing   II-312
Zhang, Junyi   II-209
Zhang, Lei   I-91, II-101
Zhang, Peixi   I-411
Zhang, Quan   I-411
Zhang, Rui   II-356
Zhang, Shanshan   I-178
Zhang, Shu   II-232
Zhang, Tong   I-22
Zhang, Wenjie   II-1
Zhang, Xian   I-46, I-249
Zhang, Xingming   II-243
Zhang, Yanning   I-58, I-238
Zhang, Yigong   I-68
Zhang, Yuan   II-369
Zhang, Yuduo   II-325
Zhang, Zhenxi   I-285
Zhang, Zipeng   I-517
Zhao, Junhui   II-78
Zhao, Lin   II-40
Zhao, Peize   II-78
Zhao, Ruidong   II-78
Zhao, Yao   I-399
Zhao, Ziyu   II-89
Zheng, Jian   I-517
Zheng, Shaomin   II-442
Zheng, Shengjie   I-165
Zheng, Yuhui   I-472, I-506
Zhong, Zhusi   I-285
Zhou, Changxin   I-312
Zhou, Kai   II-175, II-187
Zhou, Xuan   I-103
Zhu, Bowen   II-312
Zhu, Qi   II-356
Zhu, Tianqi   II-78
Zhu, Yi-Heng   II-126
Zhu, Yue   I-495
Zou, Zhipeng   II-151

Printed in the United States
By Bookmasters

Printed in the United States
By Bookmasters